DYNAMICS OF SMART STRUCTURES

DYNAMICS OF SMART STRUCTURES

Dr. Ranjan Vepa

A John Wiley and Sons, Ltd., Publication

This edition first published 2010

Registered Office
John Wiley & Sons Ltd, The Atrium, Southern Gate, Chichester, West Sussex, PO19 8SQ, United Kingdom

For details of our global editorial offices, for customer services and for information about how to apply for permission to reuse the copyright material in this book please see our website at www.wiley.com.

Library of Congress Cataloging-in-Publication Data

Vepa, Ranjan.
Dynamics of smart structures / Ranjan Vepa.
p. cm.
Includes bibliographical references and index.
ISBN 978-0-470-69705-4 (cloth)
1. Smart materials. 2. Smart structures. I. Title.
TA418.9.S62V47 2010
624.1–dc22

2009050259

A catalogue record for this book is available from the British Library.

ISBN 978-0-470-69705-4 (H/B)

Typeset in 9/11pt Times by Aptara Inc., New Delhi, India.
Printed and bound in Singapore by Markono Print Media Pte Ltd.

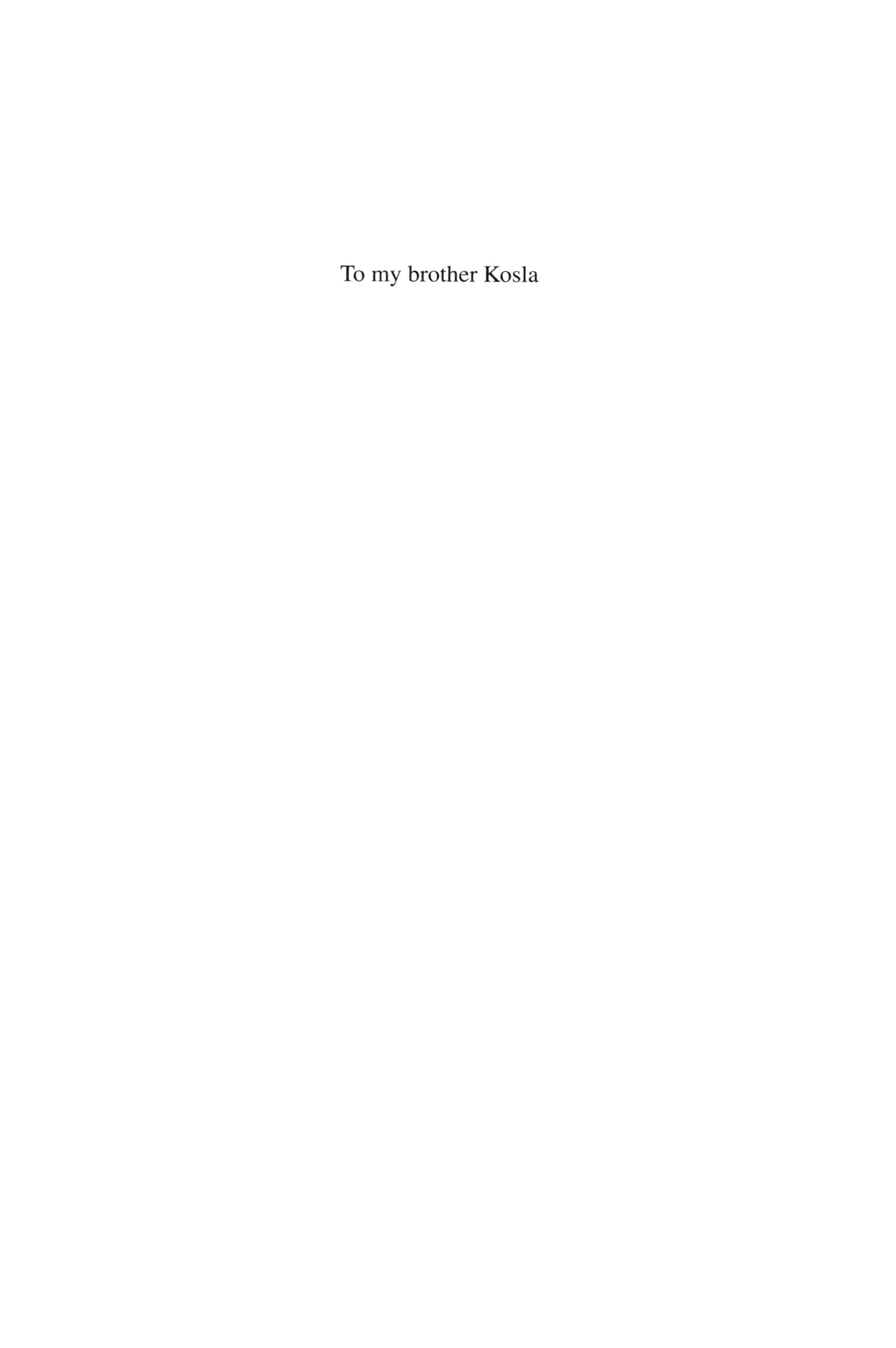

To my brother Kosla

Contents

Preface

The dynamics of smart structures has evolved over the years and has now matured into an independent and identifiable subject. The objective of this text is to provide an introduction to the fundamental principles of dynamics of smart structures. The goal is to provide a practical, concise and integrated text on the subject matter.

In this book fundamental principles are stressed. The treatment in detail on structures assembled from specific smart material types is based on the need to illustrate basic principles and to introduce the properties of a variety of commonly used smart structure prototypes. The text extends considerably beyond the needs of a single specific two-semester or one-year course. There are two compelling reasons for this: the first being the need for flexibility in structuring a course, which might be based on this text, while the second is due to a desire to achieve a degree of completeness that makes the text a useful reference to contemporary practising engineers. The result is a text that should meet the needs of a senior undergraduate or postgraduate course. It will also be possible to offer two one-semester courses based on the text: the first on the dynamics of smart structures, covering substantial portions of the first four chapters in the book, and the second on the active control of smart structures where review of the earlier chapters is covered with a control flavour followed by the remaining chapters. Relevant application examples from across the book may be included in each course.

The first chapter provides an introduction to make a smooth transition from a general review of smart materials to smart structures. The second chapter addresses the dynamical issues related to smart sensors and actuators. The topics covered are transducers and sensors for smart structures, piezoelectric actuators, shape memory alloy actuators and fibre-optic sensors, as well as other conventional transducers. The last section of this chapter is dedicated to fibre-optic transducers, as these transducers are expected to play a key role in the evolution of smart structures. The next chapter introduces the basic concepts of structural control. Only the basics of control theory and topics that are relevant to structural control are presented here. As the fundamentals of acoustic waves and transmission lines play a basic role in understanding the latter concepts, the fourth chapter begins with this topic. Following an introduction to waves in solids, the dynamics of continuous structures, particularly the mechanics of elastic media and waves in elastic media, are covered. The focus of the fifth chapter is vibrations in plates and plate-like structures. The sixth chapter is devoted to the dynamics of piezoelectric media. After providing an introduction to piezoelectric crystalline media, the topics covered are wave propagation in piezoelectric crystals, vibrations in piezoelectric plates, transmission line modelling, discrete element modelling of transducers and the generation of acoustic waves in piezoelectric media. The next chapter deals with mechanics of composite laminated media and flexural vibrations in laminated composite plates. The modelling of flexible laminated plates employing one of several theories such as the classical theory of Kirchhoff, and the theories of Reissner and Mindlin, with the inclusion of the shear, as well as the more recent theories of composite zig-zag laminates, are extensively discussed. Furthermore, the chapter covers the mechanics of composite laminated media, failure of fibre composites, flexural vibrations in laminated composite plates, dynamic modelling of flexible structures, including the finite element method and equivalent circuit modelling, and active composite laminated structures. The eighth chapter covers the dynamics of thermo-elastic and magneto-elastic media and includes sections on fundamentals of thermoelasticity, creep in materials, the shape memory effect, thermal control of shape memory alloys, constitutive relations, dynamics of hysteretic media and the modelling and control of shape memory alloy-based actuators. The last chapter is devoted entirely to the design of active controllers for flexible structures, and the sections include controller synthesis for structural control, optimal design of structronic systems, the analysis of structures with repeated components and application case studies.

The last chapter is particularly important, as this chapter is the one that brings together the concepts, techniques and systems presented in the earlier chapters. Rather than present the general principles of control engineering, this chapter focuses on the application of control theory to problems related to the active feedback control of flexible structures. Thus, it is assumed that the reader is familiar with the basic principles of control engineering and no attempt is made to provide an in-depth explanation of these basic tools. The validation of controller designs could be achieved by the application of advanced validation tools, which is covered in most advanced textbooks on control systems design. These topics are not covered here. However, the reader is also assumed to be familiar with the matrix analysis package for the PC, MATLAB/Simulink (MATLAB® and Simulink® are registered trademarks of the The Mathworks Inc., MA, USA) and some of its tool boxes.

I would like to thank my wife, Sudha, for her love, understanding and patience. Her encouragement was a principal factor that provided the motivation to complete the project. Finally, I must add that my interest in the subject of structural dynamics was nurtured by my brother Kosla, when I was still an undergraduate. His encouragement and support throughout my academic life motivated me substantially in this fascinating project.

R. Vepa

1

From Smart Materials to Smart Structures

1.1 Modern Materials: A Survey

One of the most challenging and intellectually satisfying endeavours of materials scientists during the last decade has been the application of the well-developed methodologies of materials science research to the study of smart materials. Smart materials may be described as materials that can sense an external stimulus (e.g. stress, pressure, temperature change, magnetic field, etc.) and initiate a response. Passively smart materials can only sense an external stimulus. Actively smart materials have both sensing and actuation capabilities. Smart materials, like materials themselves, may belong to one of four classes: metals or alloys, polymers ceramics, or composites (Figure 1.1). Metals and alloys of different metals are the classical materials. It is other materials that are unusual and most important as smart materials (Askeland and Phule, 2003).

1.1.1 Polymers

Polymers are chained molecules that are built up from simple units called monomers (Schultz, 1974; Hearle, 1982). Polyethylene is an archetypal example of a polymer. Polyethylene is produced from ethylene molecules by an addition reaction. The unsaturated double bond in ethylene (C_2H_4) is broken, and this acts as an active agent, attracting self-similar units at either end to produce a chain molecule. Some examples of polymers are silicone, polydimethyl siloxane, polyurethane, polyethylene (PE), polyvinylchloride (PVC), hydrogels, polyester, polytetrafluoroethylene (PTFE), acetal, polyethylene, polymethyl methacrylate (PMMA), polylactic acid, polyglycolic acid, nylon and ultra-high molecular weight polyethylene (UHMWPE).

1.1.2 Structure and Classification of Polymers

PMMA is an example of a hydrophobic linear chain polymer. It is known by the trade names Lucite and Plexiglas and possesses excellent light transmittance properties. It is therefore used for the manufacture of intra-ocular and hard contact lenses. Polypropylene is another polymer that possess high rigidity, good chemical resistance, good tensile strength and excellent stress cracking resistance. PTFE, known by the trade name Teflon, is a very hydrophobic polymer with good lubricity and low wear resistance.

Dynamics of Smart Structures Dr. Ranjan Vepa

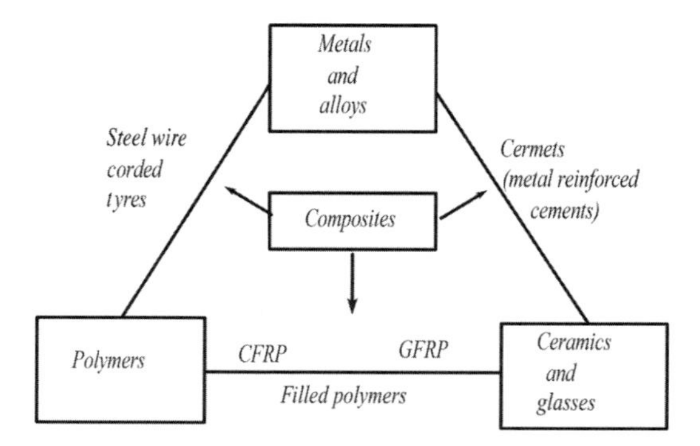

Figure 1.1 Classification of smart materials

Polyethylene is a polymer that can be compacted to a high density (HD) form with an ultra-high molecular weight and has good toughness and high wear resistance. Polyethylene terephthalate (PET), known by the trade name Dacron, is a high-melting crystalline polymer with a very high tensile strength and is therefore used in fabrics for various applications. Polylactic glycolic acid (PLGA) is a random copolymer used in medical applications.

A polymer chain layout has a macroscopic structure to it and two types of structures, amorphous and crystalline, are common. The molecular structure could be chained, linear, branched, cross-linked or networked. Polymers may also be classified as homopolymers, with all repeating units in the chain of the same type (PE, PVC, PTFE, PMMA, PET and nylon), or copolymers, which consists of two or more dissimilar 'mer' units along the chain or branches (Acrylonitrile, Butadiene, and Styrene or ABS, dimethylsiloxane). Copolymers can be random, alternating, graft or block type.

Based on their thermal characteristics, polymers may be classified as thermoplastics, which are linear or branched polymers that can be melted and remelted with heat, or thermosetting plastics, which are cross-linked or networked and are therefore rigid structures that cannot be remelted and degrade on heating. Thermoplastics are recyclable, e.g. polyethylene, while thermosets are not recyclable, e.g. polyurethane.

Based on the chain structure, polymers could be classified as linear, branched or cross-linked. In linear polymers, the mer units form a single continuous chain. The mer units in the chain are bonded together by weak van der Waals forces. In the case of branched polymers, additional side branches result from further reactions that occur during synthesis. A consequence of the existence of the side branches is that the packing efficiency of the polymers is reduced, resulting in a lower density (LD). In cross-linked polymers, the side branches join up with adjacent branched chains. These bonds are essentially covalent bonds formed during the synthesis of the polymer or at elevated temperatures, and the net result is a cross-linked chain polymer. A typical example is vulcanized rubber.

1.1.3 Characteristic Properties of Polymers

Polymers are generally elastic at relatively low temperatures below the melting point, although they are not as stiff as metals and are viscous fluids near the melting point. At intermediate temperatures they are viscoelastic. Some other key characteristics of polymers are good thermal stability, high tensile strength, high impact resistance, excellent ultraviolet resistance and extremely high purity. Thermosetting plastics are liquid or low-melting-point solids that 'cure' with heat, which is an irreversible process. Thermoplastics soften to viscous fluids when heated and regain the solid state when cooled. This process is reversible. The main properties of polymers that set them apart from other materials are molecular weight, glass transition, crystallinity, melting point, mechanical behaviour and viscoelasticity, fatigue and fracture resistance, and crazing. The glass transition temperature is one below which all segmental motion of polymer chains ceases. The glass transition temperature varies from polymer to polymer because it is a function of compound chemistry. Due to the alignment and ordering of chain segments in thermoplastics, 100% crystallinity is seldom achieved; polyvinylidene fluoride (PVDF) is typically 50–60% crystalline. The extent of crystallinity achieved is dependent on the complexity of side groups and leads to higher densities: $\rho(\text{HDPE}) = 0.97$ g/cc; $\rho(\text{LDPE}) = 0.92$ g/cc.

Elastic deformation is due to stretching of the constituent chains in the polymer. Viscoelasticity refers to instantaneous elastic strain followed by time-dependent strain. Mechanical deformation is highly time, rate and temperature dependent. The fatigue strength of polymers is more sensitive to load characteristics/temperature than metals/ceramics. Thermosets exhibit brittle fracture, while thermoplastics can exhibit either brittle or ductile fracture. Ductile to brittle transition is possible in thermoplastics, e.g. PMMA is brittle at 4°C and completely ductile at 60°C. Crazing refers to the narrow zones of highly deformed and voided polymer. Typically a 'craze' contains from about 20% to about 90% voids.

1.1.4 Applications of Polymers

An unusual ferroelectric polymer is PVDF, which is a highly crystalline polymer that belongs to a class of materials based on the vinylidene fluoride monomer [–CH_2–CF_2–]. Depending on the structure, in some crystal lattices, the centres of the positive and negative charges do not coincide even in the absence of an external electric field. In these cases, there exists spontaneous polarization in the crystal, which is referred to as ferroelectricity. When the polarization of the material can be altered by an electric field, it is said to be ferroelectric. Ferroelectric crystals are spontaneously polarized only below a certain temperature known as the Curie temperature. A ferroelectric polymer such as PVDF is one with groups of molecules linked as orderly crystallites. The crystallites form in an amorphous matrix of chemically similar, but differently structured, material. The piezoelectric behaviour of the material, which refers to the generation of charge due the application of mechanical stress, is determined by the relative population of crystallites.

PVDF is a piezoelectric polymer material used for vibration sensing. PVDF, which is known by the trade name Kynar®, is a high molecular weight, semi-crystalline, semi-opaque and white thermoplastic polymer that is melt processable. It belongs to a class of materials based on the vinylidene fluoride monomer [–CH_2–CF_2–]. In practice, both uniaxial and biaxial mechanical orientation is used in applications. A process known as poling or subjecting the material to high electric field in a heated condition gives a different balance of piezo/pyroelectric properties. Poled PVDF has excellent corrosion and chemical resistance and outperforms other piezoelectric materials in many applications up to 300°F (149°C). PVDF has a glass transition temperature of about −35°C. It is suitable for chemical processing applications because of its unique combination of properties. Poled PVDF possesses excellent chemical resistance, is tough and durable, and is easily fabricated into finished parts.

Large applied AC fields ($\sim$200 MV/m) can induce non-linear electrostrictive strains of the order of $\sim$2%. Co-polymerization of trifluoroethylene (TrFE) with VDF produces a random co-polymer, P(VDF-TrFE), a PVDF polymer that has been subject to electron radiation and has shown electrostrictive strain as high as 5% at lower-frequency drive fields (150 V/μm). These polymers belong to a class known as electroactive polymers (EAPs) and may be used in the construction of actuators.

1.2 Ceramics

Ceramics (Boch and Niepce, 2005) are inorganic and non-metallic materials, the term 'ceramics' originates from the Greek word 'keramikos' or 'burnt stuff'. Ceramic materials are intrinsically hard, strong and stiff, but brittle, as determined by their atomic structure. Ceramics are different from metals, which are characterized by metallic bonding, and polymers, which are characterized by organic or carbon covalent bonding. Most ceramics usually contain both metallic and non-metallic elements (with a combination of ionic and covalent bonding).

The structure of a ceramic depends on the structure of the constituent metallic and non-metallic atoms, and the balance of charges produced by the valence electrons. Examples of ceramics are glass ceramics that are based on a network of silica (SiO_2) tetrahedrons, nitride ceramics (TiN) and other silicate ceramics such as quartz and tridymite. A classic example of a ceramic is cement. There are a number of different grades of cement, but a typical Portland cement will contain 19–25% SiO_2, 5–9% Al_2O_3, 60–64% CaO and 2–4% FeO.

1.2.1 Properties of Ceramics

Ceramics are characterized by some unique properties due to the coupling of their chemistry with the mechanics of cracking. These properties include thermo-electrical properties, non-linear optical properties and ferroelectric (piezoelectric) properties as well as several others. A typical ceramic ferroelectric is barium titanate, $BaTiO_3$. Lead zirconate titanate (PZT) powders are processed in rolling mills to form a visco-plastic dough. The dough may be extruded or processed by calendering or lamination. PZT powder is manufactured by calcinating an appropriate mix of at least three different lead oxides, titanium and zirconium oxide. PZT with the formula $PbZr_{1-x}Ti_xO_3$ is available in several forms. Particle size has a direct influence on both the direct and the inverse piezoelectric response. The latter effect, the inverse piezoelectric effect, refers to the generation of stress by the application of an electric field. Complex dielectric properties are also controlled by adding dopants. In typical examples of acceptor doping Fe^{3+} replaces Ti^{4+}/ Zr^{4+} or Na^+ replaces Pb^{2+}, while in the case of donor doping Nb^{5+} replaces Ti^{4+}/Zr^{4+} or La^{3+} replaces Pb^{2+}.

Ceramics and polymers do not exhibit piezoelectric properties in their natural state but only after the temporary application of a strong electric field. The process of making polycrystalline ceramics and polymers piezoelectric is called poling. It has been described as a process analogous to magnetizing a permanent magnet. Several ceramics exhibit the piezoelectric property when poled. These include barium titanate, PZT, lead metaniobate, bismuth titanate, sodium potassium niobate and lead titanate.

After a ceramic has been fired, the material will be isotropic and will exhibit no piezoelectric effect because of this random orientation. Before polarization, the ceramic material is thus a mass of minute, randomly oriented crystallites. The ceramic may be made piezoelectric in any chosen direction by poling, which involves exposing it to a strong electric field. Following poling and the removal of the field, the dipoles remain locked in alignment, giving the ceramic material a permanent polarization and a permanent piezoelectric property.

This poling treatment is usually the final stage in the manufacture of a piezoelectric ceramic component. The manufacturing process involves the dry mixing and ball milling of the components, calcining above 900°C, followed by milling to a powder, spray drying and the addition of a binder. The dried, calcined powder with the binder added is then compacted. In this state it is still 'chalky' and in a soft condition. The binder is then burnt out by heating to between 600 and 700°C. It is then machined if necessary and sintered at 1800°C. During the sintering or firing process, high energy is provided to encourage the individual powder particles to bond or 'sinter' together, thus removing the porosity present in earlier compaction stages.

During the sintering process, the 'green compact' shrinks by around 40% in volume. The shrinkage is predictable and can be accommodated. In the final stages of manufacture, after cutting, grinding and polishing as required, electrodes are applied either by screen printing or by chemical plating or vacuum deposition. Poling then is carried out by heating in an oil bath at 130–220°C, and applying an electric field of 2–8 kV/mm to align the domains in the desired direction.

1.2.2 Applications of Ceramics

Ferroelectric ceramics are very promising for a variety of application fields such as piezoelectric/electrostrictive transducers and electro-optics. PZT is a piezoelectric ceramic that is suitable for incorporation in composites in the form of ceramic fibres. PZT fibres are an active piezoelectric ceramic material and are particularly suitable for actuating and sensing applications for vibration control in smart and adaptive structures. Innovative fabrication technologies, which enable the realization of very high volume fraction fibre composites, have been developed for the manufacture of high-performance active and novel flexible composites for smart structure applications (Srinivasan and McFarland, 2000).

1.3 Composites

A composite is a combination of two or more materials (reinforcement, resin, filler, etc.) (Agarwal and Broutman, 1990; Vijaya and Rangarajan, 2004; Daniel and Ishai 2005). The component materials may differ in form or composition on a macroscale. A unique feature of composites is that the materials retain their identities, i.e. they do not dissolve or merge into each other, and act cooperatively. The components can be physically identified and exhibit an interface between each other. A typical composite is a two-phase composite consisting of a matrix, which is the continuous phase, that surrounds a second phase known as reinforcement, which in turn is a dispersed phase that normally bears the majority of stress.

There are several ways to classify composites and a broad classification of composites is shown in Figure 1.2. Based on the classification certain main types of composites can be identified:

1. micro- and macrocomposites
2. fibre-reinforced composites
3. continuous fibre composites
4. short-fibre composites
5. fibre-matrix composites.

1.3.1 Micro- and Macrocomposites

Micro- and macrocomposites can be considered as polymer combinations, metal combinations and ceramic combinations. Typical examples of polymer combinations are polymer–polymer fibre-reinforced

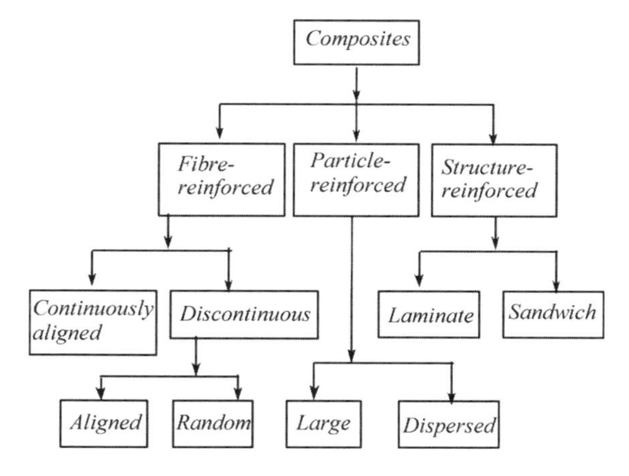

Figure 1.2 Classification of composites

plastic (FRP), such as composites manufactured from epoxy and aramid, polymer–ceramic fibre glass, polymer–metal composites and polymer–carbon composites such as carbon-fibre-reinforced plastic or CFRP. Typical examples of metal composites are metal–metal with iron and aluminium, metal–ceramic with aluminium and silicon carbide (SiC), metal–polymer, metal–carbon such as carbon-fibre-reinforced magnesium, graphite–carbon fibre and reinforced carbon–carbon (RCC). Typical examples of ceramic combinations are ceramic–ceramic such as concrete, ceramic–polymer such as flexible cement and ceramic–metal consisting of a whole family of cermets.

Particulate reinforced composites are manufactured from a variety of particulates and matrix materials. The particulates are harder and stiffer than the matrix material. The particulates are of macro-, micro- or nanoscopic scale. The benefits of particulate reinforcements in terms of mechanical behaviour depend on the interface bonding. Typical examples of particulate reinforced composites are silica-epoxy composites used in electronic moulding applications, Cermets (tungsten carbide (WC) or titanium carbide (TiC), reinforced cobalt or nickel), concrete (aggregate-gravel and sand-reinforced cement) and silicon carbide (SiC), alumina (Al_2O_3) particle-reinforced Al matrix.

1.3.2 Fibre-reinforced Composites

Fibre-reinforced polymer (FRP) composites are defined as 'a matrix of polymeric material that is reinforced by fibres or other reinforcing material'. A 'reinforcement' is the strong, stiff, integral component that is incorporated into the matrix to achieve desired properties. The term 'reinforcement' usually implies enhancement of strength. Different types of fibres or filaments, such as continuous fibres, discontinuous fibres, whiskers and particulate reinforcements, may be of any shape, ranging from irregular to spherical, plate-like and needle-like nanoparticles. The matrix materials are made from metals, polymers or ceramics and constitute a continuous phase. In fibre-reinforced composites, the applied load is transmitted and distributed within the fibres via the matrix. Some ductility is desirable within the matrix, considering that its main function is to bind the reinforcements (fibres/particulates) together while mechanically supporting and transferring the load to them. A secondary function of the matrix is to protect the reinforcements from surface damage due to abrasion or chemical attacks. A high bonding strength between fibre and matrix is vitally important for effective load transference. Fibre-reinforced composites may be classified as continuous-fibre and short-fibre composites.

1.3.3 Continuous-fibre Composites

In continuous-fibre composites, the reinforcing fibres are long, extending over the entire length of the cut. They are highly anisotropic as their properties, such as tensile strength and elastic modulus, in the direction parallel to the orientation of the fibre differ markedly from those in the other two directions. Thus, it is important to differentiate between these two cases and analyse them independently. For example, the elastic modulus in the direction of fibre orientation can be found by weighted addition, while the compliance is found by weighted addition in the transverse directions. Consequently, the elastic modules are quite low in the transverse directions. The weights are linear functions of the volume fraction of the fibre in the composite.

1.3.4 Short-fibre Composites

In short-fibre composites, the reinforcements are generally less effective than in the continuous-fibre case. Consequently, they are not as anisotropic, which can be advantageous in applications where isotropy is desirable. The fibre length is a critical parameter in short-fibre composites. The critical length is defined as the minimum length above which the load is transferred to the fibre effectively, so the reinforcements are effective in bearing the load. Short-fibre composites themselves could be classified according to the parallel or random orientation of the fibres.

1.3.5 Fibre-matrix Composites

A wide range of fibre-matrix composites are manufactured for a variety of engineering applications. These include polymer fibre-, metal fibre- and ceramic fibre-matrix composites. Commonly used fibre materials are aramid (Kevlar), carbon and glass. Modern active fibre-matrix composites use piezoelectric ceramic fibres such as PZT and shape memory alloy (SMA) wires such as Nitinol. The advantage of using metal fibres is that their operating temperatures can be high and they have higher specific strength. They also have higher creep resistance and thermal conductivity. Ceramic fibre-based active composites are suitable for high-temperature applications, although they suffer from a major disadvantage because of the need for a high-voltage source to generate the control forces (Taya, 2005).

1.4 Introduction to Features of Smart Materials

Several features are characteristic of smart materials and provide the fundamental mechanisms that set apart smart materials from others. They are briefly discussed in this section.

1.4.1 Piezoelectric, Piezoresistive and Piezorestrictive

Piezoelectricity refers to the generation of charge due the application of mechanical stress. It is a feature of certain crystalline materials where the crystals do not possess reflection symmetry. *Acousto-electricity* refers to the generation of an electric current when an acoustic wave travels through the material. Piezoelectric materials are used to construct *surface acoustic wave* (SAW) devices, which use an AC signal (30–300 MHz) applied to a set of inter-digitated input electrodes to generate a surface (Rayleigh) wave. The propagation delays to output electrode are affected by changes in the surface properties, which can in turn be influenced by a number of factors. The phase shifts of the output electrode signal are used as a measure of the response to the features of interest. Piezoelectric materials are also used to construct *quartz crystal microbalances* (QMB), which are also known as *bulk acoustic wave* (BAW) devices. These devices are operated in a resonating oscillator circuit. The feature changes in the sensing membrane affect the resonant frequency (5–20 MHz) of the device.

Piezoresistivity is the change in the conductivity of a material that is induced by the application of mechanical stresses.

1.4.2 Electrostrictive, Magnetostrictive and Magnetoresistive

Electrostriction refers to the elastic deformation of all electrical non-conductors or dielectrics induced by the application of an electric field; specifically, the term applies to those components of strain that are not dependent on the direction of the applied field. Electrostriction is a property of all electrical non-conductors or dielectrics, and is thus distinguished from the inverse piezoelectric effect, a field-induced strain that changes sign on field reversal and occurs only in piezoelectric materials. It is the property of all electrical non-conductors or dielectrics, which manifests itself as a relatively slight change of shape under the application of an electric field. Reversal of the electric field does not reverse the direction of the deformation.

The classical electromagnetic effect is the change in magnetization due to an electric current.

All magnetic materials generally fall into four broad categories: (1) diamagnetic, (2) paramagnetic, (3) ferromagnetic and (4) ferrimagnetic. If one considers magnetic materials on a microscopic scale, the total contribution to atomic magnetic moment is from three distinct sources: (1) the magnetic orbital moment due to electrons orbiting the atomic nucleus, (2) the magnetic spin moment due to the spin of the electrons about their axis (related to the so-called Bohr magneton) and (3) the magnetic moment of the nucleus because of the spin of the nucleus about its own axis. In *diamagnetic* materials, the total or net magnetic moment adds up to zero. *Diamagnetism* is the property of a material that induces in it a magnetic field in opposition to an externally applied magnetic field, thus causing a repulsive effect. It is a form of magnetism that is only exhibited by a substance in the presence of an externally applied magnetic field. The force acting on such materials in a strong electromagnetic field is repulsive and proportional to the square of the current. By contrast, *paramagnetic* materials are attracted to magnetic fields and hence have a relative magnetic permeability >1 (or, equivalently, a positive magnetic susceptibility).

Iron, nickel, cobalt and some of the rare earths (gadolinium, dysprosium, alloys of samarium and neodymium with cobalt) exhibit a unique magnetic behaviour, called *ferromagnetism*, when a small externally imposed magnetic field, say from an electric coil with a DC current flowing through it, causes the magnetic domains to line up with each other: the material is said to be magnetized. In a ferromagnetic material, the magnetic moments of the atoms on different sublattices are in the same direction. (By contrast, a *ferrimagnetic* material is one in which the magnetic moment of the atoms on different sublattices are opposed, as in *antiferromagnetism*; however, in ferrimagnetic materials, the opposing moments are unequal and a spontaneous magnetization remains.) All ferromagnets have a maximum temperature where the ferromagnetic property disappears as a result of thermal agitation. This temperature is called the Curie temperature. *Magnetostriction* is the property of all ferromagnetic materials, which manifests itself as a relatively slight change of shape under the application of a magnetic field. Conversely, application of stresses with the subsequent change in shape or deformation can cause a magnetic field to be generated in these materials.

Permittivity, also called electric permittivity, is a constant of proportionality that exists between electric displacement and electric field intensity. This constant is equal to approximately 8.85×10^{-12} *farad* per *metre* (F/m) in free space (a vacuum). In other materials it can be much different, often substantially greater than the free-space value, which is symbolized by ε_o. Permittivity, which is symbolized by ε, is a dielectric constant that determines the extent to which a substance concentrates the electrostatic lines of flux. *Ferroelectric* materials are named after ferromagnetic ones because they behave in a similar way, but under the application of an electric field. The main difference between ferroelectric and ferromagnetic materials is that ferroelectric materials are not magnetic like ferromagnetic materials but are permanently electrically polarized.

1.4.3 The Shape Memory Effect

SMAs are novel materials that have the ability to return to a predetermined shape when heated. When an SMA is cold, or below a certain critical transformation temperature, it has a very low yield strength and can be deformed quite easily into any new shape, which it will retain. However, when the material is heated above its transformation temperature it undergoes a change in crystal structure, which causes it to return to its original shape. If the SMA experiences any resistance during this transformation, it generates extremely large forces. However, this extreme force is proportional to the size of the SMA used. In the microworld, a micron-sized SMA is just good enough to push a micro-sized beam or diaphragm. The most common shape memory material is an alloy of nickel and titanium called *Nitinol*. This particular alloy has very good electrical and mechanical properties, and microvalves made of this nickel and titanium alloy, *Ni–Ti*, are currently available in the market. The voltage required to heat the SMA alloy is generated by relying on the Joule effect. The heat thus generated is capable of heating the actuating SMA element, the one usually available in integrated circuits (ICs). SMA technology is most used in micro-electro-mechanical systems (MEMS). When heated, the valve made of SMA pops open and allows the fluid to go through a sealed ring. When the temperature decreases, the valve goes back to its original shape and seals the orifice.

SMAs in their quiescent low-temperature state have a martensite (M) phase structure, while in a high-temperature state they have an austenite (A) phase structure (Callister, 1999). The property of shape memory facilitates dramatic shape changes when transitioning between these two crystallographic material phase changes in a structured manner. Mechanical work is done as the shape changes from the low-strength, lower-temperature, martensite phase to the high-strength, higher-temperature, parent austenite phase. The austenite phase exhibits a high stiffness Hooke's law-type constitutive behaviour, whereas the martensite is in a lower-stiffness pseudo-plastic domain. Although transitions are completely reversible, they exhibit hysteresis when the paths are reversed. Moreover, when the transitions are such that the phase transformations are only partially complete, the materials exhibit minor loop hysteresis, which is relatively harder to model.

The ability of SMAs to memorize their original configuration after they have been deformed in the martensitic phase, by heating the alloys above their austenitic characteristic transition temperature (A_f), hence recovering large strains, is known as the shape memory effect and is utilized in the design and development of SMA actuation applications. A shape memory material can be easily deformed when it is in the low-temperature martensitic state and can be returned to its original configuration by heating through the reverse transformation temperature range. This type of shape memory effect is referred to as free recovery. Conversely, in a constrained recovery configuration, the shape memory element is prevented from recovering the initial strain and consequently a large tensile stress (recovery stress) is induced. When the shape memory element performs work by deforming under load, the recovery is restrained. It is this feature that can be exploited in using the material in an actuating mechanism.

1.4.4 Electro- and Magnetorheological Effects

Rheology can be defined as the study of the flow of matter, particularly solids that can flow rather than deform elastically. An electrorheological (ER) fluid is a kind of colloid consisting of dielectric particles suspended in a non-conducting liquid. The dielectric rheological model is defined as a colloidal dispersion of dielectric particles whose electrical response is governed by linear electrostatics. The ER fluid can change from liquid to solid state in a relatively short time (1–10 ms). When no external electrical field is present, the electric dipoles within the dielectric particles are randomly oriented. Under the action of a mild external electrical field, the dipoles align themselves and form chains of dipoles. Under the action of a strong external electrical field, the chains close in towards one another in small groups and form columns. While the chains are easily broken and deformed by a small shear force, it is relatively more difficult to break or deform the columns. Thus, a structural behavioural transition always accompanies the rheological transformation. The dipole interaction model can be used effectively to qualitatively explain the structural behavioural transition.

Like an ER fluid, a magnetorheological (MR) fluid is a suspension of carbonyl iron and soft iron particles in a carrier fluid that is typically a hydrocarbon-based oil. The MR fluid can change from liquid to solid and vice versa by the application of a strong magnetic field. It makes the structures in which it is embedded particularly flexible so that they can adapt to relatively large external forces that would normally make some common rigid structures snap.

1.5 Survey of Smart Polymeric Materials

One of the first plastic transistors was conceived and built by Francis Garnier at the Centre National de la Recherche Scientifique's Molecular Materials Laboratory, in the late 1990s. The device was feasible due to the discovery of a group of organic semiconductors called conjugated oligomers. These organic materials belonged to a group of plastics. While conventional transistors rely on silicon, the new semiconductor devices were made from organic semiconductors, with polymer-based conducting ink replacing the metal electrodes.

1.5.1 Novel Inorganic Thin Film Materials

The controlled generation of novel thin film materials is another technological development, leading to the development of new composite materials. The new developments involve meso-structured thin film synthesis with control of structural features over several length scales. These controlled methods involve a number of techniques specifically designed for thin films as well as other processing methods such as electron beam-induced deposition, pulsed laser deposition, ion beam-assisted deposition, chemical vapour deposition, chemical solution deposition and other vapour deposition techniques, vacuum-based deposition processes and deposition using a sputtering process and heat treatment, which are specially adapted to produce high-quality thin films. A number of new and novel measurement and control techniques applied at atomic and molecular levels have emerged involving ellipsometry, plasma analysis, scanning probe microscopy, Raman spectroscopy, and scanning and transmission electron microscopy. The synthesis of controllers at molecular and atomic levels involves the modelling of quantum level dynamical processes and the synthesis of controllers for these dynamical systems. Stochastic models are therefore incorporated as a matter of course, and one must necessarily rely on probabilistic control synthesis techniques with a high degree of predictability.

1.5.2 Integrative Polymeric Microsystems

The development of organic semiconductors paved the way for the development of ICs and microsystems in plastics. In classical MEMS technology, electronic components are fabricated using IC process sequences, while the micromechanical components are fabricated using compatible 'micromachining' processes that selectively etch away parts of the silicon wafer or add new structural layers to form the mechanical and electromechanical devices. By combining MEMS technology, which is the integration of mechanical elements, such as sensors, actuators and electronic components etched on a common silicon substrate by microfabrication, with organic semiconductors and polymer-based conductors, prototype-integrated active polymeric microsystems are currently being built which will facilitate the design and construction of smart structures.

1.5.3 Electroactive Polymers

EAPs (Bar-Cohen, 2002) can be divided into two major categories based on their activation mechanism: ionic and electronic. Coulomb forces drive electronic EAPs, which include electrostrictive, electrostatic, piezoelectric and ferroelectric types. These EAP materials can be made to hold the induced displacement while activated under a DC voltage, allowing them to be considered for robotic applications. They have a greater mechanical energy density, and they can be operated in air with no major constraints. However, electronic EAPs require a high activation field ($>100\,\mathrm{V/\mu m}$), which may be close to the breakdown level.

In contrast to electronic EAPs, ionic EAPs are materials that involve mobility or diffusion of ions. They consist of two electrodes and an electrolyte. The activation of ionic EAPs can be made as low as 1–2 V, and mostly a bending displacement is induced. Examples of ionic EAPs include gels, polymer–metal composites, conductive polymers and carbon nanotubes. Their disadvantages are the need to maintain wetness and difficulties in sustaining constant displacement under the activation of a DC voltage (except for conductive polymers). SAW-IDT actuators also represent a class of piezoceramic or EAP-based smart-materials actuators driven by a modulated microwave.

1.6 Shape Memory Materials

There are a number of materials that are capable of changing their shape when presented with an external stimulus. Such materials also possess a certain level of shape memory ability, although there is often no common mechanism governing this phenomenon. Materials with the shape memory property include certain alloys, magnetic materials, polymers and gels.

1.6.1 Shape Memory Alloys

SMAs are materials capable of changing their crystallographic structure, following changes in temperature and/or stress, from a high-temperature state or austenite phase to a low-temperature state or martensite phase. These changes, referred to as martensitic phase transformations, are reversible. In general, characterization of phase transformation is done by four transformation temperatures: martensite start (M_s), martensite finish (M_f), austenite start (A_s) and austenite finish (A_f) temperatures. The SMAs used in the design of actuators for lifting surfaces are K-type NiTiCu alloys from Memry Corporation and Nitinol (*Ni-Ti*), which was the first SMA to be commercially developed when it was discovered in 1965.

The SMA specimen in the austenitic phase will expand to the maximum amount when the temperature drops below the martensite finish temperature M_f without a load on the specimen. Thus, the formation of detwinned martensite will take place directly from the austenitic phase without any externally applied load. Since $M_f < M_s < A_s < A_f$ for certain materials such as *K*-type *NiTiCu* alloys and $M_f < A_s < M_s < A_f$ for *Nitinol*, and typically ranges from about 30 to 85°C for certain *K*-type *NiTiCu* alloys and from $-12°$ to $49°$ for *Nitinol* wires, we will need to raise the temperature by about 50–60° to A_f to contract the material followed by cooling to just below M_f to expand it fully. It is this feature that is used to actuate other structures. SMA heating, which triggers the martensite-to-austenite (M to A) phase change, is normally achieved electrically by passing a high current through the SMA wire. However, the cooling process cannot be achieved by similar means and convective heat transfer-based methods must be adopted. These generally require more time, and therefore the actuator time constants and bandwidths involved are critically dependent on the cooling times. The high-energy density of fuel-based combustors compared to typical electrical batteries, or even fuel cells, allows for a dedicated the fuel combustor to be used as a heat source inside the actuator system. Although the thermoelectric heat transfer mechanism utilizing semiconductors, based on the Peltier effect, can result in a higher bandwidth, this method has a very low efficiency and convective cooling is still the preferred approach for reducing the SMA's temperature below M_f.

1.6.2 Magnetically Activated Shape Memory Alloys

Magnetically activated SMAs are novel smart materials that exhibit a magnetic field-induced shape memory effect of up to 10% strain. The magnetic shape memory effect is only possible if the martensite -to-austenite transformation is feasible in the alloy and when the material is also magnetic. Several alloys containing manganese (Mn) in addition to Ni and third element, such as aluminium (Ni_2MnAl) or gallium (Ni_2MnGa), are found to have the property. Certain alloys of iron and palladium (Fe_xPd_y) also exhibit the magnetic shape memory effect. The magnetic moment in the material originates mainly from Mn in the Mn-containing alloys, and the existence of different structural phases is due to a band level Jahn–Teller effect in the Ni band. The Jahn–Teller effect refers to a distortional instability of the martensitic tetragonal structure due to a magnetocrystalline anisotropy, which induces the phase transformational changes in the material under the influence of a magnetic field.

1.6.3 Shape Memory Polymers

There exist a number of self-standing polymeric materials capable of changing shape under the influence of specific external stimuli (electrical, chemical, thermal and radiation). This is an essential feature of several liquid crystalline polymers, liquid crystalline elastomers and polymer gels synthesized from cross-linked polymers. A gel is simply a solid but jelly-like material, constructed from a cross-linked polymer that additionally contains a solvent.

Shape memory polymers are based on the minimum energy principle that the polymeric configuration at the point of cross-linking is the one with the lowest energy. Any deformation away from that configuration increases the energy. Thus, when the polymer settles to a minimum energy configuration it regains the shape associated with that particular configuration.

1.7 Complex Fluids and Soft Materials

A number of complex fluids have been developed, such as colloidal fluid suspensions of non-adsorbed polymers and polyelectrolytes, as well other particle-based suspensions that have resulted in new fluid stability and control problems relating to inter-particle potentials. Polyelectrolytes are charged, water-soluble polymers that can be processed at room temperatures. Control of the inter-particle potentials is essential if one wishes to control the properties of the complex fluids, which in turn could be used in the synthesis of a number of other smart structures.

1.7.1 Self-assembled Fluids

Biologically produced composites and fluids are fabricated by a highly coupled synthesis process involving self-assembly. These structures are formed by a template-assisted self-assembly where a primary organic material such as a protein or lipid forms an initial scaffold, which then provides a structure for the deposition of inorganic materials. The process of self-assembly is a multi-step one. The first step is the selection and operation of a mechanism for the assembly of the organic scaffold. The second is the orientation and shaping of the scaffold, which must be accomplished while the deposition of the inorganic material is on-going. The third step is the controlled nucleation and growth of inorganic material in specific locations. The final step involves the assembly of the hierarchically organized structure.

Biomimetically inspired methods of synthesis of such materials is feasible due to the availability of methods of controlling the microstructure of the materials by applications of methods used in the control of thin films. These synthesis techniques are collectively known as biomineralization and involve atomic level control and sensing, including molecular recognition and purpose-designed thin film processing techniques.

1.7.2 Electro- and Magnetorheological Fluids

ER and MR effects were introduced in an earlier section. It is often necessary to tune and improve ER and MR properties to match the requirements of a particular application. This can be done by coating the dielectric particles (generally a particular kind of glass particle) suspended in a carrier fluid by a metallic film. A suitable metal for such a coating is nickel, and a further coating with titanium oxide (TiO_2) results in substantial improvements in the ER property. However, the maximum yield stress of the ER fluid saturates at some high level, typically in the order of 10 kPa. Thus, the ER property of the fluid can be matched to a particular requirement and therefore finds applications in the design of several smart devices such as smart clutches and dampers involving active feedback.

1.7.3 Smart Polyelectrolyte Gels

Some gels have the ability to swell when immersed in a multi-component aqueous solution. Gel swelling may be used to force braided structures to change shape from one form from to another, hence generating useful work. Polyelectrolyte gels possess mechanical and actuation properties very similar to living matter. For example, one polyelectrolyte gel, polyacrylamide gel, is known to swell in solution, increasing its volume by up to 100%. The swelling may be induced by a change of temperature, by controlling the acetone content of the solution or by controlling the acidity. The swelling is completely reversible, and the gel may be shrunk back to its original shape.

Hydrogels are colloidal gels with water as the dispersion medium. They are used in gas and oil exploration drilling. Temperature-sensitive hydrogels prepared from a mixture of monomer, a polymer of the monomer and its cross-linked counterpart are known to have fast deswelling and swelling rates. They may be used for the design and development of shape-changing actuators. However, gels are characterized by both hysteresis and creep, which must be properly compensated for during the design of the actuators.

1.8 Active Fibre Composites

Active fibre composites (AFCs) have been developed by embedding active fibres such as piezoelectric ceramic or SMA fibres into a polymer matrix, and have several advantages over monolithic active materials such as piezoelectric ceramics and SMAs. AFCs are thin composite plies comprising unidirectional active fibres embedded in a thermoset polymer matrix, as illustrated in Figure 1.3.

An electric field is supplied to the fibres embedded in the outer layers via an inter-digitated set of electrodes bonded on either side of the AFC. AFCs facilitate anisotropic planar actuation, with the matrix providing the load transfer and an even load distribution, thus making it an ideal actuator for shape control. Although there is a marked reduction in the loads generated due to a mismatch of dielectric properties of the fibres and the bonding materials, recent research has shown that by appropriate impedance matching techniques and by using functionally graded materials the mismatch may be minimized and the actuator's performance maximized. The impedance matching is provided for by interfacing an inductive impedance in series with the input electrodes. The entire circuitry is integrated by employing very large system integration (VLSI) techniques.

1.9 Optical Fibres

Although John Tyndall demonstrated as early as 1870 that light could be trapped within in a narrow jet of water by virtue of the property of total internal reflection, and William Wheeling actually patented a

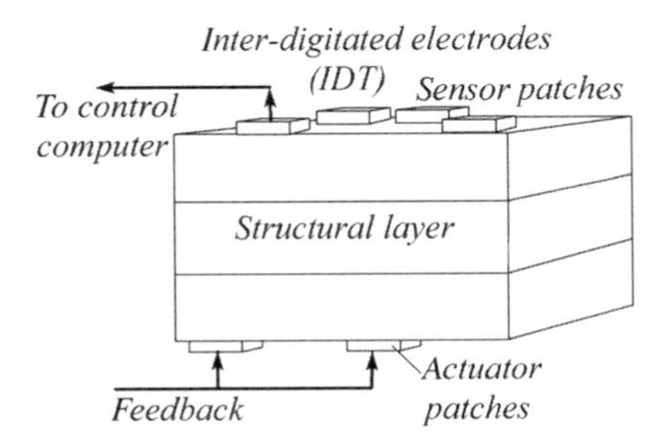

Figure 1.3 Concept of an active composite material

Table 1.1 Attenuation of light in the visible and near infra-red regions

Transparent material	λ_0 (μm)	Attenuation in decibels per kilometre (db/km)
Window glass		10 000
Glass fibres (1960)		1000
Optical glass (spectacles)		300
Pure water	0.5	90
Silica glass optical fibres		
Corning (USA) 1970		20
Corning (USA) 1972		4
Corning (USA) 1973		2
Fujikawa (Japan)	0.83	1
Fujikawa (Japan)	1.10	0.5
Several	1.30	<0.5
Approximate present limit		0.1–0.2

method for piping light, the concept of an optical fibre being employed to pipe light was concurrently conceived in 1956 by Brian O'Brien at the American Optical Company and by Narinder Kapany and colleagues at the Imperial College of Science and Technology in London. In 1970, the fact that some amorphous materials such as glass have a high transparency and allow for the relative lossless guidance of light prompted Drs Robert Maurer, Donald Keck and Peter Schultz of Corning Glassworks in the USA to fabricate several hundreds of metres of glass fibre with a minimal attenuation as low as 20 db/km. This achievement was a significant breakthrough that led to the subsequent development of optical fibres with substantially lower attenuation, as shown in Table 1.1.

Although it was known that light could be guided with a very small loss in a cylindrical rod made from optically transparent non-dissipating dielectric material, the identification of the precise causes for light attenuation and the progressive elimination of these causes have resulted in the development of optical fibres with minimal attenuation. A primary cause for the absorption of optical energy is the presence of free electrons in the fibre material as a result of impurities. Thus, the minimization of the impurity content in the fibre material has led to the development of fibres with minimal attenuation. Yet the overall development of optical fibre-based transmission systems was also in part due to developments in semiconductor technology, which resulted in the fabrication of compact semiconductor-based light sources, light sensors and other components that essentially make up a fibre-optic transmission link.

1.10 Smart Structures and Their Applications

Structures with added functionality over and above the conventional purpose of providing strength by reinforcement or stiffness may be regarded as smart. Smart or adaptive structures, based on using a small change in the structure geometry at critical locations induced by internally generated control signals, can result in a non-linear amplification of the shape, stiffness or strength, and so the structure will adapt to a functional need. In practice, smart structures may be classified depending on their functionality and adaptation to the changing situation:

1. Passive smart: providing additional features-related stiffness and strength in a passive mode, involving no form of internal energy generation or inputs.
2. Active smart: tuning functionality to specific inputs such as changing the stiffness or strength in response to a controllable input.
3. Intelligent: adapting their functionality to changing environment and also requiring some form of 'intelligence' of the actuation control in the form of an internal self-adapting feedback loop or mechanism.

An adaptive or smart structure is a structure using a distribution of sensors, control filters and actuators to generate the essential control forces and moments at critical locations in the structure (Wagg *et al.*, 2007). Measurements relating to the structure's deformation or forces and moments acting on it are used in a control algorithm that is based on robust control theory and structural dynamics. Depending on what the operator or controller commands, the control algorithm sends command signals to the actuator to elongate, twist, etc. to generate the desired control forces that change the shape, stiffness or strength of the structure.

1.10.1 Medical Devices

Smart structures are currently being employed in the manufacture of a number of medical devices (Machado and Savi, 2003), including orthopaedic applications such as bone clips and plates; spinal spacers; actively controlled catheters and stents; general surgical tools such as removal baskets, lachrymal probes, suction tubes and epicardial retractors; surgical tools for laparoscopic and endoscopic surgical operations; a variety of cardiovascular devices and implants such as Simon filters and septal occlusion plugs; bionic prosthetic devices; and a host of other biomimetic robotic structures (Cohn *et al.*, 1995; Chonan *et al.*, 1996; Fatikow and Rembold, 1997; Sastry *et al.*, 1997; Vepa, 2009). Actively controlled grasping structures are another area of development that would benefit from emerging and innovative smart structures.

Endoscopy involves a minimally invasive surgical procedure inside the human body. Present-day endoscopes used in minimal access surgery are rigid, inflexible and not controllable. Laparoscopy not only is a term given to a group of operations that are performed with the aid of a camera placed in the abdomen but also refers to minimally invasive surgery in the abdomen via a small incision. Instruments used in endoscopic surgery via natural incisions have severe limitations, such as rigidity and inflexibility, and are often not adequately controllable. The difficulties faced in laparoscopy are the requirement for coordinated handling of instruments (hand–eye coordination) and the limited workspace, which increases the possibility of damage to surrounding organs and vessels either accidentally or through the complexity of the procedures.

A novel approach that can be applied to endoscopy and laparoscopy is to employ actively controlled micro-manipulators. Such micro-manipulators employ flexible linkage systems that can be completely controlled to any shape or geometry as desired. The control actuation is performed by smart actuators employing smart composites with layers of piezopolymers or piezoceramics or embedded SMA wires, in addition to conventional devices. The embedded smart actuators are complemented by a matching set of embedded smart sensors. Together these systems can be employed to achieve the closed loop servo control of the micro-manipulator.

1.10.2 Aerospace Applications

Aerospace applications of smart structures are primarily related to the development of adaptive flow control devices that induce large aerodynamic changes due to drag reduction, increase in propulsive efficiency or increase in lift forces by introducing small shape changes or changes in geometry at key locations in the flow. There also a class of actively controlled couplers, fasteners and fixtures used for remote assembly of structures. The monitoring of aircraft structures is fast emerging as a vital application area and is briefly discussed in the next section.

1.10.3 Structural Health Monitoring

An emerging application of smart structures is the monitoring of the 'health' of a structure in a critical application, such as an aircraft structure, where it is essential to continuously assess its ability to withstand the effects of aerodynamic loading without suffering any damage due to fatigue and fracture.

Structural health monitoring (SHM) of aircraft structures, especially composite structures, is now considered an essential element of the suite of on-board monitoring systems (see for example Sohn *et al.*, 2003; Staszewski *et al.*, 2004, for extensive overviews of the field). In ageing aerospace systems, a retrofitted SHM system would alert users of incipient damage, preventing catastrophic failure. For newer systems, incorporating an SHM can reduce life-cycle costs. Central to such SHM systems is the ability to detect damage in a structure. With the extensive use of a variety of co-cured structures, wherein bonded joints are replacing bolted joints, there is a concern about a number of de-bonding, delamination and crack propagation issues. It is hence necessary to be able to detect and assess potentially damaging events before they achieve criticality. For aerospace structures, the primary sensing issues are (1) measuring the reaction of the structure to external loads and (2) determining the internal state of health of the structure. For these purposes, sensors are embedded and spatially distributed in the structure. Sensors have been developed to detect variations in crack formation, strain, temperature and corrosion. A significant advantage of several recent sensor designs is the ability to multiplex a number of continuous or discrete sensors on one fibre to form a distributed sensor system. The availability of distributed SHM offers a means of predicting fatigue in real time based on the distributed measurements. However, most of the current implementations are limited to passive structures. In the case of active structures, there is not only the distinct possibility of integrating the SHM function with the existing suite of transducers but also the fact that custom SHM sensors could also support the active vibration or shape control functions.

Recent advances in fibre-optic sensing have led to the availability of several technologies in place of the classical electrical, mechanical or vibrating wire strain gauges for strain monitoring. Thus, one family of health monitoring strategies is through strain monitoring using fibre-optic strain sensors based on measuring changes to an optical path or a secondary feature due to changes in strain, rather than the more conventional strain gauges. Among them are a variety of fibre-optic-based techniques, including small-diameter Fibre–Bragg grating (FBG) sensors and interferometric Fabry–Perot sensors, which are available for direct local strain measurements; laser displacement sensors can provide a very accurate and reliable measurement of the relative displacement of any two points chosen in a structure at large distances from a reference station. The working principles of a selection of these sensors are explained in Chapter 2. Other fibre-optic sensors are based on polarimetry or Raman or Brillouin scattering. Although experimental and field monitoring of several fibre-optic sensors show that extrinsic Fabry–Perot fibre-optic sensors perform linearly and show good response to thermal variations and mechanical loading conditions, intra-core FBG sensors are associated with a number of advantages related to the stability of the measurements, the potential long-term reliability of the optical fibres and the possibility of performing distributed and remote measurements. Notwithstanding these advantages, there are indeed a number of practical issues related to the implementation of the technique, including the characterization, performance in an embedded environment, and cost of the FBG sensors and the associated signal processing.

An alternative strategy is to employ piezoelectric embedded thin film actuators and sensors for SHM. While there are several advantages in employing thin film piezosensors, a third strategy that effectively realizes the advantages of using both FBGs and piezosensors is to employ a variety of functionally different and distributed sensors. Such an approach not only provides for complementary redundancy but also offers a relatively low-cost approach to SHM. Moreover, such an approach would permit the incorporation of global positioning system (GPS) measurements on board an aircraft to simultaneously monitor aircraft structural vibrations, which in turn would allow for the motion measurements to be correlated in real time with the damage developing in the structure. Thus, it is possible to develop and implement a multi-functional SHM system and integrate it with existing active control loops.

Exercises

1. Identify the various transducers and transduction mechanisms that are particularly important for measurement and control applications at a molecular level.

2. List the key features that characterize the dynamics of molecular systems with particular emphasis on quantum dynamical principles. How would the modelling and synthesis of controllers for quantum mechanical systems differ from conventional modelling and synthesis techniques?

3. Discuss the primary sources of noise and interference in measurements associated with thin films. Describe and distinguish between these noises sources.

4. Write a brief survey of the main techniques for damage detection in structures. Compare and contrast these techniques with reference to the nature of measurements that must be made, stimuli that may have to be supplied and the signal processing that must be performed.

References

Agarwal, B. D. and Broutman, L. J. (1990) *Analysis and Performance of Fiber Composites*, 2nd edn, John Wiley & Sons Inc., New York.

Askeland, D. R. and Phule, P. P. (2003) *The Science and Engineering of Materials*, Thomson Books Inc, WA.

Bar-Cohen, Y. (2002) Electro-active polymers: current capabilities and challenges, Paper 4695-02, *Proceedings of the SPIE Smart Structures and Materials Symposium*. EAPAD Conference, San Diego, CA, March 18–21.

Boch, P. and Niepce, J.-C. (eds) (2007) *Ceramics Materials: Processes, Properties, and Applications*, ISTE Ltd, London.

Callister, W. D. (1999) *Materials Science and Engineering: An Introduction*, 5th edn, John Wiley & Sons Inc., New York.

Chonan, S., Jiang, Z. W. and Koseki, M. (1996) Soft-handling gripper driven by piezoceramic bimorph strips, *Smart Materials & Structures* 5, 407–414.

Cohn, M., Crawford, L., Wendlandt, J. and Sastry, S. (1995) Surgical applications of milli-robots, *Journal of Robotic Systems* 12(6), 401–416.

Daniel, I. M. and Ishai, O. (2005) *Engineering Mechanics of Composite Materials*, 2nd edn, Oxford University Press, Oxford.

Fatikow, S. and Rembold, U. (1997) *Microsystem Technology and Microrobotics*, Springer-Verlag, Berlin, Heidelberg.

Hearle, J. W. S. (1982) *Polymers and Their Properties, Fundamental and Structures and Mechanics*, Vol. 1, John Wiley & Sons Inc., New York.

Machado, L. G. and Savi, M. A. (2003) Medical applications of shape memory alloys, *Brazilian Journal of Medical and Biological Research* 36, 683–691.

Sastry, S., Cohn, M. and Tendick, F. (1997) Millirobotics for remote minimally invasive surgery, *Robotics and Autonomous Systems* 21, 305–316.

Schultz, J. M. (1974) *Polymer Materials Science*, Prentice-Hall, Inc., New York.

Sohn, H., Farrar, C. R., Hemez, F. M., Shunk, D. D., Stinemates, D. W. and Nadler, B. R. (2003) A review of structural health monitoring literature: 1996–2001, Los Alamos National Laboratory Report, LA-13976-MS.

Srinivasan, A. V. and McFarland, D. M. (2000) *Smart Structures: Analysis and Design*, Cambridge University Press, Cambridge.

Staszewski, W. J., Boller, C. and Tomlinson, G. R. (eds) (2004) *Health Monitoring of Aerospace Structures: Smart Sensor Technologies and Signal Processing*, John Wiley & Sons Ltd, Chichester.

Taya, M. (2005) *Electronic Composites*, Cambridge University Press, Cambridge.

Vepa, R. (2009) *Biomimetic Robotics*, Cambridge University Press, Cambridge.

Vijaya, M. S. and Rangarajan, G. (2004) *Materials Science*, Tata McGraw-Hill, India.

Wagg, D., Bond, I., Weaver, P. and Friswell, M. (eds) (2007) *Adaptive Structures – Engineering Applications*, John Wiley & Sons Ltd, Chichester.

2

Transducers for Smart Structures

2.1 Introduction

Active or 'smart' structures usually contain embedded and/or surface-mounted patches of piezoelectric materials in a polymer base. Vibration suppression and shape control of lightweight structures based on feedback control principles may be achieved by the use of distributed actuation and sensing. Conventional transducers are essentially discrete dynamic systems and are capable of sensing or actuation only at certain discrete locations. Because of their capacity to convert mechanical deformation into electric voltages, and electric voltages into mechanical motion, piezoelectric materials are used in surface-mounted patches. The surface-mounted patches of piezoelectric materials may be used as distributed actuator patches to apply distributed forces on a structure or as distributed sensor patches to measure the structural deformation over a distributed region.

The piezoelectric effect, a property of several classes of materials, and discovered by Jacques and Pierre Curie in 1880, refers to the fact that certain materials become electrically polarized when a mechanical stress is applied across two faces of the material. Consequently, there is a net electric moment and an electric charge appears across the two faces. This feature was first observed in quartz, which also features the converse (or inverse) piezoelectric effect: an electric field applied across the quartz crystal generates mechanical strains as well. Thus, the piezoelectric effect manifests itself in two forms: the direct piezoelectric effect and the converse piezoelectric effect. The direct piezoelectric effect is the appearance of an electric potential across certain faces of a crystal when it is subjected to mechanical pressure. When an electric field is applied on certain faces of the crystal, the crystal undergoes mechanical distortion, and this is the converse piezoelectric effect. The latter is a considerably smaller effect than the former and is also not that common. The piezoelectric effect occurs in several crystalline substances. The effect is caused by the displacement of ions in the simplest polyhedron that makes up the crystal structure, the unit cell, due to its inherent lack of symmetry. When the crystal is subjected to compression, the ions in each unit cell are displaced, causing the electric polarization of the unit cell. Because of the regularity of crystalline structure, these effects are coherent and, as a consequence, they accumulate rather than cancel each other. This causes the appearance of an electric potential difference between certain faces of the crystal. When an external electric field is applied to the crystal, the ions in each unit cell are displaced by electrostatic forces, resulting in the mechanical deformation of the whole crystal.

Certain ceramics and polymers also exhibit piezoelectric properties, although not in their natural state, but rather only after the temporary application of a strong electric field. The process of making polycrystalline ceramics and polymers piezoelectric is known as poling, a process that is analogous to magnetizing a permanent magnet.

Dynamics of Smart Structures Dr. Ranjan Vepa

Piezoelectricity has also been observed in certain ferroelectric crystals; in a ferroelectric crystal, each cell of the crystal lattice is known to spontaneously polarize along one of a set of preferred directions. This spontaneous polarization disappears at a critical temperature (the Curie temperature or Curie point), above which the crystal remains 'paraelectric' or non-polar. If the crystal is cooled through its Curie point in the presence of an external electric field, the dipoles tend to align in a preferred direction that is most nearly parallel to the direction of the applied field. This 'poling' process, as it is known, involves heating the crystal well beyond the Curie temperature, followed by the application of a strong electric field of the order of 2000 V/mm or more. The result is the appearance of groups of ferroelectric, or Weiss, domains within which the electric dipoles are parallel to each other. The alignment is preserved after the material is slowly cooled through the Curie point. When such a 'poled' crystal is subjected to mechanical stress, the aligned lattice structure will distort and there will be a resultant change in the dipole moment of the crystal. Within certain characteristic limits of applied stress, this change in the dipole moment results in a charge that is directly proportional to the applied stress. The effect is also reversible, without exception, in all poled crystals that exhibit the piezoelectric effect. All ferroelectric materials exhibit piezoelectricity, albeit to a different extent, depending on the nature of the crystalline or semi-crystalline structure. Piezoelectricity is the result of interactions between the force and electric charge within a material and depends on the symmetry of the material. As the symmetry of the material is a consequence of the crystal class to which it belongs, piezoelectricity can be directly associated with 10 of the 32 crystal classes. Typical examples of such ferroelectric crystals are tourmaline and Rochelle salt. The property is also commonly observed in several ceramics that are generally classed as piezoceramics. These include lithium niobate, barium titanate ($BaTiO_3$), lead titanate ($PbTiO_3$) and lead zirconate ($PbZrO_3$), which can be modified by additives. They belong to a general class of ceramics referred to as lead zirconate-titanate (PZT) and are used extensively to build commercial-grade piezoactuators.

Synthetic polymeric materials have also been observed to exhibit a lasting and substantial electric dipole polarization. Two types of polymers are known to exhibit the property of ferroelectricity: semi-crystalline and amorphous. The latter, including materials such as polyvinyl chloride (PVC) and polyacrylonitrile (PAN), belong to a class of non-crystalline non-ferroelectric materials. Typical examples of the crystalline polymers are polyvinyl fluoride (PVF) and polyvinylidene fluoride (PVF_2 or PVDF), which are termed as electrets since they acquire a permanent polarization when crystallized under an electric field. The discovery of these piezopolymers led to the discovery of others such as difluoroethylene (VF_2) and its co-polymers with related fluorinated vinyl monomers such as vinyl fluoride (VF), trifluoroethylene (VF_3) and tetrafluoroethylene (VF_4), which tend to exhibit the piezoelectric effect to a substantially greater extent. These semi-crystalline polymers are known to belong to one of four classes. Of these, the properties of PVF_2 are extensively documented and are the basis of several commercial-grade piezopolymeric actuators and sensors. They are made of long-chain molecular structures involving the unit [CF_2–CH_2], which repeats itself in the chain. The chemistry and processing of this class of polymers have been considerably optimized, and this results in materials with robust piezoelectric properties.

Physically in a piezoelectric material patch there is a coupling between the electric and elastic fields. As a consequence, there is an interactive behaviour between the electrical and mechanical responses of the structure and the patch. While there are many approaches to modelling such systems, all the modelling methods rely on either an electrical energy balance relation or a constructive application of the material constitutive relations. The linear piezoelectric material constitutive equations coupling the elastic and electric field can be expressed as the direct and converse piezoelectric equations:

$$\mathbf{D} = \mathrm{E}\,\mathbf{E} + \mathbf{e}\,\mathbf{S}, \tag{2.1a}$$

$$\mathbf{T} = -\mathbf{e}^T\,\mathbf{E} + \mathbf{C}^E\,\mathbf{S}, \tag{2.1b}$$

where $\mathbf{D}$ is the electric displacement vector, $\mathbf{E}$ is the electric field vector derived from the negative of the gradient of the electric potential, E is the dielectric permittivity matrix evaluated at constant elastic strain, $\mathbf{e}$ is the piezoelectric stress matrix, $\mathbf{S}$ is the elastic strain vector, $\mathbf{T}$ is the elastic stress vector and $\mathbf{C}^E$ is the elastic matrix (Hooke's law) at constant electric field. These relationships form the basis for deriving the governing equations describing the behaviour of actuator and sensor patches. By assuming that these relationships are essentially static, the first of these may be integrated over the depth of a typical actuator patch to derive a relationship between the input voltage and the elastic and piezoelectric moments acting on the structure. When the second equation is integrated over the depth of a sensor patch, the result is a relationship between the output voltage and strain in the patch to the deformation of the structure.

2.2 Transducers for Structural Control

A transducer is defined as a device that transforms one form of signal or energy into another form and is generic term that is used to refer to both sensing and actuating transducers. There are many ways of classifying transducers, and we have already classified them as being either passive or active, depending on the need for an external energy source to drive them. Passive transducers generally employ only passive circuit elements and require an external energy source to drive them, while active transducers involve some form of internal energy generation, which implies energy transfer from the environment or an 'unlisted' source. When measuring a mechanical entity, it is desirable to first convert to some form of angular position by employing a kinematic mechanism and then measuring the angular position by a transducer. The primary types of passive transducers employed in most actuator and sensor packages are resistive, inductive or capacitive types. To interface to a controller, a typical transducer is usually connected to a signal conditioning device such as an encoder or an analogue-to-digital converter (*A*-to-*D* converter). The *A*-to-*D* converter is itself, in many cases, a feedback device that employs a digital-to-analogue converter.

2.2.1 Resistive Transducers

Possibly the most common way of measuring angular position is with a rotary potentiometer. This consists of an uninsulated resistance element with movable contact that is capable of making contact with the resistance element at either end or at any conducting point along the element. The ends of the rotary potentiometer are connected to opposite-polarity power supplies, and the output of the moving contact or wiper gives an indication of its angular position relative to either of the ends. Thus, when the wiper is in contact with one end the output across it is 0 V, while at the other end it is the maximum voltage applied across the terminals. When the wiper is at any intermediate point, the output across it is linearly proportional to its position relative to the 0 V end. Resistance elements in common use are wire wound, carbon film or conducting plastic.

Linear potentiometers may also be constructed on the same principle and may be employed to measure the relative position of the wiper. A typical example is illustrated schematically in Figure 2.1.

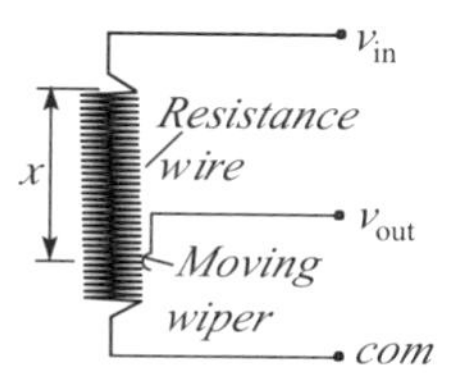

Figure 2.1 Schematic illustration of a linear potentiometer

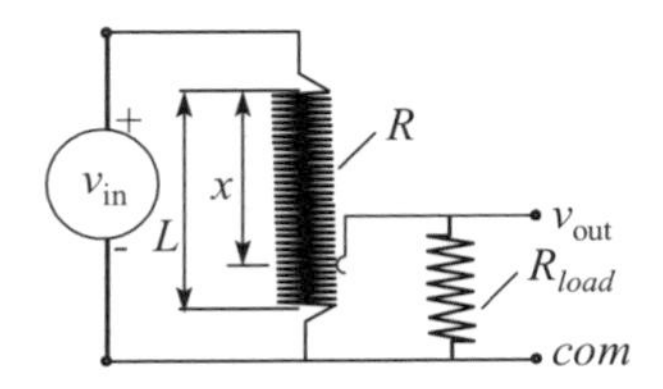

Figure 2.2 Typical potentiometer circuit

Figure 2.2 illustrates a typical application of the potentiometer where the output is connected to a resistive load. Applying Kirchoff's voltage law to the loops and Kirchoff's current laws to the junctions, the analysis of the circuit in Figure 2.2 gives

$$i_{total} = \frac{v_s}{R}\left(\frac{R\left(1-\frac{x}{L}\right)+R_{load}}{R\left(1-\frac{x}{L}\right)\frac{x}{L}+R_{load}}\right), \tag{2.2a}$$

$$i_{load} = i_{total}\frac{1}{1+\frac{R_{load}}{R\left(1-\frac{x}{L}\right)}} = i_{total}\frac{R\left(1-\frac{x}{L}\right)}{R\left(1-\frac{x}{L}\right)+R_{load}} \tag{2.2b}$$

and

$$i_{load} = \frac{v_s}{R}\left(\frac{R\left(1-\frac{x}{L}\right)}{R\left(1-\frac{x}{L}\right)\frac{x}{L}+R_{load}}\right). \tag{2.3}$$

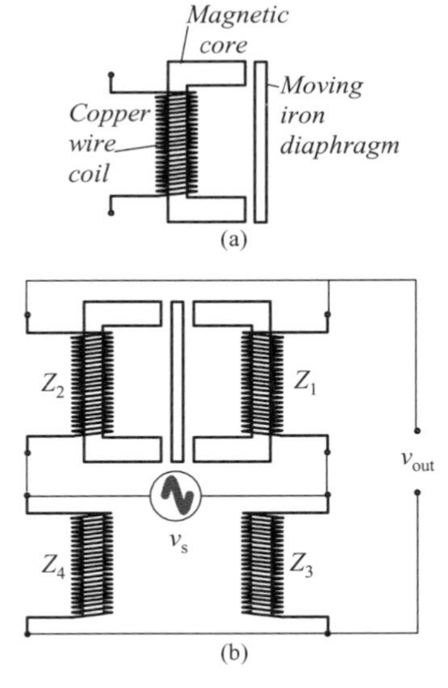

Figure 2.3 (a) Schematic diagram of a typical inductive EI pickoff with moving iron diaphragm. (b) Typical example of differential inductive EI pickoff with a Whetstone's bridge circuit used to measure position of the iron diaphragm

Thus, as a result of the voltage divider effect, the output voltage is considerably lower and the behaviour is non-linear, which is definitely not desirable.

When the assumption is made, the above expression reduces to

$$i_{load} = \frac{v_s}{R_{load}}\left(1 - \frac{x}{L}\right), \tag{2.4}$$

and the potential difference across the load is a linear function of the position x of the wiper. Thus, the output voltage is a linear function of the position x of the wiper only when the load impedance is relatively much larger than the total resistance of the potentiometer R. In practice, it is naturally important to ensure that this condition is satisfied. This is usually done by introducing an operational amplifier (which has a very large input impedance) at the output stage as a buffer.

A major limitation of potentiometers, or pots, is that the resolution of these devices is always limited. Typically, these are constructed by winding a high-resistance wire onto a circular or straight tubular insulating core. The wiper moves over the wire wound core, from one turn to the next. The resistance therefore increases in small finite increments, and this limits the resolution of the pot. The situation is considerably improved when carbon film or conductive plastic-deposited cores are used. The presence of mechanical contacts results in noise due to contact-related effects and also severely limits the useful life of these elements.

Strain gauges are based on the principle of resistance measurement. Strain gauges are assembled from elements that change resistance when small mechanical displacements result in a tensile stress in them. However, there are no mechanical parts and the variation of the resistance of a length of conducting wire is caused by the application of the tensile stress to it. They also require a well-regulated power supply or a stable battery. There are a number of materials that exhibit the property of variation of resistivity due to the application of a tensile stress. For example, consider a wire of length L and area of cross-section A with a coefficient of resistivity given by ρ. The resistance of this wire is given by $R = \rho(L/A)$. When a tensile stress σ is applied there is a change in the resistance, and it therefore follows that

$$\frac{1}{R}\frac{\mathrm{d}R}{\mathrm{d}\sigma} = \frac{1}{\rho}\frac{\mathrm{d}\rho}{\mathrm{d}\sigma} + \frac{1}{L}\frac{\mathrm{d}L}{\mathrm{d}\sigma} - \frac{1}{A}\frac{\mathrm{d}A}{\mathrm{d}\sigma}. \tag{2.5}$$

Since the area of cross-section of a circular section of wire is proportional to the square of the diameter D:

$$\frac{1}{A}\frac{\mathrm{d}A}{\mathrm{d}\sigma} = 2\frac{1}{D}\frac{\mathrm{d}D}{\mathrm{d}\sigma}. \tag{2.6}$$

assuming that the Poisson's ratio is almost a constant,

$$\frac{1}{D}\frac{\mathrm{d}D}{\mathrm{d}\sigma} = -\nu\frac{1}{L}\frac{\mathrm{d}L}{\mathrm{d}\sigma}.$$

it follows that

$$\frac{1}{R}\frac{\mathrm{d}R}{\mathrm{d}\sigma} = \frac{1}{\rho}\frac{\mathrm{d}\rho}{\mathrm{d}\sigma} + \frac{1}{L}\frac{\mathrm{d}L}{\mathrm{d}\sigma}(1+2\nu). \tag{2.7}$$

The above expression may be written as

$$\frac{1}{R}\frac{\mathrm{d}R}{\mathrm{d}\sigma} = \frac{1}{\rho}\frac{\mathrm{d}\rho}{\mathrm{d}\sigma} + \frac{1}{L}\frac{\mathrm{d}L}{\mathrm{d}\sigma}(1+2\nu) = \frac{1}{L}\frac{\mathrm{d}L}{\mathrm{d}\sigma}G_f, \tag{2.8}$$

where G_f, the strain sensitivity or gauge factor, is defined as

$$G_f = 1 + 2\nu + \left(\left(\frac{1}{\rho}\frac{\mathrm{d}\rho}{\mathrm{d}\sigma}\right)\Big/\left(\frac{1}{L}\frac{\mathrm{d}L}{\mathrm{d}\sigma}\right)\right). \tag{2.9}$$

Table 2.1 Thermoresistivity coefficients

Tungsten	$b_1 = 52$	$b_2 = 70$
Platinum	$b_1 = 35$	$b_2 = -55$

Thus, it is customary to choose a resistance wire with a value as large as possible, and in the case of most metal strain gauges this is limited to $G_f \approx 2$. Another drawback of metal strain gauges is their temperature sensitivity, which is quite significant in comparison to strain sensitivity. Semiconductor strain gauges, although very expensive, have a G_f value in the region of 100–150 and are therefore eminently suitable for several applications. However, these also have temperature sensitivities that are comparable. Using two gauges with similar temperature sensitivities but almost equal but opposite gauge factors alleviates the problem.

Semiconductor strain gauges are also used as strain or force sensors to measure a force on a solid. There are two types of strain or force sensors: quantitative and qualitative. Quantitative strain and force sensors, such as strain gauges and load cells, assess the force and proportionally represent its value into an electric signal. Qualitative strain and force sensors are a Boolean type of output signal and do not represent the force value accurately. They detect if there is a sufficient force applied, and the output signal indicates when the pre-determined threshold is reached. Computer keyboards employ this type of sensor. For example, this distinction is relevant for textile applications: the former finding an application as yarn tension detection and the latter being used for stop motion replacement.

While temperature sensitivity is a problem with strain gauges, it is also the basis of temperature-sensitive resistance thermometers and thermistors. These are based on the change of mobility inside the conductor with temperature. As a general cubic or quadratic relation in the absolute temperature can approximate the temperature dependence of most resistance thermometers, the resistance measured may be interpreted in terms of an equivalent temperature. For example, in the case of tungsten and platinum we may write

$$R_T = R_a \left[1 + b_1 \frac{T - T_a}{10^4} + b_2 \frac{(T - T_a)^2}{10^8} \right], \tag{2.10}$$

where T_a is the ambient temperature in °C, R_a is the resistance at the ambient temperature and R_T is the resistance at the temperature T. The coefficients b_1 and b_2 are as in Table 2.1.

Unlike resistance thermometers, thermistors have a sharp, non-linear, exponentially increasing dependence with temperature and are employed as switching elements when it is desired to switch a circuit at or beyond a critical temperature. They are produced commercially by processing oxides of manganese, nickel and cobalt. Pieces of doped semiconductor materials exhibit a great variation in their resistance with temperature. Thus, with proper biasing semiconductor p–n junctions may also be employed as thermistors.

2.2.2 *Inductive Transducers*

The inductance of a coil indicates the magnitude of flux linkages in the circuit due to the current. Various types of inductive transducers, both for translational and for angular motion, are in common usage. They operate on the principle that the voltage drop v_L across an inductance L, which produces a magnetic field, is proportional to the rate of change of current i, with respect to time, and is $v_L = L(di/dt)$.

The archetypal inductive transducer is the EI pickoff, schematically shown in Figure 2.3. It consists of an electrical inductance coil in a magnetic circuit. The reluctance of the magnetic circuit is varied with a moving piece of iron in such a way that the inductance of the coil is altered whenever there is a change in the position of the iron piece. The change in the inductance is detected by means of an AC bridge circuit.

In the EI pickoff shown in Figure 2.3(a), the flux path includes the moving iron diaphragm as well as air gaps. The inductance L may be expressed in terms of the relative permeability of the magnetic material (μ_r), the number of windings in the coil (N) and the geometry of the iron and air circuits as

$$L = \mu_0 N^2 / \left(\frac{l_{air}}{A_{air}} + \frac{l_{iron}}{\mu_r A_{iron}} \right), \tag{2.11}$$

where (l_0/A_0) is the ratio of the length of the iron/air circuit to the area of cross-section of the iron/air circuit and μ_0 is the permeability of air. In this type of transducer, a single coil is used and the input displacement to a moving iron diaphragm changes the self-inductance of the coil, thus generating a voltage across the coil.

A common application of the principle of the EI pickoff is in a differential mode, as shown in the circuit in Figure 2.3(b). This type of practical implementation employs two coils. The motion of iron diaphragm alters the mutual inductance by altering the mutual coupling between the two coils.

A Wheatstone's bridge circuit is employed to measure the position of the iron diaphragm. In a real practical circuit, additional potentiometers and impedances are included in series and in parallel with the voltage input terminals and in parallel with the EI pickoff for the purpose of adjusting the sensitivity, zero adjust and calibration of the instrument. Amplifiers, summing amplifiers and pre-amplifiers are employed on a routine basis to condition the measurand in a typical feedback loop.

Considering a general inductive bridge circuit and applying Kirchoff's voltage law to the loops and Kirchoff's current laws to the junctions or by replacing the bridge circuit by its Thevenin equivalent gives, for the voltage across the current measuring AC galvanometer,

$$v_{out} = v_s \left(\frac{Z_3}{Z_3 + Z_4} - \frac{Z_1}{Z_1 + Z_2} \right), \tag{2.12}$$

where v_s is an AC supply voltage, obtained typically from an oscillator.

Thus, when $Z_1 = Z_2$, i.e. when the moving iron diaphragm is in its nominal equilibrium position, the wiper is adjusted so that $Z_3 = Z_4$ and there is no current in the galvanometer. When the moving iron diaphragm moves from its normal equilibrium position, the wiper is adjusted so that there is no current in the galvanometer and its position is directly calibrated on an indicating dial to directly give the displacement of the diaphragm.

When only the impedance Z_1 changes by an amount $Z_1 + \Delta Z$, v_{out} will change by an amount Δv. Assuming that the bridge was initially balanced this is given by

$$\frac{\Delta v}{v_s} = \left(\frac{1}{2} - \frac{1 + \dfrac{\Delta Z}{Z_1}}{2 + \dfrac{\Delta Z}{Z_1}} \right)$$

or by

$$\frac{\Delta v}{v_s} = \frac{1}{2} \left(1 - \left(1 + \frac{\Delta Z}{Z_1} \middle/ 1 + \frac{\Delta Z}{2Z_1} \right) \right). \tag{2.13}$$

Thus, the resistance bridge is linear only when $\Delta Z \ll Z_1$, and in this case equation (2.13) may be approximated as

$$\frac{\Delta v}{v_s} \approx -\frac{1}{4} \frac{\Delta Z}{Z_1}. \tag{2.14}$$

When a current galvanometer is used to measure the current across the terminal, the resistance of the galvanometer also influences the non-linearity. Assuming that $Z_1 = Z_2 = Z_3 = Z_4$, the corresponding relationship for the increase in the current is

$$\Delta i = \frac{v_s}{4R_g} \frac{\left(1+\frac{\Delta Z}{4Z_1}\right)}{\left(1+\frac{\Delta Z}{4Z_1}\right)+\frac{Z_1}{R_g}\left(1+\frac{\Delta Z}{2Z_1}\right)} \left(1-\frac{1+\frac{\Delta Z}{Z_1}}{1+\frac{\Delta Z}{2Z_1}}\right). \tag{2.15}$$

When $R_g \gg Z_1$,

$$\Delta i = \frac{v_s}{4R_g}\left(1-\left(1+\frac{\Delta Z}{Z_1}\Big/1+\frac{\Delta Z}{2Z_1}\right)\right) \approx -\frac{1}{8}\frac{v_s}{R_g}\frac{\Delta Z}{Z_1} \tag{2.16}$$

the circuit in Figure 2.3(b) is a balanced differential bridge circuit, i.e. when the impedance Z_1 changes by an amount $Z_1 + \Delta Z$ and when the impedance Z_2 changes by an amount $Z_2 - \Delta Z$. Hence, in this case,

$$\Delta v/v_s = -\Delta Z/(2Z_1), \tag{2.17}$$

and with a current galvanometer we have

$$\Delta i = -\frac{v_s}{4R_g}\left(1\Big/\left(1+\frac{Z_1}{R_g}\right)\right)\frac{\Delta Z}{Z_1}. \tag{2.18}$$

These expressions are exact, indicating that the output of the bridge varies linearly with the change in the inductance. When $R_g \gg Z_1$, equation (2.18) may be expressed as

$$\Delta i \approx -\left(v_s/\left(4R_g\right)\right) \times \left(\Delta Z/Z_1\right). \tag{2.19}$$

The most widely used of all inductive transducers is the linear variable differential transformer (LVDT) and is shown in Figure 2.4. It consists of a primary coil wound over the entire length of a moving iron core and two symmetrically placed secondary coils wound over cylindrical tubes with the magnetic core free to move inside the cylinders in the axial direction. The stroke of the core of a typical LVDT varies from about 100 m to about 25 cm. The motion of the core changes the coupling between the primary and secondary coils, and as a result there is a change in the mutual inductance. The motion of the core is generally with the minimum of friction and wear due to the non-contact arrangement, which is one of the primary advantages of this sensor.

Another common example of a *magnetic pickup* is a phonograph needle. The needle moves a coil through a magnetic field established by a permanent magnet, thereby producing an induced electrical

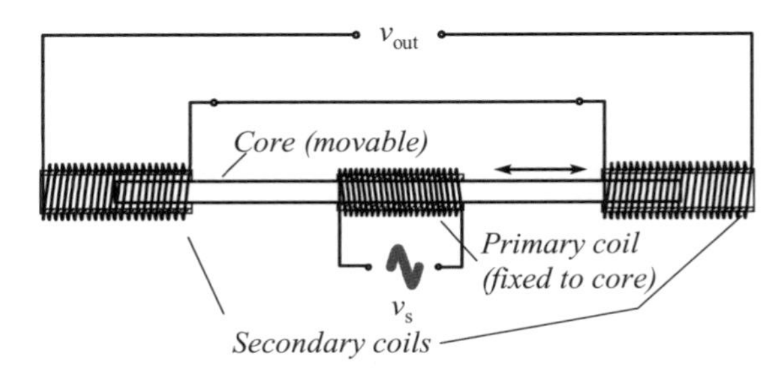

Figure 2.4 Principle of operation of a linear variable differential transformer

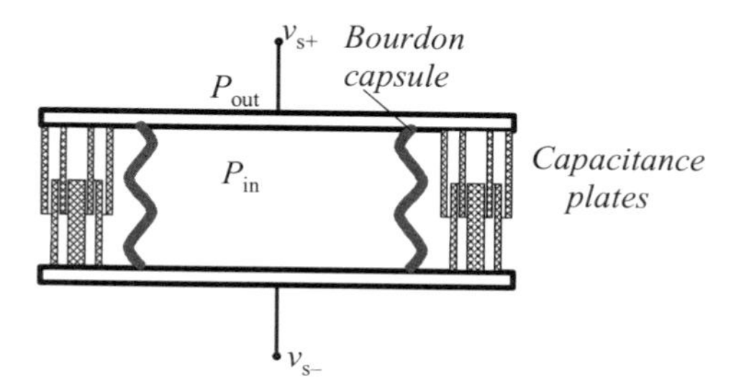

Figure 2.5 Principle of a capacitance-type differential pressure transducer, incorporating Bourdon-type capsules to provide spring restraints

voltage proportional to the velocity of the mechanical movement of the coil. A dynamic audio microphone also works on the same principle.

2.2.3 *Capacitive Transducers*

A capacitor consists of two conducting metal plates of equal area, separated by an insulator, sandwiched between the two plates. When a DC voltage is applied to the metal plates, equal and opposite electric charges appear on the two faces of the capacitor. The capacitance is the ratio of the charge to the applied voltage and is proportional to the dielectric constant ε_r of the material in the gap, the permittivity of free space ε_0 and the area of the plates overlapping to each other, and is inversely proportional to the separation distance between the two inner faces of the plates. Thus, the capacitance of a typical capacitor varies linearly with variations in the area of overlap A. Variable area capacitors, in a differential arrangement of the type shown in Figure 2.5, are most commonly employed as pickups and operate on the same principle as variable reluctance pickups. This type of pickup is biased, and the variation in the capacitance due to mechanical displacement is measured by measuring the impedance of the pickup. The measurement is usually done by employing an appropriate capacitive bridge circuit.

2.2.4 *Cantilever-type Mechanical Resonator Transducers*

A cantilever is an elastic beam or 'tongue' anchored at one end. These beams may be fabricated in silicon and in other semiconducting materials (gallium arsenide, GaAs) by a fabrication process known as micro-machining. They are used as micro-resonators for motion detection. The beams may be in an excited state due to electrostatic excitation, dielectric property changes in a sandwich beam, excitation generated due to the inverse piezoelectric effect when the application of a electric potential difference causes deformation, heating such as resistive (conductive) or infrared (radiative) heating, optical heating by focusing a laser (light amplification by stimulated emission of radiation) beam, electromagnetic excitation by using the force generated by the interaction of an electric current with an applied magnetic field or by any other mechanical means. Figure 2.6 illustrates the principle of mechanical transducer used with a scanning force microscope (Cunningham *et al.*, 1995). A laser beam is focused on to the back of a flexible cantilever beam with nanometre scale tip on the other end that can exert a force on the cantilever. The reflected laser beam is directed to a light-sensitive photo-detector, thus allowing for nanoscale motions to be detected and measured.

Figure 2.7 illustrates a typical cantilever transducer constructed within a semiconductor substrate and integrated into a capacitive sensor for detecting infrared radiation.

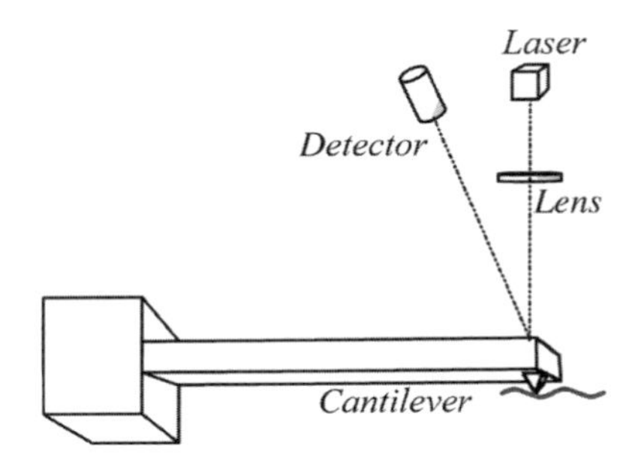

Figure 2.6 Mechanical transducer used with scanning force microscope

2.2.5 *Eddy Current Transducer*

Eddy current-based non-contact displacement transducers operate on eddy current principles. Eddy current transducers use eddy currents induced in a non-magnetic but conductive material to sense the proximity of the material dynamically. A typical eddy current transducer uses a single coil to induce the eddy current and senses the presence of a nearby conductive object. It usually consists of a probe, cable, oscillator and signal demodulator. The probe tip emits a magnetic field, and when a conductive material or strip is brought close to it, an eddy current is induced at the surface of the material or strip. The energy extracted from excitation circuitry, and consequently the excitation amplitude, varies linearly over short range of the gap between the conductive strip and the probe. The probe is excited at a frequency in the 1–10 kHz range, and this excitation produces a magnetic field radiating from probe tip. Consequently, eddy currents are induced in the conductive material close to the tip and energy is extracted from the probe, which is proportional to the distance to the conductive material. As a further consequence of the energy extraction, there is a reduction in the oscillatory output of the probe, which is proportional to the gap. As the distance from tip to conductive material is varied, the output from the probe is demodulated to extract a proportional DC voltage. As long as this voltage is much smaller in magnitude than the voltage supplied to the probe, a linearly proportional response is observed. Proper calibration and an appropriate copper-based conductive material are essential for measuring low-amplitude motion.

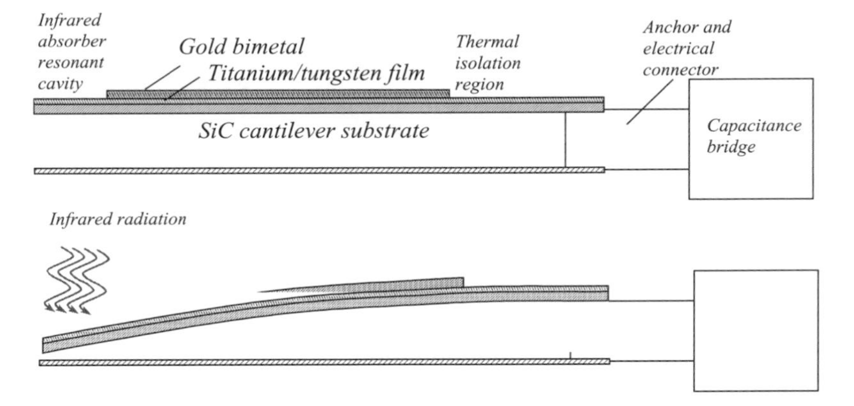

Figure 2.7 Cantilever-type mechanical transducer used in an infrared sensor

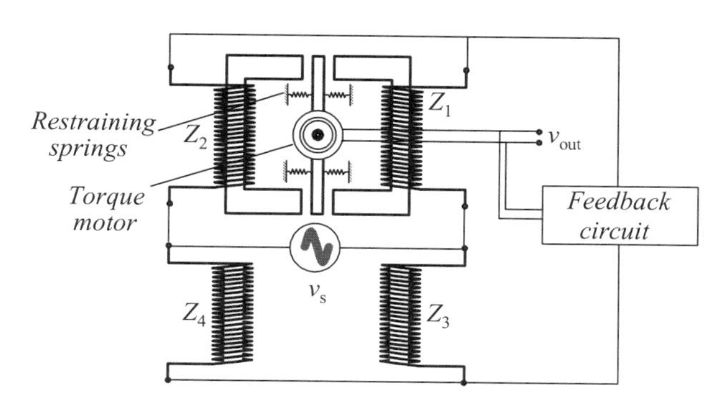

Figure 2.8 Typical example of a balancing-type differential inductive EI pickoff, incorporating a feedback circuit and a torque motor to maintain the moving iron armature in an equilibrium position

2.2.6 *Balancing Instruments*

Figure 2.8 shows a typical example of a balancing instrument for measuring angular displacement, incorporating a feedback circuit and a torque motor to maintain the spring-restrained moving iron armature in an equilibrium position. The current to the torque motor, which generates the rebalancing torque and maintains the spring-restrained moving iron armature in an equilibrium position, is proportional to its angular displacement.

2.2.7 *Transduction Mechanisms in Materials*

Several mechanisms are often exploited in the design of transducers and provide the fundamental channel in the operation of the transducer (Van Putten, 1996). Some of those that are characteristic of smart materials have already been presented in Chapter 1. Apart from these there are several others, which are briefly discussed in this section. They may be classified and grouped into four different classes for the convenience of identifying the underlying principles. Moreover, we assume that the primary materials used in the construction are generic semiconductors, which covers conductors and insulators as well as materials that are not semiconductors.

Thermoelasticity refers to generation of a potential difference across two points of a structure by the combined application of a mechanical strain and a temperature gradient across the two points. *Piezo-opticity* refers to change in the refractive index of material by the application of mechanical stresses. *Photo-elasticity* refers to the generation of birefringence or double refraction by the application mechanical force. *Tribo-electricity* refers to the generation of positive or negative surface charge due to rubbing between two materials. The *lateral photo-voltaic effect* refers to the generation of lateral potential difference across a junction by radiation that is incident on it. The *lateral photo-electric effect* refers to the generation of spatial currents across two contacts at one side of a reverse biased p–n junction by radiation that is incident on it.

The *thermoelectric or Seebeck effect* refers to the generation of current in a closed-loop coil of two dissimilar conductors by a temperature gradient across the two junctions. The *Peltier* effect is the converse effect and results in the generation of different junction temperatures when a current is flowing through a closed-loop coil of two dissimilar conductors. A related but independent effect is the *electrothermal effect*, which results in the generation of heat when a current is flowing through a conductor. *Thermoconductivity* refers to the change in electrical conductivity with temperature. *Superconductivity* refers to the almost complete absence of resistance to current flow in certain materials when cooled below a specific critical temperature. The *thermochemical effect* refers to the change in chemical structure due to temperature.

The response of many common solids to an applied electric field is such that the electric field inside the sample is less than the applied field. It appears that the solid has generated its own charge distribution, i.e. it has become polarized so that it produces an electric field that opposes and reduces the applied field. Materials that do this are called dielectrics, and we may define the polarization of the material to be the internal response of the material to the applied electric field. The *pyroelectric effect* refers to the change in polarization due to a change in temperature.

Thermodielectricity refers to the change in permittivity of a ferroelectric with temperature. The *Nernst effect* refers to the generation of an electric field due to the combined presence of a temperature gradient and mutually perpendicular magnetic field. *Incandescence* refers to the emission of radiation when a material is heated. *Thermoluminescence* refers to the emission of radiation by certain crystals due to changes in temperature.

Several electromagnetic mechanisms rely on the relation between the magnetization and the applied magnetic field. For low levels of magnetization, Curie's law is good approximation for the magnetization of paramagnets:

$$M = \chi H = (C/T)\,H, \tag{2.20}$$

where M is the resulting magnetization or magnetic moment, χ is the magnetic susceptibility, H is the auxiliary magnetic field, measured in amperes/metre, T is absolute temperature, measured in K, and C is a material-specific Curie constant. Paramagnetic materials are also characterized by a net remnant magnetic moment per unit volume M and may be described by the so-called Langevin function:

$$M = N_{dp} m_{dp} \left(\coth\left(\mu_0 m_{dp} H/kT\right) - \left(\mu_0 m_{dp} H/kT\right)^{-1}\right), \tag{2.21}$$

where N_{dp} is total number of magnetic dipoles per unit volume, m_{dp} is the magnetic dipole moment (in A m^2) of the atom, μ_0 is the permeability in a vacuum ($1.256\,637 \times 10^{-6}$ H/m), k is the Boltzmann constant[1] (1.38×10^{-23} J/K) and T is the absolute temperature. When this law is linearized for small arguments of the function *coth*(.), we obtain a linear law, which is equivalent to Curie's law. Micro-magnetic phenomena can occur in small ferromagnetic films, which are exploited to construct novel micro-magnetic sensors.

An important effect due to the presence of a magnetic field is the *Hall effect* discovered by Edwin Hall in 1879. When a current is flowing in a thin conductor, which is placed in magnetic field that is not parallel to it, an electric potential is generated in a direction mutually perpendicular to the directions of both the current and the magnetic field. Thus,

$$V_{Hall} = R_{Hall}\left(\mathbf{I} \times \mathbf{B}/A\right), \tag{2.22}$$

where V_{Hall} is the Hall voltage, R_{Hall} is the Hall constant, $\mathbf{I}$ is the current vector, $\mathbf{B}$ is magnetic induction vector and A is the cross-sectional area of the conductor. Another important effect is the phenomenon of *magneto-resistivity*, which is the increase in the resistance of a conductor due to the presence of a magnetic field that exerts a force (the Lorentz force) on the charge carriers.

When a material makes the transition from the normal to superconducting state, it actively excludes magnetic fields from its interior; this is called the *Meissner effect*. Superconductors are perfectly dia-magnetic since if one attempts to magnetize a superconductor, then by virtue of *Lenz's law*, current loops are established internally that null the imposed field. By *Faraday's law*, when there is no change in the applied magnetic field, there is no voltage generated to drive a current within the material. However, as a superconductor has zero resistance, it must also exclude the applied magnetic field from its interior.

There are a number of other laws that are used to define transduction mechanisms in the presence of a magnetic field. *Suhl's law* refers to the change in the conductivity of a semiconducting surface in

---

[1] The following electronic constants are standard: $q = 1.666 \times 10^{-19}$ C, $k = 1.38 \times 10^{-20}$ J/degree, $h = 6.62 \times 10^{-34}$ J.s, and $c = 3 \times 10^8$ m/s.

the presence of a magnetic field. The *photo-magneto-electric effect* refers to an electric potential in the presence of both a magnetic field and an incident radiation. *Faraday's effect* refers to the rotation of the plane of polarization of a polarized incident radiation in the presence of an externally applied magnetic field. The *Ettinghausen effect* is the generation of a temperature gradient due to the presence of both an electric and a magnetic field. The *Righi–Leduc effect* refers to generation of a temperature gradient due to the presence of both an external heat source and a magnetic field. The *Maggi–Righi–Leduc effect* refers to the change in thermal conductivity due to the presence of a magnetic field.

Light radiation is a form of electromagnetic radiation as the wave motion is accompanied by mutually perpendicular electric and magnetic fields in a plane perpendicular to the direction of motion. While photometry refers to the study of electromagnetic radiation in the visible band, radiometry refers to the study of electromagnetic radiation over all bands. In most radiation transducers such as Schottky barrier diodes, photo-resistors, photo-diodes and photo-transistors, PIN diodes, avalanche photo-diodes (APDs), metal–insulator semiconductor junctions, metal–oxide semiconductor junctions, laser devices and optical fibres, an interaction exists between the incident radiation and a material, whether it is a semiconductor or any other. These interactions involve certain primary phenomena such as refraction, transmission or reflection, absorption, interference or scattering and diffraction. Refraction is governed by Snell's law, which describes the relationship between incident and transmitted radiation at the interface of two media. Absorption of photons results in the attenuation of the incident radiation in a medium during the process of transmission. Interference occurs when two waves with equal frequencies, but different in phase from independent sources, coincide. This results in the generation of an interference pattern that depends on the relative amplitudes and phases of the two incident waves. An interferometer is any measurement device based on the fact that radiation from a point source arrives at different points of an aperture with a phase difference dependent on the angle of arrival. The interferometer breaks the larger aperture into small sub-apertures and facilitates the comparison of phase. The presence of several sub-apertures leads to interference, and the comparison of phase is a non-linear operation involving the recognition of the features of an interference pattern. Scattering refers to generic phenomenon whereby some forms of radiation are forced to deviate from their nominal trajectory by one or more localized non-uniformities or obstacles in the medium through which they pass. An example of this is the deviation of reflected radiation from the angle predicted by the law of reflection. Diffraction occurs when radiated waves are broken up into constituent frequency bands in the presence of a transparent or opaque obstacle.

Polarization is an important feature of electromagnetic radiation, which is not normally polarized. Looking along a ray of light, the electric vectors make all angles with the vertical. Light that is plane polarized in one plane has all its electric vectors aligned in the same plane. The plane of polarization is the plane that includes both the vibration direction and the ray path. Light may be polarized by passing it through crystals or polarizing filters, or by reflection. However, reflected light is only partially polarized, mainly along the vibration direction perpendicular to the plane of the ray path (including both the incident and reflected rays). On the other hand, polarizing filters exclude all light not vibrating in the preferred direction of the filter. Birefringence, or double refraction, is the decomposition of a ray of light into two rays (the ordinary ray and the extraordinary ray) when it passes through certain types of anisotropic materials, such as calcite crystals or silicon carbide, or in defective optical fibres, depending on the polarization of the light.

There are a number of effects that arise due to the interaction of electromagnetic radiation with conductors and semiconductors. We shall briefly describe some of the principal effects. The *photo-electric effect* is the generation of electrons and holes in a semiconductor junction area when photons (radiation) are incident on the junction. The electrons and holes subsequently migrate in different directions away from the junction. The *photo-conductivity* refers to the change in electrical conductivity due to incident radiation. *Radiation heating* refers to the process of heating a material by exposing it to incident radiation. The *photo-voltaic* effect refers to the generation of an electrical potential difference when radiation is incident at the junction of two dissimilar materials. The *photo-magneto-electric effect* is the generation of an electric potential difference when both an incident radiation and a mutually perpendicular magnetic field are present. The *photo-dielectric effect* refers to the change in the dielectric constant and dielectric loss in the presence of incident radiation. The *photo-magnetic effect* refers to the generation of a magnetic moment in the presence of incident radiation. The *photo-chemical effect* refers to the change in chemical structure due to incident radiation.

There are a number of effects related to the phenomenon of *luminescence*. This refers to the emission of light by a substance and generally occurs when an electron returns to the electronic ground state from an excited state and loses its excess energy as a photon. *Fluorescence*, a form of *photoluminescence*, occurs when the molecule returns to the electronic ground state, from the excited vibration state known as a singlet state, by emission of a photon. When an electron is excited to a vibrational state and the spin of the excited electron can also be reversed, the molecule is said to be in an excited triplet state. *Phosphorescence*, which is another form of *photoluminescence*, occurs when the electron in a triplet state could lose energy by emission of a photon. *Chemiluminescence* occurs when a chemical reaction produces an excited electron that in turn emits a photon in order to reach the ground state. These kinds of reactions could be common in biological systems; the effect is then known as *bioluminescence*. *Radioluminescence* refers to the emission of visible radiation when the material is excited by *X-rays* and *γ-rays*. *Electroluminescence* refers to the emission of radiation when the material is excited by an alternating electric field. *P–N luminescence* refers to the emission of radiation following recombination in a forward biased *p–n junction*.

There are a number of other optical effects that are used to define transduction mechanisms in the presence of an electromagnetic field. The *Kerr-electro-optic effect* refers to the generation of double refraction of incident radiation in the presence of an electric field. The *Kerr-magneto-optic effect* refers to the rotation of the plane of polarization of a polarized incident radiation in the presence of an externally applied magnetic field. *Pockel's effect* refers to the rotation of the plane of polarization of a polarized incident radiation in the presence of an externally applied electric field. The *Cotton–Mouton effect* refers to the generation of double refraction of incident radiation, within a liquid, in the presence of an electric field.

2.2.8 Hydrodynamic and Acoustic Transduction Mechanisms

Two important properties of fluids are viscosity and surface tension. Viscosity is the quantity that describes a fluid's resistance to flow, and if the flow is 'Newtonian', viscosity is defined as the ratio of the shear stress to the velocity gradient. Viscosity contributes to viscous friction and boundary layer flow. Consequently, hydrodynamic resistance is proportional to flow rate. Hydrodynamic ducts are analogous to electrical capacitance in that the flow rate is proportional to the time rate of change of pressure. *Darcy's law* in Hele–Shaw cells (shallow channels) is very similar to *Ohm's law* for electrical circuits: flow velocity is directly proportional to the pressure or head gradient and inversely proportional to viscosity. The analogy between Darcy's law and the *Hagen–Poiseulle's law* for flow through a pipe may be used to evaluate the proportionality constant. Surface tension is due to the presence of cohesive forces between liquid molecules and is what keeps bubbles together. Capillary action is the result of surface tension and the adhesion of water to the walls of a vessel or narrow pipe and will cause an upward force on the liquid at the edges, which in turn results in a meniscus that turns upwards. The capillary effect is important in narrow pipes, as the pressure drops caused by capillarity are $\sim L^{-1}$ while those due to viscosity behave like $\sim L^{\circ}$. The *Coandă* effect is the phenomenon in which a jet flow attaches itself to a nearby surface and remains attached even when the surface curves away from the initial jet direction. The Coandă effect is a direct consequence of surface tension or van der Waals forces and follows from the application of Newton's laws to the fluid flow.

Bernoulli's principle results from the application of the principle of conservation of energy to hydrodynamic flows. It is often used to obtain the flow rates in many transducers and fluidic devices.

The *Venturi effect* is a phenomenon that is described by Bernoulli's equation. When a fluid is flowing through a pipe and is forced through a narrower section of pipe, the pressure decreases while the velocity increases.

An important flow paradigm is the flow past a cavity. In the cavity, the flow turns around, with small separation zones along the boundary walls. This in turn is responsible for generating noise due the presence of very small eddies in the flow. In fact, the flow past a rectangular cavity can result in self-sustained oscillations, a phenomenon known as cavity resonance. Cavity flows are often used as flow transducers for the indirect measurement of pressures and flow rates. Cavity flows can also be controlled quite easily by introducing disturbances that influence the separation of the flow within the cavity. When cavity flows are combined with capillary effects (flows in narrow pipes), there are a number of interesting physical problems that are particularly relevant to the design of practical micro-fluidic transducers.

Sound waves in a fluid (air or water) are essentially longitudinal waves; the particles in the medium are displaced from their equilibrium position parallel to the direction that the wave propagates. There are three types of sound waves that may be differentiated by their frequency range: infrasonic waves, with frequencies $f < 20\,\text{Hz}$ produced by earthquakes and seismic disturbances; audible waves, in the frequency range $20\,\text{Hz} < f < 20\,000\,\text{Hz}$; and *ultrasonic waves* with frequencies $f > 20\,000\,\text{Hz}$, which are used in the design of transducers. The other type of waves that are particularly important in the design of transducers are the *surface acoustic waves*, which are present on the surface of an elastic half space, which were first discovered by Lord Rayleigh. There are a wide variety of surface waves that occur in elastic materials, which are also piezoelectric. When waves are established in a cavity, they undergo multiple reflections at the walls of the cavity and the net result is the appearance of a standing wave. It is these standing waves that lead to problem of cavity resonance.

A microphone is a generic term that refers to any element that transforms acoustic wave energy (sound) into electrical energy (electricity (audio signal)). It is essentially an acoustic transducer. The most common types of microphones are the dynamic or inductive pickups and variable capacitance types.

2.2.9 *Transducer Sensitivities, Scaling Laws for Example Devices*

Multiplicative or transducer scaling errors are generally influenced by errors due to sensitivity. Transducer sensitivities are in turn influenced by scaling laws. Some base scaling laws are listed in Table 2.2.

While transducers can be calibrated to minimize scaling errors, knowledge of the transducer sensitivities is useful in designing the electronic circuitry used to condition the output of a sensor. We shall briefly explain this by considering a typical case study: a piezoelectric transducer.

Table 2.2 Some fundamental length scaling laws based on forces

Intermolecular Van der Waals' forces	L^{-7}
Density of Van der Waals' forces between interfaces	L^{-3}
Cantilever natural frequency	L^{-1}
Capillary force	L^{1}
Distance	L^{1}
Thermal power by conduction	L^{1}
Electrostatic force	L^{2}
Diffusion time	L^{2}
Force of gravity	L^{3}
Magnetic force with an external field	L^{3}
Electrostatic micro-motor torque	L^{3}
Internal magnetic forces	L^{4}
Forces in a centrifuge	L^{4}

2.2.10 Modelling and Analysis of a Piezoelectric Transducer

Piezoelectric transducers are by far the most common type of transducer used in structural control. Piezoelectric ceramic and polymeric actuators are generally modelled by equivalent circuit methods or by the finite element approach. Such models are particularly important for the optimal design of the transducer when additional conditions must be met to ensure maximum power transfer or some other optimality criteria (Kim and Jones, 1991; Low and Guo, 1995; Silva *et al.*, 1997).

Circuit models are particularly useful in estimating the optimum power requirements via the maximum power transfer theorem. A typical piezoelectric transducer may be modelled as two impedances in parallel. In the first, inductance, resistance and capacitance are coupled in series. The inductance represents the inertial effect of the piezoelectric material set in vibration and is similar to the apparent mass effect of a vibrating mass of fluid. The elastic stiffness of the piezoelectric material is represented by the capacitance, while the dissipation of energy is represented by the resistance. The second impedance in parallel is purely capacitive and corresponds to the dielectric properties of the piezoelectric material, which is assumed to be encased within two parallel plates. To ensure maximum transfer of power and consequently the optimal use of the supply power, the impedance of the voltage must be properly matched to the supply voltage and the load circuit. To do this, one needs to apply the maximum power transfer theorem in circuit theory. In order to apply the theorem, it is essential to first find the total impedance of the circuit Z_{eq}. This can be estimated by applying the parallel formula to the series combination of the resistance R and the series capacitance C_s and an inductance L_s, and the capacitive impedance Z_c, which is given as

$$\frac{1}{Z_{eq}} = \frac{i\omega C_s}{i\omega C_s R + 1 - \omega^2 L_s C_s} + \frac{1}{Z_c} = \frac{i\omega C_s}{i\omega C_s R + 1 - \omega^2 L_s C_s} + i\omega C.$$

The above expression is equivalent to

$$\frac{1}{Z_{eq}} = -\frac{\omega^2 C C_s R - i\omega\left(C_s + C - \omega^2 C L_s C_s\right)}{i\omega C_s R + 1 - \omega^2 L_s C_s}. \tag{2.23}$$

Hence,

$$Z_{eq} = -\frac{i\omega C_s R + 1 - \omega^2 L_s C_s}{\omega^2 C C_s R - i\omega\left(C_s + C - \omega^2 C L_s C_s\right)}. \tag{2.24}$$

The numerator of equation (2.23) is expressed as a real function of ω to give

$$\frac{1}{Z_{eq}} = -\frac{\omega^2\left(\omega^2\left(C^2 C_s^2 R^2 - 2C L_s C_s\left(C + C_s\right)\omega^2\right) + \left(C_s + C\right)^2 + \omega^4 C^2 L_s^2 C_s^2\right)}{\left(i\omega C_s R + 1 - \omega^2 L_s C_s\right)\left(\omega^2 C C_s R + i\omega\left(C_s + C - \omega^2 C L_s C\right)\right)}$$

and it follows that

$$Z_{eq} = -\frac{\left(i\omega C_s R + 1 - \omega^2 L_s C_s\right)\left(\omega^2 C C_s R + i\omega\left(C_s + C + \omega^2 C L_s C_s\right)\right)}{\omega^2\left(\omega^2\left(C^2 C_s^2 R^2 - 2C L_s C_s\left(C + C_s\right)\right) + \left(C_s + C\right)^2 + \omega^4 C^2 L_s^2 C_s^2\right)}. \tag{2.25}$$

From the first of the above expressions for Z_{eq}, we observe that the circuit would resonate when

$$\omega = \sqrt{1/L_s C_s}, \tag{2.26}$$

as the current in the circuit tends to a peak value. However when

$$\omega = \sqrt{\frac{\left(C_s + C\right)}{C L_s C_s}} = \sqrt{\frac{1}{L_s C_{eq}}},\quad C_{eq} = 1\bigg/\left(\frac{1}{C} + \frac{1}{C_s}\right), \tag{2.27}$$

the current is a minimum, and this frequency corresponds to anti-resonance. Yet neither of these frequencies corresponds to a situation where the power transfer is a maximum.

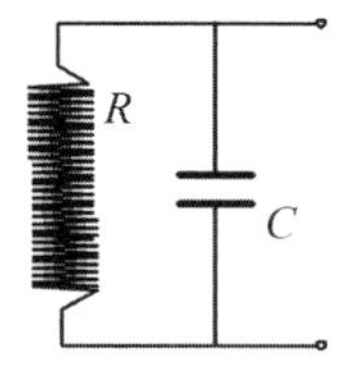

Figure 2.9 Equivalent RC circuit of a piezoelectric transducer

The inertial effect is often small and may be neglected. In this case $L_s \to 0$ and

$$Z_{eq} = -\frac{(i\omega C_s R + 1)\left(\omega^2 C C_s R + i\omega (C_s + C)\right)}{\omega^2 \left(\omega^2 C^2 C_s^2 R^2 + (C_s + C)^2\right)} \tag{2.28}$$

$$Z_{eq} = \frac{C_s^2 R}{\left(\omega^2 C^2 C_s^2 R^2 + (C_s + C)^2\right)} - i\omega \frac{\omega^2 C C_s^2 R^2 + (C_s + C)}{\omega^2 \left(\omega^2 C^2 C_s^2 R^2 + (C_s + C)^2\right)}.$$

A typical piezoelectric transducer operating near resonance may be modelled as an RC circuit, as shown in Figure 2.9, ignoring both the capacitive and the inductive impedances in series with the resistance. There are a number of issues to consider in using such a transducer. The first is the interfacing problem to ensure maximum power transfer.

Assuming $C_s \to \infty$, the equivalent impedance Z_{eq} is

$$Z_{eq} = \frac{R(1 - i\omega CR)}{\left(1 + (\omega CR)^2\right)} = \frac{R}{\left(1 + (\omega CR)^2\right)} - i\omega \frac{CR^2}{\left(1 + (\omega CR)^2\right)}. \tag{2.29}$$

For maximum power transfer the total load impedance, Z_s must be

$$Z_s = Z_{eq}^* = R_s + i\omega L_s = \frac{R}{\left(1 + (\omega CR)^2\right)} + i\omega \frac{CR^2}{\left(1 + (\omega CR)^2\right)}. \tag{2.30}$$

It follows that the load must be inductive and this can be achieved by employing an inductive chalk coil. Moreover, it depends on the frequency of operation ω. The equivalent load resistance and inductance, which are both frequency dependent, are given by

$$R_s = \frac{R}{\left(1 + (\omega CR)^2\right)}, \quad L_s = \frac{CR^2}{\left(1 + (\omega CR)^2\right)}. \tag{2.31}$$

Thus, the actual operation frequency plays a key role in the performance of the transducer.

For most transduction applications, a two-port model is most useful. This model relates the port inputs, the current I and displacement u to the port outputs; the force F; and voltage V. The model parameters are the mechanical stiffness k_p and mechanical damping b_p, the total electrical capacitance C and the electromechanical coupling coefficient α. The port transfer matrix may be expressed as

$$\begin{bmatrix} V \\ F \end{bmatrix} = \frac{1}{j\omega C} \begin{bmatrix} 1 & \alpha \\ \alpha & k_p C + j\omega C b_p + \alpha^2 \end{bmatrix} \begin{bmatrix} I \\ j\omega u \end{bmatrix}. \tag{2.32}$$

For a piezoelectric stack actuator, which consists of a number (n_p) of thin layers of piezoelectric slabs, coaxially stacked in the direction of the polarization axis with a total thickness of t_p, a Young's modulus E_{33}^E and an electric field applied in the same direction, the electrical capacitance and mechanical stiffness are, respectively, given by

$$C = n_p^2 \varepsilon_{33} \left(A/t_p\right), \quad k_p = E_{33}^E \left(A/t_p\right). \tag{2.33}$$

The electromechanical coupling coefficient α is given in terms of d_{33} and ε_{33}, which are piezoelectric material charge and dielectric constants. For a stack actuator, each layer and the total thickness of the stack are typically about 0.5 and 10 mm, respectively.

If one is intent on measuring the output voltage and the measurand is a force in the absence of any displacement, the voltage–force relationship is desirable. The total charge generated by a single piezoelectric slab of thickness t may be expressed in terms of the intrinsic *charge constant* d_{33} and the induced polarization per unit stress, both in the longitudinal direction, and the applied longitudinal force F, as

$$Q = C \times V = F \times d_{33}. \tag{2.34}$$

Since equivalent capacitance is inversely proportional to the thickness t and directly proportional to the cross-sectional area A of the slab, the voltage generated by the applied stress may be expressed in terms of the *piezoelectric voltage constant* g_{33} and the induced electric field per unit applied stress, both in the longitudinal direction, and is

$$V = F \times (d_{33}/C) = (F \times d_{33} \times t)/(\varepsilon_{33} \times A) = (F \times g_{33} \times t)/A. \tag{2.35}$$

In the presence of any parasitic capacitance, this formula is changed to

$$V/F = 1/\alpha = (g_{33} \times t)/\left(A + A_p\right), \tag{2.36}$$

where A_p is an equivalent parasitic area that one should add to the true area of cross-section of the equivalent capacitance of the piezoelectric transducer.

If one is interested in estimating the displacement of a piezoelectric actuator due to an input voltage, under no load conditions, we use the fact that, assuming negligible mechanical damping, the displacement is related to the force by the relationship

$$u = \left(s_{33}^E \times t \times F\right)/A, \tag{2.37}$$

where s_{33}^E is the elastic compliance of the piezoelectric material. Eliminating the force we obtain the relationship

$$u/V = s_{33}^E/g_{33} = d_{33}. \tag{2.38}$$

In a transverse configuration, this is modified to

$$u/V = d_{31}\left(L_p/t\right), \tag{2.39}$$

where L_p is the length of the piezoelectric slab. For a bimorph, it may be shown to be

$$u/V = (3/2)\, d_{31}\left(L_p/t\right)^2. \tag{2.40}$$

Piezoelectric disc transducers produce the largest force (~4000–5000 N) in the longitudinal direction, while bimorphs produce the largest displacements (~1 mm).

It must be noted that there is a temperature limitation in using piezoelectric devices defined by the Curie point or Curie temperature. The property of piezoelectricity may vanish altogether when one is operating above this temperature.

Finally, it must be noted that that piezoelectric transducers exhibit a hysteresis in the relationship between the electric displacement (D) and the electric field (E) similar to the B–H curve between magnetic induction B or flux density and magnetic field strength H. This curve is illustrated in Figure 2.10.

One may estimate the bending moment exerted on a beam by two piezoelectric patches symmetrically placed on two parallel lateral faces of a beam, above and below the middle surface. For a piezoelectric bimorph, the appropriate constitutive relations are

$$\varepsilon_{11} = \sigma_{11}/E_a + d_{31}E_3, \quad D_3 = d_{31}\sigma_{11} + \epsilon_{33}\, E_3, \tag{2.41}$$

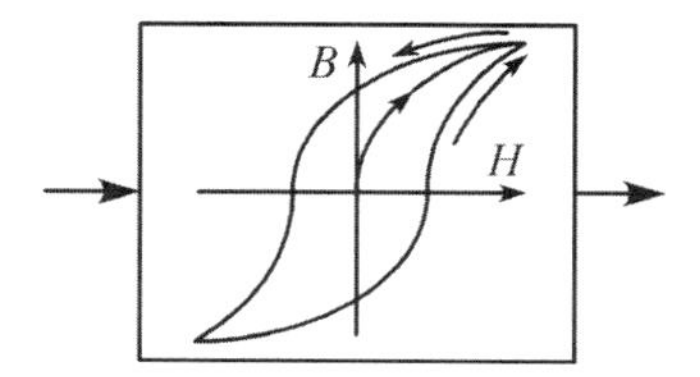

Figure 2.10 Illustration of hysteresis: the B–H curve or equivalently the D–E curve for a piezoelectric transducer

where E_3 is the electric field component, E_a is the Young's modulus of the piezoelectric material, d_{31} is the piezoelectric coefficient, D_3 is the electric displacement and $\in_{33}$ is a dielectric constant. Let us consider two piezoelectric actuators of length, width and thickness, L_p, w and t, respectively (all of them constant values), surface bonded to a beam of length, width and thickness, L, b and d. Whenever an electrostatic field $E_3 = V_a/t$ is applied out of phase to both piezoelectric actuators, the deformation in the piezoelectric materials produces a pure bending moment acting on the beam. Consequently, we assume that there is linearly varying strain across the depth of the beam and the patch, which is assumed to be equal to $\theta\ z$. Given the Young's modulus of the beam and the actuator patch materials as E_b andE_a, respectively, the bending moment acting on the beam is the equal to

$$M_b = E_b b\theta \int_{-d/2}^{d/2} z^2 \mathrm{d}z = E_b b\theta \frac{d^3}{12} = E_b I_b \theta. \tag{2.42}$$

Applying the constitutive relation, the stress in the actuator patches is

$$\sigma_{11} = E_a \theta z - E_a d_{31} E_3. \tag{2.43}$$

Hence, the total bending moment on the actuator patches is

$$M_a = 2E_a w\theta \int_{d/2}^{t+d/2} z^2 \mathrm{d}z - 2E_a d_{31} E_3 w \int_{d/2}^{t+d/2} z\mathrm{d}z. \tag{2.44}$$

The expression for M_a simplifies to

$$M_a = 2E_a \theta I_a (w, d, t) - 2E_a d_{31} E_3 w \frac{t(d+t)}{2}, \tag{2.45}$$

where

$$I_a (w, d, t) = wt\left(4t^2 + 6dt + 3d^2\right)/12. \tag{2.46}$$

Since the beam is in equilibrium,

$$M_a + M_b = 0. \tag{2.47}$$

Hence, one may solve for θ, which is given by

$$\theta = \frac{E_a d_{31} E_3 wt (d+t)}{(E_b I_b + 2E_a I_a (w, d, t))}. \tag{2.48}$$

Thus, the moment acting on the beam is given by

$$M_b = E_b I_b \frac{E_a d_{31} wt (d+t)}{(E_b I_b + 2E_a I_a (w, d, t))} \frac{V_a}{t}, \tag{2.49}$$

which may be expressed as

$$M_b = E_b I_b \frac{E_a w (d+t)}{(E_b I_b + 2E_a I_a (w, d, t))} d_{31} V_a. \tag{2.50}$$

When $E_b I_b \gg 2E_a I_a(w, d, t)$,

$$M_b = E_a w(d + t) d_{31} V_a, \tag{2.51}$$

and the bending moment is directly proportional to the applied voltage, over the length of the patch.

2.3 Actuation of Flexible Structures

Piezoactuators like stacks, benders, tubes and rings make use of the deformation of electroactive PZT ceramics (PZT: a lead (Pb), zirconate (Zr), titanate (Ti) compound) when they are exposed to an electrical field. This deformation can be used to produce motions or forces if the deformation of the piezoelement is constrained. The above effect is the complementary effect to piezoelectricity, where electrical charges are produced on application of mechanical stress to the ceramics.

Two different types of piezoelectric materials that are currently commercially available for constructing the actuator and surface patches are the piezoceramics or polymeric film-based piezoelectric materials. PVDF is a typical piezoelectric material based on polymeric films such as S028NAO and is characterized by a relatively low Young's modulus (2.3×10^9 N/m^2). The PZT piezoceramics are lead–zirconate–titanate based, light-weight, low-cost materials with a relatively high strain constant that are eminently suitable for use in material patches designed to perform as actuators and sensors for active control of structural vibrations. One such PZT material, G1195, has a density of 7600 kg/m^3, Young's modulus of 63×10^9 N/m^2 and a Poisson's ratio of 0.31, which are of the same relative order as the corresponding properties of aluminium and steel. On the other hand, the characteristic piezoelectric strain constant of G1195 is of the order of 1.667×10^{-10} m/V. For this reason, the choice of this material for constructing PZT material patches is very attractive.

Piezoelectric fibre composite actuators are manufactured using a layer of extruded piezoceramic fibres encased in a protective polymer matrix material. Layers of these piezoceramic fibres encased in a protective polymer layer are excited by a set of interdigitated electrodes and combine together to form micro-active fibre composites. The two primary disadvantages of piezoelectric fibre composite actuators are the high cost due to the difficulty of processing and handling expensive piezoceramic fibres during actuator manufacture, and the need for a high voltage source for driving the actuator. A typical actuator of this type consists of an interdigitated electrode pattern etched on a polyimide film, a layer of structural epoxy to inhibit crack propagation and to bond the actuator layers together, a layer of aligned PZT fibres encased in a protective polymer matrix followed by another layer of structural epoxy and a layer of interdigitated electrodes. Typically, the fibres are less than 200μm thick, while the spacing between the fibres is of the order of 0.25 mm. The interdigitated electrodes are etched across the direction of the fibres and are also less than 200μm in width, but the spacing between electrodes of opposite polarity is of the order of 1 mm, while it is twice that between electrodes of the same polarity.

Pre-stressed PZT piezoelectric ceramic devices are of interest in a variety of applications for the control of smart structures due to their durability and enhanced strain capabilities (Crawley and de Luis, 1987; Jenkins *et al.*, 1995). The enhanced durability and strain output is due to the combination of the piezoelectric ceramics with other materials to form a range of layered composites. Differences between the thermal and structural properties of the various materials that form the layers of the composites lead to the pre-stressing of the PZT layer, which creates internal stresses. These internal stresses, when combined with restricted lateral motion, are known to enhance axial displacement. They are then used as morphing actuators to bend another structure. Typical configurations of such actuators are unimorphs and bimorphs.

Bimorphs are an assembly of two piezoelectric layers that are bonded with their polarity in opposite directions. When opposite voltages are applied to the two piezoceramic layers, a bending moment is induced in the beam. Under the action of an electric field, one piezoelectric layer contracts in the thickness direction while the other expands. Due to the contraction and expansion in the thickness direction, one layer expands along the length and the other contracts, inducing bending of the bonded layers. Unimorphs are similar in structure to bimorphs, which are an assembly of a single piezoelectric layer that is bonded to a passive layer. Under expansion in the poling direction, the strain in the plane perpendicular to the poling direction undergoes a contraction in the active layer, leading to a bending of the whole device. Such devices can be used to induce relatively large deflections, and the amplitude increases with the lateral dimensions.

Another approach used to increase the strain that can be induced with a piezoelectric material for a given field is to drive the material at its resonance frequency. However, most piezoelectric actuators are characterized by a resonance frequency corresponding to the frequency of sound waves, and this severely limits the application of these devices. However, one could use amplifiers operating near resonance to increase the effectiveness of the excitation.

2.3.1 Pre-stressed Piezoelectric Actuators

A typical pre-stressed piezoelectric ceramic-based actuator is composed of a metallic base material (e.g. aluminium or stainless steel), a piezoceramic wafer and an adhesive employed in spray or film form. The materials are bonded under high pressures and temperatures and then cured at room temperature after the adhesive has solidified. Pre-stressed piezoelectric ceramic-based micro-positioning actuators provide a form of precise solid-state actuation, with a fast response and associated high bandwidths (Williams and Inman, 2002), and have found applications in many areas such as shape controls (Andoh *et al.*, 2001), vibration controls (Kamada *et al.*, 1997) and robot controls (Bruch *et al.*, 2000). However, like many other materials-based actuators, piezoceramic actuators possess hysteresis, a phenomenon wherein the input and the output bear a non-linear relationship, which depends on prior history. This non-linearity considerably degrades a system's performance, especially in cases that require precision positioning, such as robot manipulators in hospital applications and in aircraft applications. Recently, piezoceramics have found new applications on board an aircraft as 'active controllers' as they are able to vary the shape of the control surface and hence control the forces and moments acting on the aircraft. For such applications, curved pre-stressed piezoceramic patch actuators, which have the advantages of achieving larger displacement, longer life cycle and greater flexibility than conventional piezoelectric actuators, have been developed. However, this type of curved pre-stressed piezoceramic patch actuator is flexible and inherently damped, and exhibits severe hysteresis as compared to a typical stack-type piezoactuator. Piezoelectric patch actuators are characterized by hysteresis, and a common model employed is the Bouc–Wen model. The presence of the hysteresis in the input–output model can be dealt with by employing a Schmidt trigger in the control input, which is itself a relay with both deadband and hysteresis.

Mechanically pre-stressed piezoactuators were first developed in the 1980s and were referred to as monomorphs, as they utilized a monolithic structure without a separate bonding layer between the active and passive lamina. The first pre-stressed bending actuator was named RAINBOW (Reduced And INternally Biased Oxide Wafer) and was introduced by Haertling in 1994 (Haertling, 1997). The CERAMBOW (CERAMic Biased Oxide Wafer) actuator (Barron *et al.*, 1996) and NASA's THUNDER^{TM} (THin-layer composite UNimorph ferroelectric DrivER and sensor) (Barron *et al.*, 1996; Nolan-Proxmire and Henry, 1996) were introduced in the same year. The development of the CRESCENT actuator (Chandran *et al.*, 1997) was followed by LIPCA (LIghtweight Piezocomposite Curved Actuator) as an improved version of the THUNDER actuator, which was developed by Yoon *et al.* (2000). In these unimorph -type actuators, active and passive layers are bonded together by using solder or thermally cured adhesive at elevated temperature. Pre-stress develops during cooling due to the different thermal shrinkage of the layers. Juuti *et al.* (2005) developed a novel pre-stressed piezoelectric actuator, PRESTO (PRE-STressed electrOactive component by using a post-fired biasing layer), where a sintered piezoelectric ceramic is used as a substrate for the post-processed pre-stressing layer. A new family of actuators developed recently has interdigitized electrodes (IDE), so that both the poling (the third-axis) and the applied electric field are oriented along the length of the piezoelectric wafer. In this case, the applied electric field couples with the piezoactuation along the length of the device. These actuators, including the QuickPack (Lazarus *et al.*, 2002), POWERACT (Masters *et al.*, 2002) and the micro-fibre composite (MFC) (Williams and Inman, 2002) developed by NASA Langley, take advantage of the more efficient electromechanical coupling in the third axis to provide greater actuation performance along their length (Yoshikawa *et al.*, 1999; Warkentin, 2000).

Pre-stressed piezoceramic actuators must usually be driven at relatively high amplitudes, and as consequences these devices operate within a non-linear regime (Bryant *et al.*, 1997). This naturally raises questions of the suitability of control laws based on linear models. Moreover, the load capacity of piezoceramic actuators is limited by the presence of the non-linearities. Piezoelectric actuators are characterized by hysteresis and a common model employed is the Bouc–Wen model. The presence of the hysteresis in the input–output model can be dealt with by employing a Schmidt trigger in the control input, which is itself a relay with both deadband and hysteresis. The method and control approach adopted varies widely for different actuators from the methods published so far in the literature for linear systems (Oh *et al.*, 2001; Song *et al.* 1999, 2005).

A new class of piezoelectric ceramic devices that are capable of achieving 100 times greater out-of-plane displacements than previously available ($>1000\mu m$) and sustaining moderate pressures (0.6 MPa (85 psi)) has been developed through cooperative agreement between NASA Langley Research Center and Clemson University, Clemson, South Carolina. This new type of piezoelectric ceramic device is known as RAINBOW and is commercially available from Aura Ceramics, Inc., Minneapolis, Minnesota. First developed by Haertling in 1990, the RAINBOW actuator is composed of two different layers: specifically a reduced layer as an elastic passive layer and an unreduced or active layer, as shown in Figure 2.11.

A RAINBOW actuator is produced by chemically reducing one side of a conventional lead-containing round or rectangular piezoelectric wafer, at an elevated temperature. The actuator is manufactured by

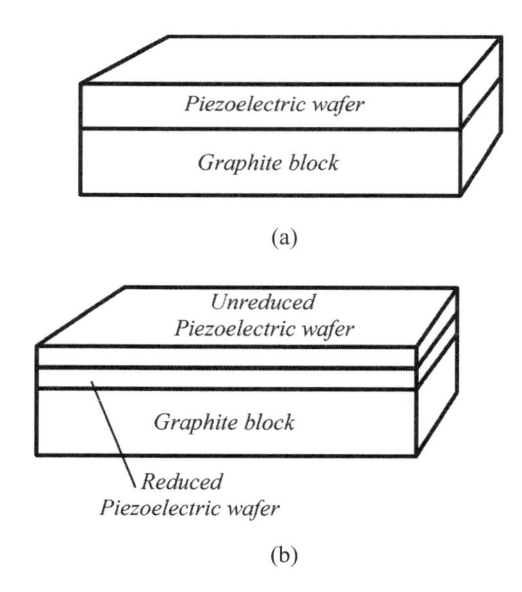

Figure 2.11 Construction of a RAINBOW actuator. (a) Piezoelectric layer and graphite block prior to reduction. (b) Piezoelectric layer and graphite block after reduction

placing a standard flat piezoelectric material, such as PZT or lanthanum-doped PLZT ($(Pb,La)(Zr,Ti)O_3$), in wafer form on top of a carbon block. Then, the whole assembly is heated in a furnace to approximately 700–1000°C. This heating process is known as a reduction process because the carbon oxidation at the ceramic–carbon interface depletes the oxygen atoms from the PZT oxide ceramics. This, consequently, leaves the ceramic wafer with two distinctive layers, like the standard unimorph. This process is depicted in the figure for a rectangular wafer. In this process, the piezoelectric wafer, which is of the order of 0.1–0.5 mm thick (5–20 mil) and up to 50 mm on a side (2 inches), is placed on graphite block and inserted into a furnace preheated to 975°C. While at high temperature, along the interface between the piezoceramic and the graphite block, oxygen is removed from the piezoelectric wafer, creating a wafer that possesses a primarily metallic lead layer (the chemically reduced, or oxygen-deficient, layer) and an unreduced piezoelectric ceramic layer. The bottom layer is a cermet layer of lead, titanium oxide or zirconium oxide, and the top layer is the intact PZT material. The thickness of the unreduced layer can be adjusted by altering the process temperature and the soaking time in the furnace. As a major effect, the reduced layer is no longer piezoelectrically active, but it is electrically conductive. Also as a result of the change in chemical component of the reduced layer from that of original PZT, a thermal expansion coefficient and elastic modulus mismatch between these two layers is generated. Specifically, there is a lack of symmetry of the layer's thermoelastic properties with respect to the actuator's geometric midplane, i.e. the actuator is an unsymmetric laminate. Thus, the wafer then has bi-material properties, i.e. differences in thermal expansion coefficients and Young's moduli between the two layers. When cooled to room temperature, because of the difference in the material properties between the reduced and unreduced layers, the RAINBOW device deforms out of the plane of the originally flat wafer, much like the bimetallic strip, creating an internally stressed structure.

A variant of a RAINBOW bender is produced by re-heating a RAINBOW bender to the cure temperature of a graphite-epoxy fibre-reinforced composite material and bonding single or multiple layers of composite material to one side of it. Once the composite layer is added to a RAINBOW actuator and cured, the temperature is returned to room temperature. Additional internal stresses are induced due to the mismatch in material properties between the unreduced ceramic, reduced ceramic and the composite layer, producing a deformed stress-biased device that can also exhibit large out-of-plane displacements. This new form of active material, known as GRAPHBOW, may have advantage over RAINBOW materials in terms of load-bearing capacity.

The THUNDER actuator is another type of curved actuator, similar to RAINBOW. THUNDER™ (Bryant *et al.*, 1997) is a typical piezoceramic-based pre-stressed actuator capable of generating significant displacements and forces in response to input voltages. These actuators were first developed to drive synthetic jets and Helmholtz resonators. They have since been utilized widely in several airflow control applications. The jets of air are usually created by the use of compressed air or an electromechanically driven vibrating wafer such as the THUNDER™ actuator. The performance capabilities of THUNDER™ actuators are tuned by appropriate choices of the component materials and process used in their construction.

The THUNDER™ actuator was developed by the NASA Langley Research Centre in 1994 and now is manufactured and distributed by Face International Corporation. In fact, it is one of the two commercially available actuators amongst the mechanically pre-stressed bender actuators that have been developed. THUNDER™ represents a piezoceramic-based actuator capable of generating significant displacements and forces in response to input voltages. However, unlike RAINBOW, which is a monolithic-layered structure, THUNDER is composed of a PZT layer sandwiched between layers of metal, such as aluminium, stainless steel, beryllium, etc., all of which are bonded together in a flat condition with a polyimide adhesive that is cured approximately at 325°C. Due to the pre-stresses that result from the differing thermal properties of the component materials during the curing period, the actuator is highly durable with respect to mechanical impacts and voltage levels. As a result of its robust construction, voltages in excess of 800 V can be applied to new actuator models without causing damage. This provides the actuators with significant displacement and force capabilities.

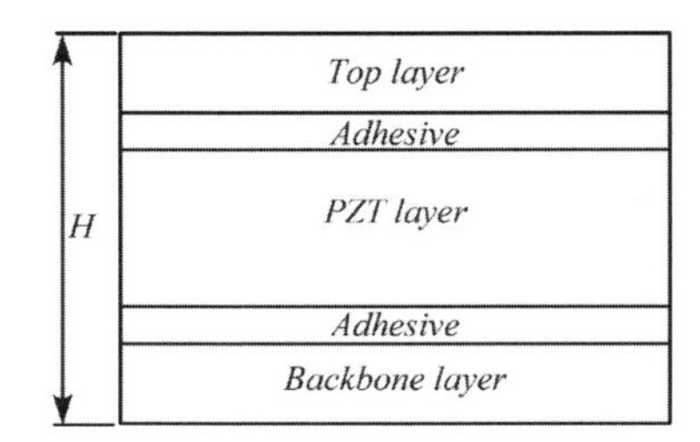

Figure 2.12 Cross-section of a THUNDER actuator

A typical cross-section of a THUNDER actuator is illustrated in Figure 2.12. The bottom layer serves as the backbone of the actuator and is used to attach the actuator to a structure. The very thin top layer, which is optional, is used for protecting the piezoceramic layer from direct exposure. All the layers are assembled in the desired order and thermally processed in an autoclave to produce the actuator. The temperature is raised to 320–325°C at 5°C/min with a full vacuum. Then, the autoclave is pressurized to 207 kPa for 30 min and cooled at a rate of 5°C/min until the temperature reaches 200°C. After this, the vacuum is released and the actuator is allowed to cool to ambient temperature. Again, because the actuator is an unsymmetric laminate, residual stresses and large out-of-plane deformations develop, so the cooled, or manufactured, shape is not flat. The performance capabilities of THUNDER^{TM} actuators are tuned by appropriate choices of the component materials and process used in their construction.

The LIPCA family of piezoceramic actuators is a recent variant of THUNDER introduced by Yoon *et al.* The LIPCA employs an epoxy resin as a pre-preg material, and it follows that the actuator can be manufactured without any adhesives for bonding the layers together. Yoon *et al.* put forward the idea of using fibre-reinforced composite materials as passive layers in order to save weight without losing the capabilities for generating high force and large displacement during electrical actuation. Three main devices have been developed, which are called LIPCA-C1, LIPCA-K and LIPCA-C2. LIPCA-C1 consists of three different materials, namely a glass-epoxy layer, a piezoceramic wafer with electrode surfaces and silver-pasted copper strip wires, and a carbon-epoxy pre-preg layer. All of the materials are stacked in that order, from bottom to top, resulting in an unsymmetric laminate. The stacked layers are vacuum bagged and cured at 177°C for 2 h in a cure oven. After removal from the flat mould, the device possesses residual stresses and curvatures.

LIPCA-K is very similar to LIPCA-C1, except Kevlar49-epoxy is used for the top layer instead of carbon-epoxy. It has been reported that LIPCA-K generates a larger manufactured curvature than LIPCA-C1. Nevertheless, the comparison of actuation responses between the two actuators has not been published. A schematic cross-section of LIPCA-C1 or LICA-K is illustrated in Figure 2.13.

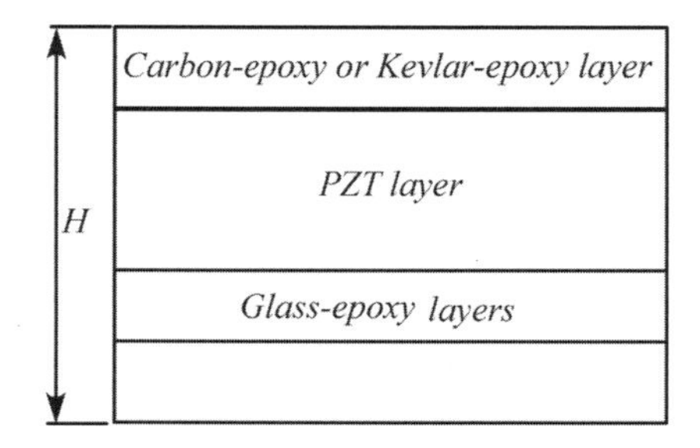

Figure 2.13 Cross-section of a LIPCA -K actuator

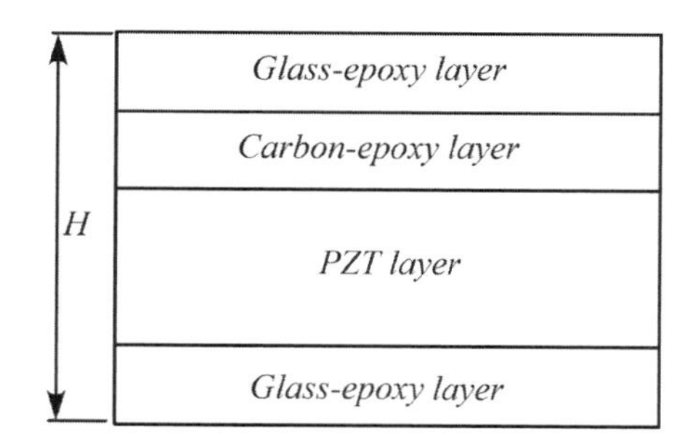

Figure 2.14 Cross-section of a LIPCA -C2 actuator

Recently, Yoon *et al.* have reported the development of the LIPCA -C2 actuator. The key objective of the LIPCA-C2 design is to place the neutral axis of the actuator laminate above the piezoelectric ceramic layer in order to produce compressive stresses in that layer. Based on the work of Barrett and Gross, Yoon and coworkers significantly increased the actuation displacement and force by placing the ceramic layer on the compressive side of laminate at the service temperature. The actuator has five layers, namely glass-epoxy as a bottom layer, piezoceramic, glass-epoxy, carbon-epoxy, and again glass-epoxy as a top layer. Figure 2.14 shows a schematic cross-section of a LIPCA-C2 actuator. After hand lay-up, the stacked laminate is vacuum bagged and cured at 177°C for 2 h in a cure oven. Performance tests have shown that a LIPCA-C2 beam generates twice as large a displacement as a LIPCA-C1 beam, for a simply supported configuration.

In multi-layered laminated actuators, also known as active constrained layer dampers, two piezoceramic layers are employed, one as an actuator and the other as a precise sensor. This permits the multi-layered system to function as a closed-loop servo that can be employed as an active material damper. A typical schematic of such an actuator is illustrated in Figure 2.15.

2.3.2 Shape Memory Material-based Actuators

Shape memory-based actuators are also used extensively for the active control of vibration. Shape memory alloys (SMAs) have the unique ability of being 'switched' from one shape to another pre-defined

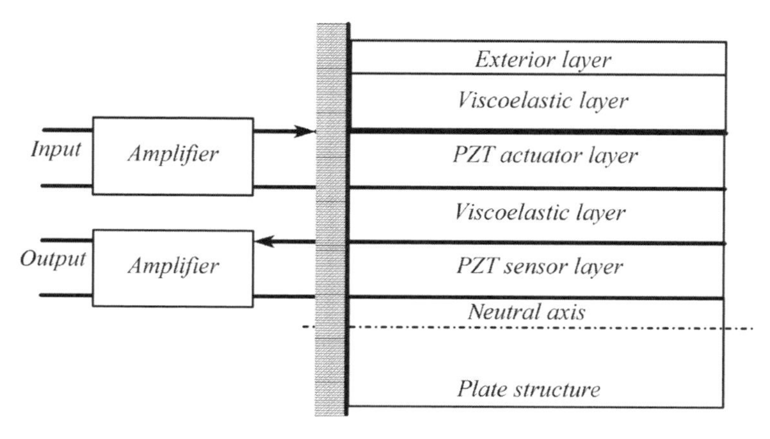

Figure 2.15 Principle of an active damper

shape, by uniformly heating it, so its temperature exceeds the critical switching value. When the alloy is deformed and then heated above this switching or phase transformation temperature, it not only recovers its original shape but also exerts a substantial force during the recovery phase. The discovery of *Nitinol*, a high-strength nickel–titanium (Ni–Ti) alloy, with this property in the 1960s led to the development of several Ni–Ti-based shape memory actuators for vibration control applications. Shape memory-based actuators are under development for actuating plate structures (Oh *et al.*, 2001), hydrofoils (Rediniotis *et al.*, 2002) and adaptive wing structures (Kudva *et al.* 1996), all of which are applications requiring large control forces.

A heated SMA wire can only provide a tensile force. Most mechanisms require cyclic motions and a bias force is needed to return the mechanism to its equilibrium position when pulled by the SMA wire. This bias force must be supplied by stored potential energy (gravity or a spring) or be provided by another SMA actuator working antagonistically.

The shape memory effect has been attributed to the distinct phase transformation that alloy's crystal structure exhibits as its temperature goes through a critical transformation temperature value. Below this critical value of temperature, the alloy is in a soft (martensite) phase and its electrical resistivity falls by a factor k, approximately equal to 0.7 for Ni–Ti alloys, while its Young's modulus and thermal conductivity fall by a factor of k^2. Above the critical temperature, the alloy returns to the stronger (austenite) phase. The relay-like switching characteristic exhibits a certain hysteresis in practice, so the critical temperature is different when the temperature falls and when the temperature rises. Nevertheless, the alloy could be deformed up to approximately 8% in the soft phase and recovers from the deformed state when the temperature rises and the alloy returns to the austenite phase. Thus, the shape memory-based actuators are particularly suitable for on–off vibration control and active control of stiffness.

2.4 Sensors for Flexible and Smart Structures

Sensors are particular examples of transducers, and a number of sensor examples have already been discussed in Section 2.1. Yet there are two classes of sensors that are extremely relevant to structural control: resonant sensors and fibre-optic sensors. These sensors are discussed in the following sections.

2.4.1 Resonant Sensors

Linear accelerometers determine the acceleration of a body by measuring the inertia force that is always present even when the motion is non-uniform and arbitrary. A typical mass–spring–dashpot arrangement is the most commonly used system for measuring linear accelerations. A simple idealized vertical axis accelerometer is illustrated in Figure 2.16. Its action is based on the measurement of the movement of an elastically suspended seismic mass, installed within a housing, rigidly attached to the motion system being investigated. The dashpot is used to restore the mass to its equilibrium position after it responds to the displacement inputs.

2.4.2 Analysis of a Typical Resonant Sensor

A typical acceleration- or displacement-measuring instrument consists of a mass–spring–dashpot system encased in a suitable casing, which could be rigidly attached to any body or surface. The instrument then provides a readout of acceleration or displacement of the body or surface. An accelerometer may be modelled as a single-degree-of-freedom vibrating mass, excited through an isolating spring or protection isolator due to the vibratory motion of the support. A model of one such system is illustrated in Figure 2.16, where two coordinates, $x(t)$ and $y(t)$, are used to describe the state of the system. The number of coordinates required defines the degrees of freedom of the system. Thus, this system is one that

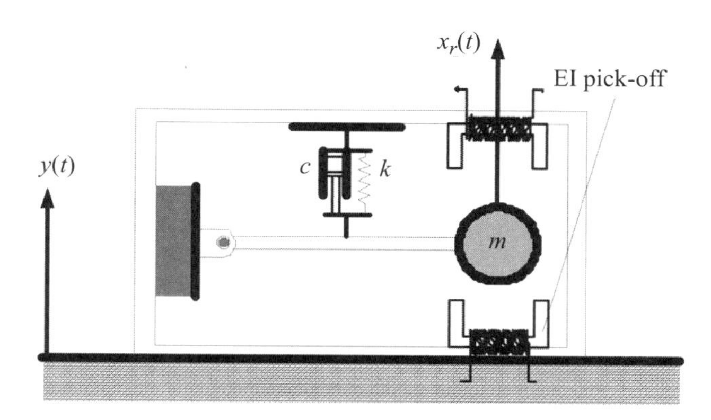

Figure 2.16 Principle of operation of accelerometers and seismographs

apparently possesses two degrees of freedom. Both the coordinates represent absolute motion relative to a fixed coordinate system. However, one of these describes the rigid body motion of support, which is the external excitation ($y(t)$), and the other the motion of the vibrating mass.

The rigid body motion of the support, in this system, may be prescribed as a specified function of time so that the system may be described by just one coordinate or degree of freedom. In effect, the rigid body motion is assumed to be imposed due to an external agency, and consequently the system is reduced to having a single vibratory degree of freedom, as shown in Figure 2.17. Thus, the mass has just one degree of freedom, and this is evident from the equations of motion for the mass.

The free body diagrams for the spring, dashpot and the mass with arbitrary time-dependent support excitation are illustrated in Figure 2.18. We introduce the relative motion coordinate

$$x_r(t) = x(t) - y(t). \tag{2.52}$$

The spring and damper forces are functions of x_r rather than on x, hence, applying Newton's second law

$$ma = \sum F, \tag{2.53}$$

to the mass particle identified in the free body diagram shown in Figure 2.18,

$$ma = -kx_r - cv_r \tag{2.54}$$

where a, v and v_r satisfy the kinematic relations $a = \ddot{x}$, $v = \dot{x}$ and $v_r = \dot{x}_r$. Since $\ddot{x} = \ddot{x}_r + \ddot{y}$, rewriting the equation, we have

$$m\ddot{x}_r + c\dot{x}_r + kx_r = -m\ddot{y}(t). \tag{2.55}$$

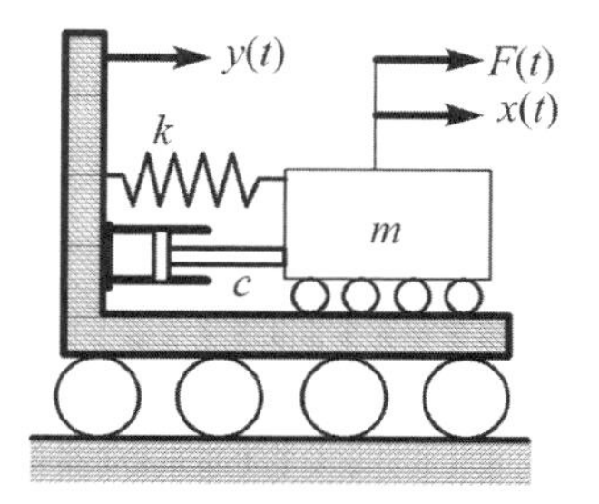

Figure 2.17 Single-degree-of-freedom system with support excitation

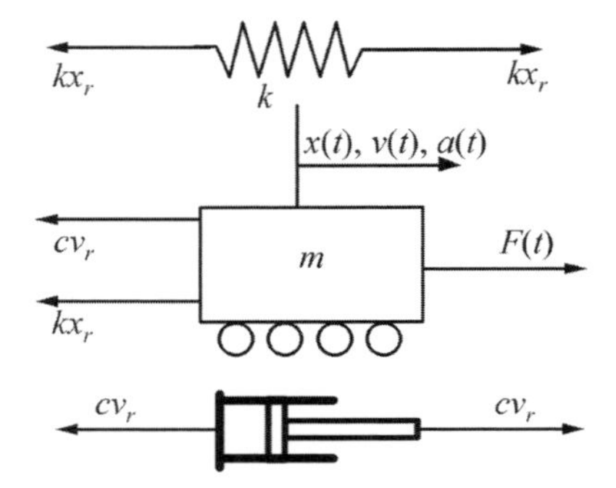

Figure 2.18 Free body diagrams for spring, dashpot and mass with arbitrary time-dependent support excitation

Equation (2.55) may also be written as

$$m\ddot{x}_r + c\dot{x}_r + kx_r = F(t) \tag{2.56}$$

where $F(t) = -m\ddot{y}(t)$. In terms of the coordinate $x(t)$,

$$m\ddot{x} + c\dot{x} + kx = c\dot{y}(t) + ky(t). \tag{2.57}$$

Equation (2.57) may also be written as

$$m\ddot{x} + c\dot{x} + kx = F(t), \tag{2.58}$$

where $F(t) = c\dot{y}(t) + ky(t)$.

Hence, the equations of motion in both the above cases are similar in form to the one where we have arbitrary time-dependent external excitation. The support motion is assumed to be in the form of a sinusoidal function, given by

$$y(t) = y_0 \sin(\omega t). \tag{2.59}$$

Hence, the equation of motion may be written as

$$m\ddot{x} + c\dot{x} + kx = ky_0 \sin(\omega t) + i\omega c y_0 \cos(\omega t). \tag{2.60}$$

Considering the case of non-zero damping, the steady-state vibration is given by

$$x_p(t) = \frac{y_0\sqrt{\left(\omega_n^2 + 4\zeta^2\omega^2\right)}}{\sqrt{\left(\omega_n^2 - \omega^2\right)^2 + 4\zeta^2\omega^2}} \sin(\omega t + \varphi - \phi), \tag{2.61}$$

where the phase angles φ and ϕ are

$$\varphi = \tan^{-1}\left(\frac{2\zeta\omega}{\omega_n}\right), \quad \phi = \tan^{-1}\left(\frac{2\zeta\omega\omega_n}{\omega_n^2 - \omega^2}\right). \tag{2.62}$$

The phase angle φ is the phase lead of the excitation while the phase angle ϕ is the phase lag of the response. The phase lag of the response to the input displacement is

$$\psi = \phi - \varphi = \frac{2\zeta\omega^3}{\omega_n^3\left(\omega_n^2 - \omega^2\right) + 4\zeta^2\omega^2}. \tag{2.63}$$

The amplitude of steady-state vibration is given by

$$x_{p0} = \frac{y_0\sqrt{(\omega_n^2 + 4\zeta^2\omega^2)}}{\sqrt{\left(\omega_n^2 - \omega^2\right)^2 + 4\zeta^2\omega^2}} \tag{2.64}$$

and the *displacement transmissibility ratio* x_{p0}/y_0 is

$$\frac{x_{p0}}{y_0} = \frac{\sqrt{\left(1 + 4\zeta^2\dfrac{\omega^2}{\omega_n^2}\right)}}{\sqrt{\left(1 - \dfrac{\omega^2}{\omega_n^2}\right)^2 + 4\zeta^2\dfrac{\omega^2}{\omega_n^2}}} = \sqrt{\left(1 + 4\zeta^2\frac{\omega^2}{\omega_n^2}\right)}M, \tag{2.65}$$

which may be written in terms of the frequency ratio, $r = \omega/\omega_n$ as

$$\frac{x_{p0}}{y_0} = \frac{\sqrt{(1 + 4\zeta^2 r^2)}}{\sqrt{\left(1 - r^2\right)^2 + 4\zeta^2 r^2}} = M\sqrt{(1 + 4\zeta^2 r^2)}. \tag{2.66}$$

If we consider the case of relative motion with respect to the support

$$m\ddot{x}_r + c\dot{x}_r + kx_r = -m\ddot{y}(t) \tag{2.67}$$

and assume that the support motion is in the form of a sinusoidal function, given by $y(t) = y_0 \sin(\omega t)$, the equation of motion may be written as

$$m\ddot{x}_r + c\dot{x}_r + kx_r = m\omega^2 y_0 \sin(\omega t). \tag{2.68}$$

Considering the case of non-zero damping, the steady-state vibration and the associated phase angle ϕ are given by

$$x_{pr}(t) = \frac{\omega^2 y_0}{\sqrt{\left(\omega_n^2 - \omega^2\right)^2 + 4\zeta\omega^2\omega_n^2}} \sin(\omega t - \phi) \tag{2.69a}$$

$$\phi = \tan^{-1}\left(\frac{2\zeta\omega\omega_n}{\omega_n^2 - \omega^2}\right). \tag{2.69b}$$

The amplitude of this steady-state vibration is given by

$$x_{pr0} = \frac{\omega^2 y_0}{\sqrt{\left(\omega_n^2 - \omega^2\right)^2 + 4\zeta\omega^2\omega_n^2}}, \tag{2.70}$$

and the *relative displacement ratio* x_{pr0}/y_0 is

$$\frac{x_{pr0}}{y_0} = \frac{r^2}{\sqrt{\left(1 - r^2\right)^2 + 4\zeta^2 r^2}} = r^2 M. \tag{2.71}$$

An interesting feature of the displacement ratio, x_{pr0}/y_0, is the fact that when the frequency ratio r approaches ∞, $x_{pr0}/y_0 \rightarrow 1$. When the frequency ratio approaches the resonant frequency ratio, i.e. $\omega = \omega_n\sqrt{1 - 2\zeta^2}$ and $r = \sqrt{1 - 2\zeta^2}$, the displacement ratio approaches a maximum. This is the phenomenon of resonance. A lightly damped spring-mass accelerometer driven by a periodic acceleration of varying frequency shows a peak mass displacement or resonance in the vicinity of the natural frequency.

When this theory is applied to the accelerometer illustrated in Figure 2.16, the *displacement ratio* of the amplitude of steady vibration of the mass relative to the casing to the amplitude of the surface vibration x_{r0}/y_0 is

$$\frac{x_{r0}}{y_0} = \frac{r^2}{\sqrt{\left(1 - r^2\right)^2 + 4\zeta^2 r^2}} = r^2 M \tag{2.72}$$

and when $r \ll 1$ and $\zeta \ll 1$, $x_{r0}/y_0 = r^2$ or $x_{r0} = r^2 y_0$. Thus, the amplitude of steady-state vibration of the mass relative to the casing is directly proportional to the acceleration of the surface to which the casing is attached. The displacement of the mass, relative to the casing, is measured by the EI pickoff. This displacement is directly proportional to the applied acceleration. The system now behaves like an accelerometer and is capable of measuring accelerations in a particular direction over a certain range of frequencies. To measure steady accelerations, corrections must be incorporated to account for the acceleration due to gravity. Mass sprung systems can be used to measure velocity or displacement in addition to acceleration. In fact, these pickups respond as accelerometers in the region below their natural frequencies and as vibrometers or displacement meters in the region well above the natural frequencies.

When $r \gg 1$ and $\zeta < 1$, $x_{r0}/y_0 = 1$ or $x_{r0} = y_0$ and the amplitude of the mass is proportional to the amplitude of the motion of the vibrating body. Over this frequency range, the sprung mass in the accelerometer behaves like a free mass and registers the displacement of the vibrating body. The system acts as seismograph or as a displacement meter over this part of the frequency spectrum. The displacement of the mass relative to the casing is directly proportional to the displacement of the surface on which the instrument is mounted. Thus, seismographs or vibrometers are built with extremely soft springs and moderately low damping constants.

2.4.3 Piezoelectric Accelerometers

Piezoelectric and piezoresistive accelerometers are extensively used in the measurement of shock and vibration. Variable capacitance devices with resolutions less than 0.1 nm, ranges up to 0.5 mm and bandwidths of 10 are also used for accurate position measurement. Alternatively, there a range of optoelectronic position-sensitive devices including photo-diodes and scanning and laser vibrometers, which are now coming into vogue for vibration measurement. Piezoresistive accelerometers have the advantage that they are useful in the measurement of constant acceleration (at zero frequency), which is an essential aspect in the measurement of long-duration shock motions. They are particularly useful in replacing strain gauge accelerometers in existing applications, resulting in an increased bandwidth as well as output power.

A typical seismic accelerometer consists of a case, which is attached to a moving part. The case contains a high-density tungsten alloy mass element connected to a case by a sensing element whose stiffness is less than that of the mass element. The sensing element is usually a cylindrical piezoelectric ceramic or a piezoresistive strain gauge, which is cemented to a heavy sleeve, forming the mass element. The cylindrical sensing element is also attached to a central post. Sometimes, several sensing elements are networked both mechanically and electrically in order that it has the desired stiffness properties. Apart from the inherent damping present, filling the accelerometer with a viscous fluid provides additional viscous damping. The damping constant is such that $\zeta^2 = 1/2$, so as to produce a relatively flat response over wide range of frequencies. As the resonance frequencies may range from 2500 to 30 000 Hz, the bandwidths vary from about 500 to 7000 Hz.

A piezoelectric material is usually either a quartz crystal or a ceramic that exhibits the piezoelectric property. In these materials, a potential difference appears across opposite faces of it, when a mechanical force is applied and the associated strain results in dimensional changes. The effect is only possible in those crystals with asymmetric charge distributions, so lattice deformations result in relative displacements of positive and negative charges within the lattice. The effect is therefore also reversible, in that a potential difference applied across two opposite faces of the crystal or ceramic results in change in dimensions. The charge is measured by attaching electrodes to the same two faces across which the force

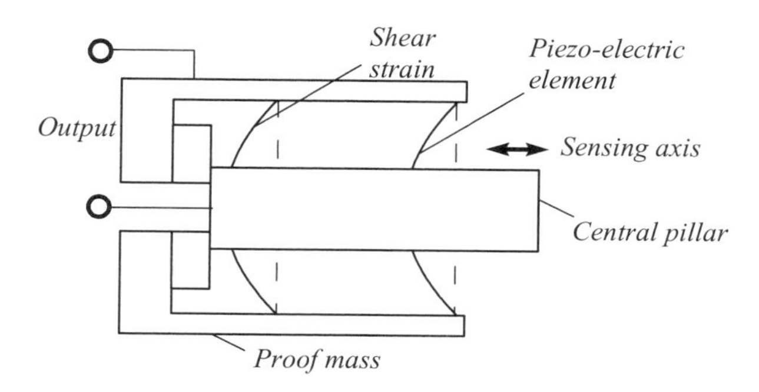

Figure 2.19 Shear-type piezoelectric transducer (the casing is not shown)

or pressure is applied. Piezoelectric accelerometers are of the compression or shear type, depending on the applied force being a compressive or shear force. The accelerometer measures shock and vibration in the direction of its axis of symmetry, which is perpendicular to the base of the accelerometer.

A shear-type piezoelectric transducer is illustrated in Figure 2.19. The spring and damper are replaced by the piezoelectric material. The motion of the mass element is opposite in phase to the motion of the accelerometer. Thus, when the accelerometer moves upwards the mass moves downwards, relative to its base.

For the shear-type of piezoelectric ceramic element, the result of the motion is a change in the shear stress acting on it. For the compression-type accelerometer, a compressive static preload is applied to the piezoelectric ceramic. The preload can be adjusted by tightening a nut provided at the top of the central post. The preloads are selected such that they greatly exceed the dynamic stresses resulting from the motion of the accelerometer. The piezoelectric constant depends on the accelerometer type as well as on the material used: quartz, lead–zirconate–titanate or other recently developed proprietary ceramics. Although quartz has the lowest piezoelectric constant, it is natural crystal and is still widely used when the advantages (cost and low capacitance) outweigh the disadvantages (lowest acceleration sensitivity).

Piezoelectric accelerometers are generally characterized by a net residual capacitance across its two terminals in the range of 50–1000 pF. The sensitivity of the accelerometer is generally expressed in terms of the charge q, on the capacitor, for an input of 1g rms acceleration. Although charge sensitivity is adequate to describe the sensitivity of the accelerometer, voltage sensitivity is often also quoted as it appears to be more meaningful to most users. The charge sensitivity is the product of the voltage sensitivity and the residual capacitance across the two terminals of the accelerometer. The voltage and charge sensitivities of typical piezoelectric ceramics are shown in Table 2.3. Typically, the charge output of a piezoelectric accelerometer is amplified by a charge to voltage converter before it is displayed by an oscilloscope or recorded by a data logger.

Table 2.3 Typical characteristics of piezoelectric accelerometers

Crystal material	Type	Typical capacitance (1000 pF)	Acceleration sensitivity	
			Charge (pC/g)	Voltage (mV/g)
Lead–zirconate–titanate	S	1	1–10	1–10
Lead–zirconate–titanate	C	1–10	10–100	10–100
Quartz	C	0.1	1	10
Endevco Piezite – P-10	S	0.1	0.1	1
Endevco Piezite – P-10	C	1	1–10	1–10

S, shear; C, compression; pF, pica Farad; pC, pica Coulomb; mV, milliVolt.

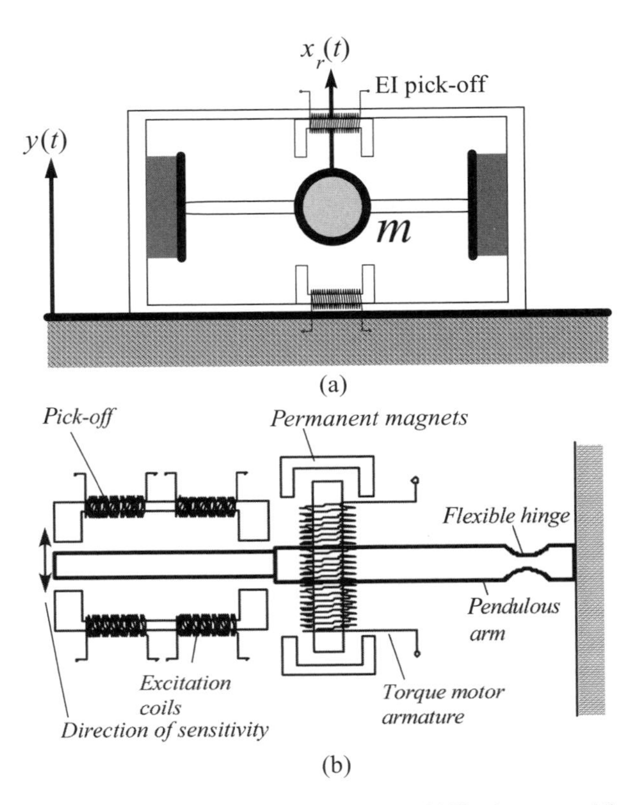

Figure 2.20 Principle of operation of vibrating beam accelerometers (a) Simply supported 'beam' type. (b) Principle of operation of a pendulous force feedback-type accelerometer

The analogue device ADXL150 integrated chip is an accelerometer that is fabricated by surface machining of polysilicon. It contains a micro-machined proof mass or shuttle suspended on a simply supported folded leaf spring, which is attached to the polysilicon substrate only at the anchor points (Senturia, 2001).

The principle of operation of a vibrating beam accelerometer is shown in the schematic diagram in Figure 2.20(a). In the polysilicon version, the EI pickoff is replaced by a capacitive pickup. A number of cantilevered electrodes are attached to the shuttle. A number of these units are fabricated in parallel, and the displacements of the proof mass are estimated by measuring the lateral differential capacitance of the entire assembly. The ADXL150 is revolutionary compared to the accelerometers that were available in the 1990s and is available at relatively low price. This very low price is possible because of way they are produced by a polysilicon batch process. Yet in terms of performance, they are as good as some of the most expensive electrodynamic accelerometers.

The principle of operation of a pendulous force feedback-type accelerometer is illustrated in Figure 2.20(b). The displacement of the pendulum is sensed by the two pickoffs in differential mode, which is then conditioned and fed back to the linear torque motor. The electromagnetic force generated by the motor acts to counter the displacement of the pendulum and maintain the pickoff output at zero. The current to the motor armature is directly proportional to the externally applied acceleration. In the polysilicon version of this device, piezoelectric pickoffs are used in place of the EI pickoffs and a pair of piezoelectric actuators replace the torque motor.

2.4.4 The Sensing of Rotational Motion

The motion of a single rigid body has six independent degrees of freedom, three of which represent translational motions of a reference point (usually the centre of mass of the rigid body) along three mutually orthogonal reference directions, while the other three represent the orientation of the body relative to the reference directions. To define the equations of motion, we begin with definition of the moment of momentum vector. The moment of momentum vector of a rigid body is given by

$$\mathbf{h} = \int_V \mathbf{r} \times (\omega \times \mathbf{r})\ \mathrm{d}m = \int_V [\omega (\mathbf{r}.\mathbf{r}) - \mathbf{r}(\mathbf{r}.\omega)]\ \mathrm{d}m = \mathbf{I}\omega, \tag{2.73}$$

where $\mathbf{I}$ is the moment of inertia matrix and ω is the angular velocity vector in a reference frame rigidly attached to the body at a point fixed in the body given as

$$\omega = \begin{bmatrix} p & q & r \end{bmatrix}^T \quad \text{or as } \omega = p\,i + q\,j + r\,k \tag{2.74}$$

in terms of the mutually perpendicular unit vectors i, j and k in the three body axes.

If the axes along which $\mathbf{h}$ is resolved are defined to be coincident with the physical principal axes of the body, then $\mathbf{I}$ is a diagonal matrix. Thus, when $\mathbf{h}$ is not resolved along principal body axes we get

$$\mathbf{h} = \mathbf{I}\omega = \begin{bmatrix} I_{xx} & -I_{xy} & -I_{xz} \\ -I_{xy} & I_{yy} & -I_{yz} \\ -I_{xz} & -I_{yz} & I_{zz} \end{bmatrix} \begin{bmatrix} p \\ q \\ r \end{bmatrix}, \tag{2.75}$$

where p, q and r are the three body components of angular velocity.

The characterization of the motion of a rigid body in a non-inertial coordinate system (i.e. a coordinate system in which the reference axes rotate and accelerate linearly) is treated in most textbooks on advanced dynamics. The rate of change of vector $\mathbf{p}$ (such as the translational velocity or angular momentum), with its components defined in a rotating reference frame, is obtained as

$$\left.\frac{\mathrm{d}\mathbf{p}}{\mathrm{d}t}\right|_{inertial} = \left.\frac{\mathrm{d}\mathbf{p}}{\mathrm{d}t}\right|_{body} + \omega \times \mathbf{p}, \tag{2.76}$$

where the symbol $\times$ denotes the *vector cross product* and the *body derivative* implies that the derivatives are taken as if the body axes are inertially fixed and ω is the angular velocity vector of the non-inertial reference frame. Thus, the Newtonian equations of motion governing the translational motion of a rigid body are

$$\mathbf{F} = m\mathbf{a}_O = \left.\frac{\mathrm{d}(m\mathbf{v})}{\mathrm{d}t}\right|_{body} + \omega \times (m\mathbf{v}) = m\left(\frac{\mathrm{d}\mathbf{v}}{\mathrm{d}t} + \omega \times \mathbf{v}\right), \tag{2.77}$$

where $(\dot{})$ represents the derivative taken as if the axes are inertially fixed, $\mathbf{a}_O$ the acceleration vector of the origin of the reference frame fixed in the body and $\mathbf{v}$ is the velocity vector in the body fixed frame given by

$$\mathbf{v} = U\,i + V\,j + W\,k \quad \text{or as } \mathbf{v} = \begin{bmatrix} U & V & W \end{bmatrix}^T \tag{2.78}$$

and $\mathbf{F}$ is the three component external force vector in the same frame given by

$$\mathbf{F} = X\,i + Y\,j + Z\,k \quad \text{or as } \mathbf{F} = \begin{bmatrix} X & Y & Z \end{bmatrix}^T. \tag{2.79}$$

The cross product is:

$$\omega \times \mathbf{v} = \begin{vmatrix} i & j & k \\ p & q & r \\ U & V & W \end{vmatrix} = i(qW - rV) - j(pW - rU) + k(pV - qU). \tag{2.80}$$

Equation (2.80) may be expressed in matrix notation as

$$\omega \times \mathbf{v} = \begin{bmatrix} 0 & -r & q \\ r & 0 & -p \\ -q & p & 0 \end{bmatrix} \begin{bmatrix} U \\ V \\ W \end{bmatrix} = \begin{bmatrix} qW - rV \\ rU - pW \\ pV - qU \end{bmatrix}. \tag{2.81}$$

The subscript O is dropped from the vector $\mathbf{v}$ for brevity. Thus, we have the three scalar equations governing the translational motion:

$$m\left(\dot{U} + qW - rV\right) = X, \tag{2.82a}$$
$$m\left(\dot{V} + rU - pW\right) = Y, \tag{2.82b}$$
$$m\left(\dot{W} + pV - qU\right) = Z. \tag{2.82c}$$

Similarly for rotational motion, we have

$$\mathbf{M} = \mathbf{r}_O \times \mathbf{F} = m\left(\mathbf{r}_{CM} \times \mathbf{a}_O\right) + \left.\frac{d\mathbf{h}}{dt}\right|_{body} + \omega \times \mathbf{h} = \dot{\mathbf{h}} + \omega \times \mathbf{h}, \tag{2.83}$$

where $\mathbf{M}$ is the three-component external torque vector in the body fixed frame, which is obtained by taking moments of all external forces about the origin of the body fixed frame and may be written as

$$\mathbf{M} = L\,i + M\,j + N\,k \quad \text{or as } \mathbf{M} = \begin{bmatrix} L & M & N \end{bmatrix}^T. \tag{2.84}$$

$\mathbf{r}_{CM}$ is the position vector of the centre of mass relative to the origin of the body fixed frame and $\boldsymbol{h}$ is the moment of momentum vector for the rigid body. The external torque is given by evaluating the sum of the moments of the forces acting on the rigid body. Equations (2.83) are the famous *Euler equations* and describe how body axes components of the angular velocity vector evolve in time in response to torque components in body axes.

For our purposes we consider a set of body axes fixed to the body at the centre of mass and aligned in the directions of the principle axes. In this case

$$\mathbf{h} = \mathbf{I}\omega = \begin{bmatrix} I_{xx}p & I_{yy}q & I_{zz}r \end{bmatrix}^T. \tag{2.85}$$

The Euler equations then reduce to

$$I_{xx}\dot{p} + (I_{zz} - I_{yy})qr = L, \tag{2.86a}$$
$$I_{yy}\dot{q} + (I_{xx} - I_{zz})rp = M, \tag{2.86b}$$
$$I_{zz}\dot{r} + (I_{yy} - I_{xx})pq = N. \tag{2.86c}$$

The most interesting application of the Euler equations is the gyroscope. In an ideal gyroscope, a rotor or wheel is kept spinning at constant angular velocity. The axis of the wheel is assumed to be coincident with the z axis. It is assumed that Oz axis is such that $\dot{r} = 0$, i.e. such that

$$h_z = I_{zz}r = I_{zz}\dot{\psi} = \text{constant}, \tag{2.87}$$

where I_{zz} is the polar moment of inertia of the wheel. The other two moments of inertia of the wheel are assumed equal to each other and equal to I_d, and hence the first two of the Euler equations may be written as

$$I_d\dot{p} + Hq = L, \; I_d\dot{q} - Hp = M, \tag{2.88}$$

where $H = (I_{zz} - I_d)\,r$.

These are the general equations of motion for two-degrees-of-freedom gyroscopes or gyros. If one assumes the gyro to be gimballed with a single gimbal, with the input angular velocity in the Ox direction, then for the Oy axis,

$$I_d \dot{q} - Hp = M, \tag{2.89}$$

where p = input angular velocity.

If the restoring torque is provided by a damper, then

$$M = -Bq. \tag{2.90}$$

Under steady conditions, the change in the output angular velocity q is proportional to the change in the input angular velocity,

$$q \approx \frac{H}{B} p, \tag{2.91}$$

and the gyro behaves like a *rate integrating gyro*.

On the other hand, if the restoring torque is given by a lightly damped torsional spring force,

$$M = -Bq - K\theta = -B\dot{\theta} - K\theta, \; q = \dot{\theta} \tag{2.92}$$

and then

$$I_d \ddot{\theta} + B\dot{\theta} + K\theta = Hp. \tag{2.93}$$

Under steady-state conditions, the output angular displacement is proportional to the input angular velocity

$$\theta \approx \frac{H}{K} p, \tag{2.94}$$

and the gyro behaves like a rate gyro.

2.4.5 The Coriolis Angular Rate Sensor

The Coriolis angular rate sensor uses the Coriolis effect in the sensor element to sense the speed of rotation. A radially vibrating element (vibrating resonator), when rotated about an axis perpendicular to the direction of vibration, is subjected to Coriolis effect, which causes secondary vibration in a direction mutually orthogonal to the original vibrating direction and the direction of rotation. By sensing the secondary vibration, the rate of turn can be detected. For detection of the vibration, a sensor based on the piezoelectric effect is often employed. The vibrating rate gyros are often called 'piezo', 'ceramic' or 'quartz' gyros, although in fact the vibration is independent of the piezo effect. This type of gyro is almost free of maintenance. The main drawback is that when it is subjected to external vibration, it cannot distinguish between secondary vibration and external vibration. This is especially true for those using tuning fork- or beam (often triangular)-shaped vibrating element, which has solid support to the base/case. Dampers around it will not solve the problem, since dampers will affect rotational motion, making the gyro's response worse.

A solid-state Coriolis rate gyro has overcome this problem using a ring-shaped element vibrating in a squeezed oval motion up and down while the ring is suspended by spokes around it; external vibration will not cause the oval squeeze vibration mode, and the exclusive lateral suspension by spokes around it enables the element to be insensitive to linear vibrations/shocks, no matter how firmly the rate gyro is strapped down to the base.

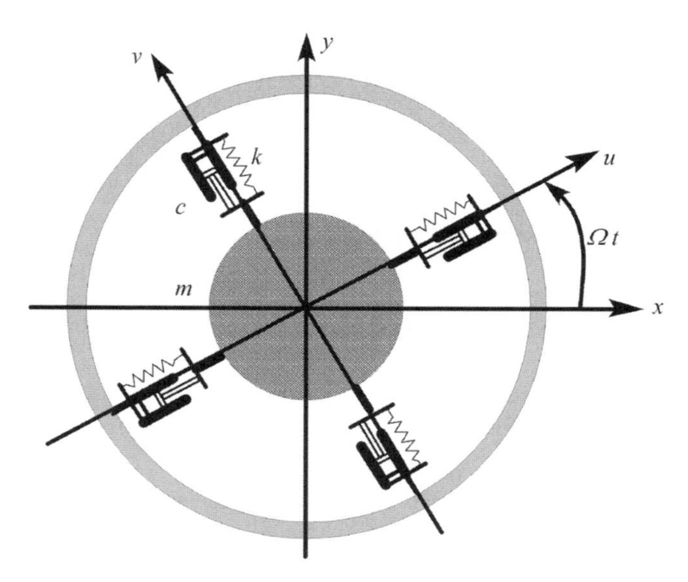

Figure 2.21 The principle of operation of the Coriolis gyro

To understand the principle of the Coriolis acceleration rate gyro, we consider shaft suspended within a ring rotating with a fixed angular velocity Ω, as illustrated in Figure 2.21. Consider an inertial reference frame (x, y), and the equations of motion in this frame are

$$m\ddot{x} + c\dot{x} + kx = 0, \quad m\ddot{y} + c\dot{y} + ky = 0. \tag{2.95}$$

However, considering a rotating frame of reference (u, v) rotating with the angular velocity Ω, the acceleration in the rotating frame may be expressed as

$$a_i = \ddot{\mathbf{r}} + \dot{\Omega} \times \mathbf{r} + 2\Omega \times \dot{\mathbf{r}} + \Omega \times (\Omega \times \mathbf{r}). \tag{2.96}$$

Since Ω is a constant, the equations of motion in the rotating frame are

$$m\ddot{u} + c\dot{u} + ku = 2m\Omega v + m\Omega^2 u, \tag{2.97a}$$

$$m\ddot{v} + c\dot{v} + kv = -2m\Omega u + m\Omega^2 v. \tag{2.97b}$$

If we assume that the motion $u(t)$ is imposed and is a sinusoidal vibration, and assuming that the spring force kv is very large relative to the damping and inertia forces,

$$v(t) = -\left(2\dot{u}(t)/\left(\omega_v^2 - \Omega^2\right)\right)\Omega. \tag{2.98}$$

Thus, the amplitude of the $v(t)$ motion is directly proportional to Ω, provided $\omega_v^2 \gg \Omega^2$. This is the basis and the principle of operation of the Coriolis rate gyro. It is in fact the same principle that is observed in the well-known Focault pendulum. Focault demonstrated that a vibrating pendulum does not actually maintain its plane of vibration but precesses around at a rate equal to the angular velocity of the earth. Thus, the rotating shaft in the Coriolis rate gyro is usually a cantilever with a flexible hinge or is constructed in the form of a tuning fork. The imposed motion $u(t)$ is in the plane of the tuning fork while the sensed output motion $v(t)$ is perpendicular to it. A typical example of such a gyro is the Systron Donner QRS rate gyro. The entire devise has been implemented in a piezoelectric material implanted in a semiconductor substrate. The amplitude in the sensed direction is measured by a charge amplifier.

2.5 Fibre-optic Sensors

2.5.1 Fibre Optics: Basic Concepts

Fibre-optic transmission of low-power data signals over long distances is characterized by attenuation losses that are less than copper wire transmission, thus making it possible to transmit signals over longer distances with the use of fewer repeater stations. While there are several advantages in employing fibre-optic transmission, there are also several disadvantages. The advantages are:

1. A very high digital data rate, in the range of several mega bits per second (Mb/s) to several giga bits per second (Gb/s), limited by multi-path and material dispersion, could be achieved.
2. Optical cables are relatively thinner than copper and other conventional cables.
3. Optical fibre transmission could cover large distances by employing repeater stations to receive and re-transmit the signals.
4. Optical fibres allow for secure communications since they cannot be intercepted either by inductive or other pickups or by contact. Extraction of the light signal without detection is virtually impossible.
5. Optical-fibre transmissions are immune to both electromagnetic interference and noise.
6. The safe usage of optical fibres can be guaranteed provided certain procedural steps are followed during the installation of light sources, which can be hazardous to the human eye.
7. They allow for the use of optical multiplexers, which could increase the data capacity, making optical fibre transmission a cost-effective alternative for data transmission.
8. Optical-fibre transmission systems are economical both to install and to use.

A major disadvantage of optical-fibre transmission is that the total power that can be transmitted is restricted as it cannot exceed certain limits.

The operation of optical-fibre-based systems in general and fibre-optic transducers in particular relies on the fundamental physical laws, which are briefly reviewed in the next section.

2.5.2 Physical Principles of Fibre-optic Transducers

2.5.2.1 Refraction and Reflection

A fundamental aspect of light propagation is that it travels in a straight line, in the direction of propagation of the energy, and such a line is usually referred to as a ray of light. A bundle of rays emerging from a single point source is known as a pencil, while a collection of pencils emerging from an extended source is referred to as a beam of light. An alternative way to imagine light is to think of it as a wave. Christiaan Huygens formulated a principle, known as Huygens' principle, that provides a geometrical recipe for deducing the propagation path of a wave and may be applied to deduce the laws of refraction and reflection from the wave theory of light. Light is in fact an electromagnetic wave, with the direction of propagation coinciding with the direction of the light ray. Whether it is perceived simply in terms of rays or in terms of electromagnetic waves, it undergoes refraction and reflection. Following the work of James Clerk Maxwell, the modern theory of fibre optics treats refraction and reflection as a boundary value problem involving electromagnetic waves. In the case of simple planar boundaries, the solutions to the boundary value problem can be explained in terms geometrical and wave theories of optics. It is important to understand the physics behind refraction and reflection so that we can predict what will take place in spectral regions that are not observable.

Refraction is the bending of light at an interface between two transmitting media. A transparent (dielectric) medium having a refractive index greater than that of its surroundings could act as a light guide. The key principle on which this is based is the refraction of light at an interface. Refraction at an

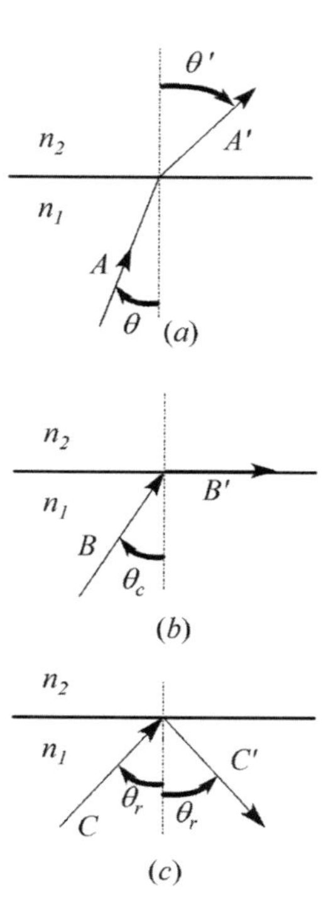

Figure 2.22 Refraction at a dielectric interface

interface between two uniform media is governed by Snell's law, formulated in 1621 and illustrated in Figure 2.22.

In Figure 2.22(a), a ray of light is shown passing from a medium of higher refractive index n_1 into a medium of lower refractive index n_2. For $0 \leq \theta \leq \theta_c$, where θ_c is limiting critical angle and $0 \leq \theta' \leq \pi/2$,

$$n_1 \sin\theta = n_2 \sin\theta', \tag{2.99}$$

where θ and θ' are the angles of incidence and refraction, as defined in the figure. When the angle θ is increased to the so-called limiting critical angle θ_c, the angle θ' increases to $\pi/2$ and in this case, as illustrated in Figure 2.22(b),

$$n_1 \sin\theta_c = n_2. \tag{2.100}$$

The critical angle at which the reflection becomes total is given by

$$\theta_c = \arcsin(n_2/n_1) \tag{2.101}$$

and when $\theta > \theta_c$, there is total internal reflection with no losses at the boundary, as illustrated in Figure 2.22(c).

The phenomenon of total internal reflection allows for the propagation of light rays, in a broken line, within a layer of higher refractive index n_1 sandwiched between two layers of material of lower refractive

Table 2.4 Critical incident angle θ_c for total internal reflection

Transparent material	Critical incident angle θ_c (degrees)
Water	48.6
Ordinary glass	41.8
Crystal glass	31.8
Diamond	24.4

index n_2. Assuming that the medium of lower refractive index is air, so $n_2 = 1$, the critical angle θ_c, is estimated for several transparent materials, is indicated in Table 2.4.

It can be seen that diamond is a real light trap and is the real reason why diamonds sparkle when they are cut so as to have several plane faces. Only light normal to each of the faces passes through while all other light is reflected, thus causing the faces to sparkle.

The laws of refraction are complemented by Fresnel's laws concerning reflectivity, which are also based on electromagnetic theory. Even those rays within the critical region $\theta < \theta_c$ suffer from partial reflection at the interface because of the change in the refractive index. This is known as Fresnel's reflection. Considering the radiation incident, perpendicular to the surface, a fraction R is reflected, which is given by

$$R = (n_1 - n_2)^2/(n_1 + n_2)^2. \tag{2.102}$$

It applies to a beam of light travelling from a medium of lower refractive index to one with a higher refractive index and vice versa. When $\theta > \theta_c$, the coefficient of reflectivity is equal to 1. The above formula applies for polarized light. When considering non-polarized light, it is essential to distinguish between two extreme cases: case I when the electric field of the light wave is perpendicular to the plane of incidence and case II when the electric field of the light wave is parallel to the plane of incidence. The coefficient of reflectivity in the first case becomes zero when

$$\theta_1 + \theta_2 = \pi/2, \tag{2.103}$$

where θ_2 is angle of the ray in the refracted or second medium. This equation has a solution when $\theta_1 = \theta_p$, which is known as the *Brewster* angle. The Brewster angle is given by

$$\theta_p = \arctan(n_2/n_1) \tag{2.104}$$

Table 2.5 compares the Brewster angle θ_p, with the limiting incidence angle θ_c. When non-polarized beam of light is travelling from a medium of lower refractive index to one with a higher refractive index and is incident on the interface at the Brewster angle, both components of the light beam are refracted but only the reflected light contains a single component and is polarized.

When total internal reflection is taking place, the electromagnetic wavefield extends into the low refractive index medium. This wavefield decreases exponentially in the direction perpendicular to the surface. It is called the evanescent wave. It is associated with an electromagnetic field moving parallel to the surface. It is the basis of several fibre-optic sensors, particularly electrochemical and biosensors.

Table 2.5 Reflection coefficients and Brewster angles

Transparent material	n_1	R	θ_p (degrees)	θ_c
Water	1.333	0.02	37	48.6
Ordinary glass	1.500	0.04	33.7	41.8
Crystal glass	1.900	0.10	27.8	31.8
Diamond	2.417	0.17	22.5	24.4

2.5.2.2 Diffraction and Interference

Apart from refraction and reflection, light is also associated with phenomenon of diffraction, and the related phenomenon of scattering and interference. If a beam of light illuminates an obstacle in which there is a small aperture, a decaying pencil of rays that have penetrated the aperture is observed behind the obstacle.

Thus, luminous energy exists in what may normally be considered as the geometric shadow of the obstacle. This is known as the phenomenon of diffraction. It is a very general phenomenon that is a consequence of the wave nature of light and is observed in all wave propagations.

When the incident beam of light is a coherent beam, an expression for the angular divergence of the pencil transmitted through a circular aperture is obtained by observing that the field at a point can have a significant intensity only if the ray from the side of the aperture is approximately in phase with ray from the centre.

Three regions can be distinguished behind the obstacle: (1) the near field, (2) the intermediate or Fresnel zone and (3) the far or Fraunhofer zone. The diameter of the diffracted pencil becomes significantly different from that of the aperture only at a distance d, which is of the order of $a \times (a/\lambda)$. The centre of the intermediate region is situated here where $d \approx a^2/\lambda$. At greater distances, the wavefront is approximately spherical and the bean then resembles a pencil of concentric rays. In the far or Fraunhofer zone, the rays propagate almost parallel to each other. Thus considering parallel rays at an angle α to the direction of the beam and an aperture of width a, we have the condition for significant luminous intensity as

$$a \sin(\alpha) \approx a\alpha = m\lambda, \tag{2.105}$$

i.e. the path difference between the two rays is an integral multiple of the wavelength. Thus

$$\alpha = \lambda/a, \tag{2.106}$$

which is only applicable when a is much greater than λ.

The fibre Bragg grating (FBG) sensor is a classic example of a sensor that exploits the phenomenon of diffraction. It is based on a diffraction grating, which is a series of regularly spaced perturbations in the form of parallel lines that are etched on an optical surface and so generate a sequence of interfering diffracted light waves.

The phenomenon of interference may be examined by considering a typical practical example such as the interference of light undergoing multiple refractions in a thin film. The thin film is a classic example where a beam of light passes from a region of low refractive index to a region of high refractive index at the first interface and vice versa at the second interface. The beam of right undergoes refraction at the first interface, while it is partially refracted and partially reflected at the second. The reflected component undergoes a second reflection when it returns to the first interface, and a further component of it is refracted a second time at the second interface. Thus, when one compares the primary refracted component with the component that undergoes a double reflection before being refracted into the medium of lower refractive index, there is a path difference that is proportional to twice the film thickness t, the film material refractive index n, and the cosine of the angle of the refracted ray makes with the surface normal cos r. When this path difference is an integral multiple of the wavelength, the interference is constructive. If the beam is captured on a screen, one observes a series of bright and dark fringes. The bright fringes correspond to constructive interference while the dark one to destructive interference. The conditions for the occurrence of bright fringes are

$$2t \times n \times \cos r = m\lambda, \quad m = 0, 1, 2, 3 \ldots, \tag{2.107a}$$

while the corresponding conditions for dark fringes are

$$2t \times n \times \cos r = (2m + 1)\lambda/2, \quad m = 0, 1, 2, 3 \ldots. \tag{2.107b}$$

Fibre-optic transducers based on the principle of interference are known as interferometric transducers. Examples of such transducers are the Michelson's two-mirror interferometer, the Mach–Zehnder refractometer and the Fabry–Perot cavity-based interferometer.

2.5.2.3 Optical Path Length and Fermat's Principle

So far in our discussion of the physical principles of fibre optics, we have confined our attention to the propagation of light in uniform media, with sudden changes in the refractive index at an interface. It is essential to consider the general principles of optical propagation in a medium with continuous variation of the refractive index even if these variations are small. The concept of *optical path length* is one that is central to a different way of looking at light propagation that takes into account the slowing of light by the refractive index of the medium through which the light is passing. The optical path length in any small region is the physical path length multiplied by the refractive index. When this quantity is summed over the entire path of propagation, taking into consideration the local variations of the refractive index, one obtains the total optical path length. The concept is employed to determine the propagation time, the number of wavelengths and the phase changes as an optical wave propagates through a medium. The time for a wave to propagate along an optical path through a medium is determined by the optical path length divided by the speed of light in vacuum. The phase change in cycles, along an optical path, is determined by the optical path length divided by the vacuum wave number. Furthermore, the importance of the optical path length stems from that fact that a key principle associated with the name of Fermat is based on it. Fermat's principle is a variational principle, which states that, of all the geometrically possible paths that light could take between two points, the optical path length is stationary along the actual path taken. When the optical path length is stationary, then so is the time of travel for light, and consequently the actual path taken by light is the one that takes the minimum time. Snell's law follows, and the angles of incidence and reflection are equal as required by the law of reflection.

2.5.2.4 The Quantum Theory of Light

Although the corpuscular theory of light was the earliest of the theories of light to be formulated, it was not until the laws of equivalence of wave and quantum descriptions of light were established that the particle theory of light was recognized. Thus, light energy is considered to be encapsulated within particles known as photons, each with energy E that depends on the frequency ν of the light, $E = h\nu$ where $h = 6.626^{\mathrm{e}-34}$ is Planck's constant. Furthermore, the momentum p of a photon is related to its wavelength λ, also through Planck's constant $p = h/\lambda$. The evidence to support the photonic or quantum description of light was provided by Einstein's photo-electric effect.

2.5.2.5 Optical Dispersion and Spectral Analysis

A final physical property of transparent materials is the variation of refractive index with wavelength. This feature is known as optical dispersion and is the basis for the analysis of the spectral content of a light beam. As the refractive index is dependent on the wavelength, it is usually indicated by the symbol n_λ or as μ_λ. As optical dispersion increases more rapidly at shorter wavelengths, Cauchy provided an empirical formula for it, given by

$$n_\lambda = 1 + \left(A/\lambda^2\right) \times \left(1 + B/\lambda^2\right). \tag{2.108}$$

To determine the constants A and B, four standard wavelengths are specified and denoted by letters of the alphabet, as indicated in Table 2.6. The first and the last of these (C, F) are employed to determine the constants A and B, while the other two are employed to define an average coefficient of dispersion

Table 2.6 Standard wavelengths used to define the *Abbe* number

Letter	Element and origin of spectral line	Wavelength (nm)
C	Hydrogen, red	656.27
D	Sodium, yellow	589.4
d	Helium, yellow	587.56
F	Hydrogen, blue	486.13

known as the *Abbe* number, given by

$$V_d = (n_d - 1)/(n_F - n_C) \text{ or } V_D = (n_D - 1)/(n_F - n_C). \tag{2.109}$$

As a consequence of optical dispersion and the refraction of a beam of light, the angle of refraction is different for each spectral component. Thus, dispersion will tend to split the beam into its various spectral components. This in turn allows for the spectrum of the light beam to be analysed. A number of different spectrometers have been devised, and the most common of these employ a prism or a diffraction grating to refract and decompose a pencil of light. Spectral decomposition with a diffraction grating is used for diffracting the signals in an optical fibre prior to performing the multiplexing and de-multiplexing functions.

2.5.3 *Optical Fibres*

2.5.3.1 Numerical Aperture

Consider a cylindrical glass fibre consisting of an inner core of refractive index n_1 and an outer cladding of refractive index n_2, where n_1 is greater than n_2. The end face of the fibre is cut at right angles to the fibre axis. Figure 2.23 shows a ray entering the end face from the air outside (refractive index, n_a). The ray will propagate unattenuated along the fibre as a consequence of multiple internal reflections, provided the incidence angle θ on to the core-cladding interface is greater than the critical angle θ_c.

To propagate unattenuated, the angle of obliqueness of the ray to the fibre axis $\phi = 0.5\pi - \theta$ must be less than $\phi_m = 0.5\pi - \theta_c$ and the angle of incidence α of the incoming ray to the end face of the fibre must be less than a certain maximum value α_m.

In order to estimate α_m and ϕ_m, we assume that $n_a = 1$ and apply the law of refraction (Snell's law) to obtain

$$\sin\alpha = n_1 \sin\phi = n_1 \cos\theta, \tag{2.110}$$

which in the case of a critical ray reduces to

$$\sin\alpha_m = n_1 \sin\phi_m = n_1 \cos\theta_c. \tag{2.111}$$

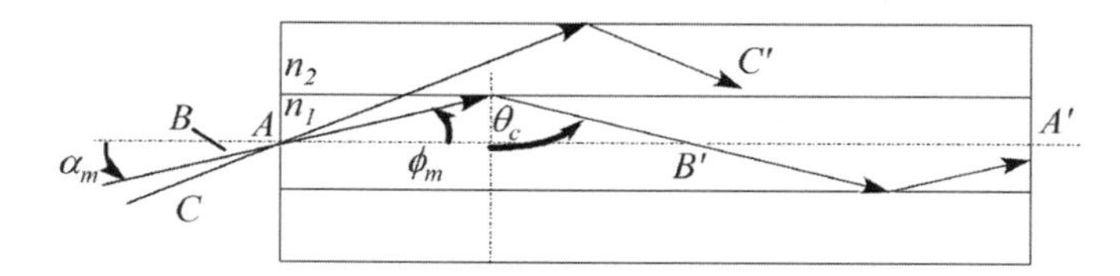

Figure 2.23 Propagation of a light ray in an optical fibre

Applying Snell's law to the core-cladding interface,

$$n_2 = n_1 \cos\theta_c, \quad \cos\theta_c = \sqrt{n_1^2 - n_2^2/n_1}. \tag{2.112}$$

Hence it follows that

$$\sin\alpha_m = n_1 \cos\theta_c = \sqrt{n_1^2 - n_2^2} = \sqrt{(n_1 - n_2)(n_1 - n_2)} = \sqrt{2\bar{n} \times \Delta n}, \tag{2.113}$$

where $\bar{n} = 0.5 \times (n_1 + n_2)$ and $\Delta n = (n_1 - n_2)$.

The greater the value of α_m, the greater is the proportion of light incident on the end face that can be collected by the fibre and transmitted by total internal reflection. Hence, the quantity $\sin\alpha_m$ is referred to as the *numerical aperture* (*NA*) and is a measure of the ability of the fibre to gather as much light as is possible.

2.5.3.2 Optical-Fibre Modes and Configurations

An optical fibre has the transmission properties of an optical waveguide, which confines the electromagnetic energy in the form of light to within the surface and guides the light in a direction parallel to the axis. The structure and characterization of waveguides have a major effect on determining how an optical signal is affected as it propagates along it. The propagation of waves along a waveguide can be described in terms of a set of modes of guided electromagnetic waves. These modes are referred to as the bound or trapped modes of the waveguide. Each of the trapped modes is a pattern of electric and magnetic field lines that are repeated along the fibre at intervals equal to the wavelength. Only a certain discreet number of modes are capable of propagating along the waveguide or fibre. A single-mode fibre can sustain just the first mode while multi-mode fibres are designed to propagate several modes. Multi-mode fibres are constructed by varying the material properties of the core. In the first case, the refractive index of the fibre core is uniform throughout a section of the fibre and undergoes an abrupt change at the cladding boundary. It is again uniform in the cladding and is referred to as a step-index fibre. In the second case, the refractive distance of the core is made to vary as a function of the radial distance from the centre of the fibre while the refractive index of the cladding continues to be uniform from the edge of the core. This type of fibre is a graded index fibre.

Multi-mode fibres have several advantages in terms of construction, cost and operation. However, for high data rates and in most sensor applications, single-mode fibres are preferable as the transmissions in a multi-mode fibre ate limited by signal degradations.

2.5.3.3 Signal Degradation in Optical Fibres

The primary causes of the degradation of a signal in an optical fibre are (1) dispersion, (2) absorption and (3) scattering.

When a short pulse of light is launched into a fibre, the transmitted rays consist of some rays that propagate along the fibre axis and some that have trajectories in the most oblique directions. An axial ray will travel a distance l along the fibre in a time $t_a = n_1 l/c$, while the most oblique ray component will travel the same axial distance in time given by

$$t_o = \frac{n_1 l}{c\cos\phi_m} = \frac{n_1 l}{c\sin\theta_c} = \frac{n_1}{n_2} \times \frac{n_1 l}{c} = \frac{n_1}{n_2} \times t_a. \tag{2.114}$$

It can be seen that t_o is greater than t_a. The time difference is

$$\Delta T = t_a - t_0 = \frac{n_1 l}{c}\left(\frac{n_1 - n_2}{n_2}\right) \equiv \frac{\Delta n}{c}\frac{n_1}{n_1} l. \tag{2.115}$$

The pulse containing rays at all possible angles between the two extremes will spread out during propagation by an amount given by $\Delta T/l$, a phenomenon known as multi-path time dispersion in the fibre that sets a limit on the maximum data rate through it.

Material dispersion is another dispersive phenomenon that changes the time of arrival of a pulse. A light ray travels normally at its phase velocity v_p, which is reduced from its maximum value in vacuum c to the value c/n in a medium with a refractive index n. However, a superposition of two waves at nearly the same frequency and wave number travels at the group velocity. If a group of two waves with slightly different frequencies and wave numbers are considered, the envelop is wave propagating with the speed

$$v_g = \lim_{\Delta k \to 0} \frac{\Delta \omega}{\Delta k} = \frac{\mathrm{d}\omega}{\mathrm{d}k}. \tag{2.116}$$

The waves themselves travel with the speed

$$v_p = \frac{\omega}{k}. \tag{2.117}$$

Hence,

$$v_p \frac{\mathrm{d}k}{\mathrm{d}\omega} = 1 - k\frac{\mathrm{d}v_p}{\mathrm{d}\omega} = 1 - \frac{\omega}{v_p}\frac{\mathrm{d}v_p}{\mathrm{d}\omega} \tag{2.118}$$

and the group velocity may be expressed in terms of the phase velocity as

$$v_g = \frac{\mathrm{d}\omega}{\mathrm{d}k} = v_p \Big/ \left(1 - \frac{\omega}{v_p}\frac{\mathrm{d}v_p}{\mathrm{d}\omega}\right). \tag{2.119}$$

It is the velocity of travel of the envelope of a group of waves within a light pulse. Thus, there is a spread in the duration of the impulses received at the output of a multi-mode step index fibre caused by modal and material dispersion when the light source is not precisely monochromatic.

When light is considered to be a collection of photons, optical transparency refers to the ability of the photon propagating through a medium without losing any of its energy. The transparency is uniform if all photons in a band of frequencies enjoy this property. However, this does not imply that a coherent beam of photons that propagate will continue to be coherent. In the case of air, it is non-dissipative to green light but the coherence of the light is not preserved. Absorption refers to the energy loss process by which photons of certain frequency bands are lost in the medium. It is the result of photon–material interactions that are characterized by inverse processes to those occurring during emission. Absorption is caused by three discreet mechanisms:

1. absorption due to atomic defects in the composition of the glass;
2. extrinsic absorption by impurity atoms in the glass;
3. intrinsic absorption by the basic constituent atoms of the fibre material.

The first is due to atomic defects or imperfections in the atomic structure of the fibre material such as missing molecules in high-density clusters of atom groups. The second is due to defects in the oxygen atoms with the glass material. Metallic traces and water that change the dielectric properties of glass are a major cause for absorption. Intrinsic absorption is associated with basic fibre material and defines the transparency of the medium to certain bandwidths.

Simple scattering refers to a loss process by which a photon of a particular frequency or frequency band is deviated from the nominal path of the beam it belongs to. This scattering is linear in that a certain percentage of the incident photons in the beam are scattered. Scattering involving the exchange of energies between a photon and a material molecule following an elastic impact is referred to as Rayleigh scattering. Raman scattering, on the other hand, involves inelastic collisions between the photon and the molecule. In such collisions, the molecule could gain or lose energy following the impact. Raman scattering is generally present when the intensities of the light are high.

2.5.3.4 Auxiliary Components in a Fibre-Optic Link

There are a number of auxiliary components that make up a fibre-optic link. These include optical transmitters, optical receivers, source fibre couplers, fibre–fibre connectors, multiplexers, switches, directional coupler repeaters as well as number of optical signal processors.

A primary optical source for fibre optics is the semiconductor laser diode. Laser is an acronym for light amplification for stimulated emission of radiation. It is a source of coherent electromagnetic waves at infrared and optical frequencies. The electrons within the atom of a substance are known to exist at various energy levels corresponding to different orbital shells in each individual atom. At a very low temperature, most of the atoms remain in the lowest possible energy level. They could be raised to higher-energy levels by the amounts of energy. One form of energy is optical energy, and this could be provided in packets of energy known as quanta. A specific number of quanta may provide the energy required to raise the level of the electron from one energy level to another. When an atom is excited by the absorption of a quantum of energy, it will remain in that state for a few micro-seconds and then re-emit the photons at a frequency proportional to the energy absorbed. The atom then returns to its original state. This assumes that the re-emission of energy has been stimulated at the expense of absorption. This could be achieved in practice by providing an optical cavity that is resonant at the desired frequency.

In a ruby laser, optical pumping of energy to the ruby is provided by a flash lamp. In the case of a ruby, the existing energy levels are already suitable for laser action. Two parallel mirrors are used, one fully silvered and the other only partially silvered, to enable the coherent light to be emitted through that end. The mirrors are made parallel to very high degree of accuracy and must be separated by a distance that is an exact number of half wavelengths apart with the ruby in place between the mirrors. A spiral flash tube pumps energy into the ruby in pulses, and coherent light is emitted from the partially silvered mirror end.

The semiconductor laser operates on a very similar principle. Optical pumping is by the DC pumping action by forward biasing a p–n junction of a layered semiconductor. The laser works on the injection principle where electrons and holes originating in the gallium arsenide-based semiconductor (GaAs(p)-GaAlAs(n)) junction give up their excess recombination energy in the form of light. Stimulation of emission is done by using two very highly polished slices of p-type gallium arsenide to ensure that total internal reflection takes place between them and a continuous beam of light is emitted from one end. The pairs of mirrors within the semiconductor laser diode form a Fabry–Perot cavity resonator. The mirror faces are constructed by making the two parallel cleaves along natural cleavage planes of the semiconducting crystal. Thus, the radiation is generated within the resonator cavity, which is then output as a collimated monochromatic beam.

Light-emitting diodes (LEDs) are used as suitable optical sources for fibre-optic sensors. They work on a principle similar to that of the laser diode. Electrons and holes are injected along a p–n junction, and optical energy is given off during recombination. Semiconducting material similar to those in laser diodes are used in LEDs. However, the structure is much simpler, as there are no polished mirror-like surfaces and stimulated emission does not take place. Consequently, not only is the power output much lower but also a much wider beam of light is emitted that is not monochromatic or collimated. The output of the LED or laser diode is coupled to an optical fibre by a small lens.

At the receiving end of an optical fibre, two different types of photo-detectors are used to accumulate the light and convert it back to an electrical signal. The first is the PIN photo-detector, which is a layered semiconductor device with p-type and n-type layers of semiconductor wafers separated by lightly *n*-doped intrinsic region. In normal operation, a sufficiently large reverse bias voltage is applied across the junction, so the depletion layer fully covers the intrinsic region and depletes it of carriers. Thus, the n-type and p-type carrier concentrations are negligible in comparison with the impurity concentration in this region. Consequently, there is no flow of current. However, when the diode absorbs energy from incident photons, each quantum will cause an electron–hole pair to be created in the intrinsic region and a corresponding current will flow in the external circuit.

The problem with PIN photo-diodes is that they are not sensitive enough. The problem is resolved in the APD. In the APD, the photo-sensitive current is amplified within the diode by a process of carrier multiplication, by making the photo-generated carriers to traverse a region where a very high electric field is present. The carriers gain enough energy within this region to ionize bound electrons within the valency band and collide with them. The ionization process followed by the impact culminates in the multiplication of the carriers. It is regenerative process and hence known as the avalanche effect. To facilitate the avalanche effect, the diode must operate close to the breakdown voltage and just below it. Only a finite number of carriers are created just below the breakdown voltage, and the output is proportional to the light energy input. APDs are capable of withstanding sustain breakdown and are about 10–150 times more sensitive than PIN diodes.

2.5.4 *Principles of Optical Measurements*

There is a range of measurements that must be performed in order that the optical fibre is used as a sensor (Senior, 1993; Kaiser, 2000). These include measurements of attenuation and changes in fibre geometry including bends, dispersion measurements and refractive index measurements. One technique that is very versatile and can be used to measure not only attenuation characteristics but also used monitoring changes in fibre geometry (due to bending) is the optical time-domain reflectometry. The optical time-domain reflectometer (OTDR) is based on the principles of Fresnel reflections and provides a method for measuring the attenuation per unit length.

2.5.4.1 Optical Fibre Measurements: OTDR

The OTDR operates very much like an optical radar. A laser source is used to launch narrow optical pulses into one end of fibre using a beam splitter or a directional coupler. The back-scattered light due to the input pulse is analysed and correlated with the changes in the geometry. The back-scattered light is principally produced by Fresnel reflections and Rayleigh scattering. Fresnel reflections refer to the reflection of light at the interface of two optically transparent materials with differing refractive indices. The analysis of the back-scattered light could involve a range of signal-processing techniques such as the fast Fourier transform to determine its spectral properties. The OTDR is now integrated into a number of different fibre-optic sensors.

2.5.5 *Fibre-optic Transducers for Structural Control*

Recent advances in fibre-optic sensing have led to the availability of several technologies in place of the classical electrical, mechanical or vibrating wire strain gauges for strain monitoring and measurement. Thus, one approach to control of flexible structure is through strain measurement using fibre-optic strain sensors rather than the more conventional strain gauges. Amongst them are a variety of fibre-optic-based techniques including small-diameter FBG sensors and interferometric Fabry–Perot sensors, which are available for direct local strain measurements, while laser displacement sensors can provide a very accurate and reliable measurement of the relative displacement of any two points chosen in a structure at large distances from a reference station. Other fibre-optic sensors are based on polarimetry or Raman or Brillouin scattering. Although experimental and field monitoring of several fibre-optic sensors show that extrinsic Fabry–Perot fibre-optic sensors perform linearly and show good response to thermal variations and mechanical loading conditions, intracore FBG sensors are associated with a number of advantages related to the stability of the measurements, the potential long-term reliability of the optical fibres and the possibility of performing distributed and remote measurements. Not withstanding these advantages, there are indeed a number of practical issues related to the implementation of the technique, including the characterization, performance in an embedded environment and cost of the FBG sensors, and the associated signal processing.

The use of fibre optics provides an alternative to employing piezoelectric-embedded thin film actuators and sensors for the control of flexible structures. While there are several advantages in employing thin film piezosensors, a strategy that effectively realizes the advantages of using both FBGs and piezosensors is to employ a variety of functionally different and distributed sensors. Such an approach not only provides for complementary redundancy but also offers a relatively low-cost approach to integrated structural control and health monitoring. Moreover, such an approach would permit the incorporation of GPS measurements on board an aircraft to simultaneously monitor aircraft structural vibrations, which in turn would allow for the motion measurements to be correlated in real time, with the damage developing in the structure. Thus, it is possible to develop and implement a multi-functional structural monitoring and control system and integrate it with existing active control loops.

The optical fibre is one of the most promising sensors for structural control applications. It is also one that can be easily embedded into composite materials. Optical fibres are small, lightweight and resistant to corrosion and fatigue, not influenced by electromagnetic interference and compatible with composites. Their geometric composite fibre-like flexibility and small size have the added advantage that they can be easily placed into composite materials. Optical fibres have been shown to effectively measure force, pressure, bending, density change, temperature, electric current, magnetic fields and changes in chemical composition, and to accurately monitor internal properties of composite materials during manufacture. However, inclusion of the optical fibre leads to a degradation of the mechanical properties of the host composite material (modulus, strength, fatigue life, etc.) unless it is embedded in directions conforming to the direction of material fibres. Yet the spectral response to strain changes can be employed not just for control applications but also for structural health monitoring to identify cracks by appropriate signal processing.

Fibre-optic sensors are often loosely grouped into two basic classes referred to as extrinsic or hybrid fibre-optic sensors and intrinsic or all fibre sensors. The intrinsic or all fibre sensor uses an optical fibre to carry the light beam, and the environmental effect impresses information onto the light beam while it is in the fibre. In the case of the extrinsic or hybrid fibre-optic sensor, the optical fibre is interface to another 'black box', which then impresses information onto the light beam in response to an environmental effect. An optical fibre then carries the light with the environmentally impressed information back to an optical and/or electronic processor. The output fibre could be the same as input fibre or could be an independent fibre. In many structural monitoring and control applications cases, it is desirable to use the fibre-optic analogue of a conventional electronic sensor. An example is the electrical strain gauge that is used to measure elastic displacements for several applications for the feedback control of flexible structures. Fibre grating sensors can be configured to have gauge lengths from 1 mm to approximately 10 mm, with sensitivity comparable to conventional strain gauges.

Amongst the sensors used for measurements associated with the control of flexible structures, there are two that are of primary importance. The extrinsic Fabry–Perot interferometric sensors have two semi-reflective mirrors that form the Fabry–Perot cavity. The mirrors are realized by the deposition of semi-reflective material inside the fibre or at straight cleaved fibre ends. The principle of operation is that the reflections from the two mirrors interfere and consequently the intensity at the detector varies as sinusoidal function of the distance between the mirrors. The distance between the mirrors is made sensitive to strain and temperature. This could be achieved, for example, by placing these mirrors on either side of flat plate structure, to measure its deflection at a certain location.

2.5.5.1 The Fibre Bragg Grating Sensor

An FBG is an optical sensor installed within the core of an optical fibre. An optical fibre has the ability to transmit a narrow bandwidth beam of light by the mechanism of total internal reflection, which is passed through a diffraction grating. A diffraction grating is an optical filter that diffracts light by an amount varying according to its wavelength. Diffraction gratings are used to separate, or disperse, polychromatic light into its monochromatic components. A diffraction grating device in which Bragg-type index modulations are recorded can be used to selectively reflect a light component having a specific wavelength, which is a linear function of the product of period of the grating and the fibre's effective refractive index. One method of fabricating such a fibre sensor is by 'writing' a fibre grating onto the core of a germanium-doped optical fibre using an angled laser beam to form an interference pattern. The interference pattern consists of bright and dark bands that represent local changes in the index of refraction in the core region of the fibre.

Since the period of the grating and the fibre's effective refractive index are both sensitive to changes in temperature and strain, it may be used as an optical narrow-band filter that responds linearly to changes in temperature and strain. Hundreds of FBG sensors can be installed in a single optical fibre by recording several diffraction grating patterns within it. The sensors may be monitored simultaneously with a single optical receiver, thus providing for the distributed monitoring of strain and/or temperature within large structures. The sensors are particularly suited for structural health monitoring and for monitoring a structure during the design, validation and test stages.

2.5.5.2 The Fabry–Perot Interferometer

The Fabry–Perot interferometer uses two closely spaced partially silvered surfaces to create multiple reflections resulting in multiple offset beams, which can interfere with each other. The large numbers of interfering rays behave like the multiple slits of a diffraction grating and increase its resolution to produce an extremely high-resolution interferometer. An etalon is a device that is typically made of a transparent plate with two reflecting surfaces. Intrinsic fibre etalons are formed by in-line reflective mirrors that can be embedded into the optical fibre. Extrinsic fibre etalons are formed by two mirrored fibre ends in a capillary tube. A fibre etalon-based spectral filter or demodulator is formed by two reflective fibre ends that have a variable spacing. High-resolution Fabry–Perot interferometers using such etalons are ideally suited for Doppler shift determination and strain measurement due to their high spectral resolution.

2.5.5.3 The Fibre-Optic Gyroscope

For strapped-down vehicle navigation systems, which relies on the direct measurement of the vehicle angular velocity by a rate gyro, the gyro drift rates are required to be very low. This requirement can be met by modern laser gyros that are based on the Sagnac principle and illustrated in Figure 2.24(a). To understand the principle of operation of the ring laser gyro, the circular model illustrated in Figure 2.24(b) is considered. Consider light entering at x and being directed by some means around the ring, then combined back at x'; it is supposed that light could travel around the circumference of the circle in both clockwise (cw) and counter clockwise (ccw) directions. Then the transit time:

$$t = 2\pi R/c, \;\; c = \text{velocity of light.} \tag{2.120}$$

Now suppose the gyro is rotated at an angular speed Ω, then the transit times for the two opposite beams change. This is because, relative to inertial space, the point x has moved to x' in the time taken for the light to transit around the ring. The speed of light is assumed to be invariant. The times taken for transits around the ring in both clockwise (cw) and counter clockwise (ccw) directions are

$$t_{cw} = \frac{2\pi R}{(c + R\Omega)}, \;\; t_{ccw} = \frac{2\pi R}{(c - R\Omega)}, \tag{2.121}$$

and since $c^2 \gg R^2\Omega^2$, the time difference is

$$\Delta t = \left(4\pi R^2/c\right) \times (\Omega/c)\,. \tag{2.122}$$

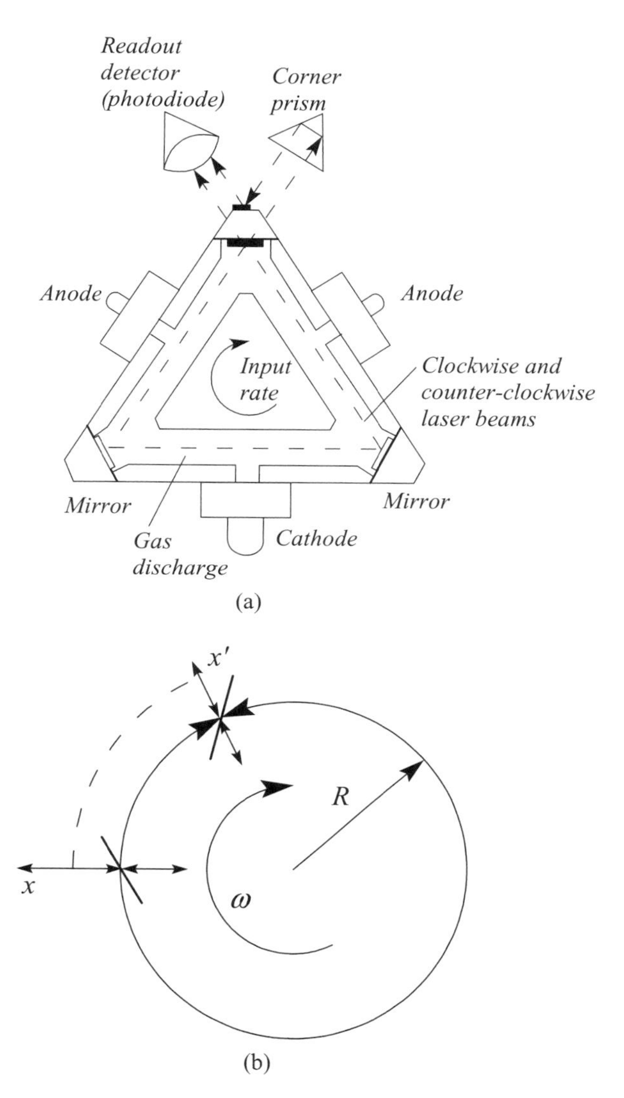

Figure 2.24 Ring laser gyro. (a) Schematic diagram. (b) Circular ring mode

The equivalent optical path difference is

$$\Delta L = c\Delta t = 4\pi R^2 (\Omega/c). \tag{2.123}$$

This can be generalized for any non-circular enclosed area A as

$$\Delta L = 4A\Omega/c. \tag{2.124}$$

This is equivalent to an apparent frequency change, which is proportional to the angular speed of the ring laser gyro.

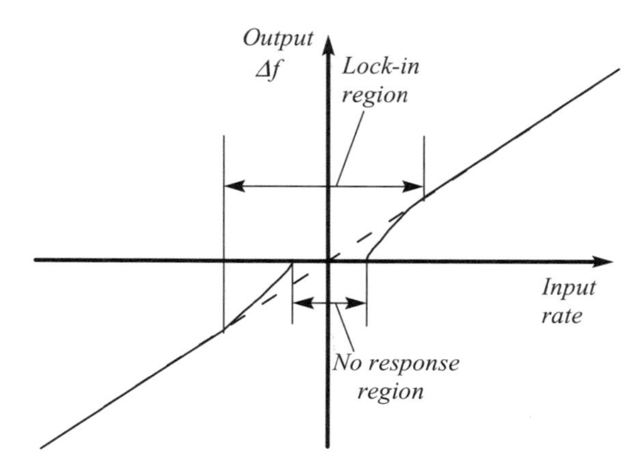

Figure 2.25 Gyro lock-in

The path difference is of the order of just 1 nm, and so a source with high spectral purity and stability such as a helium–neon gas laser is employed to ensure that this path difference is practically error free. The laser itself employs active feedback control of the gain and path length to maintain the coherence (constant phase) of the source and a stationary path length. The path difference is measured by employing appropriate photo-diodes to measure the light returns in the two directions.

At very low rotation rates, because the laser cavity resonance is essentially non-linear, the two laser beams travelling in the opposite directions oscillate synchronously and as a result lock together. In this lock-in mode, they oscillate at the same frequency and phase and there is no tangible output from the gyro. This feature of the output of the ring laser gyro is illustrated in Figure 2.25. This phenomenon of frequency synchronization is caused by back scattering of the light within the cavity, a fundamentally non-linear phenomenon. From the point of view of the application to the ring laser gyro, one must apply a suitable control technique to inhibit this synchronization. One method of effectively controlling this phenomenon, which is a feature of a number of chaotically behaving non-linear systems, is to mechanically oscillate the entire gyro assembly with a very low rotational amplitude and at a very high frequency. This process, known as 'dithering', is a standard technique for controlling lock-in-type phenomenon (it is also employed in servo valves). When the dither product, the product of the dither amplitude and dither frequency, is chosen to be as close as possible to the velocity of light, the lock in problem is completely alleviated.

The principle of operation of a fibre-optic gyro is exactly the same as the ring laser gyro and relies also on the Sagnac principle. However, in fibre-optic gyro, illustrated in Figure 2.26, a differential phase shift is induced in light that travels clockwise and anti-clockwise around a rotating optical system. In a realistic and practical arrangement, light from a coherent light source is first passed through a beam splitter to select a single optical mode. This is then passed through a second beam splitter allowed to propagate around a fibre-optic coil, in both clockwise and anti-clockwise directions. In the absence of any rotation, the propagation times are identical and when the two beams travelling along the fibre coil arrive back at the beam splitter they are completely in phase, which results in constructive interference. When the coil is set in rotation, as a result of the Sagnac effect, there is a difference in the propagation time in the two beams, which results in a phase difference. Light returning from the coil passes back through the phase modulator, interferes at the Y-junction-type beam splitter and is directed to a detector via a polarizer and coupler.

In practice, the fibre-optic gyro employs a phase-balancing method. Light propagating around the coil has its phase modulated by a signal that is synchronous with the delay time around the coil. The

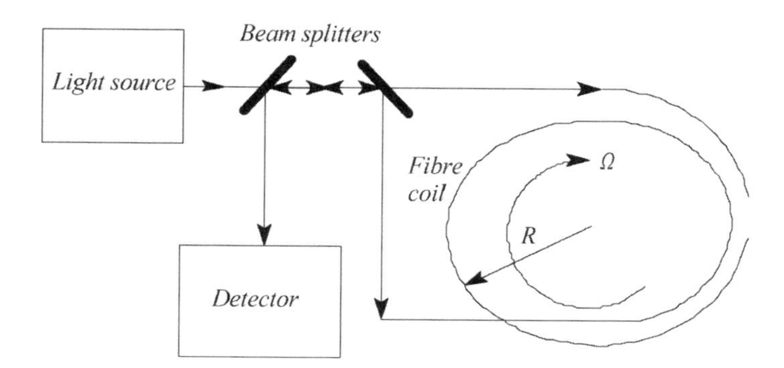

Figure 2.26 Basic concept of the fibre-optic gyro

modulation biases the gyro to the point of maximum sensitivity. Rotation induces an angular velocity-dependent output, which is demodulated, filtered and integrated. This signal is used to drive a serrodyne voltage-controlled oscillator, which outputs a sawtooth waveform. This waveform is, in turn, employed to modulate the light propagating around the coil and so null the Sagnac phase shift induced by the applied angular light.

An indication of the angular rate is obtained by counting pulses from the serrodyne oscillator between resets, each of which corresponds to a positive or negative increment depending on the polarity of the ramp. A micro-processor performs the counting and other associated operations and outputs a digital signal directly indicating the angular rate. The entire device is usually encapsulated into a single opto-electronic chip, incorporating electrosensitive polymers as sensors for the closed-loop operation. The performance of the fibre-optic gyro is limited by photon shot noise within the fibre, which can be modelled as a coloured noise corrupting the measurement and appropriately filtered out by employing a Kalman filter. The custom Kalman filter is tuned to the fibre-optic gyro, and by this process really accurate estimates of the input angular velocities may be obtained.

Interferometric fibre-optic gyros and ring laser gyros are prime candidate gyros for autonomous angular velocity measurement for applications in space structural control.

Recent advances in fibre-optic sensing have led to the availability of several technologies in place of the classical electrical, mechanical or vibrating wire strain gauges for strain monitoring and measurement. Thus, one approach to control of flexible structure is through strain measurement using fibre-optic strain sensors rather than the more conventional strain gauges. Amongst them are a variety of fibre-optic-based techniques, including small-diameter FBG sensors and interferometric Fabry–Perot sensors, which are available for direct local strain measurements, while laser displacement sensors can provide a very accurate and reliable measurement of the relative displacement of any two points chosen in a structure at large distances from a reference station. Other fibre-optic sensors are based on polarimetry or Raman or Brillouin scattering. Although experimental and field monitoring of several fibre-optic sensors show that extrinsic Fabry–Perot fibre-optic sensors perform linearly and show good response to thermal variations and mechanical loading conditions, intracore FBG sensors are associated with a number of advantages related to the stability of the measurements, the potential long-term reliability of the optical fibres and the possibility of performing distributed and remote measurements. Not withstanding these advantages, there are indeed a number of practical issues related to the implementation of the technique, including the characterization, performance in an embedded environment and the cost of the FBG sensors, and the associated signal processing. The use of fibre optics provides an alternative to employing piezoelectric-embedded thin film actuators and sensors for the control of flexible structures. While there are several advantages in employing thin film piezosensors, a strategy that effectively realizes the advantages of using both FBGs and piezosensors is to employ a variety of functionally different and distributed sensors.

Such an approach not only provides for complementary redundancy but also offers a relatively low-cost approach to integrated structural control and health monitoring. Moreover, such an approach would permit the incorporation of GPS measurements on board an aircraft to simultaneously monitor aircraft structural vibrations, which in turn would allow for the motion measurements to be correlated in real time with the damage developing in the structure. Thus, it is possible to develop and implement a multi-functional structural monitoring and control system and integrate it with existing active control loops.

Exercises

1.

(i) A strain gauge has a sensitivity of 10 mV/mm and a temperature sensitivity of 10 μV/K. If a variation of 0.1 mV is caused by temperature fluctuations, determine the parasitic sensitivity.
(ii) A strain gauge is fed by a current source of 1 mA. The total length of the wire in the gauge is 2 m, while the diameter of the wire is 0.5 mm. The specific resistivity is $45 \times 10^{-8}\ \Omega$ m. If the applied stress causes a change in length of 4 mm and a reduction of the diameter to 0.4 mm, estimate the change in the output voltage.

2. A capacitive difference transducer has plate dimensions of 1.5 × 1 mm. Because of a change in the pressure, the distance between the plates varies by 3 μm and the corresponding change in capacitance is 4.4 pF. If the distance between the plates is reduced by 9 μm and the measured charge across the transducer is 0.667 pC, estimate the change in the potential difference across the two plates of the capacitor.

3. Figure 2.27 illustrates a typical Wheatstone's bridge circuit constructed in solid state. Nominally, the resistors R_1 and R_2 are equal to each other and $R_2 \approx R_1 \approx R_0$. However, because of imperfections in the substrate, $R_1 = R_0\,(1+\alpha)$ and $R_2 = R_0\,(1-\beta)$.
Show that the output voltage in terms of the source voltage, α and β is given by

$$v_0 = v_s\left(\frac{\alpha+\beta}{2+\alpha-\beta}\right).$$

4. Calculate the magnitude, the dB magnitude (in decibels) and the phase angle (in degrees) of

$$G(D) = 1/(D+1)$$

for $\omega = 0.01, 0.1, 0.2, 0.4, 0.5, 1, 2, 2.5, 5, 10$ and 100 rad/s.
[Note: If $G = a + i\,b = r\,e^{i\theta} = r\{\cos(\theta) + i\,\sin(\theta)\}$ then r is the magnitude and θ is the phase angle and the dB magnitude is given by $20\,\log_{10}(r)$.]

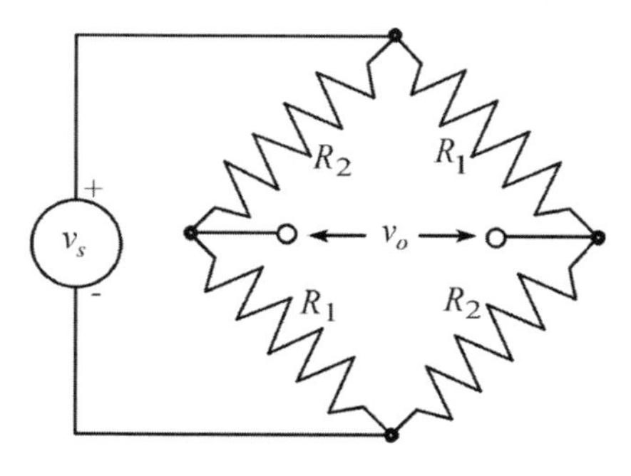

Figure 2.27 A Wheatstone's bridge circuit

5.

(i) Obtain the transfer function, the characteristic equation and its roots, the undamped natural frequency, the relative damping ratio and the damped natural frequency, of the system described by the following differential equations:

(a) $\dfrac{d^2y}{dt^2} + 4\dfrac{dy}{dt} + 6y = 6u$

(b) $3\dfrac{d^2y}{dt^2} + 9\dfrac{dy}{dt} + 15y = 4u$

(ii) Plot the roots of the characteristic equations on the complex plane (real (x axis)/imaginary (y axis) plane).

(iii) Estimate the dB magnitude and phase for the same frequencies as (1) and sketch the Bode plots for the two transfer functions from the asymptotes as ω tends to zero and to infinity. Hence, estimate the error in your plot at the corner frequency.

6.

(i) A viscously damped mass-spring system is forced harmonically at the undamped natural frequency, $\omega/\omega_n = 1$. If the damping ratio, $(\zeta = (c/c_{criticak}) = c/2m\omega_n)$ is doubled from 0.1 to 0.2, compute the percentage reduction in the steady-state amplitude.

(ii) Compare with the result of a similar calculation for the condition $\omega/\omega_n = 2$.

7.

(i) A vibrometer consists of a spring-mass system, with a mass of 50 g supported on four springs, each of stiffness 750 N/m. If the instrument is mounted on a vibrating surface, with the vibration given as, $x(t) = 0.002\cos(50t)$ m, determine the amplitude of steady-state motion of the vibrometer mass relative to the surface.

(ii) A vibrometer consists of a undamped mass-spring system encased in a rigid casing. The natural frequency of the mass-spring system is 5 Hz. When the vibrometer is fixed on to the surface of a motor that is running at 750 rpm, the absolute amplitude of vibration of the mass within the casing is observed to be 1 mm relative to the casing.

(a) Determine the amplitude of vertical vibration of the casing.

(b) Determine also the speed of the motor at which the amplitude of motor vibration be the same as the amplitude of vibration of the vibrometer casing.

(iii) Consider a seismograph, which consists of a damped mass-spring system where it is specified that when $r = 5$, $x_{r0} = y_0$.

Determine the damping ratio.

8. A very common example of support excitation is a car driving over a rough road or an aircraft taxiing over a rough runway. As an approximation, consider a vehicle driven on a rough road as depicted in Figure 2.28.

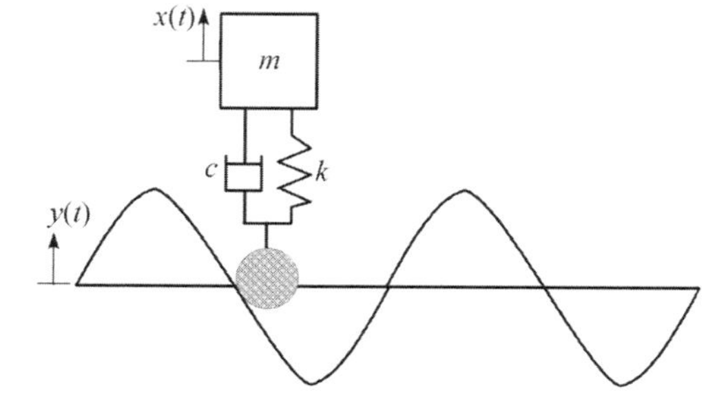

Figure 2.28 One-degree-of-freedom model with support excitation

Table 2.7 Specifications of a permanent magnet DC motor

Torque constant: 0.5 N m/A	Back EMF = 0.5 V/rad/s
Rotor inertia = 4×10^{-4} N m s^2	Dry friction torque = 0.023 N m
Weight = 1 kg	Armature resistance: 7.5 Ω
Armature inductance: 16 mH	Peak torque : 1.6 N m
Brush life: $>10^7$ revolutions	Size: 10 cm diameter, 4 cm long

The system is modelled by a one-degree-of-freedom system that is constrained to move in the vertical direction and is subjected to a sinusoidal support excitation $y(t) = Y_o \sin \omega t$.

(i) Draw a free body diagram of the mass m and set up the equation of motion.

(ii) Show that the equation of motion is given by

$$m\ddot{x} + c\dot{x} + kx = \sqrt{(kY_o)^2 + (c\omega Y_o)^2} \sin(\omega t + \beta), \ \beta = \tan^{-1}(c\omega/k),$$

where m is the mass, and c and k are the damping and stiffness coefficients, respectively.

(iii) Consider an accelerometer that has a mass of 1 kg attached to the wheel axle below the suspension system via a spring and damper, which provide an equivalent stiffness of $k =$ 400 N/m, and damping $c =$ 20 Ns/m. Assuming that the axle follows the variation of the road height perfectly and that the speed of the vehicle is 80 km/h and the sinusoidal road roughness has a wavelength of 6.0 m and an amplitude of 0.01 m, determine the amplitude of vibration measured by the accelerometer.

9.

(i) A permanent magnet DC motor with armature inertia of 5×10^{-3} kg m^2 is employed to drive a mass of 10 kg. The load is located at a mean effective radius of 0.5 m. A 100:1 gearing is used. Ignoring the inertia of the gearing and all friction, obtain the equation relating torque to angular motion.

(ii) The DC motor in part (i) is subjected to extensive testing. When stalled, it draws 10 A from a 24 V supply. When allowed to run without load, the motor shaft accelerates and reaches a maximum rotational speed of 3000 rpm.

(a) Draw an equivalent electrical circuit of the motor.

(b) Employ the equivalent circuit model to determine the electrical characteristics of the motor.

(iii) A permanent magnet DC motor is tested as follows. With a constant voltage of 15 V applied, and stalled, the motor draws 5 A. When rotating at 50 rad/s the motor draws 1 A. Employing an equivalent circuit model, find the effective armature resistance and the back EMF generated per unit rotor speed.

(iv) The motor in part (iii), when stalled, draws a current of 5 A and develops a torque of 1 N m. Given that the torque is related to the armature current by the relationship

$$T = K_{pm} I_a,$$

determine the parameter K_{pm}.

(v) The specifications of a permanent magnet DC motor are listed in Table 2.7. It is desired to employ this motor to design an electric position servo. Draw a block diagram of the closed-loop system, indicating on the diagram the main feedbacks and the type of sensors that may be employed to measure them.

Table 2.8 Properties of two piezoelectric materials

Material	Density (kg/m^3)	Young's modulus (E_1 $\times 10^{-10}$ Pa)	$d_{31} \times 10^{12}$
Lithium niobate, $LiNbO_3$	4650	8.5	5.85
Lead zirconate titanate, PZT5A	7700	6.3	171

Hence, otherwise make suitable assumptions about the closed-loop performance of the servo and select the proportional derivative gains for building an electric position servo.

10. It is desired to design a piezoelectric bimorph patch actuator. There are two piezo materials available for the purpose. The properties of these materials are given in Table 2.8. Each patch is required to occupy a volume less than 0.1 cm^3. The mechanical gain of each patch in the longitudinal direction is required to be less than 1 mm/V. The stall force with no strain is required to be 10 N at 7.5 V.

Select a reasonable thickness for the patch and determine the bending moment generated on a cantilever beam of depth d, width b and Young's modulus E. Assume a supply voltage of 75 V.

References

Andoh, F., Washington, G. and Utkin, V. (2001) Shape control of distributed parameter reflectors using sliding mode control, Proc. SPIE., vol. 4334, pp. 164–175.

Barron, B. W., Li, G. and Haertling, G. H. (1996) Temperature dependent characteristics of Cerambow actuators, *Proc. IEEE International Symposium on Applications of Ferroelectrics*, ISAF'96, IEEE, New York, vol. 1, pp. 305–308.

Bruch Jr, J. C., Sloss, J. M., Adali, S. and Sadek, I. S. (2000) Optimal piezo-actuator locations/lengths and applied voltage for shape control of beams, *Smart Materials and Structures* 9, 205–211.

Bryant, R. G., Mossi, K. M. and Selby, G. V. (1998) Thin-layer composite unimorph ferroelectric driver and sensor properties. *Materials Letters* 35, Issues 1–2, 39–49.

Chandran, S., Kugel, V. D. and Cross, L. E. (1997) Crescent: a novel piezoelectric bending actuator, *Proc. International Society for Optical Engineering, SPIE, USA*, vol. 3041, pp. 461–469.

Crawley, E. F. and de Luis, J. (1987) Use of piezoelectric actuators as elements of intelligent structures, *AIAA Journal* 25(10), 1373–1385.

Cunningham, M. J., Jenkins, D. F. L., Clegg, W. W. and Bakush, M. M. (1995) Active vibration control and actuation of a small cantilever for applications in scanning probe instruments, *Sensors and Actuators A* 50, 147–150.

Haertling, G. H. (1997) Rainbow actuators and sensors: a new smart technology. *Proc. International Society for Optical Engineering, SPIE, USA*, vol. 3040, pp. 81–92.

Jenkins, D. F. L., Cunningham, M. J. and Clegg, W. W. (1995) The use of composite piezoelectric thick films for actuation and control of miniature cantilevers, *Microelectronic Engineering* 29, 71–74.

Juuti, J., Kordás, K., Lonnakko, R., Moilanen, V.-P. and Leppävuori, S. (2005) Mechanically amplified large displacement piezoelectric actuators, *Sensors and Actuators A* 120, 225–231.

Kaiser, G. (2000) *Optical Fiber Communications*, 3rd edn, McGraw-Hill, New York.

Kamada, T., Fujita, T., Hatayama, T. *et al.* (1997) Active vibration control of frame structures with smart structures using piezoelectric actuators (vibration control by control of bending moments of columns), *Smart Materials and Structures* 6(4), 448–456.

Kim, S. J. and Jones, J. D. (1991) Optimal design of piezoactuators for active noise and vibration control, *AIAA Journal* 29(12), 2047–2053.

Kudva, J. N., Lockyer, A. J. and Appa, K. (1996) Adaptive Aircraft Wing, AGARD-LS-205; Paper No. 10.

Lazarus, K. B. *et al.* (2002) Packaged strain actuator, US Patent 6,404,107, June 2002.

Low, T. S. and Guo, W. (1995) Modeling of a three-layer piezoelectric bimorph beam with hysteresis, *Journal of Microelectromechanical Systems* 4(4), 230–237.

Masters, B. P. *et al.* (2002) Laser machining of electroactive ceramics, US Patent 6,337,465, January 2002.

Nolan-Proxmire, D. and Henry, K. (1996) NASA rolls out award-winning "THUNDER" . NASA News Release No. 96-154, NASA, Washington, DC, USA, http://www.nasa.gov/centers/langley/news/releases/1996/Oct96/96_154.html.

Oh, J. T., Park, H. C. and Hwang, W. (2001) Active shape control of a double-plate structure using piezoceramics and SMA wires, *Smart Materials and Structures* 10, 1100–1106.

Rediniotis, O. K., Wilson, L. N., Lagoudas D. C. and Khan, M. M. (2002) Development of a shape-memory-alloy actuated biomimetic hydrofoil, *Journal of Intelligent Material Systems and Structures* 13, 35–49.

Senior, J. M. (1993) *Optical Fiber Communications: Principles and Practice*, 2nd edn, Prentice-Hall International.

Senturia, S. D. (2001) *Microsystems Design*, Springer, New York.

Silva, E., Fonseca, J. and Kikuchi, N. (1997) Optimal design of piezoelectric microstructures, *Computational Mechanics* 19, 397–410.

Song, G., Jinqiang Zhao, Xiaoqin Zhou and Alexis De Abreu-García, J. (2005) Tracking control of a piezoceramic actuator with hysteresis compensation using inverse preisach model, *IEEE/ASME Transactions on Mechatronics* 10(2).

Song, J.-K. and Washington, G. (1999) Thunder actuator modeling and control with classical and fuzzy control algorithm, *Proc. SPIE*, vol. 3668, pp. 866–877.

Van Putten, A. F. P. (1996) *Electronic Measurement Systems: Theory and Practice*, Taylor and Francis Group, New Delhi.

Warkentin, D. (2000) Modeling and electrode optimization for torsional Ide piezoceramics, smart structures and integrated systems. In N. Wereley (ed.), Proc. SPIE 3985, pp. 840–854, March 2000.

Williams, R. B. and Inman, D. J. (2002) An overview of composite actuators with piezoceramic fibers, *Proc. 20th International Modal Analysis Conference, Los Angeles, CA*.

Yoon, K. J., Shin, S., Park, H. C. and Goo, N. S. (2002) Design and manufacture of a lightweight piezo-composite curved actuator, *Smart Materials and Structures* 11(1), 163–168.

Yoshikawa, S., Farrell, M., Warkentin, D., Jacques, R. and Saarmaa, E. (1999) Monolithic piezoelectric actuators and vibration dampers with interdigital electrodes, *Proc. 6th Annual International Symposium on Smart Structures and Materials, Newport Beach, CA, 1–5 March 1999*, SPIE vol. 3668, pp. 578–585.

3

Fundamentals of Structural Control

3.1 Introduction

Recent innovations in smart materials coupled with developments in control theory have made it possible to control the dynamics of structures. The feasibility of incorporating a distribution of structural actuators and sensors assembled from components made from smart materials into structures in the form of patches and lamina has made the concept of smart structure a practical reality. However, to be able design such control configured structures one must necessarily be able to design a control that can modify the dynamics of a structure. Thus, an understanding of the principles of control system design is essential. Although there are a number of excellent books on control system design (see, for example, Friedland, 1987; Utkin, 1992; Dorf and Bishop, 2004; Franklin *et al.*, 2005), the subject is briefly revisited and its application to structural control is discussed in this chapter.

3.2 Analysis of Control Systems in the Time Domain

3.2.1 Introduction to Time Domain Methods

The revolution brought about in the modern world by the invention of the digital microprocessor has been paralleled and even surpassed with the introduction of the microcomputer. Its uses today cover an unbelievable range of applications, and there is no doubt it has influenced control system design and implementation.

Control systems analysis and design in general can take place in one of two environments. These are referred to as the *time domain* and *frequency domain.* In time domain, analysis and design computations are made on the physical variables, that is, ones that are directly observable and measurable. Such methods are computationally intensive, requiring the use of computer-aided tools. Examples of measurable variables on which computations can be directly performed are voltage, current, position, velocity, temperature, flow rate, pressure, etc. In the frequency domain, however, the physical variables are subjected to some type of mathematical transformation before any analytical or design computations are initiated. Such transformations simplify the problem in some sense but make real-time calculations quite impossible. On the other hand, even though the variables are no longer physically observable, general input-output relations are more readily derived, and the properties of classes of systems can be explored and categorized. In this sense, therefore, frequency domain techniques are still invaluable and indispensable.

Dynamics of Smart Structures Dr. Ranjan Vepa

The type of time domain operations that most systems impose on physical variables can be either characterized by *differential equations* or suitably approximated by differential equations. Most dynamic system models can be described by *linear* and *time-invariant* models. In these cases, it is extremely convenient to use the *Laplace* and *Z transform* techniques. In many situations, one is primarily interested in the *steady-state* behaviour of the system, i.e. when it has been in operation for a long time and all *transients* have subsided. In these cases, the asymptotic *stability* of the system is of primary concern. Generally speaking, information about the initial conditions can be neglected for stability analysis and in these situations the *transfer function*-based methods are relatively easy to use.

A very powerful method of control systems analysis is the *state variable* method. Basically, the method requires that the variables used to formulate the system equations be chosen in such a way that the equations can write compactly in terms of *matrices*. The state variable technique is simply a method of representing an nth-order differential equation as a set of n first-order differential equations, which can be written in a standard form using matrix notation. Such a representation of system dynamics permits the formalization of many properties as well as the analysis and design techniques pertaining to the system. Furthermore, many of the methods associated with state variable representations can be generalized and applied to cases where transfer function-based techniques are not applicable. These cases include, among others, *time-varying* and *non-linear* systems, systems with *multiple inputs and outputs* and *non-stationary random* inputs.

A classic example is the multi-degree of freedom spring-mass system, which is described as a set of coupled second-order differential equations as

$$\mathbf{M}\ddot{\mathbf{d}}(t) + \mathbf{D}\dot{\mathbf{d}}(t) + \mathbf{K}\mathbf{d}(t) = \mathbf{u}(t), \tag{3.1}$$

where $\mathbf{d}(t)$ is a $q \times 1$ vector of the displacement degrees of freedom, $\mathbf{u}(t)$ is a $q \times 1$ vector of control forces generated by the actuators, $\mathbf{M}$ is a $q \times q$ mass matrix associated with the flexible structure, $\mathbf{D}$ is a $q \times q$ damping matrix associated with the structure and $\mathbf{K}$ is the associated $q \times q$ stiffness matrix.

The classical transfer function method is relatively easy to use when compared to the state variable technique, especially if the control system designer has no access to a computer. On the other hand, most computer-aided design methods are based on the state variable representation as the state variable representation is particularly suited for that purpose. State variables, however, are 'internal' variables, and they can be used to reconstruct the input–output description of the system. In the above example, the displacement and velocity vectors are completely internal to the system unless they appear in the measurement. Thus, the state variable representation is simply an alternative way of representing the behaviour of the system that makes the internal dynamics transparent.

The set of state variables is the minimum set of state variables of the system such that the knowledge of them at any initial time together with information about the inputs is sufficient to specify the states at any other time. The output is then synthesized from a linear convolution of state variable vector and the inputs.

The dynamics may be expressed in state space form as

$$\begin{bmatrix} \mathbf{M} & \mathbf{0} \\ \mathbf{0} & \mathbf{M} \end{bmatrix} \frac{\mathrm{d}}{\mathrm{d}t} \begin{bmatrix} \mathbf{d} \\ \dot{\mathbf{d}} \end{bmatrix} = \begin{bmatrix} \mathbf{0} & \mathbf{M} \\ -\mathbf{K} & -\mathbf{D} \end{bmatrix} \begin{bmatrix} \mathbf{d} \\ \dot{\mathbf{d}} \end{bmatrix} + \begin{bmatrix} \mathbf{0} \\ \mathbf{I} \end{bmatrix} \mathbf{u} \tag{3.2}$$

and reduced to

$$\frac{\mathrm{d}}{\mathrm{d}t} \begin{bmatrix} \mathbf{d} \\ \dot{\mathbf{d}} \end{bmatrix} = \begin{bmatrix} \mathbf{0} & \mathbf{I} \\ -\mathbf{M}^{-1}\mathbf{K} & -\mathbf{M}^{-1}\mathbf{D} \end{bmatrix} \begin{bmatrix} \mathbf{d} \\ \dot{\mathbf{d}} \end{bmatrix} + \begin{bmatrix} \mathbf{0} \\ \mathbf{M}^{-1} \end{bmatrix} \mathbf{u}.$$

If one assumes that certain linear combinations of the displacements, velocities and inputs are measured, the above equations may be expressed in state space form as

$$\frac{\mathrm{d}}{\mathrm{d}t}\mathbf{x}(t) = \mathbf{A}\mathbf{x} + \mathbf{B}u, \quad \mathbf{y} = \mathbf{C}\mathbf{x} + \mathbf{G}\mathbf{u}, \tag{3.3}$$

where

$$\mathbf{x} = \begin{bmatrix} \mathbf{d} \\ \dot{\mathbf{d}} \end{bmatrix}, \quad \mathbf{A} = \begin{bmatrix} \mathbf{0} & \mathbf{I} \\ -\mathbf{M}^{-1}\mathbf{K} & -\mathbf{M}^{-1}\mathbf{D} \end{bmatrix} \quad \text{and} \quad \mathbf{B} = \begin{bmatrix} \mathbf{0} \\ \mathbf{M}^{-1} \end{bmatrix}.$$

The state space description of a *linear time-invariant* system is given by the above state equation, where $\mathbf{x}$, $\mathbf{u}$ and $\mathbf{y}$ are the $n \times 1$ state vector, $m \times 1$ input vector and the $1 \times k$ output vector, and $\mathbf{A}$, $\mathbf{B}$, $\mathbf{C}$ and $\mathbf{G}$ are $n \times n$, $n \times m$, $k \times n$ and $m \times m$ matrices. The first of equations (3.3) represents the relationship between the inputs and the states, while the second represents the relationship of the states and the inputs to the outputs. Thus, in principle, the transfer function representation may be obtained by 'eliminating' the states from the second of equations (3.3) using the first. For a *single-input single-output* (SISO) system, the number of inputs and number of outputs are equal to unity, i.e. $m = k = 1$. However, we do not restrict ourselves to SISO systems, and the general case of multiple inputs and multiple outputs is considered unless otherwise stated.

The property of linearity has two important consequences among others. The first is the principle of superposition. If two independent outputs are obtained with each of two inputs applied independently to the system, then when the sum of the two inputs are applied to the system, the output is the sum of the two independent outputs. Secondly, when equations (3.3) are satisfied by two independent solutions it follows from linearity that the equations are satisfied by any linear combination of the two solutions.

3.2.2 Transformations of State Variables

The fact that the state vector is a set of variables used to describe the internal dynamics of the system implies that the state vector is essentially non-unique. Hence, it is always possible to transform the state variables from one set to another using a *transformation* that may be *linear* or *non-linear*. Linear transformations preserve the linearity properties of the system. As an example we consider the application of a transformation to equations (3.2). Consider the transformation of state variables defined by

$$\mathbf{x} = \begin{bmatrix} \mathbf{d} \\ \dot{\mathbf{d}} \end{bmatrix} = \begin{bmatrix} \mathbf{z}_2 \\ -\mathbf{z}_1 \end{bmatrix} = \begin{bmatrix} \mathbf{0} & \mathbf{I} \\ -\mathbf{I} & \mathbf{0} \end{bmatrix} \begin{bmatrix} \mathbf{z}_1 \\ \mathbf{z}_2 \end{bmatrix} \equiv \mathbf{T} \begin{bmatrix} \mathbf{z}_1 \\ \mathbf{z}_2 \end{bmatrix}, \tag{3.4a}$$

where

$$\mathbf{T} \equiv \begin{bmatrix} \mathbf{0} & \mathbf{I} \\ -\mathbf{I} & \mathbf{0} \end{bmatrix}. \tag{3.4b}$$

Equations (3.2) then take the form

$$\begin{bmatrix} \mathbf{0} & \mathbf{I} \\ -\mathbf{I} & \mathbf{0} \end{bmatrix} \frac{\mathrm{d}}{\mathrm{d}t} \begin{bmatrix} \mathbf{z}_1 \\ \mathbf{z}_2 \end{bmatrix} = \begin{bmatrix} -\mathbf{I} & \mathbf{0} \\ \mathbf{M}^{-1}\mathbf{D} & -\mathbf{M}^{-1}\mathbf{K} \end{bmatrix} \begin{bmatrix} \mathbf{z}_1 \\ \mathbf{z}_2 \end{bmatrix} + \begin{bmatrix} \mathbf{0} \\ \mathbf{M}^{-1} \end{bmatrix} \mathbf{u}. \tag{3.5}$$

Multiplying equations (3.5) by the inverse of the transformation matrix $\mathbf{T}$, i.e. by $\mathbf{T}^{-1}$ where $\mathbf{T}^{-1}$ is defined by the relationship

$$\mathbf{T}\mathbf{T}^{-1} = \mathbf{T}^{-1}\mathbf{T} = \mathbf{I},$$

where $\mathbf{I}$ is the identity matrix and $\mathbf{T}^{-1}$ is given by

$$\mathbf{T}^{-1} = \begin{bmatrix} \mathbf{0} & -\mathbf{I} \\ \mathbf{I} & \mathbf{0} \end{bmatrix} \tag{3.6}$$

and equations (3.5) are transformed to

$$\frac{\mathrm{d}}{\mathrm{d}t} \begin{bmatrix} \mathbf{z}_1 \\ \mathbf{z}_2 \end{bmatrix} = \begin{bmatrix} -\mathbf{M}^{-1}\mathbf{D} & \mathbf{M}^{-1}\mathbf{K} \\ -\mathbf{I} & \mathbf{0} \end{bmatrix} \begin{bmatrix} \mathbf{z}_1 \\ \mathbf{z}_2 \end{bmatrix} + \begin{bmatrix} -\mathbf{M}^{-1} \\ \mathbf{0} \end{bmatrix} \mathbf{u}. \tag{3.7}$$

The form of these equations is unchanged in that they continue to be of the same general form as equations (3.3) where the state vector is defined in terms of the transformed state variables. In general, when linear transformations of the type defined by equation (3.4) are applied to the state space equations (3.3), the matrices **A, B** and **C** in equations (3.3) transform as

$$\mathbf{A} \to \mathbf{T}^{-1}\mathbf{AT}, \quad \mathbf{B} \to \mathbf{T}^{-1}\mathbf{B}, \quad \mathbf{C} \to \mathbf{CT}. \tag{3.8}$$

3.2.3 Solution of the State Equations

The techniques to solve state space equations of the kind presented above are briefly reviewed. In dynamic systems of the type one is concerned with herein, the solution of the state equations generally represents the response of the system to various inputs. The calculation of the time domain response is first considered, and this is then related to the Laplace transform domain.

The solution of the state equations can be easily established using the principle of convolution or the Duhamel integral. We consider only the case when $\mathbf{G} = \mathbf{0}$ or the null matrix. Hence, the system is characterized by the triple (**A**, **B**, **C**). In order to solve the state equations, the solution for the state vector is assumed to have the form

$$\mathbf{x}(t) = \Phi(t, 0)\,\mathbf{x}(0) + \int_0^t \Phi(t, \tau)\,\mathbf{Bu}(\tau)\,\mathrm{d}\tau, \tag{3.9}$$

where the $n \times n$ matrix $\Phi(t, \tau)$ is usually referred to as the *state transition matrix* and is to be determined. The output of the system can be determined by substituting for $\mathbf{x}(t)$ in the equation for the output vector, and the output vector is then given by

$$\mathbf{x}(t) = \Phi(t, 0)\,\mathbf{x}(0) + \int_0^t \Phi(t, \tau)\,\mathbf{Bu}(\tau)\,\mathrm{d}\tau. \tag{3.10}$$

Substituting the assumed solution for the state vector in the state equations (3.3), one finds that $\Phi(t, \tau)$ must satisfy the equation

$$\frac{\mathrm{d}}{\mathrm{d}t}\Phi(t, \tau) = \mathbf{A}\Phi(t, \tau), \quad \Phi(t, t) = \mathbf{I} \tag{3.11}$$

The solution of the equation is relatively easy to find and can be shown to be

$$\Phi(t, \tau) = e^{\mathbf{A}(t-\tau)} = \Phi_r(t - \tau). \tag{3.12}$$

With equation (3.12) substituted into equation (3.9), it satisfies the state equations (3.3). In fact, one method of computing the transition matrix $\Phi(t, \tau)$, which is only a function of the single variable $t - \tau$, is apparent from equation (3.12). Other techniques of computing the transition matrix include numerical integration and the use of discrete approximations. It may be recalled that the *matrix exponential function* used in equation (3.12) is defined by first defining the scalar exponential function in the form of a series as

$$\Phi_r(t) = e^{\mathbf{A}t} = \mathbf{I} + \mathbf{A}t + \frac{\mathbf{A}^2 t^2}{2!} + \frac{\mathbf{A}^3 t^3}{3!} + \cdots + \frac{\mathbf{A}^n t^n}{n!} + \cdots \tag{3.13}$$

and replacing the variable $\mathbf{A}t$ and its powers by the corresponding matrix and its powers. The computations are generally carried out by a computer. The series is truncated when the next term in the series is smaller than a pre-defined number. There is a point when adding further terms to the summation becomes an exercise in futility. Sooner or later the $n!$ term in the denominator of the exponential coefficient becomes large enough and the term becomes so small that it becomes rounded off in any additions that occur. When that happens, there is no sense adding any more terms and the series is truncated. However, exactly when that happens depends on the size of the **A** matrix.

Consider the evaluation of $\mathbf{x}(t)$ for a sample input such as a unit step, which is defined by

$$U(t) = 1, \quad 0 \le t \le \infty, \quad U(t) = 0, \quad -\infty \le t < 0. \tag{3.14}$$

Substituting $\mathbf{u}(t) = U(t)$and the series expansion for the matrix exponential function in equation (3.9) and evaluating the integral, one obtains

$$\mathbf{x}(t) = \mathbf{B}t + \frac{\mathbf{AB}t^2}{2!} + \frac{\mathbf{A}^2\mathbf{B}t^3}{3!} + \cdots + \frac{\mathbf{A}^{n-1}\mathbf{B}t^n}{n!} + \cdots. \tag{3.15}$$

The solution may be expressed in matrix notation as

$$\mathbf{x}(t) = \begin{bmatrix} \mathbf{B} & \mathbf{AB} & \mathbf{A}^2\mathbf{B} & \cdots & \mathbf{A}^{n-1}\mathbf{B}\cdots \end{bmatrix} \begin{bmatrix} t & \frac{t^2}{2!} & \frac{t^3}{3!} & \cdots & \frac{t^n}{n!}\cdots \end{bmatrix}^T,$$

where $[.]^T$ denotes the transpose of the row vector. An important property of a matrix is that every $n \times n$ matrix raised to the power n or higher can be expressed in terms of linear combinations of lower powers of the matrix. Thus, the solution can be written as

$$\mathbf{x}(t) = \begin{bmatrix} \mathbf{B} & \mathbf{AB} & \mathbf{A}^2\mathbf{B} & \cdots & \mathbf{A}^{n-1}\mathbf{B} \end{bmatrix} \mathbf{r}(t) \tag{3.16}$$

Equation (3.16) represents the general form of the solution for an arbitrary input. Generally, the vector $\mathbf{r}(t)$ depends only on the nature of the input but not on the system properties, while the matrix

$$\mathbf{C}_c = \begin{bmatrix} \mathbf{B} & \mathbf{AB} & \mathbf{A}^2\mathbf{B} & \cdots & \mathbf{A}^{n-1}\mathbf{B} \end{bmatrix} \tag{3.17}$$

is quite independent of the input. Although equation (3.16) is not very useful for practical computations, it does provide insight into the structure of the solution. However, the matrix $\mathbf{C}_c$ plays a significant role in control systems design and is known as the *controllability* matrix. It is quite clear from equation (3.16) that if one wants to generate a set of inputs in order to obtain a desired state vector as the state response, then the matrix $\mathbf{C}_c$ must be inverted to obtain a suitable solution for $\mathbf{r}(t)$. This is the basis for the controllability condition discussed in a later section.

From a practical point of view, any numerical technique that can be used to solve a first-order differential equation may be used to solve the state equations. For example, if we use the fourth-order Runge–Kutta method to solve a set of vector first-order differential equations of the form

$$\frac{\mathrm{d}}{\mathrm{d}t}\mathbf{x}(t) = \mathbf{g}(t, \mathbf{x}) \tag{3.18}$$

and given the value of $\mathbf{x}(t_i) = \mathbf{x}_i$ at time $t = t_i$, then the state vector $\mathbf{x}(t_{i+1}) = \mathbf{x}_{i+1}$ at time $t = t_{i+1} = t_i + \Delta t_i$ may be found as

$$\mathbf{x}_{i+1} = \mathbf{x}_i + (\mathbf{g}_1 + \mathbf{g}_2 + \mathbf{g}_3 + \mathbf{g}_4)\,\Delta t_i/6, \tag{3.19}$$

where

$$\begin{aligned} &\mathbf{g}_1 = \mathbf{g}(t_i, \mathbf{x}_i), \quad \mathbf{g}_2 = \mathbf{g}\left(t_i + \Delta t_i/2, \mathbf{x}_i + \mathbf{g}_1\Delta t_i/2\right), \quad \mathbf{g}_3 = \mathbf{g}\left(t_i + \Delta t_i/2, \mathbf{x}_i + \mathbf{g}_2\Delta t_i/2\right), \\ &\mathbf{g}_4 = \mathbf{g}(t_i + \Delta t_i, \mathbf{x}_i + \mathbf{g}_3\Delta t_i). \end{aligned} \tag{3.20}$$

The value of the time step Δt_i must generally be small for accuracy and numerical stability. For more accurate calculations, the predictor–corrector methods are more suitable. In these methods, a formula of type presented above is first used to predict the state response at the next time step and it is then corrected suitably to account for the predicted characteristics.

When state space equations are small in number, then it is possible to solve them by conventional techniques without using a computer. The conventional method for solving them is the Laplace transform method. The Laplace transform may also be used to relate the transfer function of the system to the state space equations. Thus, taking Laplace transforms of equations (3.3) we have

$$s\mathbf{X}(s) = \mathbf{x}(0) = \mathbf{AX}(s) + \mathbf{BU}(s), \tag{3.21a}$$

$$\mathbf{Y}(s) = \mathbf{CX}(s) + \mathbf{GU}(s), \tag{3.21b}$$

where $\mathbf{X}(s) = \mathcal{L}(\mathbf{x}(t))$, $\mathbf{U}(s) = \mathcal{L}(\mathbf{u}(t))$, $\mathbf{Y}(s) = \mathcal{L}(\mathbf{y}(t))$ and the symbol $\mathcal{L}$ stands for the Laplace transform of the argument.

Solving equation (3.21a) for $\mathbf{X}(s)$, we have,

$$\mathbf{X}(s) = [s\mathbf{I} - \mathbf{A}]^{-1}[\mathbf{x}(0) + \mathbf{BU}(s)] \tag{3.22}$$

and substituting in the output equation (3.22a), we have

$$\mathbf{Y}(s) = \mathbf{C}[s\mathbf{I} - \mathbf{A}]^{-1}\mathbf{x}(0) + [\mathbf{C}[s\mathbf{I} - \mathbf{A}]^{-1}\mathbf{B} + \mathbf{G}]\mathbf{U}(s). \tag{3.23}$$

Comparing equation (3.23) and the first term in equation (3.9) and (3.10), it is possible to identify

$$\mathbf{R}(s) = \mathcal{L}(\Phi r\,(t)) = [s\mathbf{I} - \mathbf{A}]^{-1}. \tag{3.24}$$

The matrix $\mathbf{R}(s)$ plays a fundamental role in the evolution of the system response and is referred to as the '*resolvent*' matrix. The polynomial given by $\Delta(s) = \det(s\mathbf{I} - \mathbf{A})$ is in fact the denominator polynomial of the transfer function relating $\mathbf{Y}(s)$ to $\mathbf{U}(s)$ and is therefore the left-hand side of the characteristic equation, which is given by

$$\Delta(s) = \det(s\mathbf{I} - \mathbf{A}) = 0. \tag{3.25}$$

Thus, in order to find the response of the system to a particular input $\mathbf{u}(t)$ and for a given set of initial conditions, the best approach would be to find the resolvent matrix and then its inverse Laplace transform to find the state transition matrix. Once the state transition matrix is known, the solution can be predicted for any combination of inputs and initial conditions and for any $\mathbf{B}$ and $\mathbf{C}$ matrices.

It should be noted that an efficient algorithm is essential to calculate the resolvent matrix. Normally, it is customary to use *Cramer's* rule for evaluating the determinant of the matrix. This is generally not suitable for programming on a computer.

3.2.4 State Space and Transfer Function Equivalence

It may have already been noted by the reader that the state space and transfer function representations are equivalent for linear time-invariant systems.

Given a state space representation, it is an easy matter to obtain the transfer function and this is a unique relationship. However, given a transfer function, the internal dynamics of the system may be represented by an infinite number of state space realizations, as state space realizations corresponding to a transfer function are not unique. In the latter case, it is important to construct a *minimal realization* that is a realization based on the *minimum* number of states.

In equation (3.23), if we assume that the vector of initial conditions $\mathbf{x}(0)$ is set equal to zero, the input/output relationship in the Laplace transform domain is given by

$$\mathbf{Y}(\mathrm{s}) = [\mathbf{C}[\mathrm{s}\mathbf{I} - \mathbf{A}]^{-1}\,\mathbf{B} + \mathbf{G}]\mathbf{U}(\mathrm{s}). \tag{3.26}$$

Hence, the transfer function relating the output to the input is given by

$$\mathbf{H}(\mathrm{s}) = [\mathbf{C}[\mathrm{s}\mathbf{I} - \mathbf{A}]^{-1}\,\mathbf{B} + \mathbf{G}]. \tag{3.27}$$

The transfer function is the ratio of the Laplace transform of the output to that of the input for zero initial conditions, i.e.

$$\mathbf{H}(s) = \mathbf{Y}(s)\mathbf{U}(s)^{-1} = [\mathbf{C}\mathbf{R}(s)\mathbf{B} + \mathbf{D}]. \tag{3.28}$$

Thus, the transfer function can be computed from the resolvent matrix. Taking the inverse Laplace transform of equation (3.27) yields the impulse response matrix of the system and is given by

$$\mathbf{H}(t) = \mathbf{C}\Phi_r(t)\,\mathbf{B} = \mathbf{C}e^{\mathbf{A}t}\mathbf{B} \quad \text{for } t \geq 0 \tag{3.29a}$$

$$= \mathbf{0} \quad \text{for } t < 0 \tag{3.29b}$$

Thus, state space models are completely equivalent to transfer function models. It can be shown that classical frequency domain design techniques and time domain design techniques using state space models are also completely equivalent. The latter are particularly useful for computer-aided design and hence are preferred.

3.2.5 *State Space Realizations of Transfer Functions*

One is normally interested in constructing state space representations of the dynamics of a system using a minimum number of states. As already mentioned, such representations are not unique and when they exist, there are an infinite number of such representations.

Existence of a minimal representation depends on conditions of controllability and observability, which are discussed in the next section. Both these conditions must be satisfied if the system is to possess a minimal realization. It is usual in control system literature to combine both these conditions into a single condition for minimal realization, where it is required that a certain matrix known as the *Hankel* matrix be of rank n. For our purposes, two 'cookbook' techniques of constructing state space realizations are presented because of their significance in control and filtering theory. Although they are both applicable only to single-input or single-output systems, they may be generalized to the case of multi-input–multi-output systems.

Given a transfer function of the form

$$H(s) = \frac{b_0 s^n + b_1 s^{n-1} + b_2 s^{n-2} \cdots b_{n-1}s + b_n}{s^n + a_1 s^{n-1} + a_2 s^{n-2} \cdots a_{n-1}s + a_n}, \tag{3.30}$$

one may define the control canonic form or the phase variable form as

$$\frac{\mathrm{d}}{\mathrm{d}t}\mathbf{x}(t) = \mathbf{A}_{cc}\mathbf{x}(t) + \mathbf{B}_{cc}u(t), \quad \mathbf{y} = \mathbf{C}_{cc}\mathbf{x} + \mathbf{G}_{cc}\mathbf{u}, \tag{3.31}$$

where

$$\mathbf{A}_{cc} = \begin{bmatrix} 0 & 1 & 0 & \ldots & 0 & 0 \\ 0 & 0 & 1 & \ldots & 0 & 0 \\ 0 & 0 & 0 & \ldots & 0 & 0 \\ \ldots & \ldots & \ldots & \ldots & \ldots & \ldots \\ 0 & 0 & 0 & 0 & 0 & 1 \\ -a_n & -a_{n-1} & -a_{n-2} & \ldots & -a_2 & -a_1 \end{bmatrix}, \quad \mathbf{B}_{cc} = \begin{bmatrix} 0 \\ 0 \\ 0 \\ 0 \\ 0 \\ 1 \end{bmatrix},$$

$$\mathbf{C}_{cc} = \begin{bmatrix} c_n & c_{n-1} & c_{n-2} & \ldots & c_2 & c_1 \end{bmatrix}, \; c_k = b_k - a_k b_0, \; k = 1 \ldots n,$$

and

$$\mathbf{G}_{cc} = b_0.$$

The *observer canonic form* is the *dual* of the control canonic form defined by

$$\frac{\mathrm{d}}{\mathrm{d}t}\mathbf{x}(t) = \mathbf{A}_{oc}\mathbf{x}(t) + \mathbf{B}_{oc}u(t), \quad \mathbf{y} = \mathbf{C}_{oc}\mathbf{x} + \mathbf{G}_{oc}\mathbf{u}, \tag{3.32}$$

where

$$\mathbf{A}_{oc} = \mathbf{A}_{\mathbf{cc}}^{T}, \quad \mathbf{B}_{oc} = \mathbf{C}_{cc}^{T}, \quad \mathbf{C}_{oc} = \mathbf{B}_{cc}^{T}, \quad \mathbf{G}_{oc} = \mathbf{G}_{cc}$$

and $[\cdot]^T$ is transpose of the matrix or vector. The above state space representations are particularly useful in constructing feedback controllers and filters with desired dynamic response properties. This generally requires that one is able to transform from one state space representation to another as well as to the transfer function form. The ability to transform from one state space representation to another the conditions of controllability and observability must be satisfied. The properties of stability, controllability and observability of linear system are therefore discussed in the next section.

3.3 Properties of Linear Systems

3.3.1 Stability, Eigenvalues and Eigenvectors

Loosely speaking, we may state that a linear system is stable if the response is bounded over a finite time domain and tends to zero in steady state (i.e. as time tends to infinity).

Considering a SISO system, one may construct the impulse response by Laplace inversion of the transfer function. The general solution for the response is the linear combination of the input and a number of exponential terms. It may be observed that exponents of these terms are products of the characteristic exponents and the time t. The characteristic exponents are the roots of the characteristic equation obtained by setting the denominator of the transfer function to zero. It is quite obvious that all the exponential terms are bounded if the roots of the characteristic equation are all negative or have negative real parts. Thus, asymptotic stability is guaranteed if all the roots of the characteristic equation are negative or have negative real parts.

In the state space representation, the characteristic equation is given by equation (3.25):

$$\Delta(s) = \det(s\mathbf{I} - \mathbf{A}) = 0. \tag{3.25}$$

The roots of equation (3.25) are generally referred to as the *eigenvalues* of the matrix $\mathbf{A}$ and correspond to the poles of the transfer function. Thus, a linear time-invariant system is stable if and only if all the eigenvalues of $\mathbf{A}$ have negative real parts. This stability condition is identical to saying that the transfer function given by equation (3.26) has all its poles in the open left half of the complex 's' plane.

It is possible to view the equation

$$\mathbf{A}\mathbf{x} = \mathbf{b}$$

as a transformation of the vector $\mathbf{x}$ into the vector $\mathbf{b}$. There is one special situation that plays an important role in matrix theory, and this is the case when $\mathbf{b}$ is proportional to $\mathbf{x}$. In this situation, we have

$$\mathbf{A}\mathbf{x} = \lambda\mathbf{x}$$

or alternatively,

$$[\lambda\mathbf{I} - \mathbf{A}]\mathbf{x} = 0$$

The set of homogeneous equations has a solution if and only if

$$\det(\lambda\mathbf{I} - \mathbf{A}) = 0. \tag{3.33}$$

This equation may be written as

$$\Delta(\lambda) = 0, \tag{3.34}$$

where $\Delta(\lambda)$ is defined by the relationship

$$\Delta(\lambda) = \lambda^n + a_1\lambda^{n-1} + a_2\lambda^{n-2} \cdots a_{n-1}\lambda + a_n. \tag{3.35}$$

Equation (3.35) is in fact the characteristic equation with 'λ' replacing 's'. The polynomial $\Delta(\lambda)$ is referred to as the *characteristic polynomial*. If λ_i is an eigenvalue or a root of the characteristic equation, then it follows that there exists a vector $\mathbf{x} = \mathbf{e}_i$ such that

$$[\lambda_i\mathbf{I} - \mathbf{A}]\,\mathbf{x} = \mathbf{0} \tag{3.36}$$

and the vector $\mathbf{e}_i$ is referred to as the *eigenvector* corresponding to the eigenvalue λ_i. The set of eigenvectors corresponding to all the n eigenvalues of the matrix $\mathbf{A}$ may be arranged sequentially to form a matrix that is usually referred to as *eigenvector matrix* or simply the *eigenmatrix*. The eigenvector matrix plays a crucial role in the construction of *similarity transformation* that diagonalizes the matrix $\mathbf{A}$, which is discussed in a later section.

One consequence of equation (3.33) is that every square matrix satisfies its own characteristic equation, i.e. if we substitute $\mathbf{A}$ for λ in the characteristic equation (3.34), it is identically satisfied. Thus, we may write any matrix raised to the power n as a linear combination of the lower powers of the matrix. This is the famous *Cayley–Hamilton theorem*, which was used in deriving equation (3.16).

3.3.2 Controllability and Observability

A system whose states are characterized by an equation such as the first of equations (3.3) is said to be completely state controllable or simply controllable if, for every arbitrary state $\mathbf{x}(t_0)$ at an arbitrary time t_0 and any other state $\mathbf{x}^1$ in the state space, there exists a finite time $t_1 - t_0$ and an input $u(t)$, defined in the time interval t_0 to t_1, that will transfer the state $\mathbf{x}(t_0)$ to $\mathbf{x}^1$ at time t_1, i.e. $\mathbf{x}(t_1) = \mathbf{x}^1$. Otherwise the system is said to be uncontrollable. Thus, a system is said to be completely controllable if every initial state can be transferred to any final state in a finite time by some input vector. There is no constraint imposed on the nature of the input or the trajectory the state should follow. Furthermore, the system is said to be uncontrollable even if the system is 'controllable in part', i.e. it may be possible to transfer some of the states in the state vector to any desired final values or to transfer all the states in the state vector to certain values in restricted regions in the state space. Controllability is a property of the coupling between the inputs and the states and therefore involves the matrices $\mathbf{A}$ and $\mathbf{B}$ only. Thus, state controllability is also often referred to as the controllability of the pair $(\mathbf{A}, \mathbf{B})$.

The condition for verifying whether or not a system is controllable is given by the *controllability condition*

$$\mathrm{rank}(\mathbf{C}_c) = n, \tag{3.37}$$

where rank(...) is the rank of the argument, which is a square matrix, and $\mathbf{C}_c$ is the controllability matrix defined by equation (3.17).

The concept of observability is analogous to controllability and is its dual. Loosely, a system is said to be observable if the initial state of the system can be determined from suitable measurements of the output. This is an important property that can be exploited in the development of filters for deriving information about the internal states of the system from a measurement of the output. From a practical point of view, it is desirable to know whether the measurement of the system output would provide all the information about the system states or if there are any system modes hidden from the observation. Observability guarantees the reconstruction of state variables that cannot be measured directly due to the limitations of the sensors used for measurement.

A system is said to be completely observable if for any initial vector $\mathbf{x}(t_0)$ at time $t = t_0$ in the state space and in the absence of any input to the system, the knowledge of the output $\mathbf{y}(t_1)$ at any time $t_1 > t_0$ is sufficient to provide the initial state $\mathbf{x}(t_0)$. The simplest example of an observable system is one where the $\mathbf{C}$ matrix in equations (3.3) is $n \times n$ (as many states as there are outputs) and is invertible. Then from the measurements of the outputs at time t_1 and in the absence of any inputs ($\mathbf{u}(t) = \mathbf{0}$),

$$\mathbf{x}(t_1) = \mathbf{C}^{-1}\mathbf{y}(t_1). \tag{3.38}$$

Therefore, given any measurement $\mathbf{y}(t_1)$ one can reconstruct the state vector $\mathbf{x}(t_1)$. In general, however, if the number of measurements m are less than the number of states n, $\mathbf{C}$ is an $m \times n$ matrix and therefore not invertible. Yet one would like to reconstruct $\mathbf{x}(t_0)$ from the output at time t_1, which in the absence of any inputs over the time interval $[t_0, t_1]$ is given by

$$\mathbf{y}(t_1) = \mathbf{C}\boldsymbol{\Phi}(t_1, t_0)\,\mathbf{x}(t_0). \tag{3.39}$$

Using the infinite series expansion for the matrix exponential function and the Cayley–Hamilton theorem, it is possible to express the exponential of a matrix as a linear combination of the first $n - 1$ powers of the matrix and therefore

$$e^{\mathbf{A}t} = \left[e_0 + e_1\mathbf{A}t + e_2(\mathbf{A}t)^2 \ldots + e_{n-1}(\mathbf{A}t)^{n-1}\right]. \tag{3.40}$$

Hence, $\mathbf{y}(t_1)$ may be written as

$$\mathbf{y}(t_1) = \mathbf{C}\left[e_0 \quad e_1(t_1 - t_0) \quad e_2(t_1 - t_0)^2 \quad \ldots \quad e_{n-1}(t_1 - t_0)^{n-1}\right] \times \begin{bmatrix} 1 \\ \mathbf{A} \\ \mathbf{A}^2 \\ \ldots \\ \mathbf{A}^{n-1} \end{bmatrix} \mathbf{x}(t_0). \tag{3.41}$$

One approach to reconstructing $\mathbf{x}(t_0)$ is to differentiate $\mathbf{y}(t_1)$ as many times as required up to a maximum of $n - 1$ times to generate sufficient number of equations that can be solved for $\mathbf{x}(t_0)$. However, this would require that the matrix $\mathbf{C}_o$, defined by

$$\mathbf{C}_o = \begin{bmatrix} \mathbf{C} \\ \mathbf{CA} \\ \mathbf{CA}^2 \\ \ldots \\ \mathbf{CA}^{n-1} \end{bmatrix}, \tag{3.41}$$

is rank n. The matrix $\mathbf{C}_o$ defined by equation (3.41) is known as the *observability* matrix and the requirement that

$$\text{rank}(\mathbf{C}_o) = n \tag{3.42}$$

is known as the *observability condition.* Satisfaction of this condition guarantees that the output has all the information necessary to reconstruct all the initial states.

If the system under consideration is ‘over modelled’ using a larger number of state variables than the minimum number required, then the system would not be completely controllable or observable. In fact, a necessary and sufficient condition for the number of states to be minimal is that the state space realization defined by the triple ($\mathbf{A}$, $\mathbf{B}$, $\mathbf{C}$) be completely controllable and completely observable. Otherwise cancellations of factors in the numerator and denominator of the transfer function would take place leading to contradictions.

The two conditions discussed above can be combined into one compact condition and expressed in terms of the Hankel matrix, which is defined as

$$\mathbf{H}_{oc} = \mathbf{C}_o\mathbf{C}_c. \tag{3.43}$$

The requirement for the state space realization of a transfer function to be minimal is

$$\text{rank}\ (\mathbf{H}_{oc}) = n. \tag{3.44}$$

The concepts of controllability, observability and the Hankel matrix play a key role in the ability to construct reduced-order models of flexible structures for the purpose of designing a feedback control system.

3.3.3 Stabilizability

If one is interested in only stabilizing a system using *feedback*, it is not essential that the system be controllable. Rather it is only required that the unstable states in the system be controllable or that the uncontrollable part of the system be stable. Probably the best approach for verifying if the system is *stabilizable* is to transform the system, whenever possible, into two subsystems: one that is in the control canonic form where the pair ($\mathbf{A}_{cc}$, $\mathbf{B}_{cc}$) is controllable and the other representing an uncontrollable subsystem. Then one needs to only verify that the uncontrollable subsystem is stable.

3.3.4 Transformation of State Space Representations

If there are two different state space representations of the same transfer function, it is natural to expect that one such representation is related to the other by a linear transformation. Such transformations are referred to as *similarity transformations*.

Consider the system described by the state variable representation:

$$\dot{\mathbf{x}}_1\,(t) = \mathbf{A}\mathbf{x}_1 + \mathbf{B}u, \quad \mathbf{y} = \mathbf{C}\mathbf{x}_1 + \mathbf{G}\mathbf{u}. \tag{3.45}$$

It has already been shown that the coefficient matrices of the transformed system are related to the original coefficient matrices, according to the transformation

$$\mathbf{A} \to \mathbf{T}^{-1}\mathbf{A}\mathbf{T}, \quad \mathbf{B} \to \mathbf{T}^{-1}\mathbf{B}, \quad \mathbf{C} \to \mathbf{C}\mathbf{T}. \tag{3.46}$$

It is of interest to find the transformation relating the state vector $\mathbf{x}_1(t)$ to the state vector $\mathbf{x}_2(t)$ of an alternative valid state space representation of the same system. It is assumed that the transformation is defined by the relationship

$$\dot{\mathbf{x}}_1\,(t) = \mathbf{T}\mathbf{x}_2(t), \tag{3.47}$$

where $\mathbf{T}$ is the non-singular transformation matrix. The state space equations in terms of the state vector $\mathbf{x}_2(t)$ must be

$$\dot{\mathbf{x}}_2\,(t) = \mathbf{T}^{-1}\mathbf{A}\mathbf{T}\mathbf{x}_2 + \mathbf{T}^{-1}\mathbf{B}u, \quad \mathbf{y} = \mathbf{C}\mathbf{T}\mathbf{x}_2 + \mathbf{G}\mathbf{u}. \tag{3.48}$$

The controllability matrix for the new representation is given by

$$\mathbf{C}_{c2} = \begin{bmatrix} \mathbf{T}^{-1}\mathbf{B} & \mathbf{T}^{-1}\mathbf{A}\mathbf{T}\mathbf{T}^{-1}\mathbf{B} & \mathbf{T}^{-1}\mathbf{A}^2\mathbf{T}\mathbf{T}^{-1}\mathbf{B} & \cdots & \mathbf{T}^{-1}\mathbf{A}^{n-1}\mathbf{T}\mathbf{T}^{-1}\mathbf{B} \end{bmatrix},$$

which reduces to

$$\mathbf{C}_{c2} = \mathbf{T}^{-1}\begin{bmatrix} \mathbf{B} & \mathbf{A}\mathbf{B} & \mathbf{A}^2\mathbf{B} & \cdots & \mathbf{A}^{n-1}\mathbf{B} \end{bmatrix} = \mathbf{T}^{-1}\mathbf{C}_{c1}, \tag{3.49}$$

where $\mathbf{C}_{c1}$ is the controllability matrix corresponding to pair $(\mathbf{A}, \mathbf{B})$ defined in equations (3.45). Similarly, the observability matrices of the two representations are related as

$$\mathbf{C}_{o2} = \mathbf{C}_{o1}\mathbf{T}. \tag{3.50}$$

From equations (3.49) and (3.50), it follows that the Hankel matrix obtained using either of the two representations is invariant and identical.

Thus, if one wishes to construct a transformation of the state space representation (3.45) to the representation (3.48) when the matrices $\mathbf{A}$, $\mathbf{B}$, $\mathbf{T}^{-1}\mathbf{A}\mathbf{T}$ and $\mathbf{T}^{-1}\mathbf{B}$ are known, then $\mathbf{T}$ is obtained from equation (3.49) and is given by

$$\mathbf{T} = \mathbf{C}_{c1}\mathbf{C}_{c2}^{-1}. \tag{3.51}$$

On the other hand, if the matrices $\mathbf{A}$, $\mathbf{C}$, $\mathbf{T}^{-1}\mathbf{A}\mathbf{T}$ and $\mathbf{C}\mathbf{T}$ are known then the transformation is analogously given by

$$\mathbf{T} = \mathbf{C}_{o2}^{-1}\mathbf{C}_{o1}. \tag{3.52}$$

It may be noted, however, that the matrices may not be arbitrarily specified, as the two representations must satisfy the requirement that the Hankel matrix be invariant.

Finally, if one wishes to find the transformation that $\mathbf{A}$ is diagonal then the matrix $\mathbf{T}$ is defined by the eigenvector matrix of $\mathbf{A}$. These concepts play a fundamental role in the design of feedback controllers, compensators and filters for estimating the states from the outputs.

3.4 Shaping the Dynamic Response Using Feedback Control

One of the basic design problems of interest to a control systems engineer is to design suitable inputs $\mathbf{u}(t)$ so that the transient system response to a typical input has a desired stable characteristic, i.e. it is desired to shape the transient response of the system to suitable inputs.

A SISO system with a transfer function of the form given by equation (3.30) with a transient response to a suitable input function is assumed to be given by

$$y(t) = \sum_{i=1}^{n} y_{0i} e^{\beta_i t}. \tag{3.53}$$

One method of shaping the response is to try and alter the coefficients $\beta_1, \beta_2, \beta_3, \ldots, \beta_{n-1}, \beta_n$ in equation (3.53), i.e. design suitable inputs $u(t)$ such that the response of the system can be characterized by an equation of the same form as equation (3.53) but with a desirable set of coefficients $\beta_{d1}, \beta_{d2}, \beta_{d3}, \ldots, \beta_{dn-1}, \beta_{dn}$ in the place of the coefficients $\beta_1, \beta_2, \beta_3, \ldots, \beta_{n-1}, \beta_n$. It may be recalled that the coefficients $\beta_1, \beta_2, \beta_3, \ldots, \beta_{n-1}, \beta_n$ are the roots of the characteristic equation and can be obtained from the characteristic polynomial that is the denominator polynomial of the transfer function. Thus, one may shape the response of the system by ensuring that the characteristic polynomial has a set of coefficients such that its roots are given by $\beta_{d1}, \beta_{d2}, \beta_{d3}, \ldots, \beta_{dn-1}, \beta_{dn}$.

Consider the control canonic representation of the transfer function given by equation (3.30). The state equations have the form given by equation (3.31). The effect of an input of the form

$$u(t) = -\mathbf{K}\mathbf{x}(t) + v(t), \tag{3.54}$$

where $\mathbf{K}$ is a $1 \times n$ row vector of constants and $v(t)$ is an auxiliary control input, is considered. It may be noted that inputs described by equations of the form given above are equivalent to feeding back a linear combination of the states of the system and is the basis of *closed-loop feedback control*. The negative sign in equation (3.54) for the input is to ensure that the feedback is stabilizing (*negative feedback*) if the system is represented in the control canonic form. It has been tacitly assumed that all the states are accessible. In practice, this is generally not the case and suitable compensation has to be provided for the fact that only the outputs are available feedback.

Substituting equation (3.54) for $u(t)$ in equation (3.31) results in the closed-loop equations:

$$\frac{\mathrm{d}}{\mathrm{d}t}\mathbf{x}(t) = \mathbf{A}_{cloop}\mathbf{x}(t) + \mathbf{B}_{cc}u(t), \tag{3.55}$$

where

$$\mathbf{A}_{cloop} = \begin{bmatrix} 0 & 1 & 0 & \dots & 0 & 0 \\ 0 & 0 & 1 & \dots & 0 & 0 \\ 0 & 0 & 0 & \dots & 0 & 0 \\ \dots & \dots & \dots & \dots & \dots & \dots \\ 0 & 0 & 0 & 0 & 0 & 1 \\ -a_n - k_1 & -a_{n-1} - k_2 & -a_{n-2} - k_3 & \dots & -a_2 - k_{n-1} & -a_1 - k_n \end{bmatrix},$$

$$\mathbf{B}_{cc} = \begin{bmatrix} 0 & 0 & 0 & \dots & 0 & 1 \end{bmatrix}^T$$

where

$$\begin{bmatrix} k_1 & k_2 & k_3 & \cdots & k_{n-1} & k_n \end{bmatrix} \equiv \mathbf{K}. \tag{3.56}$$

If we now construct the equivalent transfer function, it would have the form

$$H(s) = \frac{b_0 s^n + b_1 s^{n-1} + b_2 s^{n-2} \cdots b_{n-1}s + b_n}{s^n + (a_1 + k_n)s^{n-1} + (a_2 + k_{n-1})s^{n-2} \cdots (a_{n-1} + k_2)s + a_n + k_1}.$$

Equation (3.5) is the closed-loop transfer function of the system with the state feedback. The new characteristic polynomial is given by the denominator of equation (3.5), and the characteristic equation is given by setting the characteristic polynomial to zero and is

$$\begin{aligned} &s^n + (a_1 + k_n)s^{n-1} + (a_2 + k_{n-1})s^{n-2} \cdots . \\ &\cdots (a_{n-1} + k_2)s + a_n + k_1 = 0 \end{aligned} \tag{3.57}$$

It follows therefore that by suitable choice of the elements of $\mathbf{K}$

$$\mathbf{K} \equiv \begin{bmatrix} k_1 & k_2 & k_3 & \cdots & k_{n-1} & k_n \end{bmatrix}, \tag{3.58}$$

which are the state feedback gains, it is possible to alter all the coefficients of the closed-loop characteristic equation (3.6) and hence the roots of the characteristic equation to any desired set. This is the basis of a feedback control design method.

The steps involved in the design are outlined below. Given a general state space representation of a system as

$$\dot{\mathbf{x}}(t) = \mathbf{A}\mathbf{x} + \mathbf{B}u, \quad \mathbf{y} = \mathbf{C}\mathbf{x} + \mathbf{G}u. \tag{3.59}$$

the first step is to transform to the control canonic form. The form of the matrices $\mathbf{A}_{cc}$ and $\mathbf{B}_{cc}$ are determined entirely from the characteristic equation of the $\mathbf{A}$ matrix. Thus, they can easily be determined by obtaining the characteristic equation of $\mathbf{A}$. The transformation relating the pair ($\mathbf{A}_{cc}$, $\mathbf{B}_{cc}$) to the pair ($\mathbf{A}$, $\mathbf{B}$) can then be determined from equation (3.49). This requires that we compute the controllability matrices of the two representations.

The next step is to choose the desired closed-loop poles $\beta_{d1}, \beta_{d2}, \beta_{d3}, \ldots, \beta_{dn-1}, \beta_{dn}$ (usually from previous experience), which are the used to compute the coefficients of the desired closed-loop characteristic equation $\begin{bmatrix} d_1 & d_2 & d_3 & \cdots & d_{n-1} & d_n \end{bmatrix}$. An algorithm based on the resolvent matrix algorithm, for computing them from the sums of the powers of all the roots of the characteristic is often used. The set of feedback gains are then obtained using the relationship

$$k_{j+1} = d_{n-j} - a_{n-j}, \quad j = 0, 1, \ldots n - 1. \tag{3.60}$$

It may be recalled at this stage that the above feedback gains relate to the control input, which is a linear combination of the states of the control canonic form. They must therefore be transformed back to the original representation. It can easily be shown that the states of the original state space representation are related to the states of the transformed representation by equation (3.47), therefore the desired feedback gains with reference to the original system are obtained as

$$\mathbf{K}_{cloop} = \mathbf{K}\mathbf{T}^{-1}, \tag{3.61}$$

where $\mathbf{T}$ is the transformation to the control canonic form and the design is complete. If one assumes that the principal feedback is the first state, the control gains may be expressed as

$$\mathbf{K}_{cloop} = k_g \mathbf{K} k_g^{-1} = k_g \mathbf{C}, \tag{3.62}$$

where $\mathbf{C}$ is the normalized gain vector. If the input–output system can be expressed as

$$\dot{\mathbf{x}}(t) = \mathbf{A}\mathbf{x} + \mathbf{B}u, \quad y(t) = \mathbf{C}\mathbf{x} \tag{3.63}$$

and the control law may be expressed as

$$u(t) = -k_g y(t) + v(t). \tag{3.64}$$

Thus, given an arbitrary output that is a linear combination of the states, another method of designing the controller is to plot the characteristic roots of the closed system for $0 \le k_g < \infty$ on the complex plane, a plot known as the root locus plot, and select an appropriate value for the gain k_g. Such a method will be used in a later section for constructing a feedback controller for a prototype flexible structure.

The design procedure presented above is applicable only to single input. In the case of multiple inputs, it is possible to assign arbitrary values not only to all the closed-loop poles but also to some of the components of the eigenvectors.

3.5 Modelling of the Transverse Vibration of Thin Beams

An important engineering example of the vibration of a continuous system is the transverse or lateral vibration of a thin uniform beam. This is an example where the elasticity or stiffness and the mass both are distributed. The oscillatory behaviour of beams, plates and other built-up structures such as trusses are of fundamental importance to engineers, and several books have been written on this topic (Timoshenko and Woinowsky-Kreiger, 1959; Tong, 1960; Meirovitch, 1967, 1970; Timoshenko *et al.*, 1974; Tse *et al.*, 1979; Rao, 1995; Inman, 2007). The bending-torsion oscillations of steel bridges and of aircraft wings have resulted in serious accidents in the early part of the twentieth century. Generally, these were due to the lower natural frequencies of the structural systems being relatively close to an external excitation frequency or excitation-related frequency, thereby giving rise to the possibility of some form of resonance.

Considering the transverse vibration of beams that is a primary element of structures with both distributed mass and distributed stiffness, it is the flexural or bending stiffness of the element that is of primary importance. In considering the flexural vibrations of a beam, we shall adopt the assumptions usually made in simple bending theory. This theory is referred to as the Bernoulli–Euler beam theory and the assumptions are

1. the beam is initially straight;
2. the depth of the beam is small compared to the radius of curvature of its maximum displacement;
3. plane sections of the beam are assumed to remain plane during all phases of the oscillations;
4. deformations due to the shearing of one section relative to an adjacent section are assumed to be negligible.

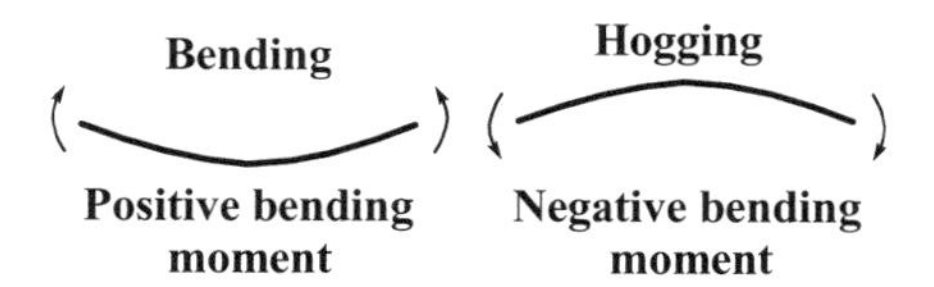

Figure 3.1 Sign convention for moments

In addition to the above assumptions, which are inherent in the Bernoulli–Euler theory of bending, we shall assume that

5. the plane of vibration is a principal plane of bending or the neutral plane and includes the principal axis of the beam;
6. the direction of vibration is perpendicular to it;
7. all the mass is concentrated along the principal axis of the beam;
8. there are no longitudinal tensile or compressive forces acting on the beam. This implies that the rotary inertia and correspondingly all moments necessary to accelerate the mass of the beam between adjacent cross-sections, as the cross-sections rotate about the principal axis in the neutral plane, are neglected. The influence of shear deformations, longitudinal forces and rotary inertia will be considered in a later section.

At the outset, it is important that a proper sign convention is adopted to distinguish between positive and negative bending moments (Figure 3.1). Moments causing positive bending will be considered to be positive, while those causing negative bending or hogging will be considered to be negative.

We now consider the transverse vibration of a slender beam. A coordinate system O'xyz is chosen, as illustrated in Figure 3.2, with the O'x axis along the centroidal axis of the undeformed beam. For simplicity, we take the O'z and O'y to be the principal axis of the undeformed cross-sections.

Thus, a bending moment distribution $Mz(x,t)$ about the O'z axis causes the deformation of the beam and produces a deflection $w(x,t)$ in the direction of y only. The deflection $w(x,t)$ varies with position x along the beam and with time, being measured along O'y from the principal axis of the undeformed state of the beam when $Mz(x,t) = 0$.

Consider the moments and forces acting on element of a beam shown in Figure 3.2. The beam is assumed to have a cross-sectional area A, flexural rigidity EI and material of density ρ. The external forces and moments acting on the element are shown in Figure 3.3. In addition, there is the total force, which by Newton's law is equal to

$$dm \times a = \rho A dx \frac{\partial^2 w}{\partial t^2}.$$

Considering the free body diagram of the element, neglecting rotary inertia and shear of the element, taking moments about O, the principal axis of the section, gives

$$M_z + Q_y \frac{\delta x}{2} + \left(Q_y + \frac{\partial Q_y}{\partial x} \delta x \right) \frac{\delta x}{2} = M_z + \frac{\partial M_z}{\partial x} \delta x$$

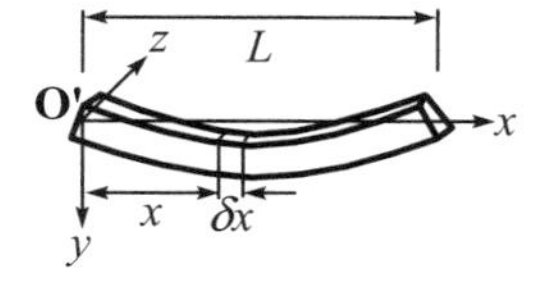

Figure 3.2 Coordinate system used in transverse vibration analysis

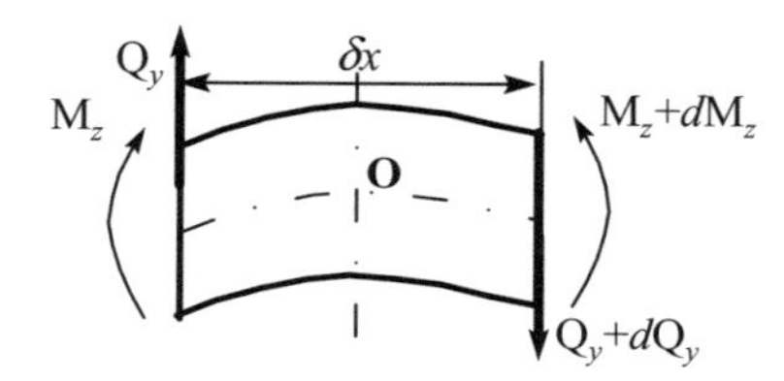

Figure 3.3 Externally applied forces and moments acting on an element

and

$$Q_y \delta x = \frac{\partial M_z}{\partial x}\delta x,$$

which simplifies to

$$\frac{\partial M_z}{\partial x} = Q_y.$$

Thus, for moment equilibrium we have

$$\frac{\partial M_z}{\partial x} - Q_y = 0. \tag{3.65a}$$

Summing all the forces in the y direction,

$$\sum F = \frac{\partial Q_y}{\partial x} dx = dm \times a = \rho A dx \frac{\partial^2 w(x,t)}{\partial t^2}.$$

Hence, for force equilibrium, we have

$$\rho A \frac{\partial^2 w(x,t)}{\partial t^2} = \frac{\partial}{\partial x} Q_y. \tag{3.65b}$$

Employing (3.65a) in equation (3.65b) it follows that

$$\frac{\partial}{\partial x}\left(\frac{\partial M_z}{\partial x}\right) = \frac{\partial}{\partial x} Q_y = \rho A \frac{\partial^2 w(x,t)}{\partial t^2}. \tag{3.66}$$

We shall assume that for a slender beam in transverse vibration the stress and strain are related according to the Bernoulli–Euler theory of bending. Furthermore, in the neutral plane, the strain is assumed to be zero and that the strain varies linearly from this plane. We shall ignore all other secondary effects. The x–z plane is then the neutral plane of the undeformed beam. For small deformations and a prismatic beam, we have the static Bernoulli–Euler bending relations:

$$M_z = -EI_{zz} \frac{\partial^2 w(x,t)}{\partial x^2}. \tag{3.67}$$

Differentiating equation (3.67) and employing equation (3.65a), we obtain

$$\frac{\partial}{\partial x} M_z = -\frac{\partial}{\partial x}\left(EI_{zz} \frac{\partial^2 w(x,t)}{\partial x^2}\right) = Q_y. \tag{3.68}$$

Differentiating equation (3.68) again,

$$\frac{\partial}{\partial x}\left(\frac{\partial}{\partial x} M_z\right) = -\frac{\partial^2}{\partial x^2}\left(EI_{zz} \frac{\partial^2 w(x,t)}{\partial x^2}\right) = \frac{\partial Q_y}{\partial x} = -q_y, \tag{3.69}$$

where qy is the distribution along the beam of the external and inertially applied forces per unit length in the Qy direction. These forces may be obtained by the application of d'Alembert's principle. From equilibrium considerations we obtain equation (3.66), and the stress strain equations reduce to

$$\frac{\partial^2 M_z}{\partial x^2} = -\frac{\partial^2}{\partial x^2}\left(EI_{zz}\frac{\partial^2 w(x,t)}{\partial x^2}\right) = \rho A\frac{\partial^2 w(x,t)}{\partial t^2}. \tag{3.70}$$

The derivatives with respect to x and t in the above equations are partial derivatives as the displacement w is a function of both x and t.

Thus, the equation for transverse oscillatory motion of a prismatic beam is

$$\rho A\frac{\partial^2 w(x,t)}{\partial t^2} + \frac{\partial^2}{\partial x^2}\left(EI_{zz}\frac{\partial^2 w(x,t)}{\partial x^2}\right) = 0. \tag{3.71}$$

This is linear fourth-order partial differential equation that may be solved in a similar manner to that of the one-dimensional wave equation, i.e. by separation of variables. When the beam vibrates transversely in one of its natural modes, the deflection at any point along the beam may be assumed to be equal to the product of two functions:

$$w(x,t) = \eta(x)P(t), \tag{3.72}$$

where $\eta(x)$ is the spatial or normal function depicting the modal displacement of the deflected beam and $P(t)$ is the harmonically varying function representing the instantaneous variation of the modal amplitude. Thus, the general solution may be written as

$$w(x,t) = \sum_{n=1}^{\infty}\phi_n(x,t) = \sum_{n=1}^{\infty}\eta_n(x)P_n(t), \tag{3.73}$$

where

$$P_n(t) = A_n\sin\omega_n t + B_n\cos\omega_n t. \tag{3.74}$$

To find the constants A_n and B_n and the functions $\eta_n(x)$, one must supplement the partial differential equation with the appropriate boundary conditions that reflect the conditions inherent in the physical situation.

For example, at a fixed (or built-in) end

$$x = x_e,\ \ w(x,t)|_{x=x_e} = 0 \quad \text{and} \quad \left.\frac{\partial w(x,t)}{\partial x}\right|_{x=x_e} = 0, \tag{3.75a}$$

reflecting the fact that the displacement and slope are zero at such an end. At a pinned end, $x = xe$, where there is no deflection or moment

$$w(x,t)|_{x=x_e} = 0 \quad \text{and} \quad M_z|_{x=x_e} = -EI_{zz}\left.\frac{\partial^2 w(x,t)}{\partial x^2}\right|_{x=x_e} = 0. \tag{3.75b}$$

When, $EI_{zz} \neq 0$ at the end $x = xe$, the second boundary condition reduces to

$$\left.\frac{\partial^2 w(x,t)}{\partial x^2}\right|_{x=x_e} = 0. \tag{3.76}$$

Boundary conditions that prescribe the geometry alone must be *imposed* to reflect the geometry of the physical situation, which represents the constraints inherent in the problem. They generally involve the displacement and slope at a particular point along the beam.

The boundary conditions involving the shear force or moment at a particular point along the beam arise naturally as a consequence of the physical conditions and are *natural* or *dynamic* boundary conditions.

In general, initial conditions must also be specified to give the complete solution. Thus, the deflected shape of the beam and its rate of change may be specified at time $t = t_0$, as

$$w(x,t)|_{t=t_0} = f(x) \quad \text{and} \quad \left.\frac{\partial w(x,t)}{\partial t}\right|_{t=t_0} = g(x). \tag{3.77}$$

In order to solve the equations of motion for the transverse vibration of the beam, we let the solution be given by equation (3.72). Thus,

$$\frac{\partial^2 w(x,t)}{\partial^2 t} = \eta(x)\frac{d^2 P(t)}{d^2 t} \quad \text{and} \quad \frac{\partial^2 w(x,t)}{\partial^2 x} = \frac{d^2\eta(x)}{d^2 x}P(t). \tag{3.78}$$

Substituting in the equation of motion (3.71), we obtain the equation

$$\eta(x)\frac{d^2 P(t)}{d^2 t} + \frac{1}{\rho A}\frac{\partial^2}{\partial x^2}\left(EI_{zz}\frac{d^2\eta(x)}{d^2 x}P(t)\right) = 0. \tag{3.79}$$

Dividing the equation by $w(x,t) = \eta(x)P(t)$ and rearranging,

$$\frac{1}{P(t)}\frac{d^2 P(t)}{d^2 t} = -\frac{1}{\rho A\eta(x)}\frac{\partial^2}{\partial x^2}\left(EI_{zz}\frac{d^2\eta(x)}{d^2 x}\right). \tag{3.80}$$

The left-hand side of the equation is entirely a function of t alone, while the right-hand side is a function of x alone. Thus, each side must be equal to a constant and we have

$$\frac{1}{P(t)}\frac{d^2 P(t)}{d^2 t} = -\frac{1}{\rho A\eta(x)}\frac{\partial^2}{\partial x^2}\left(EI_{zz}\frac{d^2\eta(x)}{d^2 x}\right) = -\omega^2, \tag{3.81}$$

which implies that we now have, for $P(t)$ and $\eta(x)$, the two equations

$$\frac{d^2 P(t)}{d^2 t} + \omega^2 P(t) = 0 \tag{3.82a}$$

and

$$\frac{1}{\rho A}\frac{\partial^2}{\partial x^2}\left(EI_{zz}\frac{d^2\eta(x)}{d^2 x}\right) - \omega^2\eta(x) = 0. \tag{3.82b}$$

We now consider the case of a uniform beam, so it follows that the flexural rigidity EI_{zz} is constant along the length of the beam. Equation (3.82b) for $\eta(x)$ reduces to

$$\frac{d^4\eta(x)}{d^4 x} - \beta^4\eta(x) = 0, \tag{3.83}$$

where

$$\beta^4 = \frac{\omega^2\rho A}{EI_{zz}}.$$

There are two separate ordinary differential equations, equations (3.82a) and (3.83) for $P(t)$ and $\eta(x)$, respectively, and ω^2 is an unspecified parameter common to both.

Considering, in the first instance, the equation for $P(t)$ when $\omega = \omega_n$, the general solution may be written as

$$P_n(t) = A_n\sin(\omega_n t) + B_n\cos(\omega_n t), \tag{3.84}$$

where ω_n is an unspecified parameter.

The general solution for $\eta(x)$ may be written as

$$\eta(x) = \sum_{n=1}^{4} A_n\exp[\lambda_n x], \tag{3.85}$$

where $\lambda = \lambda_n$ are the four roots of the equation

$$\lambda^4 = \beta^4, \quad \lambda_1 = \beta, \quad \lambda_2 = -\beta, \quad \lambda_3 = \sqrt{-1}\beta \quad \text{and } \lambda_4 = -\sqrt{-1}\beta. \tag{3.86}$$

The general solution may also be written in terms of the hyperbolic functions, as these are linear combinations of the appropriate exponential functions. Thus,

$$\eta(x) = A\sin(\beta x) + B\cos(\beta x) + C\sinh(\beta x) + D\cosh(\beta x), \tag{3.87}$$

where the constants A, B, C and D must be determined by satisfying the appropriate boundary conditions, which are either *geometric* or *dynamic*.

From the general solution, given by equation (3.87), it follows that

$$\frac{\partial}{\partial x}\eta(x) = \beta A\cos(\beta x) - \beta B\sin(\beta x) + \beta C\cosh(\beta x) + \beta D\sinh(\beta x)$$

and that

$$\frac{\partial^2 \eta(x)}{\partial^2 x} = -\beta^2 A\sin(\beta x) - \beta^2 B\cos(\beta x) + \beta^2 C\sinh(\beta x) + \beta^2 D\cosh(\beta x).$$

Applying the *geometric* boundary conditions given by equations (3.75a) arising due to the *kinematic* constraints imposed at the fixed or clamped end $x = 0$, we have

$$w(x,t)|_{x=0} = P(t)\,\eta(x)|_{x=0} = 0$$
$$\left.\frac{\partial}{\partial x}w(x,t)\right|_{x=0} = P(t)\left.\frac{\partial}{\partial x}\eta(x)\right|_{x=0} = 0.$$

Since $P(t) \neq 0$, it follows that

$$\eta(0) = B + D = 0 \quad \text{and} \quad \frac{\partial}{\partial x}\eta(0) = \beta A + \beta C = 0.$$

Hence, it follows that $D = -B$ and $C = -A$ and the general solution reduces to

$$\eta(x) = A(\sin(\beta x) - \sinh(\beta x)) + B(\cos(\beta x) - \cosh(\beta x)) \tag{3.88}$$

and there are now only two unknown constants A and B.

On the other hand, if the end $x = 0$ is pinned or hinged, the conditions are

$$w(x,t)|_{x=0} = 0 \quad \text{and} \quad \left.\frac{\partial^2 w(x,t)}{\partial^2 x}\right|_{x=0} = 0 \tag{3.89}$$

and it follows that

$$\eta(0) = B + D = 0$$

and

$$\frac{\partial^2}{\partial^2 x}\eta(0) = -\beta^2 B + \beta^2 D.$$

Hence, it follows that in this case, $B = D = 0$ and that

$$\eta(x) = A\sin(\beta x) + C\sinh(\beta x). \tag{3.90}$$

We shall now consider several special cases.

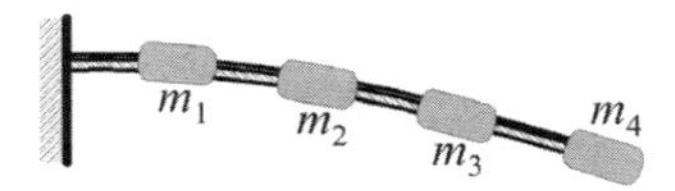

Figure 3.4 Non-uniform cantilever with distributed discrete masses

3.5.1 Vibrations of Cantilever Beam

The vibrations of a cantilever beam provide an important example in the study of the vibrations of continuous structures. Many real-life structures such as aircraft wings may be idealized as a cantilever. These may be composed of several elastic as well as inertial elements distributed along the beam, in addition to discrete masses that may be present at certain points along the beam, as shown in Figure 3.4.

In many such situations, it may be appropriate to model the beam as a non-uniform cantilever.

In certain other situations where there is large additional tip mass present at the tip, it may be appropriate to model the beam as a continuous massless elastica with a proportion of the mass of the beam $m$eff added to the tip mass $m$tip. This proportion may vary from about 20% to about 30% of the mass of the beam mb, depending on the degree of non-uniformity. The situation is illustrated in Figure 3.5.

An idealized cantilever is a long slender beam, rigidly clamped to a fixed support at one end and free at the other end, as shown in Figure 3.6.

The geometric boundary conditions at the fixed or clamped end $x = 0$ are

$$w(x,t)|_{x=0} = 0 \quad \text{and} \quad \left.\frac{\partial w(x,t)}{\partial x}\right|_{x=0} = 0. \tag{3.91}$$

The boundary conditions at the free end are obtained from the conditions of zero moment and zero shear force at that end. These boundary conditions arise as a result of the application of the condition's force and moment equilibrium and hence are *dynamic* in nature. They are also referred to as the *natural* boundary conditions, as they follow naturally from the condition's force and moment equilibrium. It follows that at $x = L$,

$$M_z|_{x=L} = -EI_{zz}\left.\frac{\partial^2 w(x,t)}{\partial x^2}\right|_{x=L} = 0 \quad \text{and} \quad Q_y|_{x=L} = \left.\frac{\partial}{\partial x}M_z\right|_{x=L} = 0. \tag{3.92}$$

Hence,

$$-\left.\frac{\partial}{\partial x}\left(EI_{zz}\frac{\partial^2 w(x,t)}{\partial x^2}\right)\right|_{x=0} = 0. \tag{3.93}$$

Considering the case of a uniform, homogeneous beam so that the flexural rigidity EI_{zz} is constant along the length of the beam, we choose the assumed mode shapes on the basis of equation (3.88) to be

$$\eta(x) = A(\sin(\beta x) - \sinh(\beta x)) + B(\cos(\beta x) - \cosh(\beta x)) \tag{3.88}$$

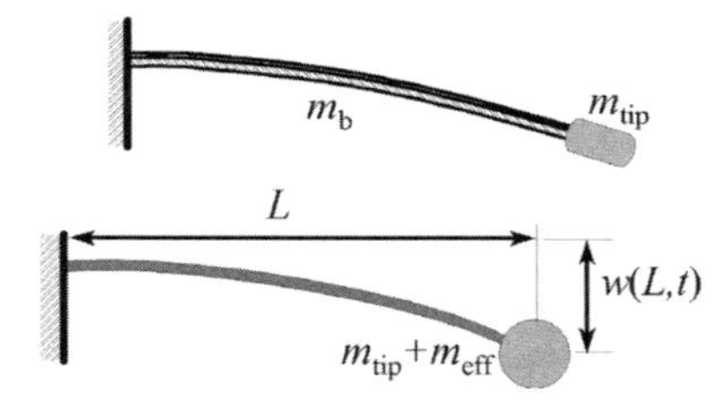

Figure 3.5 Non-uniform cantilever with a tip mass and its idealization

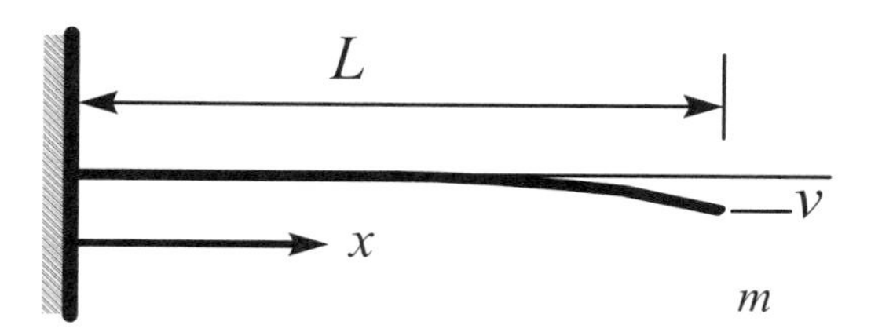

Figure 3.6 Cantilever of length L

and apply the boundary conditions at $x = L$ to it. Hence at $x = L$,

$$\left.\frac{\partial^2}{\partial^2 x} w(x,t)\right|_{x=L} = P(t)\left.\frac{\partial^2}{\partial^2 x}\eta(x)\right|_{x=L} = 0 \tag{3.89a}$$

and

$$\left.\frac{\partial^3}{\partial^3 x} w(x,t)\right|_{x=L} = P(t)\left.\frac{\partial^3}{\partial^3 x}\eta(x)\right|_{x=L} = 0. \tag{3.89b}$$

Hence, assuming that $P(t) \neq 0$, the two boundary conditions are

$$\left.\frac{\partial^2}{\partial^2 x}\eta(x)\right|_{x=L} = 0 \quad \text{and} \quad \left.\frac{\partial^3}{\partial^3 x}\eta(x)\right|_{x=L} = 0. \tag{3.90}$$

Differentiating the assumed solution for $\eta(x)$ and applying the boundary conditions (3.90), we obtain two simultaneous equations for the constants A and B:

$$-A\beta^2(\sin(\beta L) + \sinh(\beta L)) - B\beta^2(\cos(\beta L) + \cosh(\beta L)) = 0, \tag{3.91a}$$

$$-A\beta^3(\cos(\beta L) + \cosh(\beta L)) + B\beta^3(\sin(\beta L) - \sinh(\beta L)) = 0. \tag{3.91b}$$

Solving for the amplitude ratio B/A from each of the above equations gives

$$\frac{B}{A} = -\frac{\sin(\beta L) + \sinh(\beta L)}{\cos(\beta L) + \cosh(\beta L)} = \frac{\cos(\beta L) + \cosh(\beta L)}{\sin(\beta L) - \sinh(\beta L)}. \tag{3.92}$$

Hence, we obtain the *frequency equation*

$$\sinh(\beta L)^2 - \sin(\beta L)^2 = (\cos(\beta L) + \cosh(\beta L))^2,$$

which simplifies to

$$2\cos(\beta L)\cosh(\beta L) + 2 = 0. \tag{3.93}$$

The solution of the frequency equation can be obtained by writing this equation as

$$\cos(\beta L) = -\frac{1}{\cosh(\beta L)}. \tag{3.94}$$

For large values of βL, the frequency equation reduces to

$$\cos(\beta L) = 0. \tag{3.95}$$

Thus, the roots of the frequency equation are $z_n = \beta_n L$,

$$\beta_n L = 1.875,\ 4.694,\ 7.855,\ 10.996,\ 14.137,\ 17.279, \ldots, \left(n - \frac{1}{2}\right)\pi, n > 6.$$

For each of the values of $\beta = \beta_i$, the corresponding natural frequency is obtained from

$$\omega_i^2 = EI_{zz}\beta_i^4 / \rho A \tag{3.96a}$$

and the corresponding normal mode is obtained by substituting $\beta = \beta_i$ in the equation for $\eta(x)$ and is

$$\eta_i(x) = \sin(\beta_i x) - \sinh(\beta_i x) + \gamma_i(\cos(\beta_i x) - \cosh(\beta_i x)), \tag{3.96b}$$

where

$$\gamma_i = \frac{B_i}{A_i} = -\frac{\sin(\beta_i L) + \sinh(\beta_i L)}{\cos(\beta_i L) + \cosh(\beta_i L)}.$$

The first four values of γ_i are given by -1.362221, -0.981868, -1.000776 and -0.999966.

In the case of the cantilevered beam, the nodes $x = x_k$ of the ith normal mode ($i > 1$) are given by the roots of the equation,

$$\eta_i(x_k) = \sin(\beta_i x_k) - \sinh(\beta_i x_k) + \gamma_i(\cos(\beta_i x_k) - \cosh(\beta_i x_k)) = 0.$$

Hence, the nodes $x = x_k$ may be estimated from the equation

$$\frac{\sin(\beta_i x_k) - \sinh(\beta_i x_k)}{\cosh(\beta_i x_k) - \cos(\beta_i x_k)} = \gamma_i. \tag{3.97}$$

An important property of these normal mode shapes or *eigenfunctions*, corresponding to each of the natural frequencies, is *orthogonality*. It follows as a result of orthogonality that

$$\int_0^L \rho A\left(\eta_i(x)\eta_j(x)\right)\mathrm{d}x = 0, \quad i \neq j. \tag{3.98}$$

The general solution for the transverse vibration of the cantilevered beam may be written as

$$w(x,t) = \sum_{n=1}^{\infty}\phi_n(x,t) = \sum_{n=1}^{\infty}\eta_n(x)P_n(t), \tag{3.73}$$

where

$$P_n(t) = A_n \sin\omega_n t + B_n \cos\omega_n t.$$

Furthermore,

$$\frac{\partial}{\partial t}w(x,t) = \frac{\partial}{\partial t}\sum_{n=1}^{\infty}\phi_n(x,t) = \sum_{n=1}^{\infty}\eta_n(x)\frac{\partial}{\partial t}P_n(t),$$

where

$$\frac{\partial}{\partial t}P_n(t) = \omega_n(A_n\cos\omega_n t - B_n\sin\omega_n t).$$

The constants A_n and B_n may be found from the specified initial conditions at time $t = 0$,

$$w(x,t)|_{t=0} = f(x) \quad \text{and} \quad \left.\frac{\partial w(x,t)}{\partial t}\right|_{t=0} = g(x). \tag{3.77}$$

Thus, it can be shown that

$$A_i = \frac{\int_0^L g(x)\eta_i(x)\mathrm{d}x}{\omega_i\int_0^L \eta_i(x)\eta_i(x)\mathrm{d}x} \quad \text{and} \quad B_i = \frac{\int_0^L f(x)\eta_i(x)\mathrm{d}x}{\int_0^L \eta_i(x)\eta_i(x)\mathrm{d}x}. \tag{3.99}$$

The general solutions for the transverse vibrations of beams with other conditions may be obtained in a similar manner.

3.5.2 *Vibrations of Simply Supported, Slender Uniform Beam*

The case a slender cantilever beam considered in the previous section is a classic case of a beam with geometric boundary conditions at one end and forced boundary conditions at the other. In the case of a beam hinged at both ends, the boundary conditions are identical at both ends. The boundary conditions at both ends, $x = 0$ and $x = L$, are given by equations (3.75b), i.e. the condition of zero displacement

$$w(x,t)|_{x=x_e} = 0$$

and the condition of zero moment

$$M_z|_{x=x_e} = -EI_{zz}\left.\frac{\partial^2 w(x,t)}{\partial x^2}\right|_{x=x_e} = 0,$$

where $xe = 0$ or L. For a uniform beam, the later condition reduces to equation (3.76), which is

$$\left.\frac{\partial^2 w(x,t)}{\partial x^2}\right|_{x=x_e} = 0.$$

The assumed solution for the mode shape is given by equation (3.90) and takes the form

$$\eta(x) = A\sin(\beta x) + C\sinh(\beta x).$$

At the end, $x = L$,

$$\eta(L) = A\sin(\beta L) + C\sinh(\beta L) = 0$$

and

$$\left.\frac{\partial^2}{\partial^2 x}\eta(x)\right|_{x=L} = -A\beta^2\sin(\beta L) + C\beta^2\sinh(\beta L) = 0.$$

Since $\sinh(\beta L) \neq 0$ for any non-zero β, the constant C must satisfy the condition $C = 0$ and the frequency equation reduces to

$$\sin(\beta L) = 0. \tag{3.100}$$

The roots of the frequency equation are

$$\beta_i L = i\pi, \quad i = 1, 2 \ldots \infty. \tag{3.101}$$

For each of the values of $\beta = \beta_i$, the corresponding natural frequency is obtained from

$$\omega_i^2 = i^4\pi^4 EI_{zz}/\rho AL^4. \tag{3.102}$$

The corresponding normal mode is obtained by substituting $\beta = \beta_i$ in the equation for $\eta(x)$ and is

$$\eta_i(x) = \sin\left(\frac{i\pi x}{L}\right). \tag{3.103}$$

Thus, in the case of the hinged-hinged beam, the nodes of the nth normal mode ($n > 1$) are given by the roots of the equation

$$\sin\left(\frac{n\pi x}{L}\right) = 0, \tag{3.104}$$

which are

$$\frac{x}{L}=\frac{m}{n}, \quad m<n. \tag{3.105}$$

The cases for a slender uniform beam free at both ends, clamped at both ends, hinged at one end and clamped at the other or any other boundary conditions may be found in the work by Gorman (1975).

3.6 Externally Excited Motion of Beams

In most practical situations involving real engineering systems, both the bending and the torsional oscillations are present. For example, in a typical model of an aircraft wing, the first three modes correspond closely to the first two bending modes of a non-uniform beam and the first torsion mode of a non-uniform rod. The next three modes are characterized by curved nodal lines, and the beam and rod models cannot easily predict these modes. More complex models based on plate and shell theories are definitely essential. Yet the beam and rod models are quite adequate for most practical purposes. The beams are usually non-uniform and the associated boundary conditions are relatively complex and non-standard. Typically, while one end of a beam may be characterized by a standard boundary condition, the other carries a mass or an inertia or both, and may be supported by a translational or rotational spring or a combination of both. Masses and springs could also be attached between the two ends of the beam. Typical practical examples include aircraft wings with attached pylons, fuel tanks and stores, multi-span bridges, suspension bridges, tall buildings such as the John Hancock building in Chicago and large turbine shafts. The analysis of such beams is illustrated by the following four examples.

The general analysis of the forced vibration of continuous structures with arbitrary single point or distributed loading is best illustrated by considering the forced lateral vibrations of thin beams. Forced excitation of beams may arise due to concentrated forces or couples acting at discrete points along the beam or distributed forces or couples acting along the length of the beam. The equation of motion for the transverse vibration of a beam with an additional distributed external force along the length of the beam, which modifies the distributed inertial loading on the beam, is given by

$$\frac{\partial^2}{\partial x^2}\left(EI_{zz}\frac{\partial^2 w(x,t)}{\partial x^2}\right)=-\rho A\frac{\partial^2 w(x,t)}{\partial t^2}+q_y(x,t) \tag{3.106}$$

To find the general solution, we assume the solution

$$w(x,t)=\sum_{n=1}^{\infty}\phi_n(x,t)=\sum_{n=1}^{\infty}\eta_n(x)P_n(t), \tag{3.73}$$

where $P_n(t)=A_n\cos(\omega_n t+\phi_n)$.

The distributed excitation force $q_y(x,t)$ is expanded in an orthogonal series similar to the assumed solution as

$$q_y(x,t)=\sum_{n=1}^{\infty}q_{yn}(x,t)=\sum_{n=1}^{\infty}\rho A\beta_{nn}\eta_n(x)Q_n(t), \tag{3.107}$$

where the quantities β_{nn} and $Q_n(t)$ are yet to be specified. Assuming that the functions $\eta_n(x)$ are an orthogonal set, and multiplying the above equation by $\eta_j(x), j=1,2,3,\ldots,\infty$ and integrating over the length of the beam, we have

$$\int_0^L q_y(x,t)\eta_j(x)\,\mathrm{d}x=\int_0^L \rho A\eta_j(x)\sum_{n=1}^{\infty}\beta_{nn}\eta_n(x)Q_n(t)\,\mathrm{d}x. \tag{3.108}$$

Following the application of the property of orthogonality of the normal modes given by

$$\alpha_{ij} = \int_0^L \rho A \eta_i(x)\, \eta_j(x)\, \mathrm{d}x = 0, i \neq j, \tag{3.109}$$

we have

$$\int_0^L q_y(x,t)\, \eta_j(x)\, \mathrm{d}x = Q_j(t)\, \beta_{jj} \int_0^L \rho A \eta_j(x)\, \eta_j(x)\, \mathrm{d}x.$$

If we let

$$\beta_{nn} = \frac{1}{\int_0^L \rho A \eta_n(x)\, \eta_n(x)\, \mathrm{d}x} = \frac{1}{\alpha_{nn}} \tag{3.110}$$

and solve for $Q_j(t)$, we get

$$Q_j(t) = \int_0^L q_y(x,t)\, \eta_j(x)\, \mathrm{d}x. \tag{3.111}$$

Substituting the series expansions for $q_y(x,t)$ and $w(x,t)$ given by equations (3.107) and (3.73), respectively, in the equation of motion (3.106), and rearranging the resulting equation within the summation sign,

$$\begin{aligned}
&\sum_{n=1}^{\infty} \rho A \eta_n(x)\, P_n(t) \left(\frac{1}{P_n(t)} \frac{d^2 P_n(t)}{dt^2} - \frac{1}{\alpha_{nn}} \frac{Q_n(t)}{P_n(t)} \right) \\
&= -\sum_{n=1}^{\infty} \rho A \eta_n(x)\, P_n(t) \left(\frac{1}{\rho A \eta_n(x)} \frac{\partial^2}{\partial x^2} \left(EI_{zz} \frac{d^2 \eta_n(x)}{dx^2} \right) \right).
\end{aligned} \tag{3.112}$$

The terms within the brackets on the left-hand side of equation (3.112) are entirely functions of t alone, while the term within the brackets on the right-hand side is a function of x alone. Thus, the quantities within the brackets on each side must be equal to a constant and we have

$$\frac{1}{P_n(t)} \frac{\mathrm{d}^2 P_n(t)}{\mathrm{d}t^2} - \frac{1}{\rho A \alpha_{nn}} \frac{Q_n(t)}{P_n(t)} = -\frac{1}{\rho A \eta_n(x)} \frac{\partial^2}{\partial x^2} \left(EI_{zz} \frac{\mathrm{d}^2 \eta_n(x)}{\mathrm{d}x^2} \right) = -\omega_n^2$$

for each n. This implies that we now have the two equations:

$$\frac{\mathrm{d}^2 P_n(t)}{\mathrm{d}t^2} + \omega_n^2 P_n(t) = \frac{1}{\alpha_{nn}} Q_n(t) \tag{3.113a}$$

and

$$\frac{\partial^2}{\partial x^2} \left(EI_{zz} \frac{d^2 \eta_n(x)}{dx^2} \right) - \rho A \omega_n^2 \eta_n(x) = 0 \tag{3.113b}$$

for $Pn(t)$ and $\eta_n(x)$. The first of these, equation (3.113a), is a set of vibrational equations with de-coupled inertial and stiffness properties. The dependent variables $Pn(t)$ are known as the *principal coordinates*. Thus, when the equations of motion are expressed in terms of the principal coordinates and in matrix form, the mass and stiffness matrices are diagonal. The second of the above two equations is identical to the one obtained for free vibrations, and it therefore follows that it can be solved as in the free vibration case.

Assuming the beam is uniform and homogeneous, it follows that ρA is a constant. While the equations of motion in the principal coordinates are unchanged, the integrals, defined by α_{nn}, may be expressed as

$$\alpha_{nn} = \rho A \int_0^L \eta_n(x)\,\eta_n(x)\,\mathrm{d}x. \tag{3.114}$$

The equation for $P_n(t)$ may also be written as

$$\alpha_{nn}\frac{\mathrm{d}^2 P_n(t)}{\mathrm{d}t^2} + \alpha_{nn}\omega_n^2 P_n(t) = Q_n(t). \tag{3.115}$$

The quantity α_{nn} may then be interpreted as a generalized inertia in the nth mode, while the quantity $\alpha_{nn}\omega_n^2$ is the corresponding generalized stiffness. The term on the right-hand side of the equation is then interpreted as the generalized force acting on the beam in the nth mode. In the absence of any external forces, the solutions for $P_n(t)$ correspond to the *normal modes* of vibration of the structure.

Assuming that $Q_n(t) = Q_{n0}\sin(\omega t)$, the particular solution for $P_n(t)$ may be written as

$$P_n(t) = \frac{1}{\alpha_{nn}}\frac{Q_{n0}}{\omega_n^2 - \omega^2}\sin(\omega t). \tag{3.116}$$

The particular solution for the general case is obtained by the convolution principle and is

$$P_n(t) = \frac{1}{\alpha_{nn}\omega_n}\int_0^t Q_n(\tau)\sin(\omega_n t - \omega_n \tau)\,\mathrm{d}\tau. \tag{3.117}$$

The general solution may then obtained by combining the particular solution with solution for the free vibration and applying the appropriate initial conditions. Thus,

$$P_n(t) = P_n(0)\cos(\omega_n t) + \frac{\dot{P}_n(0)}{\omega_n}\sin(\omega_n t) + \frac{1}{\alpha_{nn}\omega_n}\int_0^t Q_n(\tau)\sin(\omega_n t - \omega_n \tau)\,\mathrm{d}\tau. \tag{3.118}$$

In the case of a point load acting at point $x = \chi$ along the beam, the response equations may be simplified considerably. In this case, $q_y(x, t)$ takes the form

$$q_y(x, t) = F_y(t)\,\delta(x - \chi), \tag{3.119}$$

where $\delta(x - \chi)$ is the Dirac-delta function satisfying the relationship

$$\int_0^\infty f(x)\,\delta(x - \chi)\,\mathrm{d}x = f(\chi).$$

Substituting for $q_y(x, t)$ in the equation defining $Q_n(t)$, it follows that

$$Q_n(t) = \int_0^L F_y(t)\,\delta(x - \chi)\,\eta_n(x)\,\mathrm{d}x = F_y(t)\,\eta_n(\chi). \tag{3.120}$$

3.7 Closed-loop Control of Flexural Vibration

An early text by Vernon (1967) is dedicated to the application of control theory to vibrating systems. The dynamics and control of structures (Meirovitch, 1990; Clark *et al.*, 1998), the active control of vibrations (Frolov and Furman, 1990; Preumont, 2002), the control of flexible structures (Leipholz and Rohman, 1986; Joshi, 1989; Gawronski, 1996) and the dynamics and control of distributed parameter systems

(Tzou and Bergman, 1998) have all been the subjects of textbooks. This section is dedicated to the design of a simple feedback controller for flexible structures. The subject will be dealt with in greater depth in Chapter 9. To illustrate the process of designing a closed-loop controller for an archetypal structure, consider the design of a controller for a typical cantilevered beam with a distributed set of PZT actuators.

In this section, the main steps involved in the controller design and validation process are presented. The governing partial differential equation of a uniform beam of length L, excited by an external actuator patch, is obtained by rearranging equation (3.106):

$$\frac{\partial^2}{\partial x^2}\left(EI_{zz}\frac{\partial^2 w(x,t)}{\partial x^2}\right) + \rho A\frac{\partial^2 w(x,t)}{\partial t^2} = q_y(x,t), \tag{3.121}$$

where $q_y(x,t)$ is related to the external applied bending moment by the PZT actuator patch by the relationship

$$q_y(x,t) = -\frac{\partial}{\partial x}\left(\frac{\partial}{\partial x}M_{z-pzt}\right). \tag{3.122}$$

M_{z-pzt} is the moment induced by the PZT actuator patch and may be assumed to be a constant over the extent of the patch. The model adopted here is a rather simplified model. Extensive models of such actuators have been developed and for a complete discussion of such models the reader is referred to the work by Crawley and Anderson (1990).

The moment induced by the actuator patch (Figure 3.7) may be modelled as

$$M_{z-pzt} = -\mu\rho A\left[H(x-x_1) - H(x-x_2)\right]V(t), \tag{3.123}$$

where the patch is attached from $x = x_1$ to $x - x_2$, $V(t)$ is the voltage input to the actuator patch and μ is the moment of the patch per unit linear density of the beam and per unit of applied voltage for the actuator patch.

Assuming a distribution of patches each of width Δx_k, $k = 1\ldots K$, located at $x = x_k$, the distributed exiting force is expressed as

$$q_y(x,t) = -\frac{\partial}{\partial x}\left(\frac{\partial}{\partial x}\sum_{k=1}^{K}M_{z-pzt}^{k}\right). \tag{3.124}$$

Following the procedure outlined in the preceding section to find the general solution, we assume the solution

$$w(x,t) = \sum_{n=1}^{\infty}\phi_n(x,t) = \sum_{n=1}^{\infty}\eta_n(x)P_n(t), \tag{3.73}$$

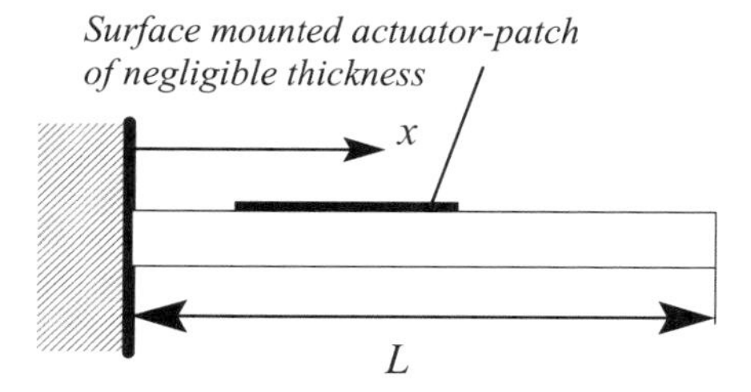

Figure 3.7 Illustration of a beam with a typical surface-mounted PZT actuator patch

where

$$P_n(t) = A_n \cos(\omega_n t + \phi_n).$$

The distributed excitation force $q_y(x,t)$, is also expanded in an orthogonal series similar to the assumed solution given by equation (3.107).

For a single patch, the equations for modal amplitudes, $P_n(t)$ are

$$\alpha_{nn}\frac{d^2 P_n(t)}{d^2 t} + \alpha_{nn}\omega_n^2 P_n(t) = \frac{1}{\rho A}\int_0^L q_y(x,t)\,\eta_n(x)\,dx = \mu F_n V(t) \tag{3.125}$$

where

$$F_n = \int_0^L (H(x-x_1) - H(x-x_2))\frac{d^2}{dx^2}\eta_n(x)\,dx = \frac{d}{dx}\eta_n(x_2) - \frac{d}{dx}\eta_n(x_1).$$

For a distribution of patches with identical characteristics, the equations for modal amplitudes $P_n(t)$ are

$$\alpha_{nn}\frac{d^2 P_n(t)}{d^2 t} + \alpha_{nn}\omega_n^2 P_n(t)$$
$$= \frac{1}{\rho A}\int_0^L q_y(x,t)\,\eta_n(x)\,dx = \mu\sum_{k=1}^{K} F_{nk}V_k(t). \tag{3.126}$$

Assuming that the patch width Δx_k is small, the voltage to force transformations coefficients may be expressed as

$$F_{nk} = \left.\frac{d^2}{dx^2}\eta_n(x)\right|_{x=x_k}\Delta x_k.$$

The equation for the modal amplitudes $P_n(t)$ may be written as

$$\alpha_{nn}\frac{d^2 P_n(t)}{d^2 t} + \alpha_{nn}\omega_n^2 P_n(t) = \mu\sum_{k=1}^{K} F_{nk}V_k(t) = Q_n(t). \tag{3.127}$$

If one requires that the control inputs only influence the first few modes, it may be assumed that

$$Q_n(t) \equiv \mu\sum_{k=1}^{K} F_{nk}V_k(t) = u_n(t), \quad \text{for} \quad n = 1\ldots m, \tag{3.128a}$$

and

$$Q_n(t) \equiv \mu\sum_{k=1}^{K} F_{nk}V_k(t) = 0, \quad n = m+1\ldots N. \tag{3.128b}$$

Only the first N modes are considered in the analysis. The vector of required exciting forces may be expressed in matrix form as

$$\mathbf{Q} = \mu\mathbf{F}\mathbf{V} = \begin{bmatrix}\mathbf{I}\\ \mathbf{0}\end{bmatrix}\mathbf{u}. \tag{3.129}$$

Assuming that the vector applied voltages, $\mathbf{V}$ may be expressed, in terms of a weighting matrix $\mathbf{W}$ and an auxiliary set of control inputs **u**, as,

$$\mathbf{V} = \mathbf{W}\mathbf{u}. \tag{3.130}$$

Thus, the voltage inputs to the *PZT* patches are constructed as weighted linear combinations of the control inputs. The same effect could be realized by shaping the patch, although there are limitations with such an approach (see for example Clark and Burke, 1996; Kim *et al.*, 2001; Preumont *et al.*, 2003). The latter approach based on shaping of the patch is particularly suitable for micro-electro-mechanical implementation of piezoelectric actuation of structures etched in silicon. A possible optimum solution for the weighting matrix **W** may be obtained as

$$\mathbf{W} = \left(\mathbf{F}^T\mathbf{F}\right)^{-1}\frac{1}{\mu}\mathbf{F}^T\begin{bmatrix}\mathbf{I}\\ \mathbf{0}\end{bmatrix}. \tag{3.131}$$

Consequently,

$$\mathbf{Q} = \mathbf{F}\left(\mathbf{F}^T\mathbf{F}\right)^{-1}\mathbf{F}^T\begin{bmatrix}\mathbf{I} & \mathbf{0}\end{bmatrix}^T\mathbf{u}. \tag{3.132}$$

The equations of motion may be expressed in matrix notation as

$$\mathbf{M}\ddot{\mathbf{P}} + \mathbf{K}\mathbf{P} = \mathbf{F}\left(\mathbf{F}^T\mathbf{F}\right)^{-1}\mathbf{F}^T\begin{bmatrix}\mathbf{I} & \mathbf{0}\end{bmatrix}^T\mathbf{u}. \tag{3.133}$$

In the above equation, **M** is the $N \times N$ diagonal generalized mass matrix, **K** is the $N \times N$ diagonal stiffness matrix and **u** is the vector of m control inputs, which predominantly influences the first m modes of vibration. It may be noted that the control inputs influence the other modes as well, and this leads to the problem of *control spillover*, particularly when one wants to construct a reduced order model. There are several approaches to reduce spillover effects, primarily by using the sensors and actuators at optimal locations (DeLorenzo, 1990; Kang *et al.*, 1996; Sunar and Rao, 1996; Balas and Young, 1999), which will be discussed in Chapter 9.

The effects of damping have not been ignored in the above analysis and are included in the modal equations by assuming the damping to be entirely due to structural effects. Thus, the generalized damping is assumed to be proportional to the generalized stiffness. A more significant aspect of the modelling technique presented here is that, using the methods discussed earlier, the equations for the modal amplitudes may be expressed in state space form as

$$\dot{\mathbf{x}} = \mathbf{A}\mathbf{x} + \mathbf{B}\mathbf{u}. \tag{3.134}$$

When this is done, a whole plethora of techniques for the synthesis of feedback control systems for controlling the behaviour of the vibration of the structure, particularly at relatively low frequencies, may be applied to design 'smart' or actively controlled vibrating systems that meet a set of specified performance requirements. These and other aspects of the design of feedback controller for flexible structures have been discussed by several authors (Balas, 1978a,b; Fanson and Caughey, 1990; Balas and Doyle, 1994a,b; Inman, 2001).

To close the loop and design a control system, a control law must first be established. A state feedback control law of the form

$$\mathbf{u} = -\mathbf{K}\mathbf{x} = -\mathbf{K}_1\mathbf{P} - \mathbf{K}_2\dot{\mathbf{P}} \tag{3.135}$$

is assumed. It is also assumed that the closed-loop damping in all the eigenvalues is increased. The structure is now endowed with sufficient damping that it will not vibrate when subjected to an oscillating force input in the vicinity of the natural frequencies. Moreover, it is specified that the feedback is allowed to marginally alter the natural frequencies of the first m modes. In the first instance, it will be assumed that m is equal to 1. This means that while the controller would be able to alter the damping in the first mode relatively easily, it is relying on the spillover effect to increase the damping in the other modes. It is natural to expect that the principal feedback would be the modal displacement amplitude and modal velocity amplitude of the first mode. To construct measurements of the modal displacement amplitude, it will be assumed that an array of displacement transducers is used. A proportional-derivative controller is assumed to be used to provide the feedback.

Assuming that L sensors are used to measure the displacements at a distributed set of locations $x = x_k$, $k = 1 \ldots L$ along the beam, the measurements may be expressed as

$$y_j = w\left(x_j, t\right) = \sum_{n=1}^{\infty} \phi_n\left(x_j, t\right) = \sum_{n=1}^{N} \eta_n\left(x_j\right) P_n(t)$$

or

$$y_j = \begin{bmatrix} \eta_1\left(x_j\right) & \eta_2\left(x_j\right) & \cdots & \eta_N\left(x_j\right) \end{bmatrix} \begin{bmatrix} P_1(t) \\ P_2(t) \\ \vdots \\ P_N(t) \end{bmatrix}.$$

In matrix form,

$$\begin{bmatrix} y_1 \\ y_2 \\ \vdots \\ y_L \end{bmatrix} = \begin{bmatrix} \eta_1(x_1) & \eta_2(x_1) & \cdots & \eta_N(x_1) \\ \eta_1(x_2) & \eta_2(x_2) & \cdots & \eta_N(x_2) \\ \cdots & \cdots & \cdots & \cdots \\ \eta_1(x_L) & \eta_2(x_L) & \cdots & \eta_N(x_L) \end{bmatrix} \begin{bmatrix} P_1(t) \\ P_2(t) \\ \vdots \\ P_N(t) \end{bmatrix} \tag{3.136}$$

or

$$\mathbf{y} = \mathbf{NP}. \tag{3.137}$$

It is assumed that there is a matrix $\mathbf{S}$ such that multiplying the measurements with $\mathbf{S}$ results in the first q modal amplitudes. Hence, a set of observations of the first few modal amplitudes can be expressed in terms of the measurements as

$$\mathbf{z} = \begin{bmatrix} \mathbf{I} & \mathbf{0} \end{bmatrix} \mathbf{P} = \mathbf{Sy} = \mathbf{SNP}. \tag{3.138}$$

It follows that

$$\mathbf{SN} = \begin{bmatrix} \mathbf{I} & \mathbf{0} \end{bmatrix}. \tag{3.139}$$

An optimum and approximate solution for $\mathbf{S}$ is given by

$$\mathbf{S} = \begin{bmatrix} \mathbf{I} & \mathbf{0} \end{bmatrix} \mathbf{N}^T \left(\mathbf{NN}^T\right)^{-1}. \tag{3.141}$$

The observation model may be expressed as a weighted linear combination of the measurements and consequently written as

$$\mathbf{z} = \begin{bmatrix} \mathbf{I} & \mathbf{0} \end{bmatrix} \mathbf{N}^T \left(\mathbf{NN}^T\right)^{-1} \mathbf{NP}. \tag{3.142}$$

As in the case of the control inputs, the measurements are a function of all the N modal amplitudes and not just the first q modal amplitudes. This leads to the problem of *observation spillover*, particularly when one wants to construct a reduced-order model, which will be discussed in Chapter 9.

The control law may now be expressed as

$$\mathbf{u} = -\mathbf{K}_{1z}\mathbf{z} - \mathbf{K}_{2z}\dot{\mathbf{z}}. \tag{3.143}$$

Assuming that $m = q = 1$, i.e. only the first mode is sensed and only the first control input is used, the control law is written as

$$u = -K\left(z - \beta\dot{z}\right). \tag{3.144}$$

The problem reduces to one of selecting the gain K for some fixed value of the constant β. The problem may be solved by using the root locus method. We shall illustrate the process with an example of a uniform cantilever beam.

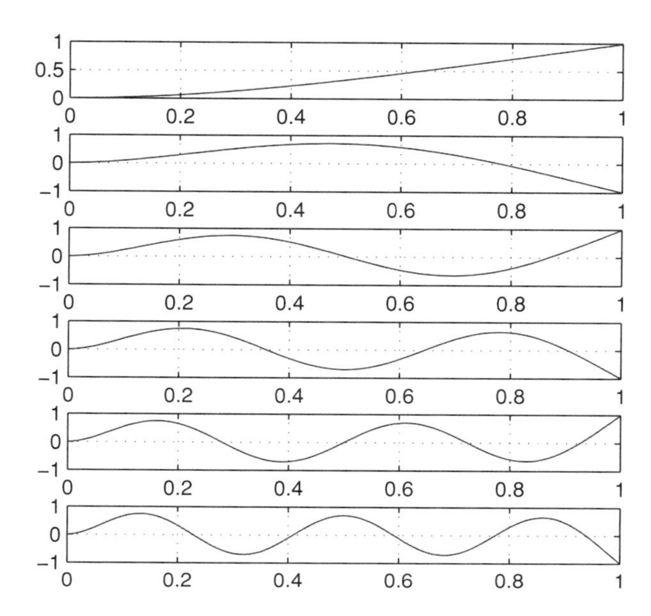

Figure 3.8 Modes of a cantilever

The exact modes of a uniform cantilever beam are well known, and it is possible to construct the relevant system matrices in MATLAB,® assuming the beam equations are in non-dimensional form both in the spatial and in the time coordinates. The first six modes were used to construct the state space model, and these are shown in Figure 3.8.

The structural damping coefficient, i.e. the ratio of the generalized damping to the generalized stiffness, was assumed to be 0.001. It was assumed that the PZT patches are located at the zeros of the fourth and fifth modes, and the width of each patch is one-hundredth of the beam length. An equal number of displacement sensors are used, but these are located at the points where the modal amplitudes are a maximum, starting with the tip of the cantilever. The number of sensor or actuator patches is seven and exceeds the number of assumed modes.

There are a number of possible choices of the design objectives. Typically for vibration control, it is necessary to shift the natural frequencies of vibration (Chen *et al.*, 1997; Leng *et al.*, 1999; Ghoshal *et al.*, 2000) in order to avoid resonance. To suppress the transients, it is essential to increase the damping in the lightly damped modes. The design objective in the above example is to seek a design with a damping ratio close to 0.7 without altering the natural frequency of the beam too much so that it is still sufficiently less than the natural frequency of the second mode. Initial plots of the root locus with β equal to 0.01, 0.1 and 1 indicated that in the first case the gains were unreasonably large for a damping ratio of close to 0.7. In the first case, it was found that the natural frequencies were increased substantially. For this reason an initial value of $\beta = 0.1$ was chosen, as in this case the root locus plot gave reasonable results.

The initial root locus plot obtained is illustrated in Figure 3.9. The plot clearly illustrates that only the first mode of the cantilever is influenced by the introduction of the feedback controller. Closing in on the region dominated by the first mode's locus, we obtain the plot shown in Figure 3.10. The chosen design point gives a damping ratio of 0.702, a non-dimensional gain of 51.7 and natural frequency of 14.7 (compared to the open loop first mode natural frequency of 3.5156 and a second mode natural

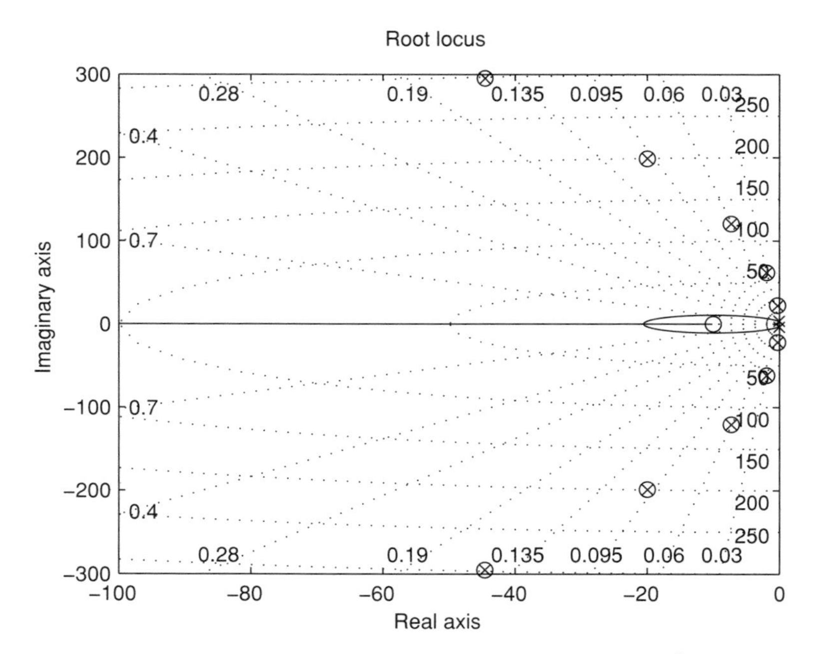

Figure 3.9 Initial root locus plot obtained by using MATLAB®

frequency of 22.0336). A lower gain of 19.2 results in a damping ratio of 0.407 and a natural frequency of 9.43. However, the overshoot corresponding to the lower gain is 24.7%, which may be just about unacceptable.

In the above example, we have used a simple displacement sensor. In practice, one may be constrained to employ a PZT sensor that gives a charge output proportional to the bending strain or curvature of the

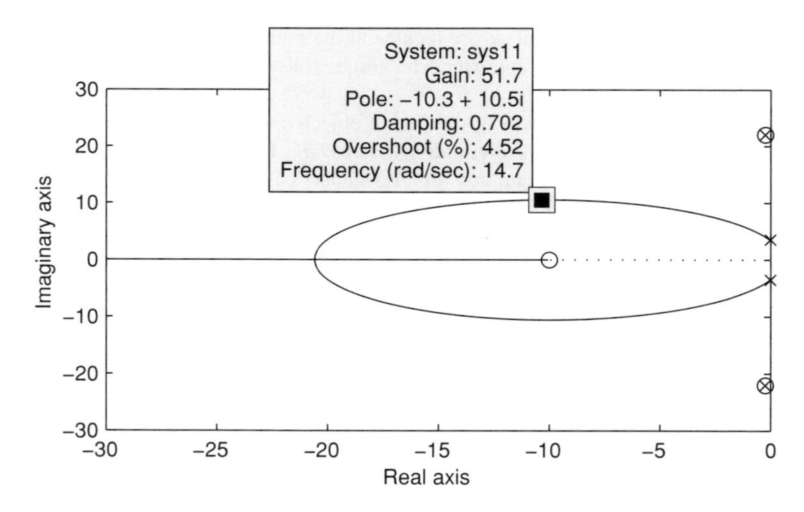

Figure 3.10 Root locus of the first mode

beam. Thus, equation (3.136) is modified and is given by

$$\begin{bmatrix} y_1 \\ y_2 \\ \vdots \\ y_L \end{bmatrix} = \begin{bmatrix} \eta_1''(x_1) & \eta_2''(x_1) & \cdots & \eta_N''(x_1) \\ \eta_1''(x_2) & \eta_2''(x_2) & \cdots & \eta_N''(x_2) \\ \cdots & \cdots & \cdots & \cdots \\ \eta_1''(x_L) & \eta_2''(x_L) & \cdots & \eta_N''(x_L) \end{bmatrix} \begin{bmatrix} P_1(t) \\ P_2(t) \\ \vdots \\ P_N(t) \end{bmatrix}. \tag{3.145}$$

The double prime $()''$, denotes double differentiation with respect to the spatial coordinate.

When using a PZT sensor, it may also be advantageous to co-locate them at the same points where the PZT actuators are located. When this is done, one may re-compute the root locus plot in Figure 3.10. The modified root locus plot is exactly identical to the one in Figure 3.10. The observation model in this case is identical to the one obtained with the displacement sensors, as the number of sensors exceeds the number of modes.

In the above synthesis example, the influence of spillover as well as several other practical issues such as the dynamics of the actuators and sensors, and the presence of disturbances was ignored. These issues will be considered in some detail in Chapter 9.

Considering the second and third mode, the same SISO methodology may be used to determine the control gain. As the modal amplitudes of higher modes are present in the observations of the first three modal amplitudes, it is not really desirable to use very high gains in the feedback. Thus, the final choices of gains for the first, second and third mode feedback were chosen as 1.5, 1.5 and 1, respectively. The complete multi-input–multi-output controller is expressed as

$$\mathbf{u} = -\mathbf{K}_{1z}(\mathbf{z} + \beta\dot{\mathbf{z}}), \quad \beta = \mathbf{I}, \quad \mathrm{K}_{1z} = \begin{bmatrix} 1.5 & 0 & 0 \\ 0 & 1.5 & 0 \\ 0 & 0 & 1 \end{bmatrix}. \tag{3.146}$$

The map of the closed-loop poles and zeros is shown in Figure 3.11. It is now assumed that the number of sensor/actuator patches is reduced to just five, i.e. one less than the number of assumed modes. The sensors located at the two zeros of the fifth mode closest to the tip were not used. It is now not possible

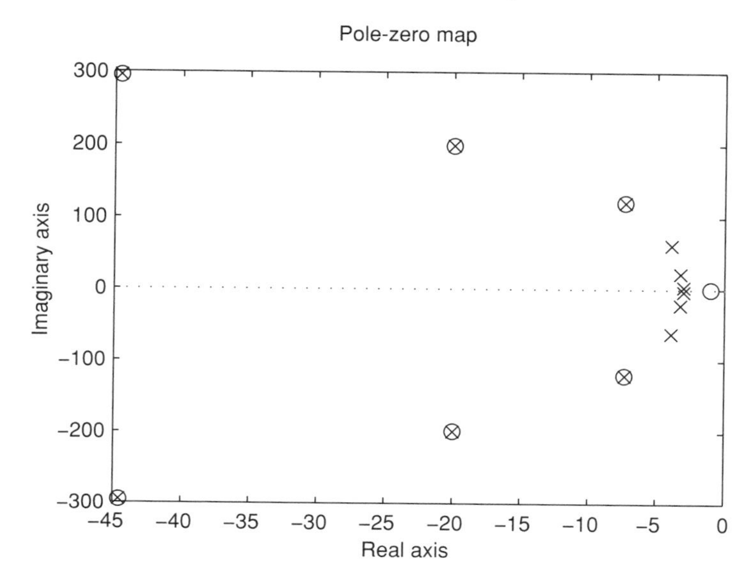

Figure 3.11 Pole-zero map of the final closed-loop system

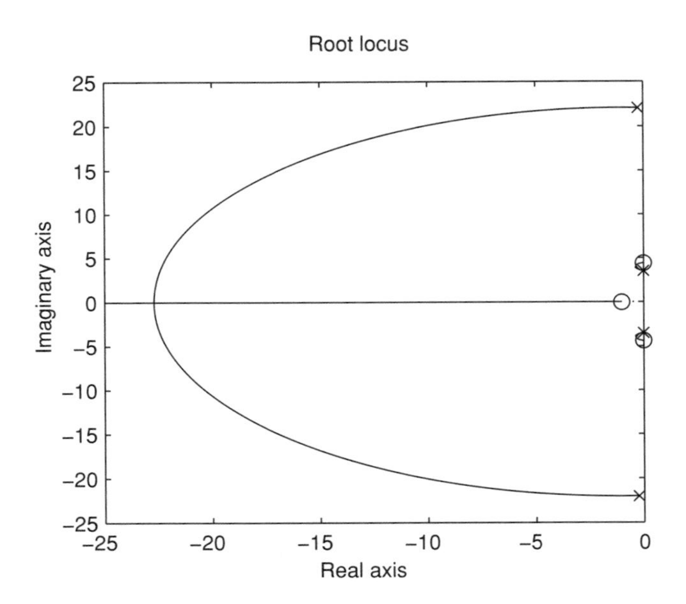

Figure 3.12 Root locus of the first mode with the number sensors reduced from 7 to 5

to design the closed-loop controller, one mode at a time, unless one ignores the control and observation spillover.

First, the root locus plot corresponding to Figure 3.10 is obtained and this is shown in Figure 3.12. It is clear that the spillover has had a profound effect on the closed-loop behaviour. In fact, the first mode is practically unaffected by the feedback. Assuming the same controller as the one given by equation (3.146), the corresponding pole-zero map is obtained, which is shown in Figure 3.13. It is

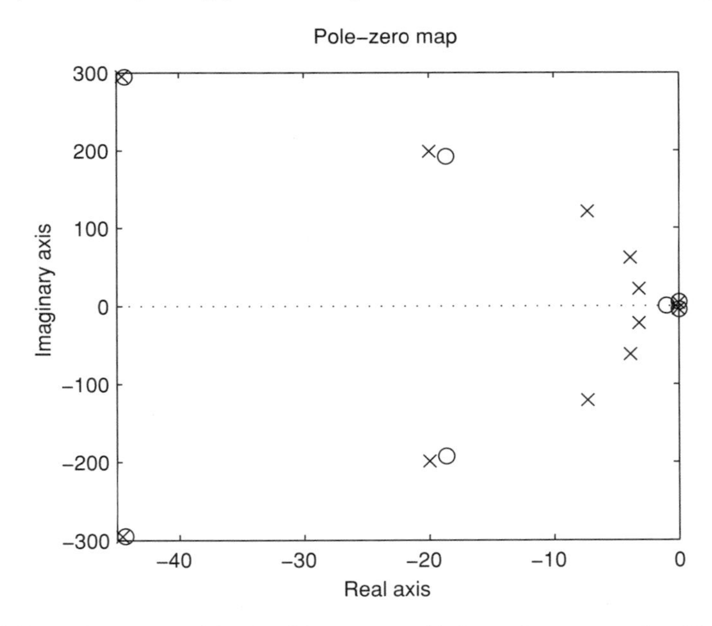

Figure 3.13 Pole-zero map of the closed-loop system with the number sensors reduced from 7 to 5

seen that the feedback does influence the two poles, which continue to remain in the vicinity of the imaginary axis and are therefore lightly damped and susceptible to external excitation. Although the two unused sensors were located at the zeros of the fifth mode, they are also relatively quite close to the tip of the cantilever. For this reason, they play a key role in the control of the first mode. Thus, it would be preferable to retain these two sensors and remove the two sensors located at the zeros of the fifth mode closest to the root, if there is a need to restrict the number of sensors to five. This example illustrates the importance of the control and observation spillover. Although we have employed the root locus technique to synthesize the controller, classical controller synthesis techniques such as Bode plots or the polar plots and the Nyquist criterion could also have been used to design and validate the controller. These are normally used to validate a controller that will be required to meet additional performance requirements.

Exercises

1. Consider a uniform beam of length L and assume that a tensile longitudinal force P is acting at the free end of the beam. The beam is assumed to have a cross-sectional area A, flexural rigidity EI_{zz} and material of density ρ. The transverse deflection of the beam along the beam axis is assumed to be $w(x, t)$. Assume that for a slender beam in transverse vibration the stress and strain are related according to the Bernoulli–Euler theory of bending. For small deformations and for a prismatic beam, we have the static bending relations:

$$M_z = -EI_{zz}\frac{\partial^2 w(x,t)}{\partial x^2}.$$

(i) Show that the transverse deflection satisfies

$$\rho A\frac{\partial^2 w(x,t)}{\partial t^2} = \frac{\partial}{\partial x}\left(P\frac{\partial w(x,t)}{\partial x}\right) - \frac{\partial^2}{\partial x^2}\left(EI_{zz}\frac{\partial^2 w(x,t)}{\partial x^2}\right).$$

(ii) Make suitable assumptions and generalize this equation for a laminated beam where some of the lamina could be made from a piezoelectric ceramic.
(iii) Assume that distribution of displacement sensors and PZT patch actuators are available and design a feedback controller so that the closed-loop natural frequencies for the first two modes of a laminated cantilevered beam with a longitudinal *compressive* force acting at the free end are equal to those of a beam without the force. State all the assumptions you have made in designing the controller.
(iv) Consider the behaviour of the closed-loop flexible beam structure with the feedback controller in place and validate the performance of the controller by considering at least the first four normal modes.
(v) Another structural engineer examining your design makes the observation 'Your design is completely equivalent to a beam on a uniform elastic foundation'. Show with suitable equations and an appropriate analysis of the first two modes of a beam on a uniform elastic foundation whether the observation is correct or incorrect. Explain your answer physically.

2.
(i) Derive the frequency equation for the transverse vibration of a uniform beam fixed or clamped at one end and sliding at the other, as shown in Figure 3.14.

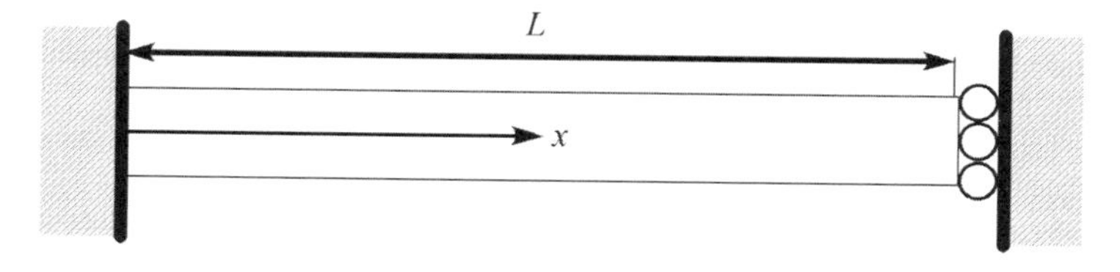

Figure 3.14 Beam fixed at one end and sliding at the other

(ii) Assume that a distribution of displacement sensors and PZT patch actuators are available and design a feedback controller so that the closed-loop natural frequencies for the first two modes of a uniform cantilever beam are equal to those of a beam with the same boundary conditions as in part (i) and with the same length, mass and elastic properties. State all the assumptions you have made in designing the controller.
(iii) Consider the behaviour of the closed-loop flexible beam structure with the feedback controller in place and validate the performance of the controller by considering at least the first four normal modes.
(iv) Discuss at least one application of the active controller designed and validated in parts (ii) and (iii).

3. Obtain the frequency equation and the lowest buckling load for the transverse vibration of a beam hinged at one end and sliding at the other, longitudinally loaded at the sliding end by a compressive force P.

4. A single-span bridge is idealized as simply supported uniform homogeneous beam and has on it a load moving with a constant velocity v, from the end $x = 0$ to the end $x = L$. The bridge may be assumed to be initially undisturbed and therefore at rest. Find the deflection of the bridge.

5. Show that the equation of motion of an element δx of an uniform bar subjected to an exciting force $g_e(x)\sin(\omega t)$, $g_e(x)$ being continuous over the whole length of the bar, is

$$\rho A \frac{\partial^2 u(x,t)}{\partial t^2} = EA \frac{\partial^2 u(x,t)}{\partial x^2} + g_e(x)\sin(\omega t).$$

Hence, show that a constant longitudinal external force acting on any cross-section of the bar has no influence on the natural frequencies of a uniform bar in longitudinal vibration.

Solve the governing equation when $g_e(x)$ is a constant for a bar with both ends fixed and determine the exciting force acting on any cross-section of the bar.

Also show that the displacement response of the bar is

$$\rho A \omega^2 u(x,t) = \left(\frac{U(x)}{\sin\left(\frac{\omega L}{a}\right)} - 1 \right) g_e \sin(\omega t),$$

where

$$U(x) = \sin\left(\frac{\omega L}{a}\right)\cos\left(\frac{\omega x}{a}\right) - \left\{\cos\left(\frac{\omega L}{a}\right) - 1\right\}\sin\left(\frac{\omega x}{a}\right).$$

6. The main propeller shafting of ships suffers from excessive axial vibrations under certain conditions. In a simple representation of the system the propeller mass M_P may be considered to be a rigid attachment to a long uniform bar of length L, area of cross-section A, Young's modulus E and mass density ρ.

To the other end of the bar is attached a rigid mass M_Q restrained on a spring of stiffness k, representing the main gear box, the thrust collar, the thrust block casing and the associated mountings.

(i) Derive the frequency equation for the system.
(ii) You are asked to design an active controller to reduce the excessive vibrations. Making suitable assumptions, design an appropriate control system to attenuate the axial vibrations.

7.
(i) Derive and solve the frequency equation for small torsional vibrations of a system consisting of two discs of moment of inertia I_1 and I_2 connected by a uniform shaft of length L, polar second moment of the area of cross-section J, modulus of rigidity G and mass density ρ.
(ii) When calculating the natural frequencies of practical systems such as a diesel engine-driven generator, it is usual to lump the inertias of the shafts with those of the rotors at each end.

Show that this is valid assumption when the lowest natural frequencies of the individual shafts are much higher than any system natural frequency of interest.
(iii) Using the results found above, derive expressions for the increases in I_1 and I_2 needed in the present case.

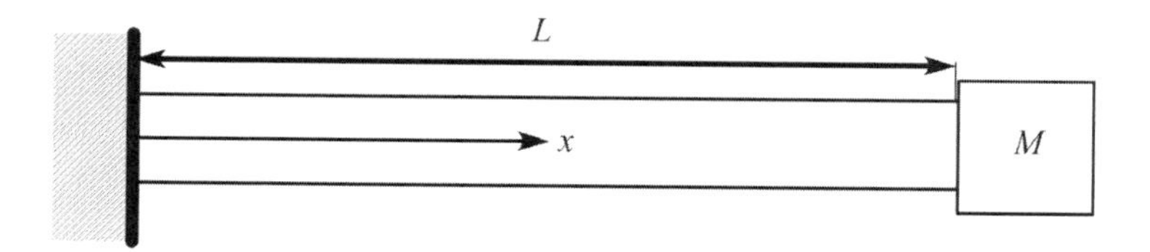

Figure 3.15 Beam free at one end and carrying a body at the other

8.
(i) Derive the frequency equation for a the transverse vibration of a beam of length L, bending rigidity EI_{zz}, mass density ρ and cross-sectional area A, which is free at one end and carries a metal block at the other. The block has a mass M and mass moment of inertia I_{M} about a principal axis passing through the end of the beam and perpendicular to the plane of the paper. The beam is illustrated in Figure 3.15. Obtain the first natural frequency and the corresponding mode shape of the beam when $I_{\mathrm{M}} = 0$ and $M = \rho AL$. Sketch the mode shape and compare with the case when $M = 0$.
(ii) Assume that a distribution of displacement sensors and PZT patch actuators are available and design a feedback controller so that the closed-loop natural frequencies for the first two modes of a uniform cantilever beam carry a rigid body of mass M at its free end and are equal to those of a uniform cantilever beam with the same length, mass and elastic properties but without the additional rigid body at its free end. State all the assumptions you have made in designing the controller.
(iii) Consider the behaviour of the closed-loop flexible beam structure with the feedback controller in place and validate the performance of the controller by considering at least the first four normal modes.

9. The equation for the longitudinal oscillation of a slender bar of length L, with viscous damping and a form of external loading, is known to be

$$\rho \frac{\partial^2 u}{\partial t^2} = E \frac{\partial^2 u}{\partial x^2} - 2\rho\alpha \frac{\partial u}{\partial t} + \frac{P_0}{L} p_e(x) f_e(t),$$

where the external loading per unit length is a separable function of x and t.
(i) Let

$$u(x,t) = \sum_{n=1,2,3..}^{\infty} \phi_n(x) q_n(t)$$

and show that

$$u(x,t) = \sum_{j=1,2,3..}^{\infty} \phi_j(x) c_j(t) + \frac{P_0}{\rho\sqrt{1-\zeta_{j^2}}} \times \sum_{j=1,2,3,...}^{\infty} \frac{b_j \phi_j(x)}{\omega_j} \int_0^t f(t-\tau) \exp(\zeta_j \omega_j \tau) \sin\left(\omega_j \tau \sqrt{1-\zeta_j^2}\right) \mathrm{d}\tau,$$

where

$$b_j = \frac{1}{\beta_j L} \int_0^L p(x) \phi_j(x) \,\mathrm{d}x, \quad \beta_j = \int_0^L \phi_j^2(x) \,\mathrm{d}x \quad \text{and} \quad \int_0^L \phi_i(x) \phi_j(x) \,\mathrm{d}x = 0, \quad i \neq j.$$

Hence or otherwise obtain expressions for ω_j, ζ_j and the stress at any point x, and state the conditions defining the function $\phi_j(x)$.
(ii) Assuming that the bar is initially at rest and stress free at both ends and $f(t)$ is a unit step, determine the response of the bar.

(iii) Assume that $f(t)$ is a controllable input generated by a distribution of actuators and design a feedback controller to increase the first natural frequency by 20% and the second natural frequency by 10%, without influencing the third and fourth natural frequencies.

10.

(i) A uniform simply supported beam of length L and flexural rigidity EI is subjected to a concentrated force F_0 applied at a distance ξ from one end.

Show that the response of the beam when the beam is initially at rest and the load is suddenly removed is given by

$$w(x,t) = \frac{2F_0L^3}{\pi^4 EI} \sum_{i=1,2\ldots}^{\infty} \frac{1}{i^4} \sin\left(\frac{i\pi\xi}{L}\right) w_i(x,t),$$

where

$$w_i(x,t) = \sin\left(\frac{i\pi x}{L}\right) \cos(p_i t) \quad \text{and} \quad p_i = \left(\frac{i\pi}{L}\right)^2 \sqrt{\frac{EI}{\rho A}}.$$

(ii) The transverse bending deflection v(z) of a uniform cantilever at a distance z from the built-in end due to the load P at a distance ζ from the same end is given by

$$v(z) = \frac{Pz^2(3\zeta - z)}{6EI}, \quad \zeta \geq z,$$

(iii) Obtain the influence coefficients (deflections for a unit load) associated with stations one-quarter of the length L from each end.

(iv) A cantilever of mass m is represented by two particles, one of mass $m_1 = m(1+\varepsilon)/2$ placed a distance $L/4$ from the tip, the other of mass $m_2 = m(1-\varepsilon)/2$ placed at a distance $L/4$ from the built-in end. Write the equations in matrix form, find the characteristic equation of the system and hence find the fundamental natural frequency, in terms of EI/mL^3, for the case when $\varepsilon = 1/8$.

(v) Obtain the first and the second natural frequencies of a uniform cantilever with its mass lumped, at two points at equal distances $L/4$ from the built-in end and from the tip, by applying the Rayleigh–Ritz method.

11.

(i) Obtain the first and the second natural frequencies of the uniform cantilever of flexural stiffness EI, length L and mass m, supported at the free end by a spring of stiffness k and illustrated in Figure 3.16, by applying the Rayleigh–Ritz method.

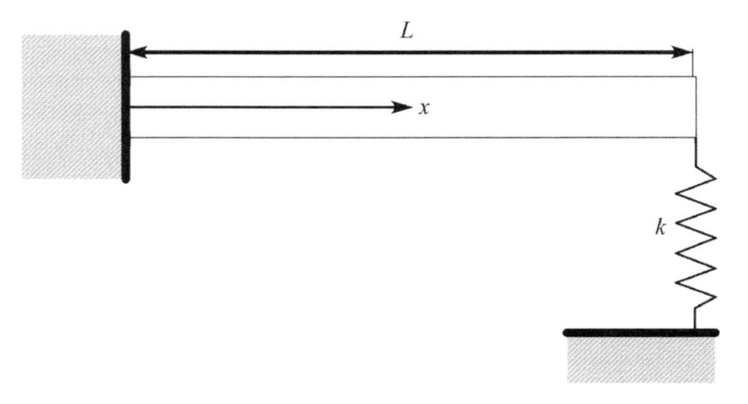

Figure 3.16 Cantilever propped up at the free end

Plot the results as a function of the parameter $\beta = kL^2/EI, 0 \le \beta \le 200$ and comment on the influence of the flexibility of the support.
(ii) Assume that a distribution of displacement sensors and PZT patch actuators are available and design a feedback controller so that the closed-loop natural frequencies for the first two modes of a uniform cantilever beam are equal to those of a beam with the same boundary conditions as in part (i) and with the same length, mass and elastic properties. State all the assumptions you have made in designing the controller.
(iii) Consider the behaviour of the closed-loop flexible beam structure with the feedback controller in place and validate the performance of the controller by considering at least the first four normal modes.
(iv) Discuss at least one application of the of the active controller designed and validated in parts (ii) and (iii).

12. Consider a non-uniform beam of length L and assume that a compressive longitudinal force P_0 is acting at one end of the beam. The beam is assumed to have a cross-sectional area $A(x)$, flexural rigidity $EI_{zz}(x)$ and material of density $\rho(x)$, which are all functions of the axial coordinate x along the beam, where the origin of the coordinate system is located at the same end where the compressive force P_0 is acting. The transverse deflection of the beam along the beam axis is assumed to be $w(x,t)$. Assume that for a slender beam in transverse vibration, the stress and strain are related according to the Bernoulli–Euler theory of bending. The beam is acted on by a distributed axial loading $q(x)$. The axial force acting at any location is given by

$$P(x) = P_0 + \int_0^x q(x)\,\mathrm{d}x.$$

(i) Show that the transverse deflection satisfies

$$\rho(x)A(x)\frac{\partial^2 w(x,t)}{\partial t^2} + \frac{\partial}{\partial x}\left(P(x)\frac{\partial w(x,t)}{\partial x}\right) + \frac{\partial^2}{\partial x^2}\left(EI_{zz}(x)\frac{\partial^2 w(x,t)}{\partial x^2}\right) = 0.$$

Hence obtain the governing equation of motion at the instant of buckling.
(ii) Write the governing equation in terms of the non-dimensional independent coordinate $\xi = x/L$, and assume that $EI_{zz}(x)$ and $P(x)$ and the mode of deflection $\eta(x)$ are all polynomial functions of ξ (Li, 2009), given by

$$EI_{zz} = \sum_{i=0}^{m} b_i\xi^i, \quad P = P_0\sum_{i=0}^{n}\beta_i\xi^i \quad \text{and} \quad \eta = \sum_{i=0}^{p} a_i\xi^i.$$

Assume that $\eta(x)$ corresponds to the approximate first mode and that $p = 4$.
Show that the coefficients b_i and β_i satisfy the following recurrence relations:

$$b_{n+2} = -P_0L^2\beta_n/(3n+12),$$

$$b_{n+1} = -\frac{a_3}{2a_4}b_{n+2} - P_0L^2\frac{(3a_3\beta_n + 4a_4\beta_{n-1})}{12(n+3)a_4},$$

$$b_i = -\frac{a_3}{2a_4}b_{i+1} - \frac{a_2}{6a_4}b_{i+2} - P_0L^2\frac{(3a_3\beta_{i-1} + 4a_4\beta_{i-2} + 2a_2\beta_i + a_1\beta_{i+1})}{12(i+2)a_4},$$

$$i = n, n-1, n-2, \cdots, 2,$$

$$b_1 = -\frac{a_3}{2a_4}b_2 - \frac{a_2}{6a_4}b_3 - P_0L^2\frac{(3a_3\beta_0 + 2a_2\beta_1 + a_1\beta_2)}{36a_4},$$

$$b_0 = -\frac{a_3}{2a_4}b_1 - \frac{a_2}{6a_4}b_2 - P_0L^2\frac{(2a_2\beta_0 + a_1\beta_1)}{24a_4}.$$

Table 3.1 Values of the coefficients a_i for beams with different boundary conditions

Case	$i=0$	$i=1$	$i=2$	$i=3$	$i=4$
C–C	0	0	1	−2	1
C–G	0	0	4	−4	1
H–H	0	1	0	−2	1
C–H	0	1	0	−3	2
H–G	0	8	0	−4	1
C–F	3	−4	0	0	1

The recurrence relations are solved sequentially for β_i, $i = n, n-1, \cdots, 1, 0$.
(iii) Assume that the boundary conditions at particular end $\xi = e$ could be one of four possibilities:
(a) clamped (C: $\eta(e) = 0$, $\mathrm{d}\eta/\mathrm{d}x|_{\xi=e} = 0$);
(b) hinged (H: $\eta(e) = 0$, $\mathrm{d}^2\eta/\mathrm{d}x^2|_{\xi=e} = 0$);
(c) free (F: $\mathrm{d}^2\eta/\mathrm{d}x^2|_{\xi=e} = 0, \mathrm{d}^3\eta/\mathrm{d}x^3|_{\xi=e} = 0$);
(d) guided (G: $\mathrm{d}\eta/\mathrm{d}x|_{\xi=e} = 0$, $\mathrm{d}^3\eta/\mathrm{d}x^3|_{\xi=e} = 0$).
Show that the coefficients a_i are given the values listed in Table 3.1.

Hence obtain the conditions for buckling.

(iv) It is proposed to strengthen the beams by actively restraining the beam in the transverse direction at a finite number of locations. Assume a 20% increase in the buckling load is desired in each case, and design and validate a suitable distributed active controller.

(Hint: Assume a distributed controller so the closed-loop dynamics takes the form

$$\rho(x) A(x) \frac{\partial^2 w}{\partial t^2} + \frac{\partial}{\partial x}\left(P(x) \frac{\partial w}{\partial x}\right) + \frac{\partial^2}{\partial x^2}\left(EI_{zz}(x) \frac{\partial^2 w}{\partial x^2}\right) + K(x) w = 0.$$

Assume that

$$K = \sum_{i=0}^{k} \kappa_i \xi^i$$

and obtain the corresponding closed-loop recurrence relations for the coefficients b_i in terms of β_i and κ_i, and P_c, the closed-loop buckling load, and solve for the controller coefficients κ_i.)

References

Balas, M. J. (1978a) Active control of flexible systems, *Journal of Optimization Theory and Applications* 25, 415–436.

Balas, M. J. (1978b) Modal control of certain flexible dynamic system, *SIAM Journal on Control and Optimization* 16, 450–462.

Balas, M. J. and Doyle, J. C. (1994a) Control of lightly damped, flexible modes in the controller crossover region, *Journal of Guidance, Control, and Dynamics* 17, 370–377.

Balas, M. J. and Doyle, J. C. (1994b) Robustness and performance trade-offs in control for flexible structures, *IEEE Transactions on Control Systems Technology* 2, 352–361.

Balas, M. J. and Young, P. M. (1999) Sensor selection via closed-loop control objectives, *IEEE Transactions on Control Systems Technology* 7, 692–705.

Chen, S. H., Wang, Z. D. and Liu, X. H. (1997) Active vibration control and suppression for intelligent structures, *Journal of Sound and Vibration* 200, 167–177.

Clark, R. L. and Burke, S. E. (1996) Practical limitations in achieving shaped modal sensors with induced strain materials, *ASME Journal of Vibration and Acoustics* 118, 668–675.

Clark, R. L., Saunders, W. R. and Gibbs, G. P. (1998) *Adaptive Structures, Dynamics and Control*. John Wiley & Sons Inc., New York.
Crawley, E. F. and Anderson, E. H. (1990) Detailed models of piezoceramic actuation of beams, *Journal of Intelligent Material System and Structure* 1, 4–25.
DeLorenzo, M. L. (1990) Sensor and actuator selection for large space structure control, *Journal of Guidance, Control, and Dynamics* 13, 249–257.
Dorf, R. C. and Bishop, R. H. (2004) *Modern Control Systems*, 10th edn, Prentice-Hall, New York.
Fanson, J. L. and Caughey, T. K. (1990) Positive position feedback control for large space structures, *AIAA Journal* 28 (4), 717–724.
Franklin, G., Powell, J. D. and Emami-Naeini, A. (2005) *Feedback Control of Dynamic Systems*, 5th edn, Prentice-Hall, New York.
Friedland, B. (1987) *Control System Design*, McGraw-Hill Book Company, New York.
Frolov, K. V. and Furman, F. A. (1990) *Applied Theory of Vibration Isolation System*. Hemisphere Publishing Corporation, New York.
Gawronski, W. (1996) *Balanced Control of Flexible Structures*, Springer-Verlag, London.
Ghoshal, A., Wheater, E. A., Kumar, C. R. A. and Sundaresan, M. J. (2000) Vibration suppression using a laser vibrometer and piezoceramic patches, *Journal of Sound and Vibration* 235, 261–280.
Gorman, D. J. (1975) *Free Vibration Analysis of Beams and Shafts*, Wiley-Interscience, New York.
Inman, D. J. (2001) Active modal control for smart structures, *Philosophical Transactions of the Royal Society, Series A* 359 (1778), 205–219.
Inman, D. J. (2007) *Engineering Vibrations*, 3rd edn, Prentice-Hall, New York.
Joshi, S. M. (1989) *Control of Large Flexible Space Structures*, Springer-Verlag, Berlin.
Kang, Y. K., Park, H. C., Hwang, W. and Han, K. S. (1996) Optimum placement of piezoelectric sensor/actuator for vibration control of laminated beams, *AIAA Journal* 34, 1921–1926.
Kim, J., Hwang, J. S. and Kim, S. J. (2001) Design of modal transducers by optimizing spatial distribution of discrete gain weights, *AIAA Journal* 39, 1969–1976.
Leipholz, H. H. E. and Rohman, M. A. (1986) *Control of Structures*, Kluwer Academic, Boston.
Leng, J., Asundi, A. and Liu, Y. (1999) Vibration control of smart composite beams with embedded optical fiber sensor and ER fluid, *ASME Journal of Vibration and Acoustics* 121, 508–509.
Li, Q. S. (2009) Exact solutions for the generalized Euler's problem, *Journal of Applied Mechanics* 76, 041015-1–041015-9.
Meirovitch, L. (1967) *Analytical Methods in Vibration*, MacMillan, New York.
Meirovitch, L. (1970) *Elements of Vibration Analysis*, 2nd edn, McGraw-Hill Book Company, New York.
Meirovitch, L. (1990) *Dynamics and Control of Structures*, John Wiley & Sons Inc., New York.
Preumont, A. (2002) *Vibration Control of Active Structures, an Introduction*, 2nd edn, Kluwer Academic Publishers.
Preumont, A., François, A., De Man P. and Piefort, V. (2003) Spatial filters in structural control, *Journal of Sound and Vibration* 265, 61–79.
Rao, S. S. (1995) *Mechanical Vibrations*, 3rd edn, Prentice-Hall.
Sunar, M. and Rao, S. S. (1996) Distributed modeling and actuator location for piezoelectric control systems, *AIAA Journal* 34, 2209–2211.
Thomson, W. T. (1993) *Theory of Vibration with Applications*, 4th edn, Chapman & Hall, London.
Timoshenko, S. P. and Woinowsky-Kreiger, S. (1959) *Theory of Plates and Shells*, 2nd edn, McGraw-Hill Book Company, New York.
Timoshenko, S., Young, D. H. and Weaver, W. (1974) *Vibration Problems in Engineering*, 4th edn, John Wiley & Sons Inc., New York.
Tong, K. N. (1960) *Theory of Mechanical Vibrations*, John Wiley & Sons Inc., New York.
Tse, F. S., Morse, I. E. and Hinkle, R. T. (1979) *Mechanical Vibrations: Theory and Applications*, 2nd edn, Allyn and Bacon Inc., Boston.

Tzou, H. S. and Bergman, L. A. (1998) *Dynamics and Control of Distributed Systems*, Cambridge University Press, Cambridge.
Utkin, V. I. (1992) *Sliding Modes in Control and Optimization*, Springer-Verlag, Berlin.
Vernon, J. B. (1967) *Linear Vibration and Control System Theory with Computer applications*, John Wiley & Sons Inc., New York.

4

Dynamics of Continuous Structures

4.1 Fundamentals of Acoustic Waves

Acoustic waves are produced by the sinusoidal variation in pressure due to sinusoidal changes in particle displacement or velocity that are propagated in a medium. Sound is the auditory sensation of acoustic waves, perceived within the human ear. The sensation of sound in humans is due to the vibrations induced by acoustic waves in a thin membrane under tension, known as the *tympanic* membrane. The audible frequency bandwidth is in the 15–15 kHz range. The human ear cannot perceive acoustic wave frequencies beyond the upper limit. The sympathetic vibrations of tympanic membrane, in sympathy with the pressure variation external to it, is conveyed by a three-link mechanical amplifier, the *ossicles*, and converted into neural stimuli within the *cochlea*. These are then conveyed to the brain by the *auditory* nerve, where the sound is interpreted meaningfully. This completes the sequence of events leading to the perception of sound.

4.1.1 Nature of Acoustic Waves

One of the simplest forms of a sound wave is produced when a balloon of compressed air is released. The air that was confined under pressure is transmitted outwards in the form of a pressure pulse. A rarefaction pulse of pressure then follows this. Thus, the sound wave consists of a condensation or high-pressure pulse followed by rarefaction or low-pressure pulse. The sound wave travels in the atmosphere with a finite velocity that is approximately 331 m/s. It is a bit lower in pure oxygen (316 m/s). The speed of a sound wave in water is about 1498 m/s while it is about 1284 m/s in hydrogen. It is generally much higher in metals and is about 5100 m/s in aluminium and 5000 m/s in iron. Although the speed of a sound wave can differ widely in different materials, the mechanism of the propagation of sound is always the same and involves the spatial motion of a cyclic variation of the pressure.

Any vibrating body in contact with the atmosphere is capable of generating sound waves. A common example of this is a vibrating piston that is hurled back and forth by a mechanical arrangement such as a rotating crank and connecting rod. Thus, there exists a close relationship between the vibration of a body in the atmosphere and the generation and propagation of acoustic waves. By acoustic waves we mean the propagation of temporal variations in the local pressure, which are also propagated spatially. These pressure variations are above and below the normal undisturbed pressure in the atmosphere. Pressure is itself an omni-directional state of uniform compressive stress. Acoustic wave propagation can therefore be considered to be synonymous with the propagation of this omni-directional state of compressive stress. Acoustic waves, however, may also be directed, in that they may propagate in a preferred direction. When

Dynamics of Smart Structures Dr. Ranjan Vepa

the direction of pressure variation and the direction of propagation remain invariant with time and bear a specific invariant relationship to each other, the waves are said to be *polarized*.

4.1.2 Principles of Sound Generation

Any vibrating body in contact with atmosphere will produce a sound wave. It is this principle that is employed in the design and construction of all string-type musical instruments, drums and cymbals as well as instruments involving soundboards, such as the piano and the harp. Another mechanism that results in the generation of sound involves the conversion of a steady stream of gas into a pulsating one. Although there are many methods of achieving this type of conversion, they all involve, in some form or the other, the throttling of the air stream. This is the principle of sound generation in a siren, and the human voice, the trumpet, the trombone, the clarinet, the saxophone and other similar musical instruments operate in the same way.

4.1.3 Features of Acoustic Waves

There are a number of features of acoustic waves that permit one to distinguish between different types of sound waves:

1. The spectrum is the relative magnitudes of different single-frequency sinusoidal wave components of an acoustic wave.
2. The intensity of an acoustic wave at any point refers to the energy flux content, the energy transmitted per unit time in a specified direction through a unit area normal to the direction, at the point.
3. The wavefront is usually characterized by a specific shape (plane, spherical or other).
4. The wave is usually polarized in that the direction of wave propagation is aligned at a fixed angle to the direction of particle motion. For example, in the case of a longitudinal wave the two directions coincide, while in the case of transverse wave they are mutually perpendicular.

The wavelength of a periodic acoustic wave is the distance the acoustic travels to complete one cycle. The frequency of an acoustic wave is the number of complete cycles that pass a certain observation point in one second. In the case of a single periodic wave, the temporal frequency of the wave f, in cycles per second, is related to the wavelength λ and the wave velocity a according to the relationship

$$f = \frac{a}{\lambda}. \tag{4.1}$$

The frequency also relates the amplitude of the displacement of the particle d to the amplitude of the velocity of the particle v by

$$d = \frac{1}{2\pi}\frac{v}{f}. \tag{4.2}$$

Acoustic waves suffer from the problems of refraction, diffraction, reflection and absorption. Refraction is the variation of the direction of acoustic wave transmission due to a spatial variation in the velocity of sound transmission in the medium. Thus, acoustic waves are refracted when the velocity of sound varies across the wavefront and then result in a bending of the acoustic waves. Diffraction is the change in direction of propagation of an acoustic wave due to the passage of the wave around an obstacle. When an acoustic wave encounters a large heavy and rigid wall, the wave experiences reflection and propagates in the opposite direction. The absorption of sound is the process by which the total energy of a group of acoustic waves incident on a surface or passing through a medium is diminished. This is usually due to the conversion of a part of the acoustic energy into other forms of energy such as heat and non-acoustic mechanical energy.

4.2 Propagation of Acoustic Waves in the Atmosphere

The propagation of waves in the atmosphere is a quintessential model of acoustic wave propagation with all its features and properties. It is also the most common example of acoustic wave propagation known. Hence, we consider this model in some detail and employ the model of one-dimensional wave propagation in the atmosphere to enunciate the most important features of acoustic waves in any medium. When an acoustic wave travels in the atmosphere, the medium is compressed and rarefied cyclically as the particles in it move to and fro in the direction of propagation. It is this motion that results in a cyclic change in the pressure over and above the normal steady pressure, and it is this excess pressure that is responsible for the sensation of sound in the human ear.

When no energy transfer to the atmosphere is present, the adiabatic gas law typifies the behaviour of the atmosphere and is given by

$$\frac{p}{\rho^{\gamma}} = \text{constant}, \tag{4.3}$$

where p is the local pressure, ρ is the local density of air and γ is the ratio of specific heats and equal to about 1.4 for air at sea level. The adiabatic gas law being primarily a non-linear relation, the differential equation of particle motion in the atmosphere and hence the governing equation for the pressure is also non-linear. However, by rearranging the adiabatic gas law it is possible to establish a dynamic model governing the dynamics of particle motion in the atmosphere, and this may then be employed to study the excess pressure that is responsible for the sensation of sound.

4.2.1 Plane Waves

A plane wave is a type of wave that propagates in just one fixed direction, with pressure being uniform laterally in a plane perpendicular to the direction of propagation. In what follows, the analysis of acoustic waves is restricted to a class of one-dimensional plane waves. To establish the governing differential equation of one-dimensional wave motion, we consider a stratum of undisturbed air of thickness δx in the direction of propagation, as illustrated in Figure 4.1.

A particle originally at a position x is displaced to a position $x + u$. This holds for all particles in the stratum as the stratum is infinitesimally narrow. Hence, it follows that the thickness of the stratum

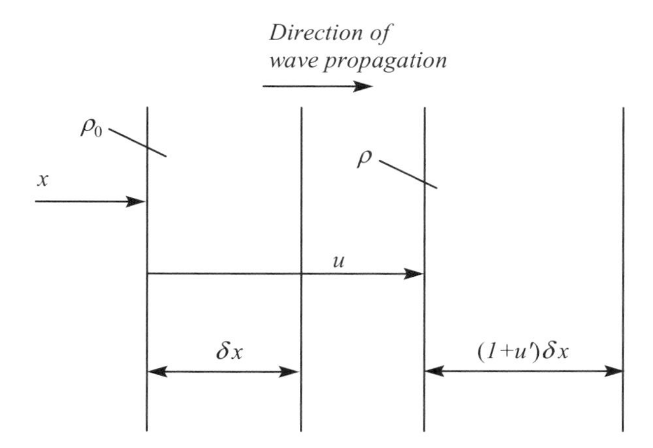

Figure 4.1 Motion of a one-dimensional plane stratum of air

changes from δx to

$$t = \delta (x + u) = \delta x + \frac{\partial u}{\partial x}\delta x = \left(1 + \frac{\partial u}{\partial x}\right)\delta x.$$

Assuming the density of the atmospheric stratum to be initially equal to ρ_0, the mass of a unit area of the stratum before and after the displacement is unchanged, and it follows that

$$\rho_0 \delta x = \rho\left(1 + \frac{\partial u}{\partial x}\right)\delta x,$$

where ρ is the density of the displaced stratum of air. Thus, it follows that

$$\rho = \rho_0\left(1 + \frac{\partial u}{\partial x}\right)^{-1} \tag{4.4}$$

and after differentiating the above we obtain,

$$\frac{\partial \rho}{\partial x} = -\rho_0 \frac{\partial^2 u}{\partial x^2}\left(1 + \frac{\partial u}{\partial x}\right)^{-2} = -\rho \frac{\partial^2 u}{\partial x^2}\left(1 + \frac{\partial u}{\partial x}\right)^{-1}.$$

Furthermore, it follows that the excess density approximately satisfies

$$\rho - \rho_0 = \rho_0\left(1 + \frac{\partial u}{\partial x}\right)^{-1} - \rho_0 = -\rho_0 \frac{\partial u}{\partial x}\left(1 + \frac{\partial u}{\partial x}\right)^{-1} \approx -\rho_0 \frac{\partial u}{\partial x}.$$

Assuming that the adiabatic gas law applies to the atmosphere,

$$\frac{p}{\rho^{\gamma}} = \frac{p_0}{\rho_0^{\gamma}},$$

which may be rearranged as

$$\frac{p}{p_0} = \left(\frac{\rho}{\rho_0}\right)^{\gamma}.$$

Hence it follows that

$$\frac{\partial p}{\partial \rho} = \frac{\gamma\, p}{\rho_0}\left(\frac{\rho}{\rho_0}\right)^{\gamma - 1} = \frac{\gamma\, p}{\rho_0}\left(1 + \frac{\partial u}{\partial x}\right)^{1-\gamma} = a_0^2\left(1 + \frac{\partial u}{\partial x}\right)^{1-\gamma},$$

where

$$a_0^2 = \frac{\gamma\, p}{\rho_0}. \tag{4.5}$$

It will be seen later that this is equal to the square of the speed of acoustic wave propagation in the ambient atmosphere. The pressure gradient may be expressed as

$$\frac{\partial p}{\partial x} = \frac{\partial p}{\partial \rho}\frac{\partial \rho}{\partial x} = -a_0^2 \rho \frac{\partial^2 u}{\partial x^2}\left(1 + \frac{\partial u}{\partial x}\right)^{-\gamma}.$$

The pressure difference on the two sides of the stratum is

$$p + \frac{\partial p}{\partial (x + u)}\left(1 + \frac{\partial u}{\partial x}\right)\delta x - p = \frac{\partial p}{\partial x}\delta x.$$

By the application of Newton's law,

$$-\frac{\partial p}{\partial x}\delta x = \rho\left(1 + \frac{\partial u}{\partial x}\right)\delta x \times \frac{\partial^2}{\partial t^2}u.$$

Hence, it follows that

$$\rho \ddot{u} = -\frac{\partial p}{\partial x}\left(1 + \frac{\partial u}{\partial x}\right)^{-1}$$

and eliminating the pressure gradient from the above equation we obtain

$$\rho \ddot{u} = \rho a_0^2 \frac{\partial^2 u}{\partial x^2}\left(1 + \frac{\partial u}{\partial x}\right)^{-1-\gamma}. \tag{4.6}$$

Assuming that

$$\frac{\partial u}{\partial x} \ll 1,$$

it follows that for small-amplitude particle motion, the governing equation for the motion of the particles of the atmosphere is

$$\ddot{u} = a_0^2 \frac{\partial^2 u}{\partial x^2}, \tag{4.7}$$

which is the archetypal equation for longitudinal wave motion. The velocity of longitudinal wave propagation may be seen to be a_0.

The most general solution for the equation foe particle motion is

$$u(x, t) = u_{01} \exp(ik(a_0 t - x)) + u_{02} \exp(ik(a_0 t + x)) \tag{4.8}$$

and considering only transmission in the positive x direction

$$u(x, t) = u_{01} \exp(ik(a_0 t - x)). \tag{4.9}$$

We also observe that when

$$\frac{\partial u}{\partial x} \ll 1,$$

the excess pressure is given by

$$p - p_0 = \frac{\partial p}{\partial \rho}(\rho - \rho_0) \approx -a_0^2 \rho_0 \frac{\partial u}{\partial x} \tag{4.10}$$

For the type of progressive wave solution assumed and given by equation (4.9),

$$\frac{\partial u}{\partial x} = -\frac{1}{a_0}\frac{\partial u}{\partial t} \tag{4.11}$$

it follows that the excess pressure is given by

$$p - p_0 \approx -a_0^2 \rho_0 \frac{\partial u}{\partial x} = a_0 \rho_0 \frac{\partial u}{\partial t}. \tag{4.12}$$

For the case of wave propagating in the negative x direction, it can be shown similarly that

$$p - p_0 \approx -a_0 \rho_0 \frac{\partial u}{\partial t}. \tag{4.13}$$

A positive excess pressure represents a *compression* while a negative excess pressure represents a *rarefaction*. A change in sign in the above expressions for the excess pressure is associated with a change in direction. The last two relationships are extremely significant and may be interpreted physically, as is done in a later section.

4.2.2 Linear and Non-linear Waveforms

No analysis of the propagation of acoustic waves would be complete without a mention of its fundamental non-linear nature and the consequences of the non-linearities.

When the assumption of small-amplitude motion

$$\frac{\partial u}{\partial x} \ll 1$$

is not made, the governing equation of particle motion may be written as

$$\ddot{u} = \frac{a_0^2}{\left(1 + \dfrac{\partial u}{\partial x}\right)^{\gamma+1}} \frac{\partial^2 u}{\partial x^2}, \tag{4.14}$$

which may also be written as

$$\ddot{u} = a^2 \frac{\partial^2 u}{\partial x^2}, \tag{4.15}$$

where

$$a^2 = \frac{a_0^2}{\left(1 + \dfrac{\partial u}{\partial x}\right)^{\gamma+1}} \approx a_0^2 \left(1 - (\gamma + 1) \frac{\partial u}{\partial x}\right). \tag{4.16}$$

Equation (4.15) is indeed a wave equation, although it is non-linear as the wave velocity a is now a function of the local displacement gradient and not a constant, as in the linear case. Physically it implies that the local wave velocity is a function of the strain, and hence it is a maximum when the strain is a minimum. It follows that the crests of the waves where the displacement gradient is a minimum travel faster than the rest, causing the waveform to distort continually with the passage of time.

The governing equation of particle motion may also be expressed as

$$\ddot{u} - a_0^2 \frac{\partial^2 u}{\partial x^2} \approx -a_0^2 (\gamma + 1) \frac{\partial u}{\partial x} \frac{\partial^2 u}{\partial x^2}. \tag{4.17}$$

Applying the method of perturbation, a first-order solution is obtained by solving the equation

$$\ddot{u}_1 - a_0^2 \frac{\partial^2 u_1}{\partial x^2} = 0.$$

The second-order solution then satisfies the equation

$$\ddot{u}_2 - a_0^2 \frac{\partial^2 u_2}{\partial x^2} \approx -a_0^2 (\gamma + 1) \frac{\partial u_1}{\partial x} \frac{\partial^2 u_1}{\partial x^2}.$$

If we let

$$u_1 = u_{10} \cos(\omega t - kx),$$

it follows that u_2 satisfies

$$\ddot{u}_2 - a_0^2 \frac{\partial^2 u_2}{\partial x^2} \approx -\frac{k^3}{2} u_{10}^2 \sin(2(\omega t - kx)).$$

A particular solution for u_2 is given by

$$u_{2p} = -\frac{(\gamma + 1)}{8} u_{10}^2 k^2 x \cos(2(\omega t - kx)).$$

A complementary solution must satisfy the boundary condition at $x = 0$ whereas the complete solution must necessarily vanish for all time. Hence, the complementary solution is of the form

$$u_{2c} = A\cos(2(\omega t - kx))$$

where the amplitude $A \equiv 0$. The complete solution to second order is therefore given by

$$u_2 = u_{10}\cos(\omega t - kx) - \frac{(\gamma + 1)}{8}u_{10}^2 k^2 x\cos(2(\omega t - kx)). \tag{4.18}$$

The solution applies only for small amplitude waves and for x less than some value $x = x_0$. This value may be found by the requirement that the total energy everywhere, of the waves at this point, is equal to the total energy content of the waves at $x = 0$. The form of the solution demonstrates the existence of a higher harmonic due to the distortion of the initial waveform as it propagates forward or backwards. The occurrence of higher harmonic is a feature of the propagation of non-linear waves and is discussed by Stoker (1950) and Nayfeh and Mook (1979).

By considering two, initial, small-amplitude waves with differing frequencies, the solution may also be shown to be dispersive, i.e. a typical wave train emerging from a source and consisting of several waves of more than one frequency, when perceived as sound after propagating through a section of the atmosphere, has several component waves with frequencies differing from those at the source. This is a fundamental feature of acoustic waves transmitted through the atmosphere, which is inherently non-linear and dispersive. The features of linear and non-linear wave propagation are discussed by Pain (1976).

4.2.3 Energy and Intensity

The total energy content in an acoustic wave is apparent in its *intensity*. The total energy content is composed of the kinetic energy and the strain energy of the particles in the atmosphere.

The kinetic energy of an acoustic wave is found by considering the motion of particles within a stratum of undisturbed air of thickness δx in the direction of propagation. Each particle in the stratum will have a kinetic energy

$$\delta E_k = \frac{1}{2}\rho_0 dx\dot{u}(x, t)^2. \tag{4.19}$$

The expression is integrated over a distance equal to n wavelengths and then averaged over the same distance. Assuming that the displacement $u(x, t)$ is given by the expression

$$u(x, t) = u_0\cos(k(at - x)),$$

where the frequency of the wave is

$$f = \frac{\omega}{2\pi} = \frac{ka}{2\pi}$$

and

$$k = \frac{2\pi}{\lambda},$$

the particle velocity is then given by

$$\dot{u}(x, t) = -2\pi f u_0\sin(k(at - x)) = \dot{u}_m\sin(k(at - x)).$$

Following the integration and the averaging, the average kinetic energy density in the medium is

$$\Delta\bar{E}_k = \frac{1}{2}\rho_0 \times \frac{1}{2}\dot{u}_m^2 = \frac{1}{4}\rho_0\omega^2 u_0^2. \tag{4.20}$$

The strain energy stored in each particle in the stratum of the atmosphere of unit area during the passage of the acoustic wave is found by considering the work done by the pressure or compressive hydrostatic stress in changing the volume by dV of a fixed mass of the gas of volume V_0 assuming that the process is adiabatic. The work done is expressed as

$$\delta E_{pot} = -\int (p - p_0)\,\delta V = -\int (p - p_0)\, d\left(\frac{\partial u}{\partial x}\right)\delta x = \frac{a_0^2 \rho_0}{2}\left(\frac{\partial u}{\partial x}\right)^2 \delta x.$$

The potential energy is then given by

$$\delta E_{pot} = \frac{1}{2}\rho_0 dx \dot{u}\,(x, t)^2\,. \tag{4.21}$$

As a result, we observe that the average values of the kinetic and potential energy densities are equal to each other. As a consequence, the maximum and minimum of the two energy densities occur at the same instant.

The intensity of an acoustic wave is the energy flux or the rate at which the energy passes by a unit area of cross-section. Thus, it is defined as the product of the total energy density and the wave velocity, and is given by

$$I = \frac{1}{2}\rho_0 a_0 \dot{u}_m^2. \tag{4.22}$$

Intensity of sound is generally a relative quantity and is therefore expressed in decibels (dB). Intensity in decibels is defined as

$$I_{dB} = 10\log_{10} I. \tag{4.23}$$

Thus, when sound intensity increases by a factor 10 the intensity in dB goes up by 10 dB.

4.2.4 *Characteristic Acoustic Impedance*

We also observe in the preceding analysis that the product $\rho_0 a_0$ appears in most of the expressions for the excess pressure and intensity. This is in fact no accident, and we may define the characteristic acoustic impedance of a medium as the absolute ratio of the excess pressure and the particle velocity. Denoting the characteristic acoustic impedance as

$$Z_0 = \rho_0 a_0, \tag{4.24}$$

the excess pressure in the case of wave travelling in the positive x direction is

$$p - p_0 \approx Z_0 \partial u/\partial t. \tag{4.25}$$

For the case of wave propagating in the negative x direction, it can be shown similarly that

$$p - p_0 \approx -Z_0 \partial u/\partial t. \tag{4.26}$$

The acoustic impedance is governed by the inertia and elasticity of the medium. It plays a fundamental role in the transmission and reflection of acoustic waves and is discussed extensively by Mason (1964).

4.2.5 *Transmission and Reflection of Plane Waves at an Interface*

Consider the case of an acoustic wave passing a boundary separating two media with different acoustic impedance characteristics. At the interface between the two media, the acoustic wave must necessarily satisfy two boundary conditions pertaining to the particle velocity and excess pressure, which are both required to be continuous. The requirement that the two are continuous across the boundary arises from the fact that the media are in complete contact with each other across the boundary. Thus, assuming that the incident wave on arriving at the boundary is split into two components, one transmitted in the same direction and the other reflected or propagated in the opposite direction, the boundary conditions may be expressed in terms of the incident, reflected and transmitted wave components. If the subscripts i, r and t denote the incident, reflected and transmitted components, respectively, the boundary conditions may be expressed in terms of the particle velocity components as

$$\dot{u}_i = \dot{u}_t - \dot{u}_r \tag{4.27}$$

and in terms of the pressure components as

$$p_i - p_0 = p_t - p_r.$$

The latter boundary condition may also be expressed as

$$Z_i \dot{u}_i = Z_t \dot{u}_t + Z_i \dot{u}_r \tag{4.28}$$

since the medium in which the reflected wave propagates is the same as the one in which the incident wave was present. Eliminating the transmitted wave component from equations (4.27) and (4.28), we have

$$\frac{\dot{u}_r}{\dot{u}_i} = \frac{Z_i - Z_t}{Z_i + Z_t}, \tag{4.29}$$

and eliminating the reflected wave component from the same two equations

$$\frac{\dot{u}_t}{\dot{u}_i} = \frac{2Z_i}{Z_i + Z_t}. \tag{4.30}$$

Similar expressions may be derived for the ratios of the transmitted and incident excess pressures, and the reflected and incident excess pressures. They determine the relative amplitudes and phases on the transmitted and reflected wave components relative to the incident wave. When Z_t is relatively much larger than Z_i, the transmitted region is relatively rigid and no wave is transmitted through. On the other hand, when the characteristic acoustic impedances of the two media are the same, wave reflection is absent. In this case, the impedances are said to be *matched*.

4.3 Circuit Modelling: The Transmission Lines

4.3.1 The Transmission Line

The transmission line is yet another fundamental paradigm that facilitates the understanding of the features of wave propagation in continuous media. Although it is an electrical model involving the flow of currents and the effects of impedances, it may be employed by appropriate use of electrical analogies to model any wave propagation problem.

When a source of electro-motive force (*emf*) such as a battery is connected to the ends of a pair of insulated parallel wires that extend outwards for an infinite distance but are not connected to each other, electric current begins to appear in the vicinity of the source *emf* and propagates outwards. The electric field from the source tends to attract the free electrons in the wire connected to the positive battery terminal and repel the free electrons in the wire connected to the negative battery terminal, thus setting up a current flow. The speed of the current flow is finite, below but near the speed of light in free space. The current flows to charge the inter-wire capacitance, which is assumed to be continuously and uniformly distributed over the whole length of the pair of wires and continuously in parallel to each other. As the current flows, energy is also stored in the wire self-inductances, which are also assumed to

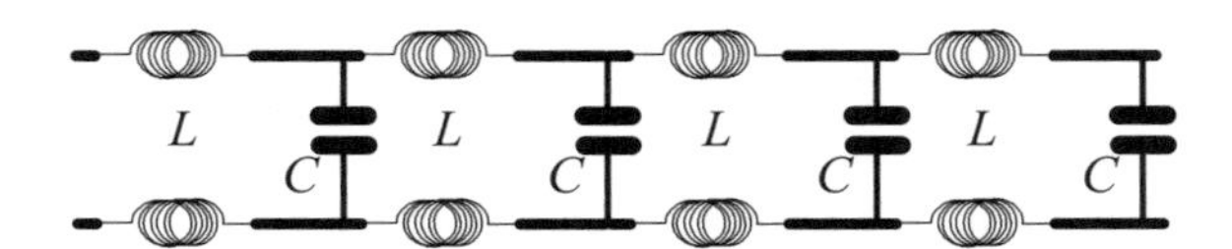

Figure 4.2 Representation of an ideal or loss-less transmission line

be uniformly distributed along the entire length of both wires and in series with each other, as illustrated in Figure 4.2.

4.3.2 The Ideal Transmission Line

Consider a short element of an ideal or loss-less transmission line of near-zero resistance and length δx, as illustrated in Figure 4.3.

The self-inductance of the element is assumed to be $L_0\delta x$ Henries and the intra-wire capacitance to be $C_0\delta x$ Farads. In terms of the rate of change of voltage per unit length at any instant time, the voltage difference between the output and input ends is

$$V_{out} - V_{in} = \frac{\partial V}{\partial x}\delta x = -\left(L_0\delta x\right)\frac{\partial I}{\partial t}$$

and it follows that

$$\frac{\partial V}{\partial x} + L_0\frac{\partial I}{\partial t} = 0. \tag{4.31}$$

In terms of the rate of change of current per unit length at any instant time, the current increment between the output and input ends is

$$I_{out} - I_{in} = \frac{\partial I}{\partial x}\delta x = -\left(C_0\delta x\right)\frac{\partial V}{\partial t},$$

where the negative sign is included as there is in fact a current loss due to the charging of the capacitance and it follows that

$$\frac{\partial I}{\partial x} + C_0\frac{\partial V}{\partial t} = 0. \tag{4.32}$$

Eliminating the current from equations (4.31) and (4.32), we obtain a pure wave equation given by

$$\frac{\partial^2 V}{\partial x^2} = L_0 C_0\frac{\partial^2 V}{\partial t^2} \tag{4.33}$$

with a velocity of propagation given by

$$c_0 = \sqrt{\frac{1}{L_0 C_0}}. \tag{4.34}$$

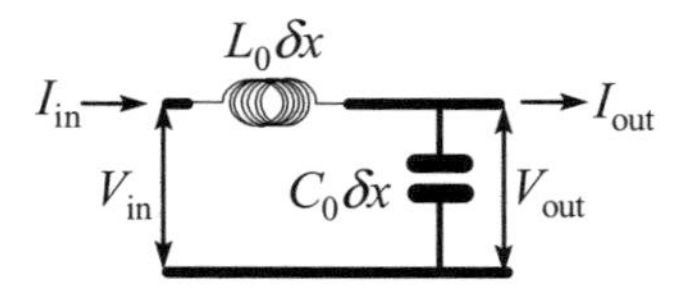

Figure 4.3 Representation of an element of an ideal or loss-less transmission line with zero resistance

Similarly, eliminating the voltage instead of the current between the same two equations, (4.31) and (4.32), we obtain another wave equation

$$\frac{\partial^2 I}{\partial x^2} = L_0 C_0 \frac{\partial^2 I}{\partial t^2} \tag{4.35}$$

with the same velocity of propagation as before. Considering wave propagation in the positive x direction, the solutions to the two wave equations, (4.33) and (4.35), may be expressed as

$$V(x,t) = V_{01} \cos\left(k\left(c_0 t - x\right)\right)$$

and

$$I(x,t) = I_{01} \cos\left(k\left(c_0 t - x\right)\right).$$

Substitution in equation (4.31) gives the relationship

$$V_{01} = c_0 L_0 I_{01}.$$

The ratio

$$\frac{V_{01}}{I_{01}} = c_0 L_0 = \sqrt{\frac{L_0}{C_0}}$$

is constant for the transmission line. It has the units of impedance (Ohms) and has a special significance for transmission lines. To an electrical input such as impulse applied at one end, such an inductive/capacitive model (non-resistive) of the transmission line has an equivalent *characteristic impedance* that is purely resistive and is given by

$$Z_c = c_0 L_0 = \sqrt{\frac{L_0}{C_0}}, \tag{4.36}$$

where L_0 and C_0 are the self-inductance and inter-wire capacitance of the line per unit length, respectively. The impedance is purely resistive, as the power from the source travels forward along the line and never returns; thus, it is all lost or dissipated along the line or at the end. Furthermore, the characteristic impedance may be employed to determine the amount of current that will flow when a fixed voltage is supplied at one end. For a wave propagating in the negative x direction, the characteristic impedance has the same magnitude but the opposite sign.

4.3.3 Matched Lines

Actual transmission lines do not extend to infinity. They are finite and are also terminated in a load at the other end, the output end. When the load is virtually purely resistive and equals the characteristic impedance, the line is said to be *matched*. To a current travelling down such a matched line, the load appears like yet another section of the same transmission line, repeating itself and thus behaving exactly like an infinite transmission line. The resistive load absorbs all of the transmitted power.

Suppose that a transmission line has a characteristic impedance Z_c and a finite length, and the output end is terminated by a load of impedance Z_L. An incident wave, with the voltage and current amplitudes of V_i and I_i, may be reflected by the load to produce a reflected voltage and a reflected current with amplitudes of V_r and I_r. The boundary conditions at the load are

$$V_i - (-V_r) = V_L \tag{4.37a}$$

and

$$I_i - (-I_r) = I_L. \tag{4.37b}$$

It follows from the first of these boundary conditions that

$$Z_c I_i - (Z_c I_r) = Z_L I_L. \tag{4.38}$$

Eliminating the load current from equations (4.37) and (4.38),

$$\frac{I_r}{I_i} \equiv \rho = \frac{Z_c - Z_L}{Z_c + Z_L} \tag{4.39a}$$

and

$$\frac{I_L}{I_i} = \frac{2Z_c}{Z_c + Z_L}. \tag{4.39b}$$

Thus, the reflected current is absent when the transmission is matched. The correspondence between equations (4.29) and (4.30), and (4.39a) and (4.39b), respectively, is no accident, and the transmission model and the transmission of longitudinal plane waves in the atmosphere are analogous to each other.

When an oscillating current or voltage source is employed at the input end rather than a battery, the transmission line continues to behave in exactly the same fashion. Yet because an alternating voltage is applied at the source and the current travels down the line with a finite speed, the current at any point in the line is the result of a voltage applied at some earlier instant, at the input terminals. Thus, the current and the voltage travel along the line as a series of waves having a length equal to the speed of travel divided by the frequency of the supply voltage. Furthermore, the reflected energy and the incident energy are directly proportional to the squares of I_r and I_i; the ratio of the reflected energy to the incident energy is

$$\frac{E_r}{E_i} = \left(\frac{I_r}{I_i}\right)^2 = \left(\frac{Z_c - Z_L}{Z_c + Z_L}\right)^2$$

and the ratio of the transmitted energy to the incident energy is

$$\frac{E_t}{E_i} = 1 - \frac{E_r}{E_i} = 1 - \left(\frac{Z_c - Z_L}{Z_c + Z_L}\right)^2 = \frac{4Z_c Z_L}{(Z_c + Z_L)^2}.$$

When $Z_c = Z_L$, no energy is reflected and the impedances are said to be matched.

4.3.4 Reflection from the End of a Transmission Line: Standing Waves

In an infinitely long transmission line pair or in a matched transmission line pair, the impedance at any point along the line is the same, as is the ratio of the voltage to the current at any point along the transmission line because it is also equal to the impedance. However with a finite transmission line that is short circuited at the output end, the impedance at the output end, the load impedance, is zero. Thus, the outgoing current is reflected at the short circuit at each of the two transmission lines. The current components in the two transmission lines are exactly out of phase with each other, as they are propagating in opposite directions, but the outgoing and reflected currents in each line are exactly in phase and a maximum at the short circuit end. Thus, the entire power incident at the short circuit end is reflected, and the reflected power travels back to the source. As far as the voltages are concerned, they are a minimum whenever the current is a maximum and vice versa. At a distance of exactly one half wavelength from the short circuit end, the current is of the same magnitude as at the short circuit end but of opposite polarity, while the voltage is again a minimum.

The reflected current travels with exactly the same speed as the outgoing current wave but in an opposite direction to the outgoing wave. The sum of the two waves constitutes a stationary oscillatory pattern. While the time variation of the current is sinusoidal, the spatial variation constitutes a standing wave or a sinusoidal normal mode of oscillation. The point of minimum line current constitutes a node of the normal mode of oscillation.

In the case of a voltage wave, the zero load impedance at the short circuit end implies a zero voltage across the load at the output end. Thus, no matter what the value of the incoming voltage wave at the short circuit end, the reflected voltage is in phase opposition at that end so that the total sum is always zero. This is equivalent to a half wavelength difference in phase. Further moving away a quarter wavelength from the short circuit end, the voltages in the outgoing and reflected waves are a further half wavelength out of phase and hence in phase with each other. The result of summing the outgoing and reflected voltage waves is again a standing wave, but of voltage. It is displaced exactly one quarter wavelength from the standing wave in the current.

Considering the case of a transmission line pair with an open-circuited output end, there is a large voltage and no current at the open circuit end. Thus, the situation is completely reversed from the short circuit case, withstanding waves of current and voltage playing exactly the opposite roles as in the short circuit case.

We now consider a pair of transmission lines terminated by an arbitrary resistive load. In this case, it is easy to surmise that a part of transmitted power is absorbed in the load while the remainder is reflected to form a standing wave. It has already been stated that when the load resistance $Z_R = Z_L$ is equal to the characteristic impedance Z_c, there is no reflection and all the transmitted power is absorbed. On the other hand, when $Z_R \to 0$ or $Z_R \to \infty$, as in the case of short-circuited and open-circuited ends, respectively, there is 100% reflection and no power is absorbed at the output end. To characterize these cases effectively we define a parameter, the *standing wave ratio* (SWR), which is the ratio of the absolute maximum to the absolute minimum along the line. The maximum current along the line will be $|I_i| + |I_r|$, while the minimum current along the line will be $|I_i| - |I_r|$ and occurs when the incident and reflected currents add in phase or out of phase, respectively. Thus, the SWR is given by

$$SWR = \frac{|I_i| + |I_r|}{|I_i| - |I_r|} = \frac{1 + \left|\dfrac{I_r}{I_i}\right|}{1 - \left|\dfrac{I_r}{I_i}\right|} = \frac{1 + |\rho|}{1 - |\rho|}. \tag{4.40a}$$

When the resistances are purely resistive,

$$SWR = \frac{1 + |\rho|}{1 - |\rho|} = \frac{1 + \rho}{1 - \rho} = \frac{Z_c}{Z_L} \tag{4.40b}$$

when $Z_c > Z_L$, and

$$SWR = \frac{Z_L}{Z_c} \tag{4.40c}$$

when $Z_c < Z_L$. The cases of the open-circuited and short-circuited output ends correspond to $SWR = \infty$, while the case of a matched load at the output end corresponds to $SWR = 1$. Thus, the SWR is an indicator of the level of mismatch between the load impedance and the characteristic impedance. Strictly it is sometimes referred to as the *standing wave impedance ratio* or simply the *impedance ratio*.

To better understand the reflection and absorption of incoming waves in a transmission line, we introduce yet another parameter, the input or source impedance Z_S. The input impedance of a transmission line is the impedance seen looking into the source end or input terminals. Thus, it is the impedance of a source of *emf* or electrical power that must be connected to the transmission line when it is operational. If the load impedance is perfectly matched to the line, the line is equivalent to an infinitely long transmission line and the input impedance may be shown to be exactly equal to the characteristic impedance. However, in the presence of standing waves, that is, when the load is not matched to the line, this is no longer true. Furthermore, the input impedance may be represented by a resistance and a capacitance, or a resistance and an inductance, and in some cases by all three types of impedances. The input impedance parameters are extremely important in determining the method by which the source *emf* is connected to the transmission line.

It is easy to observe that for a line that is exactly one half wavelength long or integral multiples of it, the magnitudes of the voltages at the two ends of the line are exactly equal to each other, no matter what the characteristic impedance. This also applies to the magnitudes of the currents. Hence, in this case the input impedance is exactly equal to the load impedance.

To derive an expression for the input impedance in the general case, consider a point on the transmission line at a distance $x = -l$. Assuming that the reflected current wave is out of phase with the incident current wave, the line current and the corresponding line voltage are

$$I(l) = I_i \exp(i\beta l) - I_r \exp(-i\beta l) \tag{4.41a}$$

and

$$V(l) = Z_c I_i \exp(i\beta l) + Z_c I_r \exp(-i\beta l). \tag{4.41b}$$

The ratio of the reflected current to the incident current at the point $x = -l$ is

$$r(l) = \frac{I_r \exp(-i\beta l)}{I_i \exp(i\beta l)} = \frac{Z_c - Z_L}{Z_c + Z_L} \exp(-i2\beta l). \tag{4.42}$$

Hence, the ratio of the line voltage to the line current at the point $x = -l$ gives the input impedance and is

$$\frac{V(l)}{I(l)} = \frac{Z_c I_i \exp(i\beta l) + Z_c I_r \exp(-i\beta l)}{I_i \exp(i\beta l) - I_r \exp(-i\beta l)} = Z_c \frac{1 + r(l)}{1 - r(l)}. \tag{4.43}$$

The quantity

$$S = \frac{1 + |r(l)|}{1 - |r(l)|} = \frac{1 + |\rho|}{1 - |\rho|},$$

where ρ, the reflection coefficient, is given by equation (4.39a), is the SWR given by equation (4.40a) and is also known as the *voltage standing wave ratio* (VSWR); it determines the maximum and minimum variation in the input impedance of the line. In the case of purely resistive impedances, it is completely equivalent to the impedance ratio given by equation (4.40b).

Substituting for $r(l)$ in equation (4.43), the input impedance may be expressed as

$$Z_S = \frac{V(l)}{I(l)} = Z_c \frac{Z_L + iZ_c \tan(\beta l)}{Z_c + iZ_L \tan(\beta l)}. \tag{4.44}$$

Thus, when the transmission line is a quarter wavelength long or an integral odd multiple of it, $\beta l = 90^0$ and the current and voltage at the load end are inverted in relation to the corresponding values at the input side; i.e. when the voltage is high the current is low at the load end, while the current is high and the voltage is low at the source end. Thus, in this case one can expect that Z_S is inversely proportional to Z_L. In fact, in this case, equation (4.44) reduces to

$$Z_S = \frac{Z_c^2}{Z_L}. \tag{4.45}$$

Thus, the quarter wavelength transmission line matches a source with impedance Z_S to a load with impedance Z_L. Matched lines are also said to be non-resonant due to the absence of standing waves. The theory of loss-less transmission lines presented in this section is also the basis for the design of antennae that are capable of radiating electro-magnetic waves. For our purposes, we shall employ the features of transmission lines to enumerate the characteristics of acoustic wave propagation in elastic and piezoelectric media. Further details on the application of the transmission line analogy may be found in the works of Auld (1973) and Campbell (1998).

4.3.5 The Mechanical Transmission Line: An Electro-mechanical Analogy

Earlier the equation governing the motion of one-dimensional acoustic waves (equation (4.7)) was derived. This may be expressed as a pair of first-order differential equations by introducing the pressure on a cross-section as an intermediate variable and relating it to the particle velocity. The equations are

$$-\frac{\partial p(x,t)}{\partial x}+\rho\frac{\partial v(x,t)}{\partial t}=0, \tag{4.46a}$$

$$-\frac{\partial v(x,t)}{\partial x}+\frac{1}{E}\frac{\partial p(x,t)}{\partial t}=0. \tag{4.46b}$$

The equations governing the transmission line, equations (4.31) and (4.32), are

$$\frac{\partial V}{\partial x}+L_0\frac{\partial I}{\partial t}=0, \tag{4.31}$$

$$\frac{\partial I}{\partial x}+C_0\frac{\partial V}{\partial t}=0. \tag{4.32}$$

Provided the pressure in the mechanical line p is interpreted as the negative of the voltage V, the particle velocity in the mechanical line v as the electrical current I, the material density of the mechanical line ρ as the electrical line inductance L_0 and Young's modulus E as the inverse of the electrical line capacitance C_0, the two pairs of equations are completely analogous to each other. Thus, an infinitely long slender rod is a *mechanical waveguide* and may be considered to be a mechanical transmission line, transmitting longitudinal plane pressure waves along its axis. On the other hand, the slender rod of finite length, considered in Chapter 5, is a *mechanical resonator* characterized by a discrete, albeit infinite, set of natural frequencies and associated normal modes of standing wave motion. Unlike resonators there are no standing waves associated with mechanical waveguides.

Since the characteristic impedance of an electrical transmission line is

$$Z_c=c_0L_0=\sqrt{\frac{L_0}{C_0}}, \tag{4.36}$$

the acoustic characteristic impedance is

$$Z_c=\sqrt{\rho E}.$$

Furthermore, since F, the force acting on a surface, may be expressed as the product of the pressure and the cross-sectional area A, in the case of simple harmonic motion, the stress–strain relationships may be invoked and equation (4.41b) may be expressed as

$$EA\frac{\partial\bar{v}(x)}{\partial x}=i\omega A\bar{p}(x)=i\omega F(x),$$

where $\bar{v}(x)$ and $\bar{p}(x)$ are the velocity and pressure amplitudes of motion. The general solutions to the equations may then be expressed in terms of the local velocity and forces as

$$\begin{bmatrix}\bar{v}(x)\\F(x)\end{bmatrix}=\begin{bmatrix}\cos kx & \frac{i}{Z_0}\sin kx\\ iZ_0\sin kx & \cos kx\end{bmatrix}\begin{bmatrix}\bar{v}_0\\F_0\end{bmatrix} \tag{4.47}$$

where $k=\omega/a$, $a=\sqrt{E/\rho}$, $\bar{v}_0=\bar{v}(0)$, $F_0=F(0)$.

$$EA\left.\frac{\partial\bar{v}(x)}{\partial x}\right|_{x=0}\equiv i\omega F(0)=ikEA\frac{F_0}{Z_m}$$

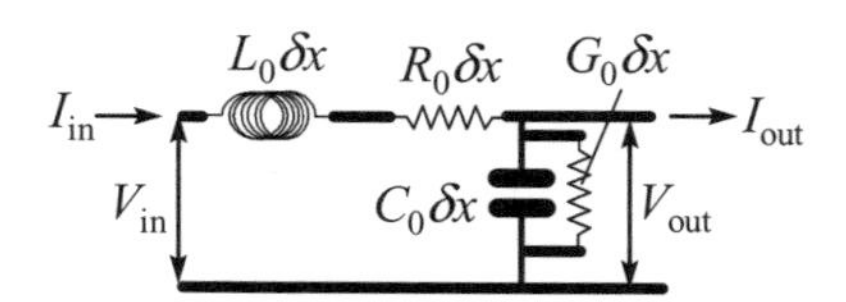

Figure 4.4 Representation of an element of a real transmission line with non-zero resistance

and Z_m, the *mechanical* characteristic impedance or *force impedance,* is related to the *acoustic* characteristic impedance or *pressure impedance* by

$$Z_m = \frac{EA}{a} = A\rho a = AZ_c.$$

Thus, the mechanical analogy, although related to the acoustic analogy, is different from it.

4.3.6 Dissipation of Waves in Transmission Lines

The analysis presented in this section thus far has been restricted to a transmission line modelled in terms of inductances and capacitances, discrete components that consume no power. In practice no such line exists and it is essential to include the effects of distributed power dissipation along the line. There is always some resistance in the wires, which is responsible for the power dissipation. We shall take this resistance into account by supposing that the transmission line has a series resistance R_0 Ohms per unit length and a shunt resistance between the wires, which is expressed as a conductance in parallel with the shunt capacitance and is given by G_0 Mhos per unit length. Consider a short element of a real transmission line of near-zero resistance and length δx, as illustrated in Figure 4.4.

In terms of the rate of change of voltage per unit length at any instant time, the voltage difference between the output and input ends is

$$V_{out} - V_{in} = \frac{\partial V}{\partial x}\delta x = -\left(L_0 \delta x\right)\frac{\partial I}{\partial t} - R_0 \delta x I,$$

and it follows that

$$\frac{\partial V}{\partial x} + L_0 \frac{\partial I}{\partial t} + R_0 I = 0. \tag{4.48}$$

In terms of the rate of change of current per unit length at any instant time, the current increment between the output and input ends is

$$I_{out} - I_{in} = \frac{\partial I}{\partial x}\delta x = -\left(C_0 \delta x\right)\frac{\partial V}{\partial t} - G_0 \delta x V,$$

where the negative sign is included as there is in fact a current loss due to the charging of the capacitance, and it follows that

$$\frac{\partial I}{\partial x} + C_0 \frac{\partial V}{\partial t} + G_0 V = 0. \tag{4.49}$$

Considering sinusoidal time variations of the current and voltage, equations (4.47) and (4.48) may be written as

$$\frac{\partial V}{\partial x} + \left(i\omega L_0 + R_0\right) I = 0 \tag{4.50}$$

and

$$\frac{\partial I}{\partial x} + (i\omega C_0 + G_0) V = 0. \tag{4.51}$$

Corresponding to equation (4.33), we have

$$\frac{\partial^2 V}{\partial x^2} - (i\omega L_0 + R_0)(i\omega C_0 + G_0) V = 0,$$

which may be expressed as

$$\frac{\partial^2 V}{\partial x^2} - \gamma^2 V = 0, \tag{4.52}$$

where

$$\gamma^2 = (i\omega L_0 + R_0)(i\omega C_0 + G_0). \tag{4.53}$$

A similar equation also holds for the current. Also, by a similar argument the characteristic impedance is given by

$$Z_c = \frac{(i\omega L_0 + R_0)}{(i\omega C_0 + G_0)}. \tag{4.54}$$

A progressive wave solution satisfies the equation

$$\frac{\partial V^+}{\partial x} + \gamma V^+ = 0, \tag{4.55}$$

while a regressive wave solution satisfies the equation

$$\frac{\partial V^-}{\partial x} - \gamma V^- = 0, \tag{4.56}$$

where the quantity γ is known as the propagation constant and is a complex number. The real part of the propagation constant is the attenuation coefficient and the imaginary part is the wave number. Thus, the propagation constant γ may be expressed as

$$\gamma \equiv \alpha + ik,$$

where

$$k = \omega / c_0$$

and

$$\alpha \approx \frac{R_0}{2} \times \frac{1}{Z_c} + \frac{G_0}{2} \times Z_c.$$

Equations (4.55), (4.56) and (4.51) or (4.52) play an important role in the modelling of acoustic wave propagation in real piezoelectric media where the amplitudes of the acoustic waves are known to decay exponentially in the direction of propagation and are also coupled to current flowing in the medium. This application of the theory of the transmission line to the design of electro-mechanical devices is discussed by Mason (1948).

Illustrative Example 4.1: Characteristic Impedance of an RF Transmission Line

The physical constants characterizing a radio frequency (RF) transmission line are as follows: R_0 is 50 mOhms per metre, L_0 is 1.6 μH per metre, C_0 is 7.5 pF per metre and G_0 is 10 (G Mhos)$^{-1}$ per metre. If the frequency of the oscillator feeding the line is 300 MHz, estimate the transmission line's characteristic impedance, the attenuation constant, the wave number, the wavelength and the wave velocity.

Solution: The characteristic impedance is

$$Z_c = c_0 L_0 = \sqrt{\frac{L_0}{C_0}} = \sqrt{\frac{1.6}{10^6} \times \frac{10^{12}}{7.5}} = 462 \text{ Ohms}$$

The attenuation constant is

$$\alpha \approx \frac{R_0}{2}\frac{1}{Z_c} + \frac{G_0}{2} Z_c = \frac{50}{2 \times 1000} \times \frac{1}{462} + \frac{10}{2 \times 10^9} \times 462,$$

which is

$$\alpha = 56.3 \times 10^{-6} \text{ nepers/m}.$$

The wave number is

$$k = \omega / c_0 = \omega \sqrt{L_0 C_0} = 2\pi \times 300 \times 10^6 \sqrt{\frac{1.6}{10^6} \times \frac{7.5}{10^{12}}} = 6.53 \text{ rads/m}.$$

The wavelength is

$$\lambda = \frac{2\pi}{k} = \frac{6.28}{6.53} = 0.96 \text{ m}.$$

The wave velocity is

$$c_0 = \lambda \times f = 0.96 \times 300 \times 10^6 = 288 \times 10^6 \text{ m/s}.$$

Illustrative Example 4.3.2: Reflection Coefficient and Input Impedance

A transmission line with a wave velocity of $c_0 = 288 \times 10^6$ m/s and a characteristic impedance of 50 Ohms is terminated by a load consisting of 40 Ohms resistance in series with a 65 Ohm capacitive impedance.

1. Calculate the reflection coefficient, the impedance ratio and the VSWR.
2. What are the maximum and minimum possible values for the input impedance?
3. If a 100 MHz RF oscillator delivers 80 W to the line, which is assumed to be loss less, estimate the power reflected and the power absorbed at the load. Also calculate the distance between adjacent nulls on the line.

Solution:

1. The load is

$$Z_L = 40 - i65 \text{ Ohms}.$$

The reflection coefficient is

$$\rho = \frac{Z_L - Z_c}{Z_L + Z_c} = \frac{40 - i65 - 50}{40 - i65 + 50} = \frac{-2 - i13}{18 - i13} = \frac{133 - i260}{493}.$$

Hence if $\rho = |\rho| \exp(i\theta)$, it follows that $|\rho| = 0.592$ and $\theta = -62.9°$.
The impedance ratio is

$$\left|\frac{Z_L}{Z_c}\right| = \left|\frac{40 - i65}{50}\right| = |0.8 - i1.3| = 1.49 \exp(-i58.4).$$

The VSWR is

$$SWR = \frac{1 + |\rho|}{1 - |\rho|} = \frac{1.592}{0.408} = 3.9.$$

2. The maximum and minimum possible values for the input impedance are

$$Z_{S\,\max} = Z_c \times SWR = Z_c \frac{1.592}{0.408} = 195 \text{ Ohms}$$

and

$$Z_{S\,\min} = Z_c \times \frac{1}{SWR} = Z_c \frac{0.408}{1.592} = 12.8 \text{ Ohms.}$$

3. The reflected power is

$$P_r = P_i \times |\rho|^2 = 80 \times (0.592)^2 = 28 \text{ W.}$$

The power absorbed by the load is

$$P_a = P_i - P_r = 80 - 28 = 52 \text{ W.}$$

The wavelength is

$$\lambda = \frac{c_0}{f} = \frac{288}{100} = 2.88 \text{ m.}$$

The distance between adjacent nulls is

$$\frac{\lambda}{2} = \frac{2.88}{2} = 1.44 \text{ m.}$$

4.4 Mechanics of Pure Elastic Media

Just as acoustic waves in the atmosphere are identified with pressure waves resulting from compressions and rarefactions of the gaseous medium, acoustic waves in an elastic solid may be associated with stress waves resulting from dilatations or volumetric strains and distortions of the medium. Thus, we consider the mechanics of pure elastic media with the aim of understanding these stress waves (Kolsky, 1963; Nadeau, 1964).

4.4.1 Definition of Stress and Strain

When a force is applied to an elastic body, there are also accompanying elastic deformations of the body in addition to the rigid body displacements and rotations. As a result of the application of a force on an elastic body, a system of stresses arises within the body. A stress on a surface element within a solid deformable body has three components: one normal to the surface and the other two mutually perpendicular to each other and in the plane of the surface element. Thus, considering a cubic body, with mutually perpendicular axes Ox, Oy and Oz attached at one corner of it, the stresses acting on three planes, each normal to one of the three axes that pass through a point P, have a total of nine components. These are denoted by σ_{xx}, σ_{yy}, σ_{zz}, σ_{xy}, σ_{xz}, etc. The first letter in the suffix denotes the direction of the stress while the second letter defines the outward normal to the plane in which it is acting. This is illustrated in Figure 4.5.

Considering an infinitesimal cubic element, with sides δx, δy and δz, to be in rotational equilibrium and taking moments about the point P, we require that

$$\sum M_y = 0 = (\sigma_{zx}\delta y\delta z)\,\delta x - (\sigma_{xz}\delta y\delta x)\,\delta z \equiv \sigma_{zx}\delta V - \sigma_{xz}\delta V,$$

$$\sum M_z = 0 = \left(\sigma_{xy}\delta x\delta z\right)\delta y - \left(\sigma_{yx}\delta y\delta z\right)\delta x \equiv \sigma_{xy}\delta V - \sigma_{yx}\delta V,$$

$$\sum M_x = 0 = \left(\sigma_{zy}\delta x\delta z\right)\delta y - \left(\sigma_{yz}\delta x\delta y\right)\delta z \equiv \sigma_{zy}\delta V - \sigma_{yz}\delta V,$$

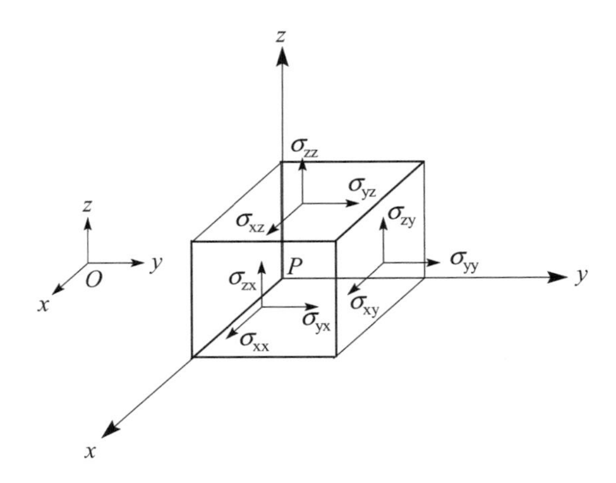

Figure 4.5 Cubic material element denoting the nine components of stresses

where δV is the volume of the infinitesimal cube. Hence, as a consequence of the requirement of rotational equilibrium of the cubic element, only six independent components of stress remain. Since the cube is infinitesimal, one may conclude that the state of stress at any point within the deformable body may be completely defined by the six components of stress.

The elastic displacement at any point P interior to the deformable body is a vector and may be resolved into three components parallel to the x, y and z directions. Denoting these components as u, v and w, it may be noted that the coordinates of the point after deformation are $x + u$, $y + v$ and $z + w$. In order to consider the strain at a point, we must consider the nature of the displacement and the resulting deformation. With this objective, consider a point P' infinitesimally close to the point P with coordinates $x + \delta x$, $y + \delta y$ and $z + \delta z$. The displacement of the point P' may be expressed as $u + \delta u$, $v + \delta v$ and $w + \delta w$. Assuming that δx, δy and δz are sufficiently small, the increments in the displacements may be expressed as

$$\delta u = \frac{\partial u}{\partial x}\delta x + \frac{\partial u}{\partial y}\delta y + \frac{\partial u}{\partial z}\delta z, \tag{4.57a}$$

$$\delta v = \frac{\partial v}{\partial x}\delta x + \frac{\partial v}{\partial y}\delta y + \frac{\partial v}{\partial z}\delta z \tag{4.57b}$$

and

$$\delta w = \frac{\partial w}{\partial x}\delta x + \frac{\partial w}{\partial y}\delta y + \frac{\partial w}{\partial z}\delta z. \tag{4.57c}$$

Thus, when the nine derivatives of the displacement components at a point are known, the displacement at any other point infinitesimally close to P may be found. The nine derivatives may be rearranged to define the nine quantities

$$\varepsilon_{xx} = \frac{\partial u}{\partial x}, \tag{4.58a}$$

$$\varepsilon_{yy} = \frac{\partial v}{\partial y}, \tag{4.58b}$$

$$\varepsilon_{zz} = \frac{\partial w}{\partial z}, \tag{4.58c}$$

$$\varepsilon_{yz} = \frac{\partial w}{\partial y} + \frac{\partial v}{\partial z}, \tag{4.58d}$$

$$\varepsilon_{zx} = \frac{\partial w}{\partial x} + \frac{\partial u}{\partial z}, \tag{4.58e}$$

$$\varepsilon_{xy} = \frac{\partial u}{\partial y} + \frac{\partial v}{\partial x}, \tag{4.58f}$$

$$\omega_x = \frac{1}{2}\left(\frac{\partial w}{\partial y} - \frac{\partial v}{\partial z}\right), \tag{4.58g}$$

$$\omega_y = \frac{1}{2}\left(\frac{\partial u}{\partial z} - \frac{\partial w}{\partial x}\right) \tag{4.58h}$$

and

$$\omega_z = \frac{1}{2}\left(\frac{\partial v}{\partial x} - \frac{\partial w}{\partial y}\right). \tag{4.58i}$$

Considering the last three of these nine quantities defined by equations (4.58g) to (4.58i), it is seen that they represent the components of the rotation vector associated with the displacement vector with components u, v and w, in a Cartesian system of coordinates x, y and z. Thus, the rotation vector with components ω_x, ω_y and ω_z is expressed as

$$\left[\omega_x \; \omega_y \; \omega_z\right]^T = \frac{1}{2}\nabla \times [u \; v \; w]^T = \frac{1}{2} curl\,[u \; v \; w]^T ,$$

where the nabla or del operator ∇ is defined as the vector

$$\nabla \equiv \left[\frac{\partial}{\partial x}\,\frac{\partial}{\partial y}\,\frac{\partial}{\partial z}\right]^T$$

and the vector cross-product $\mathbf{A} \times \mathbf{B}$ of any two vectors $\mathbf{A}$ and $\mathbf{B}$ is defined by the relationship

$$\mathbf{A} \times \mathbf{B} = \begin{vmatrix} i & j & k \\ A_1 & A_2 & A_3 \\ B_1 & B_2 & B_3 \end{vmatrix} .$$

Thus, the deformation may be considered to be pure strain when the rotational components of the displacement are zero, i.e. $\omega_x = 0$, $\omega_y = 0$ and $\omega_z = 0$.

To explain the strain–displacement relationships, it is essential to differentiate between the geometries of the undeformed and deformed states. If we define the vectors

$$\mathbf{s} = [x + u \;\; y + v \;\; z + w]^T$$

and

$$\mathbf{r} = [x \; y \; z]^T ,$$

the relationship that describes the deformation of the particles between two points, a distance $\mathbf{dr}$ apart, after the application of a force, may be expressed as

$$\delta^2 = (\mathbf{ds})^2 - (\mathbf{dr})^2 ,$$

where

$$\mathbf{ds} = [\mathrm{d}x + \mathrm{d}u \;\; \mathrm{d}y + \mathrm{d}v \;\; \mathrm{d}z + \mathrm{d}w]^T ,$$
$$\mathbf{dr} = [\mathrm{d}x \; \mathrm{d}y \; \mathrm{d}z]^T ,$$

and δ is the difference between the square of the length of an element between two points after and before deformation.

If we let, $x_1 = x, x_2 = y$ and $x_3 = z$, and $u_1 = u, u_2 = v$ and $u_3 = w$, we may approximate to second order and write

$$\delta \cong \frac{1}{2} \sum_{i,j} \in_{ij} \mathrm{d}x_i \mathrm{d}x_j,$$

where $\in_{ij}$ is a second-order strain tensor[1] defined by

$$\in_{ij}= \frac{1}{2} \left(\frac{\partial u_i}{\partial x_j} + \frac{\partial u_j}{\partial x_i} + \sum_k \frac{\partial u_k}{\partial x_i} \frac{\partial u_k}{\partial x_j} \right).$$

Assuming small displacements, the last term in the expression for the second-order strain tensor $\in_{ij}$ may be ignored and the linear strain tensor may be defined as

$$\in_{ij}\cong \frac{1}{2} \left(\frac{\partial u_i}{\partial x_j} + \frac{\partial u_j}{\partial x_i} \right). \tag{4.59}$$

The first three components of the linear geometric strain tensor defined by equation (4.59) are exactly the same as those defined in equations (4.58a) to (4.58c), while the remaining components are twice the corresponding strains defined by equations (4.58d) to (4.58f). Thus, the six equations (4.58a) to (4.58f) may be considered to define the strain components, as they relate to the deformation of the material in the presence of an external force. The first three of the nine relations defined by equations (4.58a) to (4.58c), ε_{xx}, ε_{yy} and ε_{zz} may be seen to correspond to the fractional expansions and contractions of infinitesimal line elements passing through the point P and parallel to the x, y and z axes, respectively. In fact, the scalar quantity

$$\Delta = \varepsilon_{xx} + \varepsilon_{yy} + \varepsilon_{zz} = \frac{\partial u}{\partial x} + \frac{\partial v}{\partial y} + \frac{\partial w}{\partial z} \tag{4.60}$$

is the dilatation or volumetric strain the element of material experiences. On the other hand, ε_{yz}, ε_{xz} and ε_{xy} correspond to the components of shear strain representing the distortional components of strain.

4.4.2 *Linear Elastic Materials*

An elastic material is one where all the work done by the application of an external force in deforming the material is stored as potential energy without any dissipation. The potential energy stored in an elastic body is completely recovered and converted into mechanical energy without any dissipation when the force is withdrawn. There are a number of materials that are said to be linear and elastic that obey the generalized Hooke's law. Thus, each of the six components of stress, at any point, is a linear function of the strain components. The generalized form of Hooke's law for a material may be written as

$$\sigma_{xx} = c_{11}\varepsilon_{xx} + c_{12}\varepsilon_{yy} + c_{13}\varepsilon_{zz} + c_{14}\varepsilon_{yz} + c_{15}\varepsilon_{zx} + c_{16}\varepsilon_{xy}, \tag{4.61a}$$

$$\sigma_{yy} = c_{21}\varepsilon_{xx} + c_{22}\varepsilon_{yy} + c_{23}\varepsilon_{zz} + c_{24}\varepsilon_{yz} + c_{25}\varepsilon_{zx} + c_{26}\varepsilon_{xy}, \tag{4.61b}$$

$$\sigma_{zz} = c_{31}\varepsilon_{xx} + c_{32}\varepsilon_{yy} + c_{33}\varepsilon_{zz} + c_{34}\varepsilon_{yz} + c_{35}\varepsilon_{zx} + c_{36}\varepsilon_{xy}, \tag{4.61c}$$

$$\sigma_{yz} = c_{41}\varepsilon_{xx} + c_{42}\varepsilon_{yy} + c_{43}\varepsilon_{zz} + c_{44}\varepsilon_{yz} + c_{45}\varepsilon_{zx} + c_{46}\varepsilon_{xy}, \tag{4.61d}$$

$$\sigma_{zx} = c_{51}\varepsilon_{xx} + c_{52}\varepsilon_{yy} + c_{53}\varepsilon_{zz} + c_{54}\varepsilon_{yz} + c_{55}\varepsilon_{zx} + c_{56}\varepsilon_{xy}, \tag{4.61e}$$

$$\sigma_{xy} = c_{61}\varepsilon_{xx} + c_{62}\varepsilon_{yy} + c_{63}\varepsilon_{zz} + c_{64}\varepsilon_{yz} + c_{65}\varepsilon_{zx} + c_{66}\varepsilon_{xy}. \tag{4.61f}$$

[1] A *tensor* is a vector in which the elements are themselves vectors or lower-order tensors.

In the case of an elastic material experiencing a deformation, the elastic component of the energy may be defined as

$$U = \frac{1}{\rho_0} \int \left(\sigma_{xx}\varepsilon_{xx} + \sigma_{yy}\varepsilon_{yy} + \sigma_{zz}\varepsilon_{zz} + \sigma_{yz}\varepsilon_{yz} + \sigma_{zx}\varepsilon_{zx} + \sigma_{xy}\varepsilon_{xy}\right) \mathrm{d}V.$$

Substituting for the stresses and requiring that the resulting strain energy is not a multi-valued function of the strain components but a single-valued function requires that the coefficients $c_{kl} = c_{lk}$ for all admissible values of l and k. This reduces the number of independent coefficients in the generalized form of Hooke's law from 36 to 21.

Considering first that the material is *monotropic* with respect to one plane, say the plane $z=0$, the material properties remain unchanged following a reflection about the plane $z=0$. Hence it follows that $c_{14} = c_{24} = c_{34} = 0$, $c_{15} = c_{25} = c_{35} = 0$ and $c_{46} = c_{56} = 0$. The number of independent coefficients then reduces to $21 - 8 = 13$. Consider further that the material is *orthotropic* or symmetric with respect two orthogonal planes, say the planes $x=0$ and $z=0$. In this case, it follows that the conditions $c_{16} = c_{26} = c_{36} = 0$ and $c_{45} = 0$ are also satisfied. The number of independent constants further reduces to $13 - 4 = 9$.

A material is said to have hexagonal symmetry if after an arbitrary rotation of the Cartesian frame about an axis (say the z axis) the material properties remain unchanged. In this case, it is easy to show that $c_{11} = c_{22}$, $c_{13} = c_{23}$, $c_{44} = c_{55}$ and $2c_{66} = (c_{11} - c_{12})$. The number of independent constants reduces to five. Such a material is *transversely isotropic*. When a material is completely *isotropic*, the number of constants reduces to just two. This is so because the material is transversely isotropic about all three axes and it follows that $c_{11} = c_{22} = c_{33}$, $c_{13} = c_{23} = c_{12}$, $c_{44} = c_{55} = c_{66}$ and $2c_{66} = (c_{11} - c_{12})$. It is customary in this case to denote $c_{13} = c_{23} = c_{12} = \lambda$ and $c_{44} = c_{55} = c_{66} = \mu$. It follows that $c_{11} = c_{22} = c_{33} = \lambda + 2\mu$. Thus, for an isotropic material we have

$$\sigma_{xx} = \lambda\Delta + 2\mu\varepsilon_{xx}, \tag{4.62a}$$

$$\sigma_{yy} = \lambda\Delta + 2\mu\varepsilon_{yy}, \tag{4.62b}$$

$$\sigma_{zz} = \lambda\Delta + 2\mu\varepsilon_{zz}, \tag{4.62c}$$

$$\sigma_{yz} = \mu\varepsilon_{yz}, \tag{4.62d}$$

$$\sigma_{zx} = \mu\varepsilon_{zx}, \tag{4.62e}$$

$$\sigma_{xy} = \mu\varepsilon_{xy}, \tag{4.62f}$$

where the dilatation

$$\Delta = \varepsilon_{xx} + \varepsilon_{yy} + \varepsilon_{zz}.$$

The two elastic constants λ and μ are known as Lame's constants and completely define the elastic behaviour of the isotropic solid. The constant μ may also be interpreted as the shear modulus or torsional rigidity. Thus, it is absent for an ideal fluid and only λ is relevant. Three other constants are often employed in most practical engineering calculations, and these are closely related to λ and μ. The *bulk modulus k* is defined as the ratio between the applied hydrostatic pressure and the fractional change in volumetric strain that follows as a result of the applied stress distribution. Under conditions of uniform hydrostatic compression,

$$\sigma_{xx} = \sigma_{yy} = \sigma_{zz} = -P \quad \text{and} \quad \sigma_{yz} = \sigma_{zx} = \sigma_{xy} = 0.$$

Hence it follows that

$$\lambda\Delta + 2\mu\varepsilon_{xx} = -P,$$
$$\lambda\Delta + 2\mu\varepsilon_{yy} = -P$$

and

$$\lambda\Delta + 2\mu\varepsilon_{zz} = -P.$$

Adding the three equations, it follows that

$$(3\lambda + 2\mu)\,\Delta = -3P.$$

The *bulk modulus* is defined as

$$k = \left|\frac{P}{\Delta}\right| = \lambda + \frac{2}{3}\mu. \tag{4.63}$$

For an ideal fluid $\mu = 0$ and it follows that

$$\sigma_{xx} = \sigma_{yy} = \sigma_{zz} = k\Delta = -P.$$

where P is the hydrostatic pressure. The negative sign arises since the pressure is reckoned positive when it acts inwards while the opposite is true for stress. Generally, k is large for liquids but not for gases. For an ideal incompressible fluid $k = \infty$.

Thus, the general theory of wave propagation in elastic materials is applicable to ideal fluids although fluids do not support the propagation of certain waves such as transverse or shear waves. The other two elastic constants conventionally employed in the engineering analysis of structural dynamics are the modulus of elasticity or Young's modulus E and Poisson's ratio ν. These properties of most common engineering materials as well as the densities of each of these materials are listed in Table 4.1. Young's modulus E is defined as the ratio between the applied longitudinal stress and the fractional longitudinal strain when a cylindrical or prismatic specimen is subjected to a uniform longitudinal stress over its plane ends and its lateral surfaces are maintained free from any constraint.

To derive the relationship between Lame's constants and the other elastic constants, it is assumed that the longitudinal axis of the specimen is the x axis, and σ_{xx} is the applied longitudinal stress and is the only non-zero component of stress. It then follows that

$$\sigma_{xx} = \lambda\Delta + 2\mu\varepsilon_{xx}, \quad 0 = \lambda\Delta + 2\mu\varepsilon_{yy}$$

and

$$0 = \lambda\Delta + 2\mu\varepsilon_{zz}.$$

From the last two equations, $\varepsilon_{zz} = \varepsilon_{yy}$ and hence we obtain

$$\sigma_{xx} = (\lambda + 2\mu)\,\varepsilon_{xx} + 2\lambda\varepsilon_{yy},$$
$$0 = (2\lambda + 2\mu)\,\varepsilon_{yy} + \lambda\varepsilon_{xx}.$$

Solving for the strains,

$$\varepsilon_{xx} = \frac{2(\lambda + \mu)}{2\mu(3\lambda + 2\mu)}\sigma_{xx}, \quad \varepsilon_{yy} = \varepsilon_{zz} = \frac{-\lambda}{2\mu(3\lambda + 2\mu)}\sigma_{xx}.$$

Young's modulus is then obtained as

$$E = \frac{\sigma_{xx}}{\varepsilon_{xx}} = \frac{\mu(3\lambda + 2\mu)}{\lambda + \mu} = \mu\left(2 + \frac{\lambda}{\lambda + \mu}\right). \tag{4.64}$$

Poisson's ratio ν is defined as the ratio between the lateral contraction and the longitudinal extension of the specimen with the lateral surfaces being maintained free of any stresses. Hence,

$$\nu = -\frac{\varepsilon_{yy}}{\varepsilon_{xx}} = \frac{\lambda}{2(\lambda + \mu)}. \tag{4.65}$$

Table 4.1 Density and elastic properties of materials

Material	Density (ρ) (10^3 kg/m^3)	Young's modulus (E) (10^9 N/m^2)	Poisson's ratio (ν)
Aluminium	2.71–2.8	71	0.34
Bismuth	9.8	32	0.33
Brass (Cu/Zn = 70/30)	8.5	100	0.35
Bronze (Cu/Sn = 90/10)	8.8	95	0.36
Constantan	8.88	170	0.33
Copper	8.93	117	0.35
German silver (Cu/Zn/Ni = 60/25/15)	8.7	130	0.33
Glass (crown)	2.6	71	0.235
Glass (flint)	4.2	80	0.24
Gold	19.3	71	0.44
India rubber	0.92	0.5	0.48
Invar (Fe/Ni = 64/36)	8.0	145	0.26
Iron (cast, grey)	7.15	110	0.27
Iron (wrought)	7.85	197	0.28
Lead	11.34	18	0.44
Magnesium	1.74	44	0.29
Manganin	8.5	120	0.33
Nickel	8.9	207	0.36
Nickel, strong alloy	8.5	110	0.38
Perspex	1.19	6.2	0.33
Platinum	21.45	150	0.38
Phosphor bronze	8.91	110	0.38
Quartz fibre	2.65	73	0.17
Silver	10.5	70	0.37
Steel (mild)	7.86	210	0.29
Steel (piano wire)	7.8	210	0.29
Tin	7.3	40	0.36
Tungsten	19.32	410	0.28
Zinc	7.14	110	0.25

The inverse relationships follow and are

$$\mu = \frac{E}{2(1+\nu)} \tag{4.66}$$

and

$$\lambda = \frac{E\nu}{(1+\nu)(1-2\nu)}. \tag{4.67}$$

Notes: For a homogeneous isotropic solid the following other constants may be derived:

1. $\mu = G = \dfrac{E}{2(1+\nu)}$,
2. $k = \dfrac{E}{3(1-2\mu)}$,
3. $k = \dfrac{EG}{3(3G-E)}$.

4.4.3 Equations of Wave Motion in an Elastic Medium

Consider a cuboid of the elastic material with sides δx, δy and δz. A Cartesian reference frame with its origin at point P and axes Px, Py and Pz is aligned along three mutually orthogonal edges of the cuboid. Let the components of the stress vector on the face $y=0$ be $-\sigma_{xy}$, $-\sigma_{yy}$, $-\sigma_{zy}$; on the face $z=0$ be $-\sigma_{xz}$, $-\sigma_{yz}$, $-\sigma_{zz}$; and the face $x=0$ be $-\sigma_{xx}$, $-\sigma_{yx}$, $-\sigma_{zz}$.

The components of the stress vector on the face $y = \delta y$ would then be

$$\sigma_{xy} + \frac{\partial \sigma_{xy}}{\partial y}\delta y,\ \sigma_{yy} + \frac{\partial \sigma_{yy}}{\partial y}\delta y,\ \sigma_{zy} + \frac{\partial \sigma_{zy}}{\partial y}\delta y,$$

while on the face $z = \delta z$ they would be

$$\sigma_{xz} + \frac{\partial \sigma_{xz}}{\partial z}\delta z,\ \sigma_{yz} + \frac{\partial \sigma_{yz}}{\partial z}\delta z,\ \sigma_{zz} + \frac{\partial \sigma_{zz}}{\partial z}\delta z$$

and on the face $x = \delta x$ they would be

$$\sigma_{xx} + \frac{\partial \sigma_{xx}}{\partial x}\delta x,\ \sigma_{yx} + \frac{\partial \sigma_{yx}}{\partial x}\delta x,\ \sigma_{zx} + \frac{\partial \sigma_{zx}}{\partial x}\delta x.$$

Considering the positive x direction and ignoring all other body forces, the resultant force acting in this direction is

$$F_x = \left(\sigma_{xx} + \frac{\partial \sigma_{xx}}{\partial x}\delta x\right)\delta y\delta z - \sigma_{xx}\delta y\delta z + \left(\sigma_{xy} + \frac{\partial \sigma_{xy}}{\partial y}\delta y\right)\delta x\delta z$$
$$-\sigma_{xy}\delta x\delta z + \left(\sigma_{xz} + \frac{\partial \sigma_{xz}}{\partial z}\delta z\right)\delta x\delta y - \sigma_{xz}\delta x\delta y,$$

and this reduces to

$$F_x = \left(\frac{\partial \sigma_{xx}}{\partial x} + \frac{\partial \sigma_{xy}}{\partial y} + \frac{\partial \sigma_{xz}}{\partial z}\right)\delta z\delta x\delta y.$$

By Newton's second law of motion, the mass of the cuboid $\rho\delta z\delta x\delta y$ multiplied by the acceleration in the x direction is equal to the applied external force in that direction. Hence,

$$F_x = \rho\delta z\delta x\delta y\frac{\partial^2 u}{\partial t^2},$$

which results in the equation

$$\frac{\partial \sigma_{xx}}{\partial x} + \frac{\partial \sigma_{xy}}{\partial y} + \frac{\partial \sigma_{xz}}{\partial z} = \rho\frac{\partial^2 u}{\partial t^2}. \tag{4.68a}$$

Similarly, by considering the y and z directions and applying Newton's second law of motion, it may be shown that

$$\frac{\partial \sigma_{yx}}{\partial x} + \frac{\partial \sigma_{yy}}{\partial y} + \frac{\partial \sigma_{yz}}{\partial z} = \rho\frac{\partial^2 v}{\partial t^2}, \tag{4.68b}$$

$$\frac{\partial \sigma_{zx}}{\partial x} + \frac{\partial \sigma_{zy}}{\partial y} + \frac{\partial \sigma_{zz}}{\partial z} = \rho\frac{\partial^2 w}{\partial t^2}. \tag{4.68c}$$

Equations (4.68a) to (4.68c) are the governing equations of motion of particle of an elastic medium and are quite independent of the stress–strain relations.

Considering the elastic medium to be isotropic with λ and μ constants, the equations of motion may be expressed in terms of the strains rather than the stresses. Thus, we obtain the equations

$$\frac{\partial\left(\lambda\Delta+2\mu\varepsilon_{xx}\right)}{\partial x}+\mu\frac{\partial\varepsilon_{xy}}{\partial y}+\mu\frac{\partial\varepsilon_{xz}}{\partial z}=\rho\frac{\partial^2 u}{\partial t^2}, \tag{4.69a}$$

$$\mu\frac{\partial\varepsilon_{yx}}{\partial x}+\frac{\partial\left(\lambda\Delta+2\mu\varepsilon_{yy}\right)}{\partial y}+\mu\frac{\partial\varepsilon_{yz}}{\partial z}=\rho\frac{\partial^2 v}{\partial t^2} \tag{4.69b}$$

and

$$\mu\frac{\partial\varepsilon_{zx}}{\partial x}+\mu\frac{\partial\varepsilon_{zy}}{\partial y}+\frac{\partial\left(\lambda\Delta+2\mu\varepsilon_{zz}\right)}{\partial z}=\rho\frac{\partial^2 w}{\partial t^2}, \tag{4.69c}$$

where the dilatation is given by equation (4.60).

Alternatively, eliminating the strains in equations (4.69a) to (4.69c), the equations of motion may also be expressed as

$$(\lambda+\mu)\frac{\partial\Delta}{\partial x}+\mu\left(\frac{\partial^2}{\partial x^2}+\frac{\partial^2}{\partial y^2}+\frac{\partial^2}{\partial z^2}\right)u=\rho\frac{\partial^2 u}{\partial t^2}, \tag{4.70a}$$

$$(\lambda+\mu)\frac{\partial\Delta}{\partial y}+\mu\left(\frac{\partial^2}{\partial x^2}+\frac{\partial^2}{\partial y^2}+\frac{\partial^2}{\partial z^2}\right)v=\rho\frac{\partial^2 v}{\partial t^2} \tag{4.70b}$$

and

$$(\lambda+\mu)\frac{\partial\Delta}{\partial w}+\mu\left(\frac{\partial^2}{\partial x^2}+\frac{\partial^2}{\partial y^2}+\frac{\partial^2}{\partial z^2}\right)w=\rho\frac{\partial^2 w}{\partial t^2}. \tag{4.70c}$$

4.4.4 Plane Waves in an Infinite Solid

In vector terms, since the dilatation may be expressed as the dot product,

$$\Delta=\nabla\cdot\mathbf{u}=div\,\mathbf{u},$$

where the gradient or del ∇ operator is defined in Cartesian coordinates as

$$\nabla=\frac{\partial}{\partial x}\mathbf{i}+\frac{\partial}{\partial y}\mathbf{j}+\frac{\partial}{\partial z}\mathbf{k},$$

i, **j** and **k** are unit vectors in the x, y and z directions and **u** is the displacement vector. The gradient of the dilatation may be written as

$$\nabla\left(\nabla\cdot\mathbf{u}\right)=grad\,div\,\mathbf{u}.$$

It follows that the equations of motion may be expressed, in vector form, as

$$\rho\frac{\partial^2\mathbf{u}}{\partial t^2}=(\lambda+\mu)\,grad\,div\,\mathbf{u}+\mu\nabla^2\mathbf{u}, \tag{4.71}$$

where ∇^2 is the Laplacian operator. In Cartesian coordinates, the Laplacian operator ∇^2 is given by

$$\nabla^2=\nabla\cdot\nabla=\frac{\partial^2}{\partial x^2}+\frac{\partial^2}{\partial y^2}+\frac{\partial^2}{\partial z^2}.$$

To reduce the equation to a standard wave equation, we assume that the displacement vector is the vector sum of two component vectors $\mathbf{u}_L$ and $\mathbf{u}_T$. Each of these components is assumed to satisfy the constraint equations

$$\nabla\cdot\mathbf{u}_T=div\,\mathbf{u}_T=0 \tag{4.72}$$

and

$$\nabla \times \mathbf{u}_L = curl\,\mathbf{u}_L = 0. \tag{4.73}$$

The first of these constraints implies that the dilatation due to a component of the displacement vector is zero, while the second constraint implies that the motion corresponding to another component of the displacement vector is irrotational, since the corresponding rotation components are equal to zero.

Further $\mathbf{u}_T$ is assumed to satisfy

$$\rho \frac{\partial^2 \mathbf{u}_T}{\partial t^2} = \mu \nabla^2 \mathbf{u}_T. \tag{4.74}$$

It then follows that $\mathbf{u}_L$ satisfies

$$\rho \frac{\partial^2 \mathbf{u}_L}{\partial t^2} = (\lambda + \mu)\, grad\, div\, \mathbf{u}_L + \mu \nabla^2 \mathbf{u}_L. \tag{4.75}$$

But for any vector, the vector identity

$$grad\, div\, \mathbf{u}_L = \nabla^2 \mathbf{u}_L - curl\, curl\, \mathbf{u}_L \tag{4.76}$$

is satisfied. Hence, equation satisfied by $\mathbf{u}_L$ may be simplified as

$$\rho \frac{\partial^2 \mathbf{u}_L}{\partial t^2} = (\lambda + 2\mu)\, \nabla^2 \mathbf{u}_L. \tag{4.77}$$

Thus, from equations (4.64) and (4.65) it follows that each of the two component displacement vectors $\mathbf{u}_L$ and $\mathbf{u}_T$ satisfies standard wave equations, although each of the two components is propagated with a different velocity. Waves involving no volumetric strain or dilatation satisfy the constraint defined by equation (4.71) and travel with the velocity

$$a_T = \sqrt{\frac{\mu}{\rho}}, \tag{4.78}$$

whilst irrotational waves satisfy the constraint defined by equation (4.72) and travel with the velocity

$$a_L = \sqrt{\frac{\lambda + 2\mu}{\rho}}. \tag{4.79}$$

The equi-voluminal waves involve no volumetric strain or dilatation and are essentially distortional waves, whilst the irrotational waves are purely dilatational waves involving no distortion. The latter are essentially compressional waves and are usually labelled as *P* waves (push waves), while the former are essentially shear waves (*S* waves) and are labelled *SV* or *SH waves* depending on the direction of particle motion. In the latter case, the particle motion may be either vertical or horizontal and perpendicular to the direction of wave motion, and in the former case, the directions of particle motion and wave propagation are in one and the same directions. If we let the suffix 1 denote the case of pure dilatational waves and the suffix 2 denote the case of distortional waves, the two wave equations may be expressed as

$$\frac{\partial^2 \mathbf{u}_k}{\partial t^2} = a_k^2 \nabla^2 \mathbf{u}_k, \quad k = 1 \text{ and } 2 \tag{4.80}$$

with

$$a_1 = \sqrt{\frac{\lambda + 2\mu}{\rho}} \quad \text{and} \quad a_2 = \sqrt{\frac{\mu}{\rho}}.$$

A plane wave solution is a solution of the form

$$\mathbf{u}_k = \mathbf{u}_k\,(\vec{\nu} \cdot \mathbf{r} \mp a_k t), \quad k = 1,\ 2 \tag{4.81}$$

where

$$\vec{\nu} \cdot \mathbf{r} = [\nu_1\ \nu_2\ \nu_3]\,[x\ y\ z]^T = \nu_1 x + \nu_2 y + \nu_3 z.$$

Substituting the assumed solutions into the wave equations for $\mathbf{u}_k$, we observe that the equations are satisfied when

$$[\nu_1\ \nu_2\ \nu_3]\,[\nu_1\ \nu_2\ \nu_3]^T = 1. \tag{4.82}$$

Thus, ν_k, $k = 1, 2, 3$ are the direction cosines of a line perpendicular to the plane P determined by the equation

$$\vec{\nu} \cdot \mathbf{r} \mp a_k t = \text{constant}$$

at any fixed time t. The plane waves travel with a speed $\pm a_k$ along the line, which remains perpendicular to the plane. Hence, solutions given by equation (4.24) and meeting the constraint (4.25) represent plane wave solutions of each of the wave equations.

It is instructive to see how the plane wave velocities a_k, $k = 1, 2$, compare with the velocity of sound in a uniform bar

$$a_0 = \sqrt{\frac{E}{\rho}}.$$

The ratios

$$\left(\frac{a_1}{a_0}\right)^2 = \frac{\lambda + 2\mu}{E} = \frac{(1-\nu)}{(1+\nu)(1-2\nu)}$$

and

$$\left(\frac{a_2}{a_0}\right)^2 = \frac{\mu}{E} = \frac{1}{2(1+\nu)}$$

are functions only of the Poisson's ratio. When $\nu = 0.33$,

$$\left(\frac{a_1}{a_0}\right)^2 = \frac{3}{2}$$

and

$$\left(\frac{a_2}{a_0}\right)^2 = \frac{3}{8}.$$

4.4.5 *Spherical Waves in an Infinite Medium*

In spherical coordinates, r, θ, ϕ, the Laplacian operator ∇^2 is given by

$$\nabla^2 = \frac{1}{r^2}\frac{\partial}{\partial r}\left(r^2\frac{\partial \mathbf{u}_k}{\partial r}\right) + \frac{1}{r^2 \sin\theta}\frac{\partial}{\partial \theta}\left(\sin\theta\frac{\partial \mathbf{u}_k}{\partial \theta}\right) + \frac{1}{r^2 \sin^2\theta}\frac{\partial^2 \mathbf{u}_k}{\partial \phi^2}.$$

When we restrict our study to waves that have spherical symmetry so that there is no dependence of $\mathbf{u}_k$ on θ or ϕ, then the waves are such that they are advancing along spherical wavefronts. Thus, the vector displacement components $\mathbf{u}_1$ and $\mathbf{u}_2$ are functions only of r and t; thus, $\mathbf{u}_k = \mathbf{u}_k\,(r, t)$ and it satisfies the equation

$$\frac{\partial^2 \mathbf{u}_k}{\partial r^2} + \frac{2}{r}\frac{\partial \mathbf{u}_k}{\partial r} = \frac{1}{c_k^2}\frac{\partial^2 \mathbf{u}_k}{\partial t^2}. \tag{4.83}$$

On substitution of

$$\mathbf{u}_k(r,t) = \frac{1}{r}\mathbf{F}_k(r,t),$$

the governing equation (4.83) reduces to

$$\frac{\partial^2 \mathbf{F}_k}{\partial r^2} = \frac{1}{c_k^2}\frac{\partial^2 \mathbf{F}_k}{\partial t^2} \tag{4.84}$$

and the general solution of equation (4.84) is

$$\mathbf{F}_k = \mathbf{F}_{k1}(r - a_k t) + \mathbf{F}_{k2}(r + a_k t),$$

where $\mathbf{F}_{k1}$ and $\mathbf{F}_{k2}$ are arbitrary functions. The general solutions for $\mathbf{u}_k$ are

$$\mathbf{u}_k = \frac{\mathbf{F}_{k1}(r - a_k t)}{r} + \frac{\mathbf{F}_{k2}(r + a_k t)}{r}. \tag{4.85}$$

The first term in the general solution given by equation (4.85) represents a spherical wavefront, $r - a_k t =$ constant, advancing in a radial direction from the origin $r = 0$. The second term in the general solution represents a system of spherical waves advancing towards the origin.

For a disturbance at the origin $r = 0$, such as an explosion, the waves may travel only in the $r > 0$ direction so that in this case $\mathbf{F}_{k2} = 0$. This condition is related to the problem of uniqueness in the case of steady-state motion and is known as the Sommerfeld radiation condition. It is a mathematical expression of the intuitive notion that steady-state waves of any type generated at a point must travel outwards, towards $r = \infty$, and must not radiate back.

4.4.6 Transmission Line Model for Wave Propagation in Isotropic Solids

It is possible to establish a direct analogy between the propagation of plane stress waves in an isotropic elastic solid and the propagation of waves in a transmission line. To reveal the analogy, consider the equations of particle motion in an elastic solid given by equations (4.68). Assuming that the waves propagate in the x direction and that the stress waves are purely longitudinal, we may set the partial derivatives in the y and z directions to zero. The equations of motion then reduce to

$$\frac{\partial \sigma_{xx}}{\partial x} = \rho \frac{\partial^2 u}{\partial t^2}, \tag{4.86a}$$

$$\frac{\partial \sigma_{yx}}{\partial x} = \rho \frac{\partial^2 v}{\partial t^2}, \tag{4.86b}$$

$$\frac{\partial \sigma_{zx}}{\partial x} = \rho \frac{\partial^2 w}{\partial t^2}. \tag{4.86c}$$

In terms of the particle velocities v_x, v_y and v_z in the x, y and z directions, the equations may be written as

$$\frac{\partial \sigma_{xx}}{\partial x} = \rho \frac{\partial v_x}{\partial t}, \tag{4.87a}$$

$$\frac{\partial \sigma_{yx}}{\partial x} = \rho \frac{\partial v_y}{\partial t}, \tag{4.87b}$$

$$\frac{\partial \sigma_{zx}}{\partial x} = \rho \frac{\partial v_z}{\partial t}. \tag{4.87c}$$

The relevant stress-strain equations are

$$\sigma_{xx} = (\lambda + 2\mu)\,\varepsilon_{xx}, \tag{4.88a}$$

$$\sigma_{zx} = \mu\varepsilon_{zx}, \tag{4.88b}$$

$$\sigma_{xy} = \mu\varepsilon_{xy}, \tag{4.88c}$$

and the relationships between the strain and particle velocity are

$$\frac{\partial}{dt}\varepsilon_{xx} = \frac{\partial v_x}{\partial x}, \tag{4.89a}$$

$$\frac{\partial}{\partial t}\varepsilon_{zx} = \frac{\partial v_z}{\partial x}, \tag{4.89b}$$

$$\frac{\partial}{\partial t}\varepsilon_{xy} = \frac{\partial v_y}{\partial x}. \tag{4.89c}$$

Eliminating the strains from above sets of equations,

$$\frac{\partial}{dt}\sigma_{xx} = (\lambda + 2\mu)\,\frac{\partial v_x}{\partial x}, \tag{4.90a}$$

$$\frac{\partial}{dt}\sigma_{zx} = \mu\frac{\partial v_z}{\partial x}, \tag{4.90b}$$

$$\frac{\partial}{dt}\sigma_{xy} = \mu\frac{\partial v_y}{\partial x}. \tag{4.90c}$$

Thus, the complete set of equations for the particle motion is

$$\frac{\partial \sigma_{xx}}{\partial x} = \rho\frac{\partial v_x}{\partial t}, \tag{4.91a}$$

$$\frac{\partial \sigma_{yx}}{\partial x} = \rho\frac{\partial v_y}{\partial t}, \tag{4.91b}$$

$$\frac{\partial \sigma_{zx}}{\partial x} = \rho\frac{\partial v_z}{\partial t}, \tag{4.91c}$$

$$\frac{\partial}{dt}\sigma_{xx} = (\lambda + 2\mu)\,\frac{\partial v_x}{\partial x}, \tag{4.91d}$$

$$\frac{\partial}{dt}\sigma_{zx} = \mu\frac{\partial v_z}{\partial x}, \tag{4.91e}$$

$$\frac{\partial}{dt}\sigma_{xy} = \mu\frac{\partial v_y}{\partial x}. \tag{4.91f}$$

The equations may be expressed in pairs, in transmission line form as

$$-\frac{\partial \sigma_{xx}}{\partial x} + \rho\frac{\partial v_x}{\partial t} = 0, \tag{4.92a}$$

$$\frac{\partial v_x}{\partial x} - \frac{1}{(\lambda + \mu)}\frac{\partial}{\partial t}\sigma_{xx} = 0, \tag{4.92b}$$

$$-\frac{\partial \sigma_{yx}}{\partial x} + \rho\frac{\partial v_y}{\partial t} = 0, \tag{4.92c}$$

$$\frac{\partial v_z}{\partial x} - \frac{1}{\mu}\frac{\partial}{\partial t}\sigma_{zx} = 0, \tag{4.92d}$$

$$-\frac{\partial \sigma_{zx}}{\partial x} + \rho\frac{\partial v_z}{\partial t} = 0, \tag{4.92e}$$

$$\frac{\partial v_y}{\partial x} - \frac{1}{\mu}\frac{\partial}{\partial t}\sigma_{xy} = 0. \tag{4.92f}$$

Table 4.2 Transmission line model analogies

Elastic quantity	Symbol	Transmission line equivalent	Symbol
Stress	σ	Negative voltage	$-V$
Particle velocity	v	Current	I
Density	ρ	Inductance per unit length	L
Lame's constants	$\lambda + \mu$ and μ	Inverse of capacitance per unit length	$\frac{1}{C_0}$

Each pair of these equations are coupled with each other but independent of the other pairs and may be compared with the equations of a transmission line, (4.31) and (4.32), which are

$$\frac{\partial V}{\partial x} + L_0 \frac{\partial I}{\partial t} = 0, \tag{4.93a}$$

$$\frac{\partial I}{\partial x} + C_0 \frac{\partial V}{\partial t} = 0. \tag{4.93b}$$

The analogies listed in Table 4.2 are then identified. As in the case of the mechanical transmission line considered in Section 4.3.4, the stress σ is interpreted as the negative of the voltage V, the particle velocity v as the electrical current I, the material density ρ as the electrical line inductance L_0 and Lame's constants $\lambda + \mu$ or μ as the inverse of the electrical line capacitance C_0. However, in the mechanical transmission line impedance was defined as the ratio of the force to the velocity, while in the above case it is completely analogous to the acoustic impedance.

4.4.7 Surface Waves in Semi-infinite Solids

While in an unbounded elastic medium two types of elastic stress waves may propagate in any direction, in an elastic half-space, which is characterized by a bounding surface free of any stresses, waves are attenuated everywhere except near the surface. Surface acoustic waves (SAWs) are a class of waves that propagate along the interface of two elastic media, at least one of which is a solid. Amongst the most important surface waves are Rayleigh waves, which propagate on the stress-free planar surface of a semi-infinite isotropic half-space. Lord Rayleigh showed in 1887 that *SV*-type waves degenerate to a surface wave and can propagate without undergoing any attenuation in the vicinity of a bounding surface free of any stresses, which is usually referred to as a free surface. White (1970) discussed extensively the various types of surface elastic waves and their physical features. The particle motion although predominantly transverse involves both longitudinal and vertical transverse motions relative to a horizontal free surface and results in the particle tracing an elliptic path with its major axis normal to the surface. The amplitude of the elliptically polarized particle displacement decays exponentially with depth and becomes negligible for a penetration depth of more than a few wavelengths. Therefore, typically more than 90% of the elastic energy remains concentrated within a distance of the order of one wavelength of the surface.

In 1911, A. E. H. Love showed that *SH*-type transverse waves may also propagate near the surface, provided the density of the material in a direction normal to the free surface varies such that there is a higher transverse velocity in the layers closer to the free surface, thus facilitating the transmission of the waves in these layers. The occurrence of surface elastic waves is similar to a boundary layer-type phenomenon in wave propagation and is akin to the manner in which electric current flows more freely near the surface of an electric conductor. The situation arises because of the total reflection of elastic waves at the free surface and the absence of any refraction. Waves travelling parallel to the free surface escape any reflection and are propagated without suffering any attenuation. Thus, only the surface waves prevail resulting in a sustained propagation of these type waves.

4.4.7.1 Rayleigh Waves

To consider the propagation of a plane wave through an elastic medium with a plane boundary, we introduce two displacement potentials ϕ and ψ defined by the displacement distributions

$$u = \frac{\partial \phi}{\partial x} + \frac{\partial \psi}{\partial z}, \quad v \equiv 0, \quad w = \frac{\partial \phi}{\partial z} - \frac{\partial \psi}{\partial x}.$$

The dilatation would then be

$$\Delta = \frac{\partial u}{\partial x} + \frac{\partial w}{\partial z} = \left(\frac{\partial^2}{\partial x^2} + \frac{\partial^2}{\partial z^2} \right) \phi,$$

whilst ω_y, the rotation in the x–z plane, is given by

$$2\omega_y = \frac{\partial u}{\partial z} - \frac{\partial w}{\partial x} = \left(\frac{\partial^2}{\partial x^2} + \frac{\partial^2}{\partial z^2} \right) \psi.$$

Consider the equation of motion for longitudinal waves given by equation (4.77). Applying the divergence operation to both sides,

$$\rho \frac{\partial^2 (\nabla \cdot \mathbf{u}_L)}{\partial t^2} = (\lambda + 2\mu) \nabla^2 (\nabla \cdot \mathbf{u}_L).$$

Since $\nabla \cdot \mathbf{u}_T = 0$, we have

$$\nabla \cdot \mathbf{u}_L = (\nabla \cdot \mathbf{u}_L + \nabla \cdot \mathbf{u}_T) = \nabla \cdot \mathbf{u} = \nabla^2 \phi,$$

and it can then be shown that

$$\rho \frac{\partial^2 \phi}{\partial t^2} = (\lambda + 2\mu) \nabla^2 \phi. \tag{4.94}$$

By a similar argument,

$$\rho \frac{\partial^2 \psi}{\partial t^2} = \mu \nabla^2 \psi. \tag{4.95}$$

Equation (4.94) represents the dilatational waves while equation (4.95) represents the distortional waves. Assuming that the amplitudes of the disturbance waves do not increase as they propagate to increasing depths (z – increasing), the general solutions of the equations (4.94) and (4.95) may be expressed as

$$\phi = A \exp(-qz + i(\omega t - fx)), \quad q = \sqrt{f^2 - \frac{\omega^2}{a_1^2}}, \tag{4.96}$$

$$\psi = B \exp(-sz + i(\omega t - fx)), \quad s = \sqrt{f^2 - \frac{\omega^2}{a_2^2}}, \tag{4.97}$$

with

$$a_1 = \sqrt{\frac{\lambda + 2\mu}{\rho}} \tag{4.98}$$

and

$$a_2 = \sqrt{\frac{\mu}{\rho}}. \tag{4.99}$$

The solution represents a plane wave propagating in the x direction with a velocity

$$a_R = \frac{\omega}{f}. \tag{4.100}$$

Furthermore, the solutions must satisfy the boundary condition on the free surface, which is totally stress free. Thus along the plane $z=0$, we have the boundary conditions

$$\sigma_{zx} = 0, \sigma_{zy} = 0 \quad \text{and} \quad \sigma_{zz} = 0. \tag{4.101}$$

Considering the last of the three boundary conditions in equation (4.101) and employing the fact that

$$\sigma_{zz} = \lambda\Delta + 2\mu\frac{\partial w}{\partial z},$$

we obtain

$$\sigma_{zz} = (\lambda + 2\mu)\frac{\partial^2\phi}{\partial x^2} + \lambda\frac{\partial^2\phi}{\partial z^2} - 2\mu\frac{\partial^2\psi}{\partial x\partial z}. \tag{4.102}$$

Substituting for the displacement potentials ϕ and ψ, and setting $z = 0$,

$$A\left[(\lambda + 2\mu)q^2 - \lambda f^2\right] - 2iB\mu\ s\ f = 0. \tag{4.103}$$

Considering the first of the three boundary conditions in equation (4.101) and employing the fact that

$$\sigma_{zx} = \mu\left(\frac{\partial u}{\partial z} + \frac{\partial w}{\partial x}\right),$$

we obtain

$$\sigma_{zz} = \mu\left(2\frac{\partial^2\phi}{\partial x\partial z} - \frac{\partial^2\psi}{\partial x^2} + \frac{\partial^2\psi}{\partial z^2}\right). \tag{4.104}$$

Substituting for the displacement potentials ϕ and ψ, and setting $z = 0$,

$$B\left(s^2 + f^2\right) + 2i\,A\,q\,f = 0. \tag{4.105}$$

Solving the two equations (4.102) and (4.104) for the ratio of A by B, we obtain

$$\frac{A}{B} = \frac{2i\mu\ s\ f}{(\lambda + 2\mu)q^2 - \lambda f^2} = -\frac{\left(s^2 + f^2\right)}{2i\ q\ f}.$$

Thus eliminating the ratio of A by B,

$$4\mu\ q\ s\ f^2 = \left(s^2 + f^2\right)\left((\lambda + 2\mu)q^2 - \lambda f^2\right). \tag{4.106}$$

If we denote the ratio of the transverse wave speed a_2 and the longitudinal wave speed a_1 as

$$\alpha = \frac{a_2}{a_1},$$

then we have

$$q = f^2\sqrt{1 - \frac{a_R^2}{a_2^2}\alpha^2}, \quad s = f^2\sqrt{1 - \frac{a_R^2}{a_2^2}}.$$

Both sides of the frequency equation (4.106) may be squared and the resulting equation expressed as

$$16f^4\left(1 - \frac{a_R^2}{a_2^2}\alpha^2\right)\left(1 - \frac{a_R^2}{a_2^2}\right) = f^4\left(\left(\frac{\lambda}{\mu} + 2\right)\left(1 - \frac{a_R^2}{a_2^2}\alpha^2\right) - \frac{\lambda}{\mu}\right)^2\left(2 - \frac{a_R^2}{a_2^2}\right)^2,$$

This reduces to

$$16\left(1 - \frac{a_R^2}{a_2^2}\alpha^2\right)\left(1 - \frac{a_R^2}{a_2^2}\right) = \left(2\left(1 - \frac{a_R^2}{a_2^2}\alpha^2\right) - \frac{\lambda}{\mu}\frac{a_R^2}{a_2^2}\alpha^2\right)^2\left(2 - \frac{a_R^2}{a_2^2}\right)^2$$

and to

$$16\left(1-\frac{a_R^2}{a_2^2}\alpha^2\right)\left(1-\frac{a_R^2}{a_2^2}\right)=\left(2-\frac{a_R^2}{a_2^2}\right)^4. \tag{4.107}$$

Equation is the frequency equation that determines the characteristic surface wave speed a_R, and it depends only on the material properties of the medium. Assuming that it is given as a multiple of the transverse wave speed

$$a_R = \upsilon\, a_2,$$

the frequency equation (4.99) may be expressed as

$$\upsilon^6 - 8\upsilon^4 + 8\left(3-2\alpha^2\right)\upsilon^2 - 16\left(1-\alpha^2\right) = 0. \tag{4.108}$$

The equation is a bi-cubic and may be expressed in terms of its three roots as

$$\left(\upsilon^2-\upsilon_1^2\right)\left(\upsilon^2-\upsilon_2^2\right)\left(\upsilon^2-\upsilon_3^2\right) = 0. \tag{4.109}$$

Usually only one of the three roots has a magnitude less than unity that will result in realistic values for q and s, corresponding to the surface waves.

The surface waves attenuate with depth due to the $\exp(-qz)$ term and the $\exp(-sz)$ term in the expressions for the displacement potentials. To illustrate the nature of the displacements, we assume that the displacements corresponding to the dilatational components may be expressed in the form

$$\begin{aligned} u_1 &= C_1 \exp\left(-qz + i\left(\omega t - fx\right)\right), \\ w_1 &= D_1 \exp\left(-qz + i\left(\omega t - fx\right)\right), \end{aligned}$$

and it follows that

$$\frac{D_1}{C_1} = i\frac{q}{f}.$$

Similarly, for the distortional components of the waves,

$$\frac{D_2}{C_2} = i\frac{f}{s}.$$

Thus, the displacement vector may be expressed as

$$\mathbf{u} = -A\left(if\,\mathbf{i} + q\,\mathbf{k}\right)\exp\left(-qz + i\left(\omega t - fx\right)\right) \tag{4.110}$$

for dilatational waves where $\mathbf{i}$ and $\mathbf{k}$ are unit vectors in the x and z, respectively, and as

$$\mathbf{u} = iB\left(is\,\mathbf{i} + f\,\mathbf{k}\right)\exp\left(-sz + i\left(\omega t - fx\right)\right) \tag{4.111}$$

for distortional waves where A and B satisfy equation (4.105). The total displacement is the vector sum of the two component displacement vectors.

Illustrative Example 4.2: Computation of the Rayleigh Wave Velocity

Compute the Rayleigh wave velocity for a material with $\nu = 0.25$, which is representative of glass ($a_2 = 3.35$ km/s).

Soution: The characteristic frequency equation takes the form

$$\upsilon^6 - 8\upsilon^4 + 8\left(3-2\alpha^2\right)\upsilon^2 - 16\left(1-\alpha^2\right) = 0$$

with

$$\alpha = \sqrt{\frac{1-2\nu}{2(1-\nu)}} = \sqrt{\frac{1}{3}}.$$

The roots are

$$v_1^2 = 2\left(1 - \sqrt{\frac{1}{3}}\right) = 0.8453,$$

$$v_2^2 = 2\left(1 + \sqrt{\frac{1}{3}}\right) = 3.1547$$

and

$$v_3^2 = 4.$$

Only the first root gives realistic values for the quantity s, and it follows that

$$a_R = v_1\, a_2 = 0.9194 \times 3.35 = 3.079 \text{ km/s}.$$

4.4.7.2 Love Waves

To illustrate the analysis of Love waves, we model the half-space as a two-layered medium. The outer or upper layer adjoining the free surface is above the inner layer and is a layer of depth H and with a density ρ_u and elastic constants λ_u and μ_u. The inner layer below the upper layer is assumed to extend to infinity, and together they constitute a semi-infinite medium. The inner layer has a density ρ_i and elastic constants λ_i and μ_i.

Assuming that the displacement vector is purely distortional, in the outer layer it is assumed to be of the form

$$u \equiv 0, \quad v = (C_1 \exp(is_1 z) + D_1 \exp(-is_1 z)) \exp(-i(\omega t - fx)), \quad w \equiv 0, \tag{4.112}$$

where

$$s_1 = f^2 \sqrt{\frac{\omega^2 \rho_v}{f^2 \mu_v} - 1}.$$

In the inner layer it is assumed to be of the form

$$u \equiv 0, \quad v = (C_2 \exp(-s_1 z)) \exp(-i(\omega t - fx)), \quad w \equiv 0,$$

where

$$s_1 = f^2 \sqrt{1 - \frac{\omega^2 \rho_\iota}{f^2 \mu_\iota}}.$$

The interface between the two layers is defined by the plane $z=0$, while the free-surface is defined by $z=-H$. Hence when $z=0$, at the interface between the two layers the displacement component v and σ_{zy} must be continuous. At $z=-H$, at the free surface it is stress free and $\sigma_{zy} = 0$.

Hence,

$$C_2 = C_1 + D_1, \tag{4.113}$$

$$i\mu_i (C_1 - D_1) s_1 = -s_2 \mu_u C_2 = -s_2 \mu_u (C_1 + D_1) \tag{4.114}$$

and

$$C_1 \exp(-is_1 H) - D_1 \exp(is_1 H) = 0. \tag{4.115}$$

Rearranging equation (4.114),

$$(i\mu_i s_1 + s_2\mu_u) C_1 = D_1 (i\mu_i s_1 - s_2\mu_u)$$

and eliminating C_1 and D_1 from equation (4.115),

$$i\mu_\iota s_1 (\exp(-is_1 H) - \exp(is_1 H)) = s_2\mu_u (\exp(-is_1 H) + \exp(is_1 H)),$$

which may be expressed as

$$\mu_\iota s_1 \tan(s_1 H) = s_2\mu_u. \tag{4.116}$$

Equation (4.116) is a transcendental equation that must solved for the Love wave velocity,

$$a_L = \frac{\omega}{f}. \tag{4.117}$$

A real solution for the wave velocity satisfies the inequality

$$\sqrt{\frac{\mu_i}{\rho_i}} < a_L < \sqrt{\frac{\mu_u}{\rho_u}}.$$

The actual value of the wave velocity depends not only on the wave number f but also on the thickness of the layer H, and hence is not a material constant.

4.4.7.3 Stoneley Waves

Stoneley investigated a type of surface waves that are a generalized form of Rayleigh waves by considering the more general problem of elastic waves at the surface of separation of two solid media. He not only showed that the waves were analogous to Rayleigh waves that propagate with the maximum amplitude at the interface between the two half-spaces but also established conditions for the existence of such waves. The velocity of Stoneley waves is generally less than the Rayleigh surface wave velocity and greater than the transverse wave velocities. Stoneley also investigated the existence of a generalized form of Love waves that propagate at the interface between the two media, provided the density variations facilitate a higher transverse wave velocity at the interface than at the interior. Stoneley's theory has since been generalized to stratified media, and Kennett (1983) presents an excellent overview of the subject. It has been applied to layered viscoelastic media and to a number of practical engineering problems (see, for example, Sun *et al.*, 2009).

Exercises

1. A long taut spring of mass m, stiffness k and length L is stretched to length $L + s$ and released. When longitudinal waves propagate along the spring, the equation of motion of a length δx may be written as

$$m\delta x \times \frac{\partial^2 u}{\partial t^2} = \frac{\partial F}{\partial x}\delta x$$

where u is the longitudinal displacement and F is the restoring force. Derive the equation of longitudinal motion of the spring and show that waves propagate along the length of the spring in both directions. Hence, derive an expression for the velocity of wave propagation along the spring.

2. Standing acoustic wave are formed in a gas-filled tube of length L with (a) both ends open and (b) one end open with the other closed. The particle displacement may be assumed to be

$$u(x, t) = (A\cos kx + B\sin kx)\sin\omega t.$$

Assuming appropriate boundary conditions at each of the two ends of the tube, show that for (a) $u(x, t) = A\cos kx \ \sin\omega t$ with a wave length $\lambda = 2L/n$, $n = 1, 2, 3, \ldots \infty$ and for (b) $u(x, t) = A\cos kx \ \sin\omega t$ with a wavelength $\lambda = 4L/(2n - 1)$, $n = 1, 2, 3, \ldots \infty$. Sketch the first three modes in each case.

3.

(i) In the section on the transmission line, it was shown that two impedances Z_1 and Z_2 may be matched by the insertion of a quarter wave element of impedance

$$Z_m = \sqrt{Z_1 Z_2}.$$

Repeat this derivation for the acoustic case by employing the boundary conditions for excess pressure continuity and particle velocity at the two ends and derive the equation for the coefficient of sound transmission. Hence or otherwise show that

$$Z_m = \sqrt{Z_1 Z_2} \text{ and } L = \frac{\lambda_m}{4}.$$

(ii) Show that the impedance of a real transmission line seen from a position x on the line is given by

$$Z_x = Z_0 \frac{A\exp(-\gamma x) - B\exp(\gamma x)}{A\exp(-\gamma x) + B\exp(\gamma x)},$$

where γ is the propagation constant while A and B are the current amplitudes at $x = 0$ of the waves travelling in the positive and negative x directions, respectively. If the line has a length L and is terminated by a load Z_L show that

$$Z_L = Z_0 \frac{A\exp(-\gamma L) - B\exp(\gamma L)}{A\exp(-\gamma L) + B\exp(\gamma L)}.$$

(iii) Hence or otherwise show that the input impedance of the line, that is the impedance at $x = 0$, is given by

$$Z_i = Z_0 \frac{Z_0\sinh(\gamma L) + Z_L\cosh(\gamma L)}{Z_0\cosh(\gamma L) + Z_L\sinh(\gamma L)}.$$

(iv) When the transmission line is short circuited show that the input impedance is given by

$$Z_{sc} = Z_0\tanh(\gamma L),$$

and when it is open circuited the input impedance is given by

$$Z_{oc} = Z_0\coth(\gamma L).$$

4.

(i) Show that the input impedance of a short-circuited loss-less line of length L is

$$Z_i = iZ_0\tan\left(\frac{2\pi L}{\lambda}\right),$$

where

$$Z_0 = \sqrt{L_0/C_0}.$$

Hence or otherwise, determine the lengths of the line when it is capacitive and when it is inductive.

(ii) A line of characteristic impedance Z_0 is matched to a load Z_L by a loss-free quarter wavelength line. Show that the characteristic impedance of the matching section is

$$Z_m = \sqrt{Z_0 Z_L}.$$

Hence or otherwise show that a loss-free line of characteristic impedance Z_0 and length $n\lambda/2$ may be employed to couple two high-frequency circuits without affecting other impedances.

5.

(i) A quarter wavelength loss-free transmission line is short circuited. Show that such a line has infinite impedance.

(ii) A short-circuited quarter wavelength loss-free transmission line is bridged across another transmission line. Show that this will not affect the fundamental wavelength but short circuit any second harmonic.

6. Consider a real lossy transmission line where $R_0 \ll \omega L_0$ and $G_0 \ll \omega C_0$, and show that the propagation constant

$$\gamma = \alpha + ik = \sqrt{(R_0 + i\omega L_0)(G_0 + i\omega C_0)}$$

may be expressed as

$$\alpha = \frac{R_0}{2Z_0} + \frac{G_0 Z_0}{2}, \qquad Z_0 = \sqrt{\frac{L_0}{C_0}}$$

and

$$k = \omega\sqrt{L_0 C_0} = \frac{\omega}{v_0},$$

where α is an attenuation constant and k is the wave number.

7. Consider a transversely isotropic material and show by appropriate rotation of coordinates that

$$c_{66} = \frac{1}{2}(c_{11} - c_{12}).$$

Hence, show that the number of independent elastic constants in the generalized Hooke's law can be no more than five.

8. Consider wave propagation in an aeolotropic medium such as a single crystal where Hooke's law in its most general form must be employed.

(i) Show that in this case there are three distinct velocities of propagation of a plane wave. Also show that the vibration directions accompanying the three velocities are mutually perpendicular.

(ii) Explain the consequences of any two of these three velocities being equal.

9. Tables 4.3 and 4.4 give the elastic constants and the material density for five common materials. Also given are the associated longitudinal, transverse, one-dimensional (rod), two-dimensional (plate) and surface (Rayleigh) wave velocities.

Table 4.3 Elastic constants of five common materials in N/m2

	N/m^2	Steel $\times 10^{10}$	Copper $\times 10^{10}$	Aluminium $\times 10^{10}$	Glass $\times 10^{10}$	Rubber $\times 10^{6}$
Lame's constant	λ	11.2	9.5	5.6	2.8	1000
Lame's constant	μ	8.1	4.5	2.6	2.8	0.7
Young's modulus	E	20.9	12.05	7.0	7.0	2.1
Bulk modulus	K	16.6	12.5	7.3	4.7	1000

Table 4.4 Poisson's ratio, material density and wave propagation velocities in m/s

		Steel	Copper	Aluminium	Glass	Rubber
Poisson's ratio	ν	0.29	0.34	0.34	0.25	0.5
Density (kg/m^3)	ρ	7.8×10^3	8.9×10^3	2.7×10^3	2.5×10^3	0.93×10^3
Longitudinal wave velocity	a_1	5930	4560	6320	5800	1040
Transverse wave velocity	a_2	3220	2250	3100	3350	28
One-dimensional wave velocity	a_0	5180	3680	5080	5290	48
Two-dimensional wave velocity	a_P	5410	3910	5410	5465	55
Rayleigh saw velocity	a_R	2980	2100	2900	3080	26

(i) Assuming that the values for the Lame's constants are correct and accurate, verify the values of E and k.
(ii) Hence or otherwise compute the values of the bulk wave velocities a_1, a_2 and a_0 and a_P. Verify the values given in Table 4.4.
(iii) Explain briefly why the velocities of these materials satisfy the relationship $a_1 > a_P > a_0 > a_2 > a_R$.
(iv) Employ a spreadsheet and using an appropriate method compute the Rayleigh surface wave velocity a_R from a_1 and a_2.
Obtain the percentage error in the computed value relative to the tabulated value.

10. Employ a vector notation and let the dilatation $\Delta = div\ \mathbf{u}$ where $\mathbf{u}$ is the vector displacement. Let the rotation vector

$$\omega = \frac{1}{2} curl\ \mathbf{u}.$$

(i) Show that the vector equation of motion in an isotropic solid may be expressed as

$$\rho \frac{\partial^2 \mathbf{u}}{\partial t^2} = (\lambda + 2\mu)\, grad\Delta - 2\mu \times curl\ \omega.$$

(ii) Hence or otherwise obtain the vector equations of motion of longitudinal and transverse waves in an isotropic solid.

11. Show that $\bar{\mathbf{u}} = \bar{\mathbf{A}} \exp(ika_2 t)$ where $\bar{\mathbf{A}} = \nabla \times (\mathbf{r}\psi)$ is a solution of the wave equation

$$\frac{\partial^2 \mathbf{u}}{\partial t^2} = a_2^2 \nabla \cdot \nabla \mathbf{u},$$

if ψ is solution of the Helmholtz equation

$$\nabla \cdot \nabla \psi + k^2 \psi = 0.$$

12.

(i) Consider a slender prismatic beam of infinite length and show that the speed of flexural waves based on the Euler theory of simple bending is given by

$$a_f = a_0 \frac{2\pi R}{\lambda},$$

where R is the radius of gyration of the cross-section of the beam and λ is the wavelength, and comment on the implications of the relationship.

(ii) Hence or otherwise show that the velocity of the envelope of a packet of waves, with the wavelength of the component waves being close to λ, is equal to

$$a_g = a_f - \lambda \frac{\partial a_f}{\partial \lambda}.$$

(iii) For the flexural waves in a prismatic beam, show that

$$a_g = 2a_f.$$

13.

(i) Employ cylindrical coordinates and show that the three stress–strain relationships reduce to

$$\sigma_{rr} = \lambda\Delta + 2\mu\frac{\partial u_r}{\partial r}, \quad \sigma_{r\theta} = \mu\left(\frac{1}{r}\frac{\partial u_r}{\partial \theta} + r\frac{\partial}{\partial r}\left(\frac{u_\theta}{r}\right)\right),$$

and

$$\sigma_{rz} = \mu\left(\frac{\partial u_r}{\partial z} + \frac{\partial u_z}{\partial r}\right),$$

where

$$\Delta = \frac{1}{r}\frac{\partial (ru_r)}{\partial r} + \frac{1}{r}\frac{\partial u_\theta}{\partial \theta} + \frac{\partial u_z}{\partial z}.$$

Note: In the case of curvilinear coordinates, the divergence and curl of a vector cannot be exactly expressed in terms of the gradient operator. The gradient and the divergence operators are, respectively, given as

$$\nabla = \mathbf{e}_r\frac{\partial}{\partial r} + \mathbf{e}_\phi\frac{1}{r}\frac{\partial}{\partial \theta} + \mathbf{e}_z\frac{\partial}{\partial z}, \quad \mathbf{div} = \frac{1}{r}\left(\mathbf{e}_r\frac{\partial (r)}{\partial r} + \mathbf{e}_\phi\frac{\partial ()}{\partial \theta} + \mathbf{e}_z\frac{\partial (r)}{\partial z}\right).$$

This is due to the fact that

$$\mathbf{r} = r\mathbf{e}_r + z\mathbf{e}_z$$

and since

$$\mathbf{e}_\theta = \frac{\partial}{\partial \theta}\mathbf{e}_r \quad \text{and} \quad \mathbf{e}_r = -\frac{\partial}{\partial \theta}\mathbf{e}_\theta,$$

it follows that

$$\mathrm{d}\mathbf{r} = \mathrm{d}r\mathbf{e}_r + r\mathrm{d}\theta\,\mathbf{e}_\theta + \mathrm{d}z\mathbf{e}_z.$$

Furthermore, when the deformation is expressed as

$$\mathbf{u} = u_r\mathbf{e}_r + u_\theta\,\mathbf{e}_\theta + u_z\mathbf{e}_z$$

the strain components are related to the symmetric part of the gradient of the deformation, a tensor or dyadic, defined by

$$\mathbf{S} = \frac{\mathbf{1}}{\mathbf{2}}(\nabla\mathbf{u} + \mathbf{u}\nabla)$$

and the rotation components are related to the curl of the deformation rather than to

$$\mathbf{R} = \frac{\mathbf{1}}{\mathbf{2}}(\nabla\mathbf{u} - \mathbf{u}\nabla).$$

(ii) Show also that the components of the rotation vector ω_r, ω_θ and ω_z are defined as

$$\omega_r = \frac{1}{2}\left(\frac{1}{r}\frac{\partial u_z}{\partial \theta} - \frac{\partial u_\theta}{\partial z}\right), \quad \omega_\theta = \frac{1}{2}\left(\frac{\partial u_r}{\partial z} - \frac{\partial u_z}{\partial r}\right)$$

and

$$\omega_z = \frac{1}{2r}\left(\frac{\partial (r u_\theta)}{\partial r} - \frac{\partial u_r}{\partial \theta}\right)$$

and satisfy

$$\frac{1}{r}\frac{\partial (r\omega_r)}{\partial r} + \frac{1}{r}\frac{\partial \omega_\theta}{\partial \theta} + \frac{\partial \omega_z}{\partial z} = 0.$$

(iii) Hence or otherwise show that the equations of motion in cylindrical coordinates are

$$\begin{aligned}
\rho\frac{\partial^2 u_r}{\partial t^2} &= (\lambda + 2\mu)\frac{\partial}{\partial r}\Delta - \frac{2\mu}{r}\frac{\partial \omega_z}{\partial \theta} + 2\mu\frac{\partial \omega_\theta}{\partial z},\\
\rho\frac{\partial^2 u_\theta}{\partial t^2} &= (\lambda + 2\mu)\frac{1}{r}\frac{\partial}{\partial \theta}\Delta - 2\mu\frac{\partial \omega_r}{\partial z} + 2\mu\frac{\partial \omega_z}{\partial r},\\
\rho\frac{\partial^2 u_z}{\partial t^2} &= (\lambda + 2\mu)\frac{\partial}{\partial z}\Delta - \frac{2\mu}{r}\frac{\partial (r\omega_\theta)}{\partial r} + \frac{2\mu}{r}\frac{\partial \omega_r}{\partial \theta}.
\end{aligned}$$

(iv) Consider the propagation of longitudinal waves in an infinite cylindrical bar, where the assumption that plane transverse sections remain plane is not made and lateral motion of the bar is permitted.

Show that the equations of motion of axisymmetric longitudinal waves, when both u_r and u_z are independent of θ, and ω_r and ω_z are both zero (ω_θ is the only non-zero rotation component), are given by

$$\frac{\partial^2 \Delta}{\partial r^2} + \frac{1}{r}\frac{\partial \Delta}{\partial r} = \frac{1}{a_1^2}\frac{\partial^2 \Delta}{\partial t^2},$$

and

$$\frac{\partial^2 \omega_\theta}{\partial r^2} + \frac{1}{r}\frac{\partial \omega_\theta}{\partial r} - \frac{\omega_\theta}{r^2} = \frac{1}{a_2^2}\frac{\partial^2 \omega_\theta}{\partial t^2}.$$

14.

(i) Consider the propagation of longitudinal waves in an infinite cylindrical bar, where the assumption that the plane transverse sections remain plane is not made and lateral motion of the bar is permitted. The velocity of waves propagating in such a bar is given by

$$a_P = a_0\left(1 - \nu^2\frac{\pi^2}{\lambda^2}R^2\right),$$

where ν is the Poisson's ratio, λ is the wavelength of the waves and R is the radius of the bar.

Determine the velocity of a packet of waves given that the wavelengths of the component waves in the packet are all equal to λ.

(ii) Consider the propagation of torsional waves in an infinite cylindrical rod, where the assumption that plane transverse sections remain plane is not made and lateral motions are permitted. The velocity of wave propagation in such a rod is given by

$$a_T = a_2\left(1 + k_1^2\frac{\lambda^2}{4\pi^2 R^2}\right),$$

where λ is root of the frequency equation

$$\lambda J_0(\lambda) - 2J_1(\lambda) = 0$$

and k_1 is a constant.

Determine the velocity of a packet of waves with the wavelengths of the component waves in the packet being close to λ.

15.

(i) Consider a Rayleigh surface wave and obtain expression for the elastic stresses and strains in terms of the parameters relating to the propagating wave.

(ii) The strain energy density U is related to the strain energy U by the integral relationship

$$U = \int_V \mathsf{U}\, dV.$$

Hence or otherwise obtain an expression for the strain energy density and find the average strain energy propagating in the direction of wave travel.

(Hint: The integral over a volume may be reduced to an integral over the boundary surface by integrating through the depth.)

References

Auld, B. A. (1973) *Acoustic Fields and Waves in Solids*, Vols. I & II, John Wiley & Sons Inc., New York.

Campbell, C. K. (1998) *Surface Acoustic Wave devices for Mobile and Wireless Communications*, Academic Press Inc., Boston, Chapters 1–5, pp. 3–158.

Kennett, B. L. N. (1983) *Seismic Wave Propagation in Stratified Media*, Cambridge University Press, Cambridge.

Kolsky, H. (1963) *Stress Waves in Solids*, Dover Publications Inc., New York.

Mason, W. P., (1948) *Electromechanical Transducers and Wave filters*, 2nd edn, Van Nostrand Co. Inc., New York.

Mason, W. P. (1964) *Physical Acoustics*, Harcourt Brace Jovanovich, Academic Press, New York.

Nadeau, G. (1964) *Introduction to Elasticity*, Holt, Rinehart and Winston Inc., New York.

Nayfeh, A. H. and Mook, D. T. (1979) *Nonlinear Oscillations*, John Wiley & Sons Inc., New York.

Pain, H. J. (1976) *The Physics of Vibration and Waves*, 2nd edn, John Wiley & Sons Ltd., London.

Stoker, J. J. (1950) *Nonlinear Vibrations in Mechanical and Electrical Systems*, Interscience, New York.

Sun, L., Gu, W. and Luo, F. (2009) Steady state wave propagation in multi-layered viscoelastic media excited by a moving dynamic distributed load, *Journal of Applied Mechanics* 76, 041001-1–041001-15.

White, R. M. (1970) Surface elastic waves, *Proc. IEEE*, 58(8), 1238–1276.

5

Dynamics of Plates and Plate-like Structures

5.1 Flexural Vibrations of Plates

In Chapter 3, the transverse vibration of a uniform prismatic bar was considered. Longitudinal vibrations of a bar may also be analysed in a similar fashion. The analysis is usually only approximate, in that, in deriving it, it is assumed that plane transverse sections of the bar remain plane during the passage of the stress waves and that the stress acts uniformly over each section. The longitudinal expansions and contractions of the beam will necessarily result in lateral contraction and expansions, the ratio between these lateral and longitudinal strains being given by Poisson's ratio. The effect of the lateral motion is to induce a non-uniform distribution of stress across sections of the bar and plane transverse sections no longer remain plane but are distorted. The effect of lateral motion is particularly important in the case of a plate where the wavelengths are large in comparison with the thickness and yet are of the same order of the dimension of the plate. When the wavelength of the propagating waves is relatively large in comparison with the thickness of the plate, the stress is uniform over any cross-section of the plate perpendicular to the direction of propagation in the absence of any flexure. Thus, to formally analyse the situation, let the x–y plane be parallel to the surface of the plate and let the direction of the propagating wave be the x direction. The plate is assumed to be of thickness d and density ρ.

Consider an element of the plate, unit length in the lateral direction and of width δx, as shown in Figure 5.1. Applying Newton's law to the element

$$\rho d \times 1 \times \delta x \frac{\partial^2 u}{\partial t^2} = d \times 1 \times (\sigma_{xx} + \delta\sigma_{xx} - \sigma_{xx}),$$

which reduces to

$$\rho \delta x \frac{\partial^2 u}{\partial t^2} = \delta\sigma_{xx} = \frac{\partial \sigma_{xx}}{\partial x}\delta x.$$

Hence, we have the governing equation

$$\rho \frac{\partial^2 u}{\partial t^2} = \frac{\partial \sigma_{xx}}{\partial x}. \tag{5.1}$$

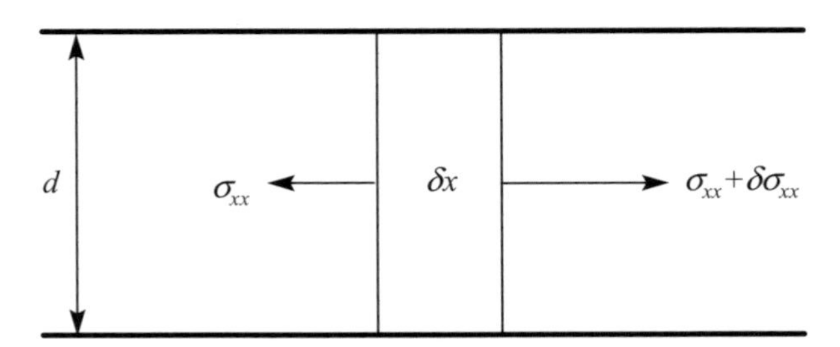

Figure 5.1 Equilibrium of a thin element of a typical plate

Similarly, in the spanwise direction,

$$\rho\frac{\partial^2 v}{\partial t^2} = \frac{\partial \sigma_{yx}}{\partial x}. \tag{5.2}$$

Considering the stress–strain relationships developed earlier for an isotropic, elastic material and recalling that σ_{zz}, the stress component perpendicular to the plate, is zero, the only non-zero stresses assuming no flexure are

$$\sigma_{xx} = (\lambda + 2\mu)\frac{\partial u}{\partial x} + \lambda\frac{\partial w}{\partial z} \tag{5.3}$$

and

$$\sigma_{xy} = \mu\frac{\partial v}{\partial x}, \tag{5.4}$$

with

$$\sigma_{zz} = (\lambda + 2\mu)\frac{\partial w}{\partial z} + \lambda\frac{\partial u}{\partial x} = 0. \tag{5.5}$$

Rearranging equation (5.5), we obtain

$$\frac{\partial w}{\partial z} = -\frac{\lambda}{\lambda + 2\mu}\frac{\partial u}{\partial x}$$

and equation (5.3) reduces to

$$\sigma_{xx} = (\lambda + 2\mu)\frac{\partial u}{\partial x} - \frac{\lambda^2}{\lambda + 2\mu}\frac{\partial u}{\partial x} = \frac{4\mu(\lambda + \mu)}{\lambda + 2\mu}\frac{\partial u}{\partial x}. \tag{5.6}$$

Substituting equations (5.6) and (5.4) into equations (5.1) and (5.2), we obtain in the longitudinal case, the governing equation

$$\rho\frac{\partial^2 u}{\partial t^2} = \frac{4\mu(\lambda + \mu)}{\lambda + 2\mu}\frac{\partial^2 u}{\partial x^2} = \rho\alpha_P^2\frac{\partial^2 u}{\partial x^2}, \tag{5.7}$$

where

$$\alpha_P^2 = \frac{\mu}{\rho}\frac{4(\lambda + \mu)}{(\lambda + 2\mu)} = \frac{E}{\rho\left(1 - \nu^2\right)}$$

and in the transverse case, the governing equation

$$\rho\frac{\partial^2 v}{\partial t^2} = \mu\frac{\partial^2 v}{\partial x^2}. \tag{5.8}$$

From the governing equation of wave motion in the transverse direction, we observe that the wave velocity is exactly the same as that given, in the general case of a medium of infinite extent, by equation (4.78). However, in the longitudinal case this is not so. In the longitudinal case,

$$\left(\frac{a_P}{a_0}\right)^2 = \frac{1}{\left(1-\nu^2\right)} \tag{5.9a}$$

and

$$\left(\frac{a_P}{a_1}\right)^2 = \frac{(1-2\nu)}{(1-\nu)^2}. \tag{5.9b}$$

The longitudinal wave motions are sometimes termed as *thickness stretch modes* because the plate is being stretched through its thickness. This is analogous to the longitudinal wave motions in slender beams, considered in Section 5.1. The transverse wave motions are sometimes termed as *thickness shear modes* and are analogous to the transverse vibrations of a string. When the wavelength of the propagating wave is relatively short in comparison with the thickness of the plate, it can be shown that the wave velocities degenerate to the case of a Rayleigh-type surface wave.

It must be said that in the above analysis it was assumed that there was no flexure. This is in fact an unrealistic assumption in so far as the wave motions in plates are concerned. Flexural effects are considered in the next section.

5.2 The Effect of Flexure

To consider the effect of flexure, plates may be separated into thee categories: thick plates, thin plates and membranes. When transverse shear deformations are of the same magnitude as bending deflections, they can no longer be ignored and the plate is considered to be a thick plate. On the contrary, in a thin plate the transverse shear deformations are relatively negligible in comparison with bending deflections. Thickness of the plates is also small in comparison with the bending deflections. In a membrane that is usually much thinner than a thin plate, the bending stiffness is negligible and the transverse loads are resisted by membrane action, i.e. the in-plane longitudinal forces and the resulting transverse forces.

To analyse the effect of flexure in a thin plate, it is essential that a number of assumptions be made. These may be grouped and described as the following:

1. The homogeneity assumption: The material of the plate is homogeneous, isotropic and obeys Hooke's law.
2. Plane sections assumption: The shear stresses σ_{xz} and σ_{yz} are negligible and as a consequence the line normal to the middle surface prior to ending remain straight and normal to the middle surface during flexure.
3. Transverse uniformity assumption: The normal stress σ_{zz} is negligible and as a consequence the transverse deflection on any point on a line normal to and away from the middle surface is equal to the transverse deflection of the corresponding point at the intersection of the normal and the middle surface.
4. The small deflection assumption: The transverse deflections are assumed to be small in comparison with the thickness of the plate and as a consequence the middle surface is assumed to be unaffected by the flexure. Thus, the in-plane stress resultants due to the flexure may be neglected.

As a consequence of assumptions (2) and (3), the stresses σ_{xz}, σ_{yz} and σ_{zz} are assumed to be zero and from Hooke's law the strains may be expressed as

$$\varepsilon_{xx} = \frac{1}{E}\left(\sigma_{xx} - \nu\sigma_{yy}\right), \tag{5.10a}$$

$$\varepsilon_{yy} = \frac{1}{E}\left(\sigma_{yy} - \nu\sigma_{xx}\right), \tag{5.10b}$$

and

$$\varepsilon_{xy} = \frac{1}{\mu}\left(\sigma_{xy}\right) = \frac{2(1+\nu)}{E}\sigma_{xy}. \tag{5.10c}$$

Solving for the stresses in terms of the strains,

$$\sigma_{xx} = \frac{E}{\left(1-\nu^2\right)}\left(\varepsilon_{xx} + \nu\varepsilon_{yy}\right), \tag{5.11a}$$

$$\sigma_{yy} = \frac{E}{\left(1-\nu^2\right)}\left(\varepsilon_{yy} + \nu\varepsilon_{xx}\right) \tag{5.11b}$$

and

$$\sigma_{xy} = \frac{E}{2(1+\nu)}\varepsilon_{xy}. \tag{5.11c}$$

To determine the governing equilibrium equations, it is essential to consider both the force and the equilibrium conditions. Thus, it is essential that we first define the forces and moments resulting from the stresses. Directly associated with the middle surface deformation are the middle surface force resultants, which are given by

$$N_{xx} = \int_{-d/2}^{d/2} \sigma_{xx} \mathrm{d}z, \tag{5.12a}$$

$$N_{yy} = \int_{-d/2}^{d/2} \sigma_{yy} \mathrm{d}z \tag{5.12b}$$

and

$$N_{xy} = \int_{-d/2}^{d/2} \sigma_{xy} \mathrm{d}z, \tag{5.12c}$$

where d is the thickness of the plate.

Similarly, the moment resultants obtained from the stresses may be defined as

$$M_{xx} = \int_{-d/2}^{d/2} \sigma_{xx} z \mathrm{d}z, \tag{5.13a}$$

$$M_{yy} = \int_{-d/2}^{d/2} \sigma_{yy} z \mathrm{d}z \tag{5.13b}$$

and

$$M_{xy} = \int_{-d/2}^{d/2} \sigma_{xy} z \mathrm{d}z. \tag{5.13c}$$

Since the plate is thin and bending deflections are small, it is assumed that the plane sections remain plane and that the middle surface remains unstrained due to the transverse deflection w. Any deflections of the middle surface are due only to externally imposed in-plane forces, and the in-plane deflections of the middle surface are therefore independent of both w and z. Consequently, these may be expressed as

$$u = u_0(x, y) - z\frac{\partial w(x, y)}{\partial x} \tag{5.14a}$$

and

$$v = v_0(x, y) - z\frac{\partial w(x, y)}{\partial y}. \tag{5.14b}$$

The strain displacement relations are given by equations (4.58a), (4.58b) and (4.58f) and are

$$\varepsilon_{xx} = \frac{\partial u}{\partial x}, \tag{4.58a}$$

$$\varepsilon_{yy} = \frac{\partial v}{\partial y}, \tag{4.58b}$$

$$\varepsilon_{xy} = \frac{\partial u}{\partial y} + \frac{\partial v}{\partial x}. \tag{4.58f}$$

Inserting equations (5.4) into equations (4.58a), (4.58b) and (4.58f) and employing equations (5.11a) to (5.11c) and (5.13a) to (5.13c), we obtain the following moment displacement relation for M_{xx}:

$$M_{xx} = -\frac{E}{1-\nu^2}\left(\frac{\partial^2 w}{\partial^2 x^2} + \nu\frac{\partial^2 w}{\partial^2 y^2}\right)\int_{-d/2}^{d/2} z^2 \mathrm{d}z.$$

On performing the integration we have

$$M_{xx} = -D\left(\frac{\partial^2 w}{\partial^2 x^2} + \nu\frac{\partial^2 w}{\partial^2 y^2}\right), \tag{5.15a}$$

where

$$D = \frac{E}{1-\nu^2} \times \frac{d^3}{12}.$$

In a similar manner,

$$M_{yy} = -D\left(\frac{\partial^2 w}{\partial^2 y^2} + \nu\frac{\partial^2 w}{\partial^2 x^2}\right), \tag{5.15b}$$

$$M_{xy} = -D(1-\nu)\frac{\partial^2 w}{\partial x \partial y}. \tag{5.15c}$$

The quantity D determines the flexural rigidity of a plate and is analogous to the flexural rigidity EI of a beam.

We shall now consider the derivation of a system of equilibrium equations governing the dynamics of thin plates. The plate is assumed to be loaded by a distribution of external in-plane forces N_{xx}, N_{yy} and N_{xy}. The transverse deflections that arise due the dynamical motions of the plate result in internal bending moments in the plate. In addition, the forces acting in the plane of the middle surface of the plate produce components of forces that act to maintain the plate in equilibrium along with the shear forces due to bending. The transverse forces generated by the in-plane forces and those generated by the bending action are considered separately.

Consider a small differential rectangular element of the plate with sides δx and δy. From equilibrium considerations, we shall sum the forces in the x, y and z directions, respectively. The transverse deflections and rotations are assumed to be small and all in-plane motion is ignored. Considering the normal and shear forces in the x direction, N_{xx} and N_{yx}, on an edge perpendicular to the x direction, we may write the condition for equilibrium of the differential element as

$$\left(N_{xx} + \frac{\partial N_{xx}}{\partial x}\delta x\right)\delta y - N_{xx}\delta y + \left(N_{yx} + \frac{\partial N_{yx}}{\partial y}\delta y\right)\delta x - N_{yx}\delta x = 0,$$

which reduces to

$$\frac{\partial N_{xx}}{\partial x} + \frac{\partial N_{yx}}{\partial y} = 0. \tag{5.16a}$$

By proceeding in a similar manner in the y direction we have

$$\frac{\partial N_{xy}}{\partial x} + \frac{\partial N_{yy}}{\partial y} = 0. \tag{5.16b}$$

In the z direction, however, there is a residual force, which can be evaluated by considering the projections of the forces on the middle surface of the plate in that direction. Thus, the normal force in the x direction, N_{xx}, contributes

$$F_{Nxx} = \left(N_{xx} + \frac{\partial N_{xx}}{\partial x}\delta x\right)\left(\frac{\partial w}{\partial x} + \frac{\partial^2 w}{\partial x^2}\delta x\right)\delta y - N_{xx}\frac{\partial w}{\partial x}\delta y \approx N_{xx}\frac{\partial^2 w}{\partial x^2}\delta x\delta y,$$

while the shear force in the same direction N_{yx} contributes

$$F_{Nyx} \approx N_{yx}\frac{\partial^2 w}{\partial x \partial y}\delta x\delta y.$$

Similarly, the normal force in the y direction N_{yy} contributes

$$F_{Nyy} \approx N_{yy}\frac{\partial^2 w}{\partial y^2}\delta x\delta y,$$

while the shear force in the same direction N_{xy} contributes

$$F_{Nxy} \approx N_{xy}\frac{\partial^2 w}{\partial x \partial y}\delta x\delta y.$$

By adding all four components and noting that

$$N_{xy} = N_{yx}$$

from the equilibrium of moments about the z axis, we obtain the total z component force due to the in-plane forces in the middle surface as

$$F_N \approx \left(N_{xx}\frac{\partial^2 w}{\partial x^2} + 2N_{xy}\frac{\partial^2 w}{\partial x \partial y} + N_{yy}\frac{\partial^2 w}{\partial y^2}\right)\delta x\delta y. \tag{5.17}$$

We now consider the bending moments and shear forces that arise due to the flexure of the middle surface of the plate.

Considering in the first instance, the net shear force in the z direction is given by comparing the shear forces along edges parallel to the x and y directions. Thus, the shear forces along the edges perpendicular to the x and y directions contribute

$$F_Q = \left(Q_x + \frac{\partial Q_x}{\partial x}\delta x\right)\delta y - Q_x\delta y + \left(Q_y + \frac{\partial Q_y}{\partial y}\delta y\right)\delta x - Q_y\delta x,$$

which reduces to

$$F_Q = \left(\frac{\partial Q_x}{\partial x} + \frac{\partial Q_y}{\partial y}\right)\delta x\delta y \tag{5.18}$$

Applying Newton's second law to the differential element,

$$F_Q + F_N = \rho d\delta x\delta y\frac{\partial^2 w}{\partial t^2}$$

and substituting for the two forces in the z direction due to the in-plane and shear forces on the edges of the plate

$$N_{xx}\frac{\partial^2 w}{\partial x^2} + 2N_{xy}\frac{\partial^2 w}{\partial x\partial y} + N_{yy}\frac{\partial^2 w}{\partial y^2} + \frac{\partial Q_x}{\partial x} + \frac{\partial Q_y}{\partial y} = \rho d\frac{\partial^2 w}{\partial t^2}. \tag{5.19}$$

Up to this point we have utilized all the three force equilibrium equations, one for each of the three directions but only one of the three moment equilibrium equations. Considering the equilibrium of moments about the x axis we have

$$-\left(M_{xy} + \frac{\partial M_{xy}}{\partial x}\delta x\right)\delta y + M_{xy}\delta y - \left(M_{yy} + \frac{\partial M_{yy}}{\partial y}\delta y\right)\delta x + M_{yy}\delta x$$
$$= -\left(Q_y + \frac{\partial Q_y}{\partial y}\delta y\right)\delta x\frac{\delta y}{2} - Q_y\delta x\frac{\delta y}{2},$$

which, after neglecting the term containing the higher-order products, reduces to

$$\frac{\partial M_{xy}}{\partial x} + \frac{\partial M_{yy}}{\partial y} - Q_y = 0. \tag{5.20}$$

By taking moments about the y axis in a similar manner, we obtain

$$\frac{\partial M_{xx}}{\partial x} + \frac{\partial M_{yx}}{\partial y} - Q_x = 0. \tag{5.21}$$

Employing equations (5.21) and (5.20) to eliminate the shear forces Q_x and Q_y in equation (5.19) and equations (5.15a), (5.15b) and (5.15c) to eliminate the moment resultants M_{xx}, M_{xy} and M_{yy}, and employing the relationship

$$D = \frac{Ed^3}{12\left(1 - \nu^2\right)},$$

we obtain the governing equilibrium equation in terms of the transverse displacement distribution and the applied in-plane forces as

$$D\left(\frac{\partial^4 w}{\partial x^4} + 2\frac{\partial^4 w}{\partial x^2\partial y^2} + \frac{\partial^4 w}{\partial y^4}\right) + \rho d\frac{\partial^2 w}{\partial t^2} = N_{xx}\frac{\partial^2 w}{\partial x^2} + 2N_{xy}\frac{\partial^2 w}{\partial x\partial y} + N_{yy}\frac{\partial^2 w}{\partial y^2}. \tag{5.22}$$

Equation (5.22) may be written in a compact form in terms of the ∇ operator in Cartesian coordinates, which is defined as

$$\nabla = \frac{\partial}{\partial x}\mathbf{i} + \frac{\partial}{\partial y}\mathbf{j} + \frac{\partial}{\partial z}\mathbf{k},$$

where **i**, **j** and **k** are unit vectors in the x, y and z directions. In Cartesian coordinates, the Laplacian operator ∇^2 is then given by

$$\nabla^2 = \nabla \cdot \nabla = \frac{\partial^2}{\partial x^2} + \frac{\partial^2}{\partial y^2} + \frac{\partial^2}{\partial z^2}$$

and the governing equation of transverse motion for an isotropic, elastic, thin plate, equation (5.22), may be expressed as

$$D\nabla^4 w + \rho d \frac{\partial^2 w}{\partial t^2} = N_{xx}\frac{\partial^2 w}{\partial x^2} + 2N_{xy}\frac{\partial^2 w}{\partial x \partial y} + N_{yy}\frac{\partial^2 w}{\partial y^2}. \tag{5.23}$$

Although equation (5.23) is not exactly of the same form as the standard equation for the propagation of waves, it has essentially all the same features and admits wave-like solutions. For most engineering applications, however, one is usually more interested in the vibration of plates of finite extent where the waves propagating along various directions are reflected at the boundaries, resulting in standing waves. Standing waves, in turn, generate the sustained vibration of the plate. Specific geometries of plates, such as rectangular and circular plates, have a number of important engineering applications and are considered in the next section.

5.3 Vibrations in Plates of Finite Extent: Rectangular Plates

Although the governing differential equations of transverse motion of thin plates in terms of the transverse displacement distribution and the applied in-plane forces have been derived for plates of infinite extent, plates of finite extent also involve the solution of these equations. To obtain the relevant solution, it is necessary to ensure that all the physical conditions at the boundaries of these plates are correctly met. To facilitate this, the mathematical description of certain important boundary conditions along the edges of the plate ought to be considered before attempting to solve the governing equations.

For rectangular plates, in which the x and y axes coincide with the directions of the edges, it is often of interest to consider solutions for which the edges are so supported that no lateral deflections occur along them. Thus, for plates that are fully supported along the edges, the transverse deflection along the edges is zero. In addition, should the plate be clamped along the edges, then no rotations will occur in a direction normal to and all along the edges. Thus,

$$\frac{\partial w}{\partial n} = 0,$$

where n is the direction normal to the edge. For rectangular plates of length a and width b, in which the x and y axes coincide with the directions of the edges,

$$\left.\frac{\partial w}{\partial x}\right|_{x=0,a} = \left.\frac{\partial w}{\partial y}\right|_{y=0,b} = 0. \tag{5.24}$$

When the edges are simply supported, $w = 0$ and the edges are free to rotate. However, the edge bending moments, M_{xx} or M_{yy}, must vanish in this case. Thus from equations (5.15a) and (5.15b), for the rectangular plate of length a and width b,

$$D\left(\frac{\partial^2 w}{\partial^2 x^2} + \nu\frac{\partial^2 w}{\partial^2 y^2}\right)\bigg|_{x=0,a} = D\left(\frac{\partial^2 w}{\partial^2 y^2} + \nu\frac{\partial^2 w}{\partial^2 x^2}\right)\bigg|_{y=0,b} = 0. \tag{5.25}$$

For an edge that is completely free of any forces or moments, both the edge bending moment and the edge shear force must vanish. Substituting the expressions for the edge bending moments defined by equations (5.15a), (5.15b) and (5.15c) into the equations (5.20) and (5.21) results in the expressions for the edge shear forces.

If we introduce the restriction that the two opposite edges of the rectangular plate along its length are simply supported, that is their deflections are zero and they experience no restraining moment, then we may assume the solutions to be of the form

$$w(x, y, t) = Y_n(y) \sin\left(\frac{n\pi x}{a}\right) \sin(\omega t), \tag{5.26}$$

where n is any integer.

Substituting the assumed solution into equation (5.23) and assuming that the in-plane shear force resultant N_{xy} is equal to zero results in the ordinary differential equation

$$D\left(\frac{\partial^4 Y_n}{\partial y^4} - 2\frac{n^2\pi^2}{a^2}\frac{\partial^2 Y_n}{\partial y^2} + \frac{n^4\pi^4}{a^4}Y_n\right) - \omega^2 \rho d Y_n = N_{yy}\frac{\partial^2 Y_n}{\partial y^2} - N_{xx}\frac{n^2\pi^2}{a^2}Y_n,$$

which may be rearranged as

$$D\frac{\partial^4 Y_n}{\partial y^4} - \left(N_{yy} + 2D\frac{n^2\pi^2}{a^2}\right)\frac{\partial^2 Y_n}{\partial y^2} + \left(D\frac{n^4\pi^4}{a^4} + N_{xx}\frac{n^2\pi^2}{a^2} - \omega^2\rho d\right)Y_n = 0. \tag{5.27}$$

Assuming further that all the coefficients of the ordinary differential equation (5.27) are constants, the general solution of this equation may be obtained in just the same way as was done in the case of the transverse vibration of a uniform beam in Section 3.4. Restricting the solution to the case of a rectangular plate, simply supported along all edges, the boundary conditions defined by equations (5.24) and (5.25) apply. The solution $Y_n(y)$ may then be expressed as

$$Y_n(y) = W_{nm} \sin\left(\frac{m\pi y}{b}\right), \tag{5.28}$$

where m is an integer.

Substituting the assumed solution given by equation (5.28) into equation (5.27) results in an expression for the frequency given by the solution of the relationship

$$\omega^2 = \frac{D\pi^4}{\rho d}\left(\frac{m^2}{b^2} + \frac{n^2}{a^2}\right)^2 + \frac{N_{yy}}{\rho d}\frac{m^2\pi^2}{b^2} + \frac{N_{xx}}{\rho d}\frac{n^2\pi^2}{a^2} \tag{5.29}$$

and the complete solution takes the form

$$w(x, y, t) = W_{nm} \sin\left(\frac{n\pi x}{a}\right) \sin\left(\frac{m\pi y}{b}\right) \sin(\omega t). \tag{5.30}$$

It follows from the solution given by equation (5.30) that it is zero when

$$\sin\left(\frac{n\pi x}{a}\right) = 0$$

or when

$$\sin\left(\frac{m\pi y}{b}\right) = 0.$$

The zeros of the solution given by equation (5.30) at any instant of time t represent nodal lines in the normal modes on the middle surface of the plate that are permanently at rest. For a plate in which the length-to-width ratio is 1.5, the nodal lines in the first six modes are illustrated in Figure 5.2. Thus, for plates simply supported on all edges, closed form solutions for the natural frequencies, normal modes and the nodal lines may be found with relative ease. However, when the plate is supported in any other manner, the situation is not as simple as it may seem.

Consider, for example, a rectangular plate of length a and width b, in which the x and y axes coincide with the directions of the edges. In what follows, we shall present a framework for employing either an exact or an approximate approach and solving a host of vibration problems associated with rectangular

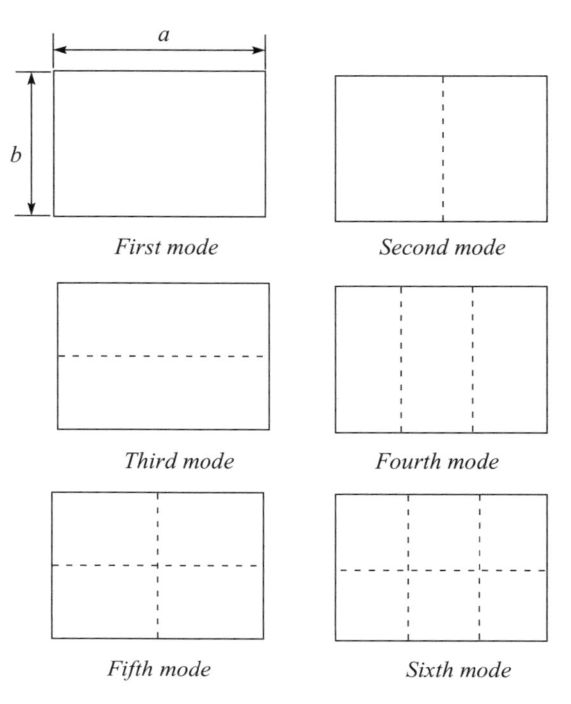

Figure 5.2 Nodal lines in the first six modes for a plate, simply supported along all its edges, in which the length-to-width ratio is 1.5

thin plates. We shall restrict the solution to the case when there are no in-plane forces or stress resultants. Hence it follows that

$$N_{xx} = N_{yy} = N_{xy} = 0.$$

Further introducing the non-dimensional coordinates,

$$\bar{x} = \frac{x}{a}, \bar{y} = \frac{y}{b}$$

the governing equation of transverse motion for a thin plate is given by

$$\frac{1}{a^4}\frac{\partial^4 w}{\partial \bar{x}^4} + \frac{2}{a^2 b^2}\frac{\partial^4 w}{\partial \bar{x}^2 \partial \bar{y}^2} + \frac{1}{b^4}\frac{\partial^4 w}{\partial \bar{y}^4} + \frac{\rho d}{D}\frac{\partial^2 w}{\partial t^2} = 0. \tag{5.31}$$

Introducing the parameter α to represent the aspect ratio of the plate given by $\alpha = b/a$ and recognizing that the area A and total mass m of the plate are given by the products ab and ρabd, respectively, the equation of transverse motion may be written as

$$\alpha^2 \frac{\partial^4 w}{\partial \bar{x}^4} + 2\frac{\partial^4 w}{\partial \bar{x}^2 \partial \bar{y}^2} + \frac{1}{\alpha^2}\frac{\partial^4 w}{\partial \bar{y}^4} + \frac{mA}{D}\frac{\partial^2 w}{\partial t^2} = 0. \tag{5.32}$$

Assuming further that the transverse displacement can be expressed as

$$w(\bar{x}, \bar{y}, t) = W(\bar{x}, \bar{y}) \exp(i\omega t),$$

equation (5.32) reduces to an eigenvalue problem given by

$$\alpha^2 \frac{\partial^4 W}{\partial \bar{x}^4} + 2\frac{\partial^4 W}{\partial \bar{x}^2 \partial \bar{y}^2} + \frac{1}{\alpha^2}\frac{\partial^4 W}{\partial \bar{y}^4} = \Omega^2 W, \tag{5.33}$$

where

$$\Omega^2 = \frac{mA\omega^2}{D}. \tag{5.34}$$

We now assume that the solution to equation (5.33) may be expressed in separable form as

$$W(\bar{x}, \bar{y}) = X(\bar{x})\, Y(\bar{y}).$$

Substituting the assumed solution into equation (5.33) and rearranging we have

$$\alpha^2 Y(\bar{y}) \frac{\partial^4 X}{\partial \bar{x}^4} + 2\frac{\partial^2 X}{\partial \bar{x}^2}\frac{\partial^2 Y}{\partial \bar{y}^2} + \frac{1}{\alpha^2} X(\bar{x}) \frac{\partial^4 Y}{\partial \bar{y}^4} = \Omega^2 X(\bar{x})\, Y(\bar{y}), \tag{5.35}$$

which may be solved for Ω^2 as

$$\Omega^2 = \alpha^2 \frac{1}{X}\frac{\partial^4 X}{\partial \bar{x}^4} + 2\frac{1}{X}\frac{\partial^2 X}{\partial \bar{x}^2}\frac{1}{Y}\frac{\partial^2 Y}{\partial \bar{y}^2} + \frac{1}{\alpha^2}\frac{1}{Y}\frac{\partial^4 Y}{\partial \bar{y}^4}.$$

We now assume that we can find functions $X(\bar{x})$ and $Y(\bar{y})$, which simultaneously satisfy

$$J_0 \frac{\partial^2 X}{\partial \bar{x}^2} = J_1 X(\bar{x}), \quad J_0 \frac{\partial^4 X}{\partial \bar{x}^4} = J_2 X(\bar{x}) \tag{5.36}$$

and

$$I_0 \frac{\partial^2 Y}{\partial \bar{y}^2} = I_1 Y(\bar{y}), \quad I_0 \frac{\partial^4 Y}{\partial \bar{y}^4} = I_2 Y(\bar{y}), \tag{5.37}$$

where

$$J_0 = \int_0^1 X^2(\bar{x})\, \mathrm{d}\bar{x}, \text{ and } Y_0 = \int_0^1 Y^2(\bar{y})\, \mathrm{d}\bar{y}. \tag{5.38}$$

Eliminating the second and fourth derivatives of $Y(\bar{y})$ in equation (5.35) by employing equation (5.37), we obtain

$$\alpha^2 I_0 \frac{\partial^4 X}{\partial \bar{x}^4} + 2I_1 \frac{\partial^2 X}{\partial \bar{x}^2} + \frac{1}{\alpha^2} I_2 X(\bar{x}) = \Omega^2 I_0 X(\bar{x}). \tag{5.39}$$

Similarly, eliminating the second and fourth derivatives of $X(\bar{x})$ in equation (5.35) by employing equation (5.37), we obtain

$$\frac{1}{\alpha^2} J_0 \frac{\partial^4 Y}{\partial \bar{y}^4} + 2J_1 \frac{\partial^2 Y}{\partial \bar{y}^2} + J_2 \alpha^2 Y(\bar{y}) = \Omega^2 J_0 Y(\bar{y}). \tag{5.40}$$

Thus to solve for the functions $X(\bar{x})$ and $Y(\bar{y})$, one must solve the system of equations (5.36) to (5.40) simultaneously. Equations (5.29) and (5.30) are ordinary differential equations with constant coefficients that could be solved by classical methods. Hence, we write equations (5.29) and (5.30) as

$$\frac{\partial^4 X}{\partial \bar{x}^4} + \beta_x \frac{\partial^2 X}{\partial \bar{x}^2} - \bar{\omega}_x^2 X(\bar{x}) = 0 \tag{5.41}$$

and

$$\frac{\partial^4 Y}{\partial \bar{y}^4} + \beta_y \frac{\partial^2 Y}{\partial \bar{y}^2} - \bar{\omega}_y^2 Y(\bar{y}) = 0, \tag{5.42}$$

and let

$$p_1^2 = \frac{\beta_x}{2} + \sqrt{\frac{\beta_x^2}{4} + \bar{\omega}_x^2},\ p_2^2 = -\frac{\beta_x}{2} + \sqrt{\frac{\beta_x^2}{4} + \bar{\omega}_x^2},$$
$$q_1^2 = \frac{\beta_y}{2} + \sqrt{\frac{\beta_y^2}{4} + \bar{\omega}_y^2} \text{ and } q_2^2 = -\frac{\beta_y}{2} + \sqrt{\frac{\beta_y^2}{4} + \bar{\omega}_y^2}. \tag{5.43}$$

Hence it follows that

$$p_1^2 p_2^2 = \omega_x^2 = \frac{\Omega^2}{\alpha^2} - \frac{I_2}{\alpha^4 I_0}, \tag{5.44a}$$

$$q_1^2 q_2^2 = \omega_y^2 = \alpha^2\Omega^2 - \frac{\alpha^4 J_2}{J_0}, \tag{5.44b}$$

$$p_2^2 - p_1^2 = -\frac{2}{\alpha^2}\frac{I_1}{I_0} = -\beta_x \tag{5.44c}$$

and

$$q_2^2 - q_1^2 = -2\alpha^2\frac{J_1}{J_0} = -\beta_y. \tag{5.44d}$$

The general solutions for $X(\bar{x})$ and $Y(\bar{y})$ are given by

$$X(\bar{x}) = A_x \sinh(p_2\bar{x}) + B_x \cosh(p_2\bar{x}) + C_x \sin(p_2\bar{x}) + D_x \cos(p_2\bar{x}), \tag{5.45}$$

and

$$Y(\bar{y}) = A_y \sinh(q_2\bar{y}) + B_y \cosh(q_2\bar{y}) + C_y \sin(q_2\bar{y}) + D_y \cos(q_2\bar{y}). \tag{5.46}$$

Also, eliminating the second and fourth derivatives of $Y(\bar{y})$ in equation (5.40) by employing equations (5.37), we obtain for Ω^2,

$$\Omega^2 = \frac{1}{\alpha^2}\frac{I_2}{I_0} + 2\frac{J_1 I_1}{J_0 I_0} + \alpha^2\frac{J_2}{J_0}. \tag{5.47}$$

From equations (5.44) and (5.47), we obtain

$$\Omega^2 = \alpha^2 p_1^2 p_2^2 + \frac{1}{\alpha^2} q_1^2 q_2^2 - \frac{1}{2}\left(p_2^2 - p_1^2\right)\left(q_2^2 - q_1^2\right). \tag{5.48}$$

Equation (5.48) together with equation (5.34) may be employed to determine the natural frequency corresponding to the normal modes.

We shall now consider two cases:

Case (i): Simply supported along all edges

In this case it is easily observed that

$$X(\bar{x}) = \sin(n\pi\bar{x})$$

and

$$Y(\bar{y}) = \sin(m\pi\bar{y}).$$

The quantities I_0, I_1 and I_2 are given by

$$I_0 = \frac{I_1}{2} = -\frac{m^2\pi^2}{2},$$

$$I_2 = 2I_0^2.$$

Similar expressions may be derived for J_0, J_1 and J_2.
Furthermore,

$$p_1 = n\pi,\ q_1 = m\pi,$$

$$p_2^2 = n^2\pi^2 + \frac{2}{\alpha^2}m^2\pi^2,$$

$$q_2^2 = m^2\pi^2 + 2\alpha^2 n^2\pi^2.$$

It may then be shown that equations (5.29) and (5.30) follow.

Case (ii): Clamped along all edges

In this case the general solution for $X(\bar{x})$ may be expressed as the sum of components, the first representing a family of symmetric modes and the second representing a family of anti-symmetric modes. A family of symmetric modes is given by

$$X_s(\bar{x}) = C_s\left[\frac{\cosh\left(\left(\bar{x}-\frac{1}{2}\right)p_2\right)}{\cosh\left(\frac{p_2}{2}\right)} - \frac{\cos\left(\left(\bar{x}-\frac{1}{2}\right)p_1\right)}{\cos\left(\frac{p_1}{2}\right)}\right]. \tag{5.49}$$

A family of anti-symmetric modes is given by

$$X_a(\bar{x}) = C_a\left[\frac{\sinh\left(\left(\bar{x}-\frac{1}{2}\right)p_2\right)}{\sinh\left(\frac{p_2}{2}\right)} - \frac{\sin\left(\left(\bar{x}-\frac{1}{2}\right)p_1\right)}{\sin\left(\frac{p_1}{2}\right)}\right]. \tag{5.50}$$

Similar expressions hold for $Y(\bar{y})$. These solution modes, for $X(\bar{x})$ and $Y(\bar{y})$, meet both the displacement boundary conditions at the edges, but only two of the four boundary conditions specifying the slope at the edges. Satisfying the remaining boundary conditions one obtains the frequency equations. Corresponding to the symmetric modes, the frequency equation relating p_1 and p_2 is

$$p_2\tanh\left(\frac{p_2}{2}\right) + p_1\tan\left(\frac{p_1}{2}\right) = 0, \tag{5.51}$$

while corresponding to the anti-symmetric modes the corresponding frequency equation is

$$p_2\coth\left(\frac{p_2}{2}\right) - p_1\cot\left(\frac{p_1}{2}\right) = 0. \tag{5.52}$$

Similar frequency equations may be derived relating q_1 and q_2. To find the natural frequencies corresponding to the mode shapes, only the quantities J_0, J_1 and the quantities I_0, I_1 are essential. Once these quantities are known, the natural frequencies are obtained numerically by solving equations (5.44c) and (5.44d), along with equations (5.51) and (5.52) and substituting the results in equations (5.48) and (5.34).

To approximately evaluate J_1, J_2, equations (5.36) are averaged. To this end they are multiplied by $X(\bar{x})$ and integrated from 0 to 1. This results in the following integral expressions for J_1 and J_2:

$$J_1 = \int_0^1 \frac{\partial^2 X}{\partial \bar{x}^2}X(\bar{x})\,\mathrm{d}\bar{x},\ J_2 = \int_0^1 \frac{\partial^4 X}{\partial \bar{x}^4}X(\bar{x})\,\mathrm{d}\bar{x}. \tag{5.53}$$

The quantity J_2 can also be evaluated by other means, and it may be shown to be related to the integrals J_0, J_1. The integral averaging method is particularly suitable for J_1.

The integrals J_0, J_1 are then evaluated as

$$J_0 = 1 - \left(\frac{p_2^2 - p_1^2}{2p_2^2 p_1^2}\right)\left(1 - \left(1 - p_2 \tanh\left(\frac{p_2}{2}\right)\right)^2\right), \tag{5.54a}$$

$$J_1 = \frac{p_2^2 - p_1^2}{2} + \left(1 - \left(1 - p_2 \tanh\left(\frac{p_2}{2}\right)\right)^2\right) \tag{5.54b}$$

in the case of the symmetric modes and

$$J_0 = 1 - \left(\frac{p_2^2 - p_1^2}{2p_2^2 p_1^2}\right)\left(1 - \left(1 - p_2 \coth\left(\frac{p_2}{2}\right)\right)^2\right), \tag{5.54c}$$

$$J_1 = \frac{p_2^2 - p_1^2}{2} + \left(1 - \left(1 - p_2 \coth\left(\frac{p_2}{2}\right)\right)^2\right) \tag{5.54d}$$

corresponding to the anti-symmetric modes.

Similar expressions may be found for I_0 and I_1. Solving equations (5.44c) and (5.44d), along with equations (5.51) and (5.52), numerically and substituting the results in equations (5.48) and (5.34), the approximate value for the square of the first natural frequency computed this way is

$$\omega^2 = \frac{36D}{mA}, \quad D = \frac{Ed^3}{12\left(1 - \nu^2\right)} \tag{5.55}$$

and is within 0.1% of other approximate values reported in several standard texts. There are also a range of other approximate methods that may be applied to the vibration analysis of rectangular thin plates, such as the Rayleigh–Ritz family of methods, which were briefly discussed in Chapter 5. These methods may also be employed to solve equations (5.27), (5.31) and (5.32), resulting in a hybrid approach with a number of potential benefits in solving complex vibration problems associated with multi-dimensional engineering structures and structural elements such as thin plates and shells. The analysis presented in this section highlights some of complexities of such problems where exact solutions are absent, and it is often almost impossible to verify the accuracy of approximate techniques such as the Rayleigh–Ritz family of methods.

Further details on the analysis and applications of plate and plate-like structures may be found in the works of Timoshenko and Woinowsky-Kreiger (1959), Leissa (1969, 1973), Donnell (1976) and Soedel (1993).

5.4 Vibrations in Plates of Finite Extent: Circular Plates

There are number of engineering applications that exploit the special features of vibrations in circular and annular plates and discs. A number of micro-scale applications such as micro-actuators and micro-generators as well as large-scale applications such as turbo-machinery involve the use of circular plates and discs. The vibration analysis of these structures often plays a crucial role in the complete design of the system.

In this section, we briefly consider the special vibrational features of circular and annular plates and discs. We have already established the governing equation of motion for the propagation of transverse wave motions in a thin plate, including flexural effects when bending moments are predominantly responsible for restoring the plate to equilibrium. In the case of a circular plate, as in the case of a rectangular plate, multiple reflections of the waves at the boundary result in sustained standing waves that manifest themselves as vibrations. We shall restrict the solution to the case when there are no in-plane forces or stress resultants. Hence it follows that

$$N_{xx} = N_{yy} = N_{xy} = 0$$

and equation (5.23) reduces to

$$D\nabla^4 w + \rho d \frac{\partial^2 w}{\partial t^2} = 0, \tag{5.56}$$

where

$$D = \frac{Ed^3}{12\left(1 - \nu^2\right)}. \tag{5.57}$$

To solve for the transverse displacements, it is essential that all the physical boundary or support conditions, usually along the circular edge of the plate, are met. It is extremely difficult to meet the mathematical constraints imposed by the physical requirements along the edge of a circular plate in Cartesian coordinates, and it is therefore necessary to transform equation (5.56) into cylindrical polar coordinates r, θ and z. The transverse direction continues to remain in the direction of the z axis, so it is only essential to transform the x and y coordinates into r and θ coordinates.

It can be shown that in cylindrical polar coordinates, the gradient or del operator may be expressed as

$$\nabla = \mathbf{i}\left(\frac{\partial}{\partial r} + \frac{1}{r}\right) + \mathbf{j}\frac{1}{r}\frac{\partial}{\partial \theta} + \mathbf{k}\frac{\partial}{\partial z}.$$

In longitudinally symmetric cylindrical polar coordinates, i.e. when

$$\frac{\partial}{\partial z} = 0,$$

the two-dimensional equivalent to the gradient operator is

$$\nabla = \mathbf{i}\left(\frac{\partial}{\partial r} + \frac{1}{r}\right) + \mathbf{j}\frac{1}{r}\frac{\partial}{\partial \theta}.$$

Hence, the Laplacian operator ∇^2 may be shown to be

$$\nabla^2 = \nabla \cdot \nabla = \frac{1}{r}\frac{\partial}{\partial r} r \frac{\partial}{\partial r} + \frac{1}{r}\frac{\partial^2}{\partial \theta^2}. \tag{5.58}$$

Assuming simple harmonic dependence with respect to time, the transverse displacement can be expressed as

$$w\left(r, \theta, t\right) = W\left(r, \theta\right) \exp\left(i\omega t\right). \tag{5.59}$$

Equation (5.56) reduces to an eigenvalue problem given by

$$D\nabla^4 W = \rho d \omega^2 W, \tag{5.60}$$

where the Laplacian operator ∇^2 is given by equation (5.58). Equation (5.60) may be expressed in product form as

$$\left(\nabla^2 + \sqrt{\frac{\rho d}{D}}\omega\right)\left(\nabla^2 - \sqrt{\frac{\rho d}{D}}\omega\right) W = 0. \tag{5.61}$$

If we let

$$\left(\nabla^2 - \sqrt{\frac{\rho d}{D}}\omega\right) W = V, \tag{5.62a}$$

equation (5.61) reduces to

$$\left(\nabla^2 + \sqrt{\frac{\rho d}{D}}\omega\right) V = 0. \tag{5.62b}$$

Applying the method of separation of variables to the pair of equations given by equations (5.62), the general standing wave-type solution for the complementary function may be expressed in polar coordinates as

$$W(r, \theta) = \sum_{n=0}^{\infty} W_n(r) \cos(n\theta + \phi), \tag{5.63}$$

where

$$W_n(r) = A_n J_n(kr) + B_n Y_n(kr) + C_n I_n(kr) + D_n K_n(kr), \tag{5.64}$$

$$k^4 = \frac{\rho d}{D}\omega^2, \tag{5.65}$$

and $J_n(kr)$, $Y_n(kr)$, $I_n(kr)$ and $K_n(kr)$ are special functions associated with the name of Bessel. The first two, $J_n(kr)$ and $Y_n(kr)$, are solutions of the so-called Bessel differential equation while the last two, $I_n(kr)$ and $K_n(kr)$, are solutions of the modified Bessel differential equation. The constants A_b, B_b, C_b and D_b are found by applying the relevant boundary conditions. The typical case of a circular plate clamped along its boundary will be considered in some detail to enunciate the method of determining the constants as well as the natural frequencies and the corresponding normal mode shapes associated with circular and annular plates (McLachlan, 1951).

In the case of a complete plate within the circle of radius R, i.e. the plate is not annular, the functions $Y_n(kr)$ and $K_n(kr)$ are inadmissible due to a singularity in these functions at $r = 0$. Non-annular circular plates have finite deflections when $r = 0$, and the general solution reduces to

$$W(r, \theta) = \sum_{n=0}^{\infty} (A_n J_n(kr) + C_n I_n(kr)) \cos(n\theta + \phi) \tag{5.66}$$

The boundary conditions at $r = R$ for a plate clamped along its edge $r = R$ are

$$W(r, \theta)|_{r=R} = 0 \tag{5.67a}$$

and

$$\left.\frac{\partial W(r, \theta)}{\partial r}\right|_{r=R} = 0. \tag{5.67b}$$

Hence, it follows that

$$A_n J_n(kR) + C_n I_n(kR) = 0 \tag{5.68a}$$

and

$$k\left(A_n \frac{\partial}{\partial R} J_n(kR) + C_n \frac{\partial}{\partial R} I_n(kR)\right) = 0. \tag{5.68b}$$

Eliminating the constants A_n and C_n in equations (5.68a) and (5.68b),

$$I_n(kR) \frac{\partial}{\partial R} J_n(kR) - J_n(kR) \frac{\partial}{\partial R} I_n(kR) = 0. \tag{5.69}$$

Table 5.1 Numerically obtained values of $\lambda_{n,m}$

m $n \rightarrow$	0	1	2
1	3.194	4.620	5.906
2	6.306	3.816	9.197
3	9.439	10.985	12.402
4	12.58	14.153	15.579
5	15.72	13.309	18.745

The derivatives of the Bessel functions of the first kind, $J_n(kR)$, and the modified Bessel function of the first kind, $I_n(kR)$, satisfy the recurrence relationships

$$\lambda \frac{\partial}{\partial \lambda} J_n(\lambda) = n J_n(\lambda) - \lambda J_{n+1}(\lambda). \tag{5.70a}$$

$$\lambda \frac{\partial}{\partial \lambda} I_n(\lambda) = n I_n(\lambda) + \lambda I_{n+1}(\lambda). \tag{5.70b}$$

The derivatives of the Bessel functions of the first kind, $J_n(kR)$, and the modified Bessel function of the first kind, $I_n(kR)$, may be eliminated from equation (5.69) by virtue of the relationships given by equations (5.70). Hence, the frequency equation (5.69) reduces to

$$I_n(kR) J_{n+1}(kR) + J_n(kR) I_{n+1}(kR) = 0. \tag{5.71}$$

The roots of the frequency equation (5.69)

$$kR = \lambda_{n,m}, \ m = 1, 2, 3, \ldots$$

for each n may be found numerically.

The values of $\lambda_{n,m}$ are tabulated in Table 5.1. Associated with each of the natural frequencies is a pattern of nodal lines.

The associated natural frequencies then are

$$\omega = \sqrt{\frac{D}{\rho d}} \left(\frac{\lambda_{n,m}}{R}\right)^2 \tag{5.72a}$$

where

$$D = \frac{E}{1 - \nu^2} \times \frac{d^3}{12}. \tag{5.72b}$$

The nodal lines associated with the first nine modes are illustrated in Figure 5.3. The ratios of the natural frequencies associated with the nine modes corresponding to m and n from 1 to 3 to that of the fundamental are 1, 2.08, 3.41, 3.88, 4.98, 5.94, 6.82, 8.26 and 8.72. It can be seen from Figure 5.3 that the numerical value of n determines the number of nodal diameters and the number of associated nodal circles is $m - 1$.

The fundamental natural frequency of clamped circular plate is often employed as a reference to quote the natural frequencies of circular plates with other boundary conditions, although it is not the lowest one associated with circular plates. For example, the natural frequencies of several modes associated with circular free plates are lower than the fundamental frequency associated with a clamped circular plate. For vibration with a nodal circle the frequency ratio is 0.88, while in the case of vibration with two nodal diameters it is 0.41. The frequency ratio corresponding to first natural frequency of a circular plate supported at its centre and vibrating in an umbrella mode has several applications in the design of turbo-machinery and is 0.33. It is much lower than the frequency ratio corresponding to the first natural

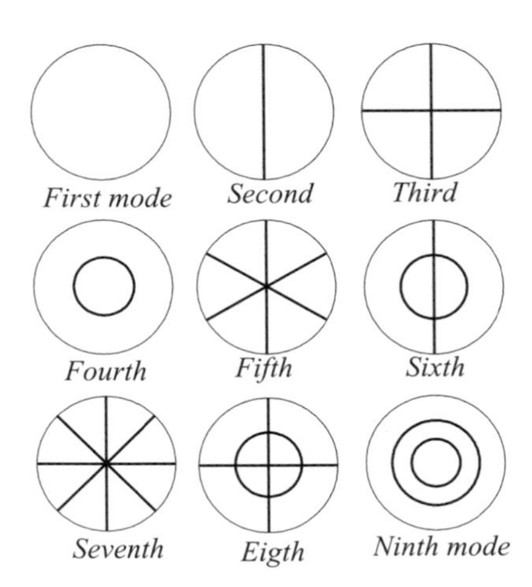

Figure 5.3 Nodal lines in the first six modes for a clamped circular plate

frequency of circular plate simply supported at its periphery with no nodal patterns within its boundary, which is 0.5.

5.5 Vibrations of Membranes

In deriving the general governing equation of the transverse motions in a thin plate, it was not assumed that the longitudinal stress resultants due to in-plane forces are negligible. In certain situations they are the predominant forces, and the flexural rigidity of thin plate is relatively negligible. This is true for membranes and as a result $D \approx 0$. When this is the case, the governing equation for the transverse motions in a thin plate reduces to

$$N_{xx}\frac{\partial^2 w}{\partial x^2} + 2N_{xy}\frac{\partial^2 w}{\partial x \partial y} + N_{yy}\frac{\partial^2 w}{\partial y^2} = \rho d\frac{\partial^2 w}{\partial t^2}. \tag{5.73}$$

A common example often encountered is the case of membrane subjected to a uniform tension in the longitudinal middle plane of the membrane. In this case, we may assume that

$$N_{xx} = N_{yy} = T$$

and that

$$N_{xy} = 0,$$

and equation (5.63) simplifies to

$$\frac{\partial^2 w}{\partial x^2} + \frac{\partial^2 w}{\partial y^2} = \nabla^2 w = \frac{\rho d}{T}\frac{\partial^2 w}{\partial t^2}. \tag{5.74}$$

Equation (5.74) is a classical example of a wave equation that admits wave-like solutions. In the case of plane waves travelling in the x direction, it reduces to the one-dimensional wave equation that was considered earlier in the chapter. The membrane behaves like a string with no stiffness.

Membranes of finite extent also admit standing wave solutions that manifest themselves in the form of sustained vibrations. Assuming simple harmonic dependence with respect to time, the solution may be expressed as

$$w = W \exp(i\omega t), \tag{5.75}$$

and equation (5.74) may be expressed as

$$\nabla^2 w + \frac{\rho d \omega^2}{T} w = 0. \tag{5.76}$$

Equation (5.76) is applicable to membranes of a variety of geometrical shapes, including rectangular, circular and polygonal geometries. A typical example of a membrane is the circular membrane, and in this case the Laplacian operator is given by equation (5.58). As in the case of a thin plate, the general solution for the complementary function may be expressed in polar coordinates as

$$W(r, \theta) = \sum_{n=0}^{\infty} W_n(r) \cos(n\theta + \phi), \tag{5.77}$$

where

$$W_n(r) = A_n J_n(kr) + B_n Y_n(kr) \tag{5.78}$$

and

$$k^2 = \frac{\rho d}{T} \omega^2. \tag{5.79}$$

Proceeding exactly in the case of the thin circular clamped plate and considering the case of a circular membrane that is stretched to a uniform tension over a circular frame and clamped at its periphery, the associated frequency equation is

$$J_n(kR) = 0. \tag{5.80}$$

The roots of the frequency equation (5.80)

$$kR = \lambda_{n,m}, \quad m = 1, 2, 3, \ldots$$

for each n may be found numerically. The associated natural frequencies then are

$$\omega = \sqrt{\frac{T}{\rho d}} \left(\frac{\lambda_{n,m}}{R} \right). \tag{5.81}$$

The values of $\lambda_{n,m}$ are tabulated in Table 5.2. Associated with each of the natural frequencies is a pattern of nodal lines, which are almost identical to those illustrated in Figure 5.3 for the case of a thin circular plate clamped along the boundary.

The fundamental vibration of a stretched circular membrane is with the circumference as a node. The ratios of the next two natural frequencies to the fundamental, with circles as the nodal lines, are 2.3 and 3.6. The ratios of the next two natural frequencies to the fundamental, with diameters as the nodal lines, are 1.59 and 2.14. The ratio of the natural frequency of vibration with one nodal circle and one

Table 5.2 Numerically obtained values of $\lambda_{n,m}$

m $n \rightarrow$	0	1	2	3	4
1	2.405	3.832	5.135	6.379	3.586
2	5.520	3.016	8.417	9.760	11.064
3	8.654	10.173	11.620	13.017	14.373
4	11.792	13.323	14.796	16.224	13.616

nodal diameter to the fundamental is 2.92. The stretched circular membrane has several applications, particularly in the design of musical instruments as well as in the design of voice transducers such as microphones and speakers.

Exercises

1. Consider a square plate of side a that is simply supported along all four edges.
(i) Obtain expressions for the first three natural frequencies of the plate.
(ii) Obtain expressions for the natural frequencies of a circular plate of radius R and simply supported along the periphery from the known roots of the associated characteristic equation.
(iii) Determine the radius of an equivalent circular plate of the same thickness and material with exactly the same first natural frequency as the square plate. Like the square plate, the circular plate is also simply supported along its periphery.
(iv) Repeat the calculations in (i), (ii) and (iii) for the case of square and circular plates clamped along the edges.

You should employ any valid and approximate method to obtain the natural frequencies of the square plate.

2. Consider a simply supported square plate subjected to a uniformly distributed compressive in-plane loading, N per unit length.
Show that that the critical in-plane compression loading, at the edges of the square plate, such that the first natural frequency of vibration is equal to zero, is given by

$$N_{Cr} = 2\pi^2 \frac{D}{a^2},$$

where D is the flexural rigidity of the plate.

3. Consider a flat rectangular panel with sides of lengths a and b and simply supported along all four edges. The longitudinal axis is assumed to be parallel to the side of length a. The panel is subjected to a uniform in-plane tension, N per unit length, such that

$$N_{xx} = N_{yy} = N \text{ and } N_{xy} = 0.$$

Furthermore, the top side of the panel is exposed on airflow. When the panel is stationary, the airflow is with a uniform velocity U_0. The excess pressure acting on the panel may be approximated as

$$\Delta p = p - p_\infty = \rho_\infty a_\infty \left(\frac{\partial w}{\partial t} + U_0 \frac{\partial w}{\partial x} \right),$$

where $w(x, y, t)$ is the transverse deflection of the panel and ρ_∞ is the density of the air.
Assume that

$$w(x, y, t) = w_1(x, t) \sin\left(\frac{n\pi y}{b}\right), \quad n = 1, 2, 3, \ldots$$

and show that

$$D\left(\frac{\partial^4 w_1}{\partial x^4} - \frac{2n^2\pi^2}{b^2}\frac{\partial^2 w_1}{\partial x^2} + \frac{n^4\pi^4}{b^4} w_1 \right) - N\left(\frac{\partial^2 w_1}{\partial x^2} - \frac{n^2\pi^2}{b^2} w_1 \right) + \rho_m h \frac{\partial^2 w_1}{\partial t^2} + p = 0,$$

where ρ_m is the density of the plate material, h the plate thickness and D the flexural rigidity of the plate. Hence explain how you would proceed to obtain the first natural frequency and the damping ratio of the first mode of oscillation of the panel.

4. Consider a rectangular panel with sides a and b and thickness h, where $h \ll a$, $h \ll b$. The panel is simply supported along all its four edges. The sides $x = 0$ and $x = a$ are restricted against any longitudinal displacements while there are no such restrictions in the y direction. The panel is initially loaded in its middle plane. The initial force N_0 is considered positive in compression and directed along the x axis. Assume that

$$w(x, y, t) = w_1(x, t) \sin\left(\frac{n\pi y}{b}\right), \quad n = 1, 2, 3, \ldots$$

and show that

$$D\left(\frac{\partial^4 w_1}{\partial x^4} - \frac{2n^2\pi^2}{b^2}\frac{\partial^2 w_1}{\partial x^2} + \frac{n^4\pi^4}{b^4} w_1\right) + N\frac{\partial^2 w_1}{\partial x^2} + \rho_m h \frac{\partial^2 w_1}{\partial t^2} = 0,$$

where

$$N = N_0 - \frac{12D}{ah^2}\int_0^a \left(\left[1 + \left(\frac{\partial w}{\partial x}\right)^2\right]^{\frac{1}{2}} - 1\right) \mathrm{d}x \approx N_0 - \frac{6D}{ah^2}\int_0^a \left(\frac{\partial w}{\partial x}\right)^2 \mathrm{d}x,$$

ρ_m is the density of the plate material, h the plate thickness and D the flexural rigidity of the plate. Assume further that

$$w_1(x, t) = W_{nm}(t) \sin\left(\frac{m\pi x}{a}\right), \quad m = 1, 2, 3, \ldots$$

and obtain an approximate non-linear differential equation for $W_{nm}(t)$ when both n and m are assumed to be equal to unity.

Comment on the features of this governing equation and its solutions, based on the experience gained with similar equations and their solutions.

5. Show that the strain energy due to flexure stored in a thin plate of thickness h, Young's modulus E and Poisson's ratio ν is given by the integral

$$U = \frac{1}{2E}\int_{-\frac{h}{2}}^{\frac{h}{2}}\int_S \mathsf{U}\, \mathrm{d}S\mathrm{d}z,$$

where

$$\mathsf{U} = \sigma_{xx}^2 + \sigma_{yy}^2 - 2\nu\sigma_{xx}\sigma_{yy} + 2(1+\nu)\sigma_{xy}^2.$$

Hence, show that for a rectangular plate

$$\mathsf{U} = \frac{D}{2}\int_0^b\int_0^a \bar{\mathsf{U}}\mathrm{d}y\mathrm{d}y,$$

where

$$\bar{\mathsf{U}} = \left(\frac{\partial^2 w}{\partial x^2}\right)^2 + \left(\frac{\partial^2 w}{\partial y^2}\right)^2 + 2\nu\frac{\partial^2 w}{\partial x^2}\frac{\partial^2 w}{\partial y^2} + 2(1-\nu)\left(\frac{\partial^2 w}{\partial x \partial y}\right)^2.$$

6. Derive the frequency equation for a rectangular plate with two opposite edges simply supported, and the other two edges (i) both clamped, (ii) both free or (iii) such that one is clamped and the other is simply supported.

7. (i) Consider a semi-infinite plate, with its middle surface in the x–y plane, infinite in the y direction and bounded by the planes $x = 0$ and $x = L$ in the x direction. The left face, $x = 0$, is subjected to a uniform, forced, longitudinal displacement $u_L = u_{L0}\cos\omega t$, while the right face is held fixed.

Show that the frequency ω at which there is resonance and the thickness stretch response of the plate are, respectively, given by

$$\omega = \frac{n\pi a_P}{L}, \ a_P = \sqrt{\frac{E}{\rho_0\left(1-\nu^2\right)}}$$

and

$$u_1(x,t) = u_{L0}\cos\omega t\left(\cos\frac{\omega x_1}{a_P} - \frac{1}{\tan\frac{\omega L}{a_P}}\sin\frac{\omega x_1}{a_P}\right).$$

(ii) Consider a semi-infinite plate, with its middle surface in the x–y plane, infinite in the y direction and bounded by the planes $x = 0$ and $x = L$ in the x direction. The left face, $x = 0$, is subjected to a uniformly applied surface traction $F = F_0\cos\omega t$, while the right face is held fixed.

Show that the frequency ω at which there is resonance and the thickness shear response of the plate are, respectively, given by

$$\omega = \frac{(2n-1)\pi a_2}{2L}, \ a_2 = \sqrt{\frac{\mu}{\rho_0}}$$

and

$$u_2(x,t) = -\frac{a_2 F_0}{\omega\mu}\cos\omega t\left(\sin\frac{\omega x_1}{a_2} - \tan\frac{\omega L}{a_2}\cos\frac{\omega x_1}{a_2}\right).$$

8. Consider the vibration of a square membrane. The membrane is clamped along all its edges. It is held in uniform tension, applied normal to the edges of the middle surface and in the direction of the undeformed middle surface.

Show that the natural frequency of the membrane is given by

$$\omega_{nm} = a\pi\sqrt{n^2+m^2}, \ n, \ m = 1, \ 2, \ 3, \ \ldots\ \infty,$$

where a is the non-dimensional velocity of transverse wave propagation in the membrane.

Obtain the mode shapes corresponding to a natural frequency equal to $a\pi\sqrt{5}$, indicating the shape and location of the nodal lines.

9. Consider the vibration of a rectangular membrane of sides a and b. The membrane is clamped along all its edges. It is held in uniform tension, applied normal to the edges of the middle surface and in the direction of the undeformed middle surface.

Determine the natural frequencies of the membrane. What is the effect of varying the aspect ratio of the plate, the ratio of b to a on the natural frequencies of the plate?

10. Consider the vibration of a square membrane. The membrane is clamped along three edges and free along the fourth. It is held in uniform tension, applied normal to the edges of the middle surface and in the direction of the undeformed middle surface.

Determine, in algebraic form, the natural frequencies of the membrane.

11. A certain elastic half-space is governed by the generalized Hooke's law. The elastic coefficients are $c_{11} = c_{22} = c_{33} = \lambda + 2\mu$, $c_{44} = c_{55} = c_{66} = \mu_T$, $c_{32} = c_{23} = \lambda$ and $c_{12} = c_{13} = c_{21} = c_{31} = \lambda$. Let

$$a_1^2 = (\lambda + 2\mu)/\rho, \ a_2^2 = \mu_T/\rho \text{ and } a_3^2 = \mu/\rho.$$

Show that a solution for a plane wave propagating in the x direction with the free surface normal to the z axis may be expressed in terms of two potential functions:

$$\begin{bmatrix}\phi\\ \psi\end{bmatrix} = \left[A\begin{Bmatrix}1\\ ie_1\end{Bmatrix}\exp(-p_1kz) + B\begin{Bmatrix}ie_2\\ 1\end{Bmatrix}\exp(-p_2kz)\right]\exp(ik(at-x)),$$

where p_1 and p_2 are roots of the characteristic equation

$$\left(p^2-1\right)^2+\left(p^2-1\right)\left(\frac{a^2}{a_1^2}+\frac{a^2}{a_2^2}+4\left(\frac{a_3^2}{a_1^2}-1\right)\left(\frac{a_3^2}{a_2^2}-1\right)\right)+\frac{a^4}{a_1^2a_2^2}+4\left(\frac{a_3^2}{a_1^2}-1\right)\left(\frac{a_3^2}{a_2^2}-1\right)=0$$

and

$$e_2=2\left(a_3^2-a_2^2\right)\frac{\left(p_2^2+1\right)}{\left(p_2^2-1\right)\left(a_2^2\left(p_2^2-1\right)+a^2\right)-4\left(a_3^2-a_2^2\right)p_2^2}p_2,$$

$$e_1=2\left(a_3^2-a_2^2\right)\frac{\left(p_1^2+1\right)}{\left(p_1^2-1\right)\left(a_1^2\left(p_1^2-1\right)+a^2\right)+4\left(a_3^2-a_2^2\right)p_1^2}p_1.$$

Furthermore, show that the transverse displacement v satisfies

$$\left(\frac{\partial^2}{\partial x^2}+\frac{\partial^2}{\partial z^2}\right)v=\frac{1}{a_2^2}\frac{\partial^2 v}{\partial t^2}$$

and that the stresses are given by

$$\sigma_{11}=\rho\left(a_1^2\frac{\partial u}{\partial x}+\left(a_1^2-2a_3^2\right)\frac{\partial w}{\partial z}\right),\quad \sigma_{13}=\rho a_2^2\left(\frac{\partial u}{\partial z}+\frac{\partial w}{\partial x}\right)$$

and

$$\sigma_{33}=\rho\left(\left(a_1^2-2a_3^2\right)\frac{\partial u}{\partial x}+a_1^2\frac{\partial w}{\partial z}\right).$$

Hence or otherwise show that the frequency equation for Rayleigh-type surface acoustic waves is

$$\begin{aligned}&\left(4p_1p_2-2p_1^2p_2^2\right)\left(1-e_1e_2\right)-2\left(p_1^2-p_2^2e_1e_2\right)\\&\quad-\left(\frac{a_1^2-2a_3^2}{a_1^2}\right)\left(\left(p_1^2p_2^2-1\right)\left(1-e_1e_2\right)+\left(p_1^2-p_2^2\right)\left(1+e_1e_2\right)\right)\\&=\sqrt{-1}\left(\left(\frac{a_1^2-2a_3^2}{a_1^2}\right)\left(e_1\left(p_2^4-1\right)+e_2\left(p_1^4-1\right)\right)+2\left(e_1\left(p_2^2-1\right)p_2^2+e_2\left(p_1^2-1\right)p_1^2\right)\right),\end{aligned}$$

where a is the velocity of the surface acoustic wave.

Show that frequency equation for Rayleigh-type surface acoustic waves may initially be approximated as

$$a_1^2\left(a^2-a_2^2\right)\left(a^2-a_1^2\left(1+\left(\frac{a_1^2-2a_3^2}{a_1^2}\right)^2\right)\right)^2=a_2^2a^4\left(a^2-a_1^2\right).$$

For a particular crystalline material ($Al_{0.3}Ga_{0.7}As$), given that $a_1^2=24.4303\times10^6\,(m/s)^2$, $a_2^2=12.1414\times10^6\,(m/s)^2$ and $a_3^2=5.9779\times10^6\,(m/s)^2$, employ an appropriate method and a computational tool such as a spreadsheet or MATLAB® and estimate the surface wave velocity. Comment on the accuracy of your estimate.

12. A certain elastic half space is governed by the generalized Hooke's law. The elastic coefficients are $c_{22}=c_{33}=\lambda+2\mu$, $c_{44}=c_{55}=c_{66}=\mu$, $c_{12}=c_{13}=c_{21}=c_{31}=c_{32}=c_{23}=\lambda$ and $c_{11}=\lambda+\mu\left(2+\beta\right)$.

Assuming β is small show that for Rayleigh-type surface acoustic waves, the *approximate* frequency equation is given by

$$\begin{aligned}&\left(4p_1p_2-2p_1^2p_2^2\right)(1-e_1e_2)-2\left(p_1^2-p_2^2e_1e_2\right)\\&\quad-\left(\frac{a_1^2-2a_2^2}{a_1^2}\right)\left(\left(p_1^2p_2^2-1\right)(1-e_1e_2)+\left(p_1^2-p_2^2\right)(1+e_1e_2)\right)\\&=\sqrt{-1}\left(\left(\frac{a_1^2-2a_2^2}{a_1^2}\right)\left(e_1\left(p_2^4-1\right)+e_2\left(p_1^4-1\right)\right)+2\left(e_1\left(p_2^2-1\right)p_2^2+e_2\left(p_1^2-1\right)p_1^2\right)\right)\end{aligned}$$

where $a_1^2=(\lambda+2\mu)/\rho$, $a_2^2=\mu/\rho$, while p_1 and p_2 are roots of the characteristic equation

$$\left(p^2-1\right)^2+\left(p^2-1\right)\left(\frac{a^2}{a_1^2}+\frac{a^2}{a_2^2}\right)+\frac{a^4}{a_1^2a_2^2}=\frac{\beta}{\left(p^2-1\right)}\left(\left(p^2-1+\frac{a^2}{a_1^2}\right)p^2+\left(p^2-1+\frac{a^2}{a_2^2}\right)\frac{a_2^2}{a_1^2}\right),$$

which are approximately defined by

$$p_1^2\approx1-\frac{a^2}{a_1^2}+\beta\left(\frac{a_2^2}{a^2}\right)\text{ and }p_2^2\approx1-\frac{a^2}{a_2^2}+\beta\left(1-\frac{a_2^2}{a^2}\right),$$

a is the velocity of the surface acoustic wave, and

$$e_2=\beta\frac{a_2^2}{\left(p_2^2-1\right)\left(a_1^2\left(p_2^2-1\right)+a^2\right)+\beta a_2^2}p_2,$$

$$e_1=\beta\frac{a_2^2}{\left(p_1^2-1\right)\left(a_2^2\left(p_1^2-1\right)+a^2\right)-\beta a_2^2p_1^2}p_1.$$

Explain why the surface wave velocity is approximate and show that when $\beta=0$, the frequency equation reduces to the same form as that for an isotropic half space.

13. A certain elastic half space is governed by the generalized Hooke's law. The elastic coefficients are $c_{22}=c_{33}=\lambda+2\mu$, $c_{44}=c_{55}=c_{66}=\mu$, $c_{32}=c_{23}=\lambda$, $c_{11}=\lambda+2\mu(1+\alpha)$ ></TB:defs> and <TB:defs><?TeX $c_{12}=c_{13}=c_{21}=c_{31}=\lambda+\alpha\mu$. Assuming α is small, show that for Rayleigh-type surface acoustic waves, the *approximate* frequency equation is given by

$$\begin{aligned}&\left(4p_1p_2-2p_1^2p_2^2\right)(1-e_1e_2)-2\left(p_1^2-p_2^2e_1e_2\right)\\&\quad-\left(\frac{a_1^2-2a_2^2}{a_1^2}\right)\left(\left(p_1^2p_2^2-1\right)(1-e_1e_2)+\left(p_1^2-p_2^2\right)(1+e_1e_2)\right)\\&=\sqrt{-1}\left(\left(\frac{a_1^2-2a_2^2}{a_1^2}\right)\left(e_1\left(p_2^4-1\right)+e_2\left(p_1^4-1\right)\right)+2\left(e_1\left(p_2^2-1\right)p_2^2+e_2\left(p_1^2-1\right)p_1^2\right)\right),\end{aligned}$$

where $a_1^2=(\lambda+2\mu)/\rho$, $a_2^2=\mu/\rho$, while p_1 and p_2 are roots of the characteristic equation

$$\begin{aligned}&\left(p^2-1\right)^2+\left(p^2-1\right)\left(\frac{a^2}{a_1^2}+\frac{a^2}{a_2^2}\right)+\frac{a^4}{a_1^2a_2^2}\\&=\frac{\alpha}{\left(p^2-1\right)}\left(\left(p^2-1+\frac{a^2}{a_1^2}\right)p^2+\left(p^2-1+\frac{a^2}{a_2^2}\right)\frac{a_2^2}{a_1^2}\left(2-p^2\right)\right),\end{aligned}$$

which are approximately defined by

$$p_1^2\approx1-\frac{a^2}{a_1^2}+\alpha\left(\frac{a_2^2}{a^2}+\frac{a_2^2}{a_1^2}\right)\text{ and }p_2^2\approx1-\frac{a^2}{a_2^2}+\alpha\left(1-\frac{a_2^2}{a^2}\right),$$

a is the velocity of the surface acoustic wave, and

$$e_2 = \alpha \frac{a_2^2}{\left(p_2^2 - 1\right)\left(a_1^2\left(p_2^2 - 1\right) + a^2\right) + \alpha a_2^2\left(2 - p_2^2\right)} p_2,$$

$$e_1 = \alpha \frac{a_2^2\left(2 - p_1^2\right)}{\left(p_1^2 - 1\right)\left(a_2^2\left(p_1^2 - 1\right) + a^2\right) - \alpha a_2^2 p_1^2} p_1.$$

Hence show that when $\alpha = 0$, the frequency equation is the same as that for an isotropic half space.

14. Consider the formation of free surface standing waves in a liquid filled container. A rigid rectangular stationary container is assumed to be partially filled with incompressible inviscid liquid to the height h.

(i) Show that the governing equations of motion of the fluid in the container can be expressed in terms of a velocity potential ϕ as

$$\nabla^2\phi = 0, \frac{\partial\phi}{\partial t} + \frac{\mathbf{V}\cdot\mathbf{V}}{\mathbf{2}} + \frac{p}{\rho} + g\left(z - h\right) = 0,$$

where

$$\mathbf{V} = \nabla\phi,$$

with the associated boundary conditions on the free surface

$$z = h + \zeta\left(x, y, t\right)$$

given by

$$\frac{\partial\zeta}{\partial t} = -\mathbf{V}\cdot\nabla\left(\zeta - z\right) \approx \frac{\partial\phi}{\partial z} \text{ and } \zeta = -\frac{1}{g}\frac{\partial\phi}{\partial t} - \frac{\mathbf{V}\cdot\mathbf{V}}{\mathbf{2}g} \approx -\frac{1}{g}\frac{\partial\phi}{\partial t}.$$

(ii) Hence show that the amplitudes of the normal modes of oscillation of the liquid free surface satisfy equations of the form

$$\ddot{q}_i + \frac{g}{L_i}q_i = 0.$$

Comment on the apparent similarity between the motion of a simple pendulum and the normal modes of motion of the liquid free surface.

References

Donnell, L. H. (1976) *Beams, Plates and Shells*, McGraw-Hill, New York.
Leissa, A. W. (1969) Vibration of Plates, NASA Special Publication SP-160, NASA, Washington D.C.
Leissa, A. W. (1973) Vibration of Shells, NASA Special Publication SP-288, NASA, Washington D.C.
McLachlan, N. W. (1951) *Theory of Vibrations*, Dover Publications, New York.
Soedel, W. (1993) *Vibrations of Shells and Plates*, Marcel Dekker Inc., New York.
Timoshenko, S. P. and Woinowsky-Kreiger, S. (1959) *Theory of Plates and Shells*, 2nd edn, McGraw-Hill Book Company, New York.

6

Dynamics of Piezoelectric Media

6.1 Introduction

For a body with dynamic disturbance propagation in three-dimensional space and with the presence of additional body forces, the stress equations of motion (4.68) (Kolsky, 1963) can be written in the Voigt's form as

$$\rho \frac{\partial v_i}{\partial t} = \sigma_{ij,j} + F_i, \tag{6.1}$$

where ρ is the mass density, F_i is the body force, σ_{ij} represents the stress component, v_i represents the particle velocity component along the ith direction and $i,\ j = 1,\ 2,\ 3$. For a linear isotropic medium, we may eliminate the stress components by introducing Hooke's law in the form

$$\frac{\partial \sigma_{ij}}{\partial t} = \lambda v_{k,k}\delta_{ij} + \mu \left(v_{i,j} + v_{j,i} \right), \tag{6.2}$$

where λ and μ are Lame's constants. In terms of displacements and body forces, the equations of motion may also be expressed as

$$\rho \frac{\partial^2 u_i}{\partial t^2} = \frac{\partial \sigma_{ij}}{\partial x_j} + \rho f_i, \tag{6.3}$$

where f_i are the body forces per unit mass or the body force density and u_i represents the component of particle displacement along the i direction. Hooke's law is essentially a constitutive relationship that relates the stress and strain components and is generally expressed by the tensorial relationship

$$\sigma_{ij} = c_{ijkl} S_{kl}, \quad S_{kl} = \frac{1}{2}\left(\frac{\partial u_k}{\partial x_l} + \frac{\partial u_l}{\partial x_k} \right), \tag{6.4}$$

where $S_{kl} = \varepsilon_{kl}$ are the second-rank strain tensor components and c_{ijkl} are the elastic stiffness constants given by a fourth-rank tensor. For a class of materials, from symmetry considerations, the number of independent constants in this relationship can be reduced to 11. Introducing a compressed notation to replace the indices ij and kl by single indices, as indicated in Table 6.1, the elastic stiffness constants may be expressed as

$$\mathbf{C} = \begin{bmatrix} \mathbf{A} & \mathbf{G} \\ \mathbf{G}^T & \mathbf{B} \end{bmatrix}, \tag{6.5}$$

Dynamics of Smart Structures Dr. Ranjan Vepa

Table 6.1 Compressed notation and index replacement scheme

Index pairs	Single index replacement	Index pairs	Single index replacement
11	1	23 or 32	4
22	2	13 or 31	5
33	3	12 or 21	6

where

$$\mathbf{A}=\begin{bmatrix}\mathbf{c}_{11} & \mathbf{c}_{12} & \mathbf{c}_{13}\\ \mathbf{c}_{12} & \mathbf{c}_{22} & \mathbf{c}_{23}\\ \mathbf{c}_{13} & \mathbf{c}_{23} & \mathbf{c}_{33}\end{bmatrix},\quad \mathbf{B}=\begin{bmatrix}\mathbf{c}_{44} & 0 & 0\\ 0 & \mathbf{c}_{55} & \mathbf{c}_{15}\\ 0 & \mathbf{c}_{15} & \mathbf{c}_{66}\end{bmatrix}\quad \text{and}\quad \mathbf{G}=\begin{bmatrix}\mathbf{c}_{14} & 0 & 0\\ -\mathbf{c}_{14} & 0 & 0\\ 0 & 0 & 0\end{bmatrix}.$$

A further reduction is possible when the properties corresponding to the 1 and 2 index pair and 4 and 5 index pair are identical to each other. Thus, there are six independent elastic constants with

$$\mathbf{A}=\begin{bmatrix}\mathbf{c}_{11} & \mathbf{c}_{12} & \mathbf{c}_{13}\\ \mathbf{c}_{12} & \mathbf{c}_{11} & \mathbf{c}_{13}\\ \mathbf{c}_{13} & \mathbf{c}_{13} & \mathbf{c}_{33}\end{bmatrix},\quad \mathbf{B}=\begin{bmatrix}\mathbf{c}_{44} & 0 & 0\\ 0 & \mathbf{c}_{44} & \mathbf{c}_{14}\\ 0 & \mathbf{c}_{14} & \mathbf{c}_{66}\end{bmatrix}\quad \text{and}\quad \mathbf{G}=\begin{bmatrix}c_{14} & 0 & 0\\ -c_{14} & 0 & 0\\ 0 & 0 & 0\end{bmatrix}.$$

Furthermore,

$$\mathbf{c}_{66}=\frac{1}{2}\left(\mathbf{c}_{11}-\mathbf{c}_{12}\right).$$

For a hexagonally crystalline material $c_{14}=0$, for a cubic crystalline material $c_{11}=c_{33}=\lambda+2\mu$, $c_{66}=c_{44}=\mu_1$ and $c_{12}=c_{13}=\lambda$, and for an isotropic material $\mu_1=\mu$.

The stress–strain relationships considered above have been tacitly applied to a non-piezoelectric, dielectric and elastic solid. In *piezoelectric materials*, the constitutive relations are modified to account for the mechanical deformation that results upon the application of electric field and vice versa. When a piezoelectric material is subjected to a load, the microscopic atomic displacements within the crystalline structure induce a residual electrical polarization in the material. Provided these electric moments combine in a constructive way, a net resultant average macroscopic electric moment is obtained, which must be accounted for in the material constitutive relationships. First, Hooke's law is generalized to account for the resulting coupling between electrical and mechanical parameters. Thus, we have

$$\sigma_{ij}=c_{ijkl}S_{kl}-e_{ijk}E_k,\quad S_{kl}=\frac{1}{2}\left(\frac{\partial u_k}{\partial x_l}+\frac{\partial u_l}{\partial x_k}\right),\tag{6.6}$$

where e_{ijk} are piezoelectric stress constants and E_k is the kth component of the electric field. There is a further relationship that relates the electric displacement vector D_i to the electric field vector and the strain tensor. This relationship is

$$D_i=e_{ikl}S_{kl}+\in_{ik}E_k,\tag{6.7}$$

where $\in_{ik}$ are constants related to the relative permittivity or dielectric constant $\in_r$ in a particular direction and to the permittivity of vacuum or the dielectric constant $\in_0$. Using the compressed notation introduced earlier, the two relationships that constitute the piezoelectric constitutive relationships may be expressed in matrix form as

$$\begin{bmatrix}\mathbf{T}\\ \mathbf{D}\end{bmatrix}=\begin{bmatrix}\mathbf{C} & -\mathbf{e}^T\\ \mathbf{e} & \mathrm{E}\end{bmatrix}\begin{bmatrix}\mathbf{S}\\ \mathbf{E}\end{bmatrix},\tag{6.8}$$

where the piezoelectric stress matrix and the permittivity matrix are, respectively, given by

$$\mathbf{e} = \begin{bmatrix} e_{11} & -e_{11} & 0 & e_{14} & 0 & 0 \\ 0 & 0 & 0 & 0 & -e_{14} & -e_{11} \\ 0 & 0 & 0 & 0 & 0 & 0 \end{bmatrix} \quad \text{and} \quad \mathrm{E} = \begin{bmatrix} \in_{11} & 0 & 0 \\ 0 & \in_{11} & 0 \\ 0 & 0 & \in_{33} \end{bmatrix},$$

for a piezoelectric material such as quartz having a trigonal (class 32) crystal classification when there are six non-zero elastic constants and $\mathbf{T}$ is the stress tensor in matrix-vector notation. Quartz belongs to a family of crystalline (such as amethyst and citrine) and non-crystalline (such as opal, flint and agate) materials that are chemically equivalent to silicon dioxide. Quartz may be cut in such a way that the strain is parallel to the applied electric field. When an alternating electric field is applied to the cut crystal, it will expand and contract at the same frequency as the field. The cut quartz crystals can be used as generators of acoustic or waves. Quartz, tartaric acid, tourmaline, Rochelle salt, ammonium dihydrogen phosphate (ADP), lithium sulphate and potassium dihydrogen phosphate (KDP) are some of the piezoelectric single crystals that have to be cut in a special way in order to maximize their piezoelectric properties.

The electric displacement vector $\mathbf{D}$ is defined to complete the formulation of Maxwell's electromagnetic wave theory, which seeks to relate the spatial dependence of the electric field intensity $\mathbf{E}$ to the temporal variations of the magnetic field intensity $\mathbf{H}$. In the theory of electromagnetic waves, the additional concept of charge is introduced. Quantity of electric charge is denoted by q, the electric current by i and the electric current density by $\mathbf{J}$. The electric current density $\mathbf{J}$ is related to the electric field intensity, the source current and the Ohmic contribution, which is given as

$$\mathbf{J} = \mathbf{J}_s + \sigma \mathbf{E} \equiv \mathbf{J}_s + \frac{1}{\rho}\mathbf{E}, \tag{6.9}$$

where $\mathbf{J}_s$ is the source current density; σ, the conductivity, is a constant characteristic of the medium; and ρ, the resistivity, is defined as its inverse. The mutual interactions of charges and currents are described in terms of the electric and magnetic field intensities. The force on an element of charge caused by all other charges is represented as the interaction of the element of charge with the fields in its vicinity, the sources of the fields being all other charges. However, rather than the magnetic field intensity it is the magnetic induction $\mathbf{B}$ that plays a role in the distribution of the force on an element of charge. The elemental force δF on element of charge δq, moving with a velocity $\mathbf{V}$, is defined by

$$\delta F = (\mathbf{E} + \mathbf{V} \times \mathbf{B})\,\delta q. \tag{6.10}$$

The magnetic induction $\mathbf{B}$ within a medium is a function of the magnetic field intensity but not necessarily a linear function. Electromagnetic wave theory also employs the physical law of conservation of electric charge, which may be stated mathematically as

$$\nabla \cdot \mathbf{J} = -\frac{\partial \rho_c}{\partial t}, \tag{6.11}$$

where ρ_c is the intrinsic charge in the material; $\rho_c = 0$ in the absence of intrinsic charge in the material.

The mutual dependence of the spatial variations of the electric and magnetic field intensities $\mathbf{E}$ and $\mathbf{H}$ may be expressed in terms of the temporal variations in the magnetic induction $\mathbf{B}$ and the electric displacement $\mathbf{D}$ by Faraday's and Ampère's laws. Faraday's law gives the relationship between an induced electric field and a time-varying magnetic field. Ampère's law describes the creation of a magnetic field due to a dielectric flux, a conductivity current and a source current. Faraday's law is stated mathematically as

$$\nabla \times \mathbf{E} = -\frac{\partial \mathbf{B}}{\partial t}. \tag{6.12}$$

In a lossy dielectric medium, a conductivity current, given by $\sigma\mathbf{E}$, will flow and Ampère's law may be expressed as

$$\nabla \times \mathbf{H} = \frac{\partial \mathbf{D}}{\partial t} + \sigma\mathbf{E} + \mathbf{J}_s. \tag{6.13}$$

Assuming that the medium is not lossy, i.e. $\sigma = 0$, applying the divergence operator to Faraday's and Ampère's laws and employing the charge conservation law, we may show that

$$div\mathbf{D} = \nabla \cdot \mathbf{D} = \rho_c \tag{6.14}$$

and

$$div\mathbf{B} = \nabla \cdot \mathbf{B} = 0. \tag{6.15}$$

The last equation expresses the physical fact that, unlike an isolated charge, a magnetic pole cannot exist in isolation.

An important relationship that can be derived from equations (6.11) and (6.14) for a boundary surface S_P with a prescribed charge distribution is

$$\int_{S_P} \frac{\partial}{\partial t}\mathbf{D} \cdot \mathbf{n}\mathrm{d}S = -\frac{\partial}{\partial t}q_P = -i_P.$$

In a seminal contribution, J. E. Maxwell was able to show that the last four equations (6.12) to (6.15), admit wave-like solutions and that the solutions propagate with a speed equal to that of light. Hence, they are known universally as Maxwell's equations.

The electromagnetic field in a given region of space depends on the kind of matter occupying the region in addition to the distribution of charge giving rise to the field. The phenomenon can be described in non-conducting, non-piezoelectric 'linear' materials by writing the magnetic induction **B** and the electric displacement **D** as products of two quantities:

$$\mathbf{D} = \vec{\kappa}_\varepsilon \cdot \in_0 \mathbf{E} \tag{6.16a}$$

and

$$\mathbf{B} = \vec{\kappa}_m \cdot \mu_0\mathbf{H}, \tag{6.16b}$$

where the $\vec{\kappa}$'s are vector quantities characteristic of the medium. In isotropic media they are scalars; $\vec{\kappa}_\varepsilon$ is the relative dielectric constant and $\vec{\kappa}_m$ is the relative permeability. They are both equal to unity in vacuum, thus defining what **D** and **B** should be in vacuum or free space. The other two constants, ε_0, and μ_0, are universal constants in these relationships. Assuming that the electric field vector is vertically polarized and that the magnetic field vector is horizontally polarized, i.e.

$$\mathbf{E} = (0, 0, E_z) \tag{6.17a}$$

and

$$\mathbf{H} = \left(0, H_y, 0\right) \tag{6.17b}$$

and assuming plane propagation of the waves, Maxwell's equations may be reduced to a pair of wave equations given by

$$\frac{\partial^2}{\partial x^2}E_z = \mu \in \frac{\partial^2}{\partial t^2}E_z \tag{6.18a}$$

and

$$\frac{\partial^2}{\partial x^2}H_y = \mu \in \frac{\partial^2}{\partial t^2}H_y. \tag{6.18b}$$

Thus, the vectors **E** and **H** both obey the same wave equation, propagating in the x direction with the same velocity,

$$c = \frac{1}{\sqrt{\mu \in}}, \tag{6.19}$$

where $\in$ and μ are the permittivity and the permeability of the isotropic propagating medium. In free space, this reduces to

$$c_0 = \frac{1}{\sqrt{\mu_0 \in_0}}, \tag{6.20}$$

where $\in_0$ and μ_0 are the permittivity of free space and the permeability of free space. Since

$$\in_0 = \frac{1}{36\pi} \times 10^{-9} \text{ Farads/m}, \quad \mu_0 = 4\pi \times 10^{-7} \text{ Henries/m},$$

the velocity of electromagnetic waves in free space is given by

$$c_0 = 3 \times 10^8 \text{ m/s}, \tag{6.21}$$

which is also equal to the velocity of light in free space.

To obtain the governing equations of wave motion in a piezoelectric medium, the equation defining the electric displacement **D**, equation (6.16a), must be replaced by to the corresponding relationship for a piezoelectric medium given by equation (6.7). Employing equations (6.8) and (6.16b) to eliminate **D** and **B** from equations (6.12) and (6.13), and assuming that the medium is loss free and that there are no electrical sources,

$$\nabla \times \mathbf{E} = -\vec{\kappa}_m \cdot \mu_0 \frac{\partial \mathbf{H}}{\partial t}, \quad \nabla \times \mathbf{H} = \mathbf{e}\frac{\partial \mathbf{S}}{\partial t} + \mathrm{E}\frac{\partial \mathbf{E}}{\partial t}. \tag{6.22}$$

Taking the curl of the first equation and eliminating the second equation results in a wave equation for the electric field given by

$$\nabla \times \nabla \times \mathbf{E} + \vec{\kappa}_m \cdot \mu_0 \frac{\partial}{\partial t}\left(\mathbf{e}\frac{\partial \mathbf{S}}{\partial t} + \mathrm{E}\frac{\partial \mathbf{E}}{\partial t}\right) = 0 \tag{6.23a}$$

that is coupled with the wave equation for particle motion

$$\rho \frac{\partial^2 u_i}{\partial t^2} = c_{ijkl} \frac{\partial^2 u_k}{\partial x_j \partial x_l} + e_{ijk} \frac{\partial E_k}{\partial x_j}, \tag{6.23b}$$

through the strain–displacement relations

$$S_{kl} = \frac{1}{2}\left(\frac{\partial u_k}{\partial x_l} + \frac{\partial u_l}{\partial x_k}\right). \tag{6.23c}$$

In the quasi-static approximation, it is assumed that the electric field intensity **E** may be expressed in terms of the gradient of a time-varying electric potential V in terms of the quasi-static relationship

$$\mathbf{E} = -\nabla \cdot V = -div V. \tag{6.24}$$

This is justified by the fact that velocity of electromagnetic waves is far greater than the velocity of acoustic waves propagating in piezoelectric media. Eliminating the components of the electric field vector in equation (6.23b), the equations governing the motion of particles in a piezoelectric medium may then be expressed as

$$\rho \frac{\partial^2 u_i}{\partial t^2} = c_{ijkl} \frac{\partial^2 u_k}{\partial x_j \partial x_l} + e_{ijk} \frac{\partial^2 V}{\partial x_j \partial x_k}, \tag{6.25}$$

and employing the constraint $\nabla \cdot \mathbf{D} = 0$, it follows that

$$0 = e_{jkl} \frac{\partial^2 u_k}{\partial x_j \partial x_l} - \in_{jk} \frac{\partial^2 V}{\partial x_j \partial x_k}. \quad (6.26)$$

In a piezoelectric medium, it can be seen that when energy is stored at potential energy the stored energy exceeds the elastic potential energy and that additional energy is stored by virtue of the electromagnetic coupling. Thus, effectively the piezoelectric effect contributes to an increase in the apparent 'elastic' energy stored and hence contributes to a piezoelectric stiffening effect, in so far as the acoustic waves are concerned. Thus, wave propagation in a piezoelectric medium is generally quite similar to wave propagation in a non-piezoelectric medium. Important engineering applications of the theory are to the vibration of piezoelectric plates of specific geometries, which are relevant to the design of vibration transducers.

6.2 Piezoelectric Crystalline Media

Before considering the features of acoustic waves propagating in piezoelectric medium, we shall briefly consider the properties of piezoelectric *crystalline* media and piezoelectric polymers. Piezoelectricity, the interaction of force and electronic charge with a material, depends on the symmetry of that material, and therefore the crystalline structure of the material plays an important role in them.

The fact that most materials had a crystalline lattice structure, that is a three-dimensional geometric arrangement of atoms in a molecule, and that this crystalline lattice structure repeated itself from molecule to molecule, with no exception, led to attempts to classify crystalline materials. The field of crystallography is a large one, and it has developed a rigorous way of classifying and identifying symmetry classes. Crystalline materials were grouped into seven divisions, as shown in Table 6.2. Cubic crystals are characterized by all equal sides and all angles being right angles. Tetragonal crystals have only two of the three sides equal, but all angles are right angles. Rhombic crystals have no sides equal but all angles are right angles. Monoclinic crystals are like rhombic crystals but are slanted in one direction, so eight of the angles are not right angles. Triclinic crystals are slanted in two directions and so have no equal sides or equal angles. Trigonal crystals are like triclinic crystals but with all sides equal. Hexagonal crystals are like tetragonal crystals, but two faces have six equal sides. Within these seven major groups are 32 subcategories, determined by the possible number of ways in which unique but symmetrical arrangements of groups of atoms can occur.

While materials belonging to 10 of these classes or subcategories exhibit piezoelectricity, the most common of these belong to six classes in the cubic, hexagonal and trigonal groups. Cubic crystals are characterized by the most regular symmetry. Monoclinic crystals are less symmetric followed by triclinic crystals. In the case of cubic crystals when the Cartesian system axes, used to define wave motion, are chosen to be parallel to the crystal axes, there are only three independent elastic coefficients. However, a unique Cartesian system of axes is employed in crystallography (see, for example, Bhagavantam (1966))

Table 6.2 Major groups and examples of crystalline geometries

No.	Name	Examples
1	Cubic	Alum, diamond, lead, copper, silver, gold and gallium arsenide
2	Tetragonal	Tin, zircon, rutile and scheelite
3	Rhombic	Topaz, sulphur, iodine and silver nitrate
4	Monoclinic	Boraz, cane sugar and gypsum
5	Triclinic	Copper sulphate and boric acid
6	Trigonal	Arsenic, quartz, ice, graphite and aluminium oxide
7	Hexagonal	Magnesium, zinc oxide, beryllium, cadmium sulphide and calcium

Table 6.3 The Miller index notation

Specific plane	$(i,\ j,\ k)$
Family of planes	$\{i,\ j,\ k\}$
Specific direction	$[i,\ j,\ k]$
Family of directions	$\langle i,\ j,\ k\rangle$

and to align the motion axes with the crystal axes one may need to rotate the reference frames through a sequence of rotations. Furthermore, when the elastic properties of a cubic crystal are the same along all three crystalline axes, it behaves like an isotropic solid and the elastic constants reduce to two. A good example of a cubic piezoelectric crystal is bismuth germanium oxide ($Bi_{12}GeO_{20}$). An example of a hexagonal crystal is piezoelectric zinc oxide (ZnO), which has five independent elastic constants, while lithium niobate ($LiNbO_3$) and lithium tantalate ($LiTaO_3$) are trigonal with as many as six independent elastic constants.

Crystallographers are particularly interested in exposing the internal structure of metallic crystalline forms so they deliberately 'cut' them in a way that exposes a large area of the specific surface of interest. The most common crystalline structures are associated with cubic and hexagonal crystals and are (1) the face-centred cubic, (2) the body-centred cubic and (3) the hexagonal close pack structures. For each of these crystal systems, although there are in principle an infinite number of possible surfaces, which can be exposed, there are in practice only a limited number of planes that are employed to reveal the internal crystalline structure. The way in which a plane intersects the crystal is very important and is defined by using Miller indices, a notation that is commonly used by crystallographers. Typical examples of this are the $(1,\ 0,\ 0)$ plane, which is a plane that cuts the x axis but not the other two axes, the $(1,\ 1,\ 0)$ plane, which cuts the x and y axes but not the z axis, and the $(1, 1, 1)$ plane, which cuts all the three axes at the same value. Thus, the Miller indices provide a method of defining a reference plane relative to the crystalline axes. The Miller index notation is summarized in Table 6.3.

The general form of stiffness and piezoelectric properties of a crystalline material may be deduced by considering microscopic symmetry properties of the medium. While it is not intended to present an exhaustive discussion of material symmetry, it is indeed essential to consider some salient features. If a crystalline medium is symmetric with respect to particular transformation of coordinates, then the stiffness and piezoelectric properties must remain unchanged by the same transformation. The transformations of coordinates that culminate in similar material symmetries define symmetry groups, which are symbolized by a unique notation in crystallography. Symmetry operations in the context of crystalline materials may be of two categories: translational operations and point transformations, which leave at least one point in the crystal lattice unchanged. The latter are particularly relevant to relations that relate two or more physical quantities at the same point in the crystal. They include *rotations*, *reflections* and *inversions* as well as their combinations. Diagrams known as *point-group diagrams* are usually employed to portray them graphically. Rotation symmetries are defined in terms of the smallest rotation under which the lattice is symmetric. Reflection symmetry refers to mirror reflection about a plane while inversion refers to reflection through a single point. In the naming of the symmetry groups, a number such as 2 or 3 indicates a 2- or 3-fold rotation axis. m indicates a plane of mirror symmetry. If we write, for example, $2m$, this means that the plane of mirror symmetry is parallel with the 2-fold rotation axis. An overbar indicates that an inversion is applied (e.g. $\bar{2}$ indicates a 2-fold axis with inversion; $\bar{1}$ is a simple inversion). It can be shown that crystalline structures that are symmetric with respect to an inversion are not piezoelectric.

To be piezoelectric, materials must also have a large and durable electric dipole polarization. Furthermore, all the crystalline directions in the material must be perfectly aligned with respect to each other so the contributions of each crystal to the overall polarization of the material combine and accumulate. Polarization is defined as the difference between the electric displacement vector in the material and the

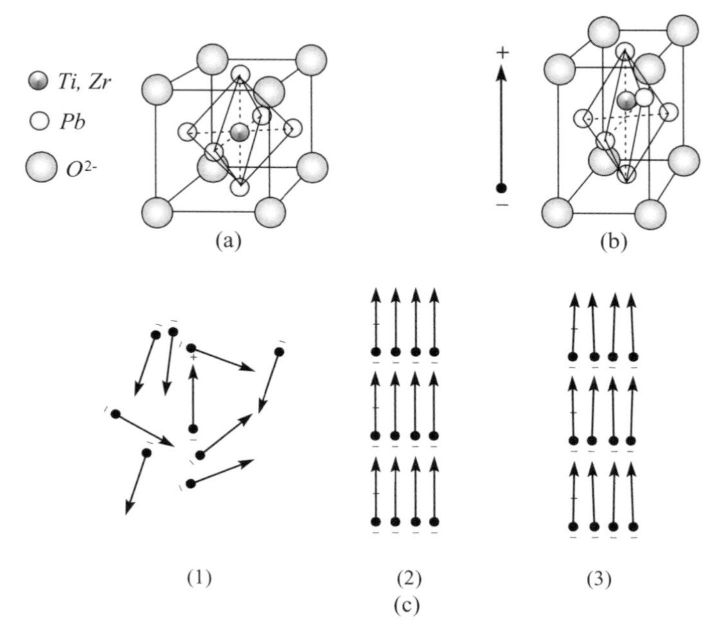

Figure 6.1 Piezoelectric elementary cell (a) before poling and (b) after poling, and (c) electric dipoles in Weiss domains corresponding to (1) an unpoled ferroelectric ceramic, (2) during poling and (3) after poling (piezoelectric ceramic)

corresponding vector in vacuum or free space. Certain materials, such as quartz, have naturally a large residual electric dipole polarization. Others such as piezoelectric ceramics and piezoelectric polymers must be processed thermally or by the application of an electric field to ensure that there is phase transition from a non-polar to a polar phase. Such materials also have a non-polar phase above a certain temperature, known as the Curie temperature, when the piezoelectric property disappears altogether. A big advantage of the polycrystalline piezoceramics and piezopolymers compared with the single crystal materials is that they do not need to be sliced in a particular direction in order to maximize the piezoelectric properties.

The most commonly used type of piezoceramics, lead zirconate titanates (PZTs), are solid solutions of lead zirconate and lead titanate, often doped with other elements to obtain specific properties. These polycrystalline ceramics are manufactured by heating a mixture of proportional amounts of lead, zirconium and titanium oxide powders to around 800–1000°C. The components in the mixture then react with each other to form the PZT powder with the Perovskite structure. This powder is mixed with a binder and sintered into the desired shape. Charge separation between the positive and negative ions results in the electric dipole behaviour. However, since the properties of the PZT crystallite are the same along all axes, the centro-symmetric cubic polycrystalline material behaves like an isotropic, elastic solid. During the cooling process, the material undergoes a non-polar paraelectric to a polar ferroelectric phase transition, and the cubic unit cell becomes tetragonal (Figure 6.1). When certain ionic crystals exhibit a spontaneous dipole moment, they are said to be ferroelectric. To be piezoelectric, the material must be ferroelectric or non-centro-symmetric. In the polar phase, the material exhibits the property of piezoelectricity but not in the non-polar phase. As a result, the unit cell becomes elongated in one direction and has a permanent dipole moment oriented along its long axis (x_3 axis). Above its Curie temperature, PZT also looses its piezoelectric property.

A volume of the material in which the polar moments of the unit cell are all oriented in the same direction is termed a Weiss domain. The 'unpoled' ceramic or raw piezoelectric material consists of many Weiss domains that orient in several random directions. Thus, they cancel each other and contribute no net polarization. Application of a high electric field greater than 2000 V/mm to the heated piezoceramic has the effect of aligning most of the domains in such a way that the polar axes of the unit cells are oriented as closely parallel to the applied field as allowed by the crystal structure. This process is called *poling* and it imparts a permanent net polarization to the ceramic. The electric dipoles align with the applied electric field and roughly continue to be in alignment with each other on cooling and after the removal of the electric field.

Materials that adopt a permanent or remanent polarization when crystallized under an electric field are termed *electrets*. Since the polar axis (x_3 axis) of the unit cell is longer than the other two axes (x_1, x_2 axes), the reorientation of the domains also creates a permanent mechanical distortion. The distortion causes an increase in the dimensions in the direction of the applied field and a contraction along normal to the electric field. The polar axes can be reoriented in certain specific directions governed by the crystal symmetry in the presence of a high electric field or mechanical stress. This process is called *switching*.

In the case of tetragonal PZTs, either 90° or 180°, switching can occur. The 90° switching is always accompanied by mechanical distortion due to the longer length of the unit cell along the polar axes, while the 180° switching can occur without any distortion. It is also known that an electric field can cause both 180° and 90° domain switching, but a mechanical stress can only cause ±90° switching. Consequently, a mechanical stress can depol a piezoelectric ceramic but cannot repol it.

A poled ceramic exhibits both the direct and the converse piezoelectric effects. When an electric voltage is applied to a poled piezoelectric ceramic material, the Weiss domains increase their alignment proportional to the voltage. The converse effect is the deformation of the piezoceramic in response to an electric field and there are three main phenomena causing it. The first of these is the piezoelectric effect of the aligned domains. This is the deformation of the unit cell caused by the electric field and is an intrinsic effect. The second of these is as a result of deformations caused by non-180° domain wall motion and related phenomena. This effect is essentially one of a category of extrinsic effects. These extrinsic effects are also the major cause of non-linearity and losses in the piezoceramic while the intrinsic contribution to such losses is small. The third is the electrostriction effect.

Electrostriction is the phenomenon of mechanical deformation of materials due to an applied electric field. This is a phenomenon present to varying degrees in all materials and occurs as a result of the presence of polarizable atoms and molecules. An applied electric field can distort the charge distribution within the material, resulting in modifications to bond length, bond angle or electron distribution functions, which in turn affects the macroscopic dimensions of the material. The electrostrictive deformation is proportional to the square of the field-induced polarization and thus proportional to the square of the electric field and independent of the direction of the field. Being inherently non-linear, the electrostrictive effects are usually much smaller than the other two effects. Yet the phenomenon of electrostriction is very similar to piezoelectricity. The relationship between the field and deformation is centro-symmetrical, in contrast to the piezoelectric effect, which is non-centro-symmetrical. One fundamental difference between the two is the closeness of transition temperature of the material to the operating temperatures. This results in the improved strain and hysteresis properties for electrostrictive materials. However, a larger number of coefficients are required to model electromechanical coupling for electrostriction. The polarization in piezoelectric materials is spontaneous, while that in electrostrictive materials is field induced. Materials that possess the following three main characteristics are termed ferroelectric:

1. for temperatures greater than the Curie temperature, the static dielectric constant is inversely proportional to the excess temperature relative to the Curie temperature, which is known the Curie–Weiss law;
2. the property of spontaneous polarization;

3. polarization due to the application of an external field in these materials is not proportional to the applied electric field as for paraelectric materials (i.e. polarization is proportional to applied electric field) but instead exhibits hysteresis.

Electrets are also ferroelectric since the direction of polarization may be reversed by the application of an electric field. All ferroelectric materials exhibit piezoelectricity but all piezoelectric materials are surely not ferroelectric. Ferroelectrics have a very large piezoelectric constant. Both the phenomena occur in non-centro-symmetrical crystals.

There are essentially three non-linear effects that are relevant to piezoelectric crystals. The first is the geometric non-linearity that manifests itself in the non-linear strain–displacement relations. The second is due to changes in the crystal volume as a result of the deformation. This change in volume induces a first-order change in the mass density that must be accounted for in a non-linear analysis. The third is due to non-linearities in the constitutive relations in the form of hysteresis effects. The hysteresis effects associated with each of the stresses and strains, and electric fields are usually different from each other, although they all have the same general structure. Thus, it is quite difficult to analyse the non-linear propagation of plane waves in a general three-dimensional piezoelectric crystal.

6.2.1 Electromechanically Active Piezopolymers

Polymers are chemical compounds or mixtures of compounds formed in chemical reactions in which two or more molecules combine to form larger molecules by a process known as polymerization. They are long-chain materials with a high molecular weight assembled from self-similar units of monomers. These are joined to each other by covalent bonds that are quite unlike metallic bonds in that they are highly directional and not as strong. Strength can be enhanced by covalent bonds between chains, ionic bonds between charged adjacent groups and by van der Waal bonds that arise due to the diffuse attraction between different varieties of groups of atoms or individual atoms. One feature of polymers is that their properties could be tailored to meet a particular functional need by combining differing monomers to form co-polymers, by appropriately controlling the extent of the polymerization and strength enhancement and by the controlled incorporation of chemical additives. They are popular for their excellent mechanical properties, for instance light weight, flexibility, malleability and processing capability. One unique feature of polymers is that they withstand high strain, which is impossible in traditional materials. The electromechanical response of polymer-based materials is low in comparison with inorganic materials. Since polymers can endure usually more than 10% strain, the idea of developing polymer-based materials with ultra-high electromechanical strain responses was conceived. Electromechanically active polymers are polymers whose electronic properties can change in response to the applied mechanical strain. They differ from inorganic materials, whose electromechanical responses are lower and are able to withstand very high strain. The level of induced strain can be as high as two orders of magnitude greater than the striction-limited, rigid and fragile electromechanically active ceramics. They are also flexible materials capable of converting energy in the form of electric charge and voltage to mechanical force and movement. As in ferroelectric ceramics, electrostriction can be considered as the origin of piezoelectricity in ferroelectric polymers. Some limiting factors of these materials are low actuation forces, mechanical energy density and the lack of robustness.

There are two major categories of electromechanically active polymers. The first is electronic and are generally driven by electric fields or Coulomb forces. The first subgroup of this category of electromechanically active polymers is the group of ferroelectric polymers. Poly-vinylidene fluoride (PVDF) and its co-polymers are the most representative of this group. PVDF polymer consists of a partially crystalline component in an inactive shapeless phase. Large applied alternating current fields of about 0.1 V/m induce electrostrictive or non-linear strains of about 1%. However, PVDF polymer that has been exposed to electron radiation is known to exhibit higher electrostrictive strain. Ferroelectric electromechanically active polymer actuators have the capability to operate in air, vacuum or water in addition to a wide temperature range. The next sub-group in this category of polymers is the dielectric electromechanically active polymer. As electrostatic fields can be applied to those polymers that show low elastic stiffness and high dielectric constants to induce large actuation strain, they are said to be electrostatically stricted polymers and are known as dielectric electromechanically active polymers. Electrostrictive graft elastomers are the third sub-group of electronic electromechanically active polymers. They are so called because they are synthesized from polymers that consist of two components, a flexible macromolecule backbone and a grafted polymer that can be produced in a crystalline form. This material is often combined with piezoelectric copolymer (polyvinylidene fluoridetrifluoro-ethylene), which is able to create a large amount of ferroelectric-electrostrictive molecular composite systems and may be operated as either piezoelectric sensors or electrostrictive actuators. The final sub-group of electronic electromechanically active polymers is liquid crystal elastomer materials. The second category of electromechanically active polymers is the ionic variety that is primarily driven by the mobility of the diffusion of ions. Although there are four different types of these materials, which include ionic polymer metallic composites and carbon nanotubes and have several unique applications in the design of transducers, they are not very useful as piezopolymers. The former have a unique property and, as a consequence, when a cantilever strip of such a material is subjected to a suddenly applied and sustained DC electric potential up to about 3 V across its faces, it bends towards the anodic face. Thus, ionic polymer–metal composites (IPMCs) are electroactive materials with the potential of being employed as artificial muscles.

6.3 Wave Propagation in Piezoelectric Crystals

The equations of motion (Mason, 1964; Auld, 1973; Ikeda 1996) in the absence of body forces and the constraint on the electric displacement, for the trigonal quartz crystal (class 32), with six non-zero elastic constants and two non-zero piezoelectric constants, e_{11} and e_{14} are ($e_{21}=-e_{11}$, $e_{25}=-e_{14}$, $e_{26}=-e_{11}$)

$$\rho\ddot{u}_1 = c_{11}\frac{\partial^2 u_1}{\partial x_1^2} + (c_{12}+c_{66})\frac{\partial^2 u_2}{\partial x_1 \partial x_2} + (c_{13}+c_{44})\frac{\partial^2 u_3}{\partial x_1 \partial x_3} + c_{66}\frac{\partial^2 u_1}{\partial x_2^2} + c_{44}\frac{\partial^2 u_1}{\partial x_3^2} + e_{11}\left(\frac{\partial^2 V}{\partial x_1^2} - \frac{\partial^2 V}{\partial x_2^2}\right) - e_{14}\frac{\partial^2 V}{\partial x_2 \partial x_3}, \tag{6.27a}$$

$$\rho\ddot{u}_2 = c_{66}\frac{\partial^2 u_2}{\partial x_1^2} + (c_{12}+c_{66})\frac{\partial^2 u_1}{\partial x_1 \partial x_2} + (c_{13}+c_{44})\frac{\partial^2 u_3}{\partial x_2 \partial x_3} + c_{11}\frac{\partial^2 u_2}{\partial x_2^2} + c_{44}\frac{\partial^2 u_2}{\partial x_3^2} - 2e_{11}\frac{\partial^2 V}{\partial x_2 \partial x_1} + e_{14}\frac{\partial^2 V}{\partial x_1 \partial x_3}, \tag{6.27b}$$

$$\rho\ddot{u}_3 = c_{44}\frac{\partial^2 u_3}{\partial x_1^2} + (c_{13}+c_{44})\frac{\partial^2 u_1}{\partial x_1 \partial x_3} + (c_{13}+c_{44})\frac{\partial^2 u_2}{\partial x_2 \partial x_3} + c_{44}\frac{\partial^2 u_3}{\partial x_2^2} + c_{33}\frac{\partial^2 u_3}{\partial x_3^2}, \tag{6.27c}$$

$$e_{11}\left(\frac{\partial^2 u_1}{\partial x_1^2} - 2\frac{\partial^2 u_2}{\partial x_1 \partial x_2} - \frac{\partial^2 u_1}{\partial x_2^2}\right) + e_{14}\left(\frac{\partial^2 u_2}{\partial x_1 \partial x_3} - \frac{\partial^2 u_1}{\partial x_2 \partial x_3}\right) - \epsilon_{11}\left(\frac{\partial^2 V}{\partial x_1^2} + \frac{\partial^2 V}{\partial x_2^2}\right) - \epsilon_{33}\frac{\partial^2 V}{\partial x_3^2} = 0. \tag{6.27d}$$

Assuming that there is a displacement in one direction only,

$$\rho\ddot{u}_1 = c_{11}\frac{\partial^2 u_1}{\partial x_1^2} + e_{11}\frac{\partial^2 V}{\partial x_1^2}, \quad \text{and} \quad e_{11}\frac{\partial^2 u_1}{\partial x_1^2} - \epsilon_{11}\frac{\partial^2 V}{\partial x_1^2} = 0. \tag{6.28}$$

Eliminating the electric potential we have, and including body forces,

$$\rho\left(\ddot{u}_1 - f_1\right) = \left(c_{11} + \frac{e_{11}^2}{\epsilon_{11}}\right)\frac{\partial^2 u_1}{\partial x_1^2} = c_{11}\left(1 + K_{11}^2\right)\frac{\partial^2 u_1}{\partial x_1^2}, \tag{6.29}$$

where f_1 is the body force density and

$$K_{11}^2 = \frac{e_{11}^2}{c_{11} \in_{11}} \tag{6.30}$$

is known as the piezoelectric stiffening coefficient and admits a maximum value of unity. The parameter K_{11}^2 is an important parameter in the design of surface acoustic wave (SAW) transducers, as it is an electromechanical coupling coefficient. It is a non-dimensional parameter that represents the ratio of the mechanical to the electrical energy stored in the transducer at any time. A letter followed by two subscripts conventionally denotes the coupling coefficient, as well as other related directionally sensitive constants or coefficients. In the case of the electromechanical coupling coefficients, the subscripts are the same as those of the elastic constants with which the electromechanical coefficient is associated. The order of the subscripts is the same, whether the stimulation is electrical or mechanical, and the value of the piezoelectric stiffening coefficient is the same.

It can be shown that in the case of a poled ferroelectric ceramic (hexagonal crystal, class 6 mm), the piezoelectric stress and permittivity matrices **e** and E are, respectively, given as

$$\mathbf{e} = \begin{bmatrix} 0 & 0 & 0 & 0 & e_{15} & 0 \\ 0 & 0 & 0 & e_{15} & 0 & 0 \\ e_{31} & e_{31} & e_{33} & 0 & 0 & 0 \end{bmatrix} \quad \text{and} \quad \mathrm{E} = \begin{bmatrix} \in_{11} & 0 & 0 \\ 0 & \in_{11} & 0 \\ 0 & 0 & \in_{33} \end{bmatrix},$$

and the piezoelectric stiffening coefficient is

$$K_{55}^2 = \frac{e_{15}^2}{c_{55} \in_{11}}. \tag{6.31}$$

The static piezoelectric d and g constants relate mechanical stresses and strain to electrical voltage (g) or charge (d) in the same manner as coupling coefficient. The g constant is the ratio of the field generated or voltage per metre thickness to the applied stress in Newtons per square metre. The d constant is the ratio of the strain developed to the field applied or voltage per metre thickness. Both coefficients are related to the piezoelectric stiffening coefficient or the electromechanical coupling coefficient, the piezoelectric coupling coefficients e_{ij} and the dielectric constants. The piezoelectric stress coefficients are defined by inverting the relationships in equation (6.8) and solving for **S** and **E**, while the piezoelectric strain coefficients are obtained from a partial inversion of equation (6.8) and solving for **S** and **D**.

Solutions to the governing equations of wave motion are found by precisely the same techniques as those enunciated for isotropic solids. A preferred approach is to replace the elastic constants of the piezoelectric material by 'stiffened' elastic constants, which are linear functions of the piezoelectric constants and reflect the influence of the piezoelectric effect on the elastic constants. The piezoelectric material may then be treated as an equivalent non-piezoelectric material. For example, in the governing equations for the one-dimensional plane wave approximation for surface wave motion, replacing c_{11} by the stiffened elastic coefficient $\bar{c}_{11}$, where $\bar{c}_{11}$ is defined as

$$\bar{c}_{11} = c_{11} \left(1 + K_{11}^2\right), \tag{6.32}$$

results in equation (6.29). Since the numbers of independent piezoelectric constants are usually only one or utmost two, the stiffened elastic constants are predominantly the function of just one electromechanical coupling coefficient. For some reason or the other, this is true of most piezoelectric materials and so it is relatively easy to relate the velocity of surface wave propagation to the stiffened elastic coefficients. Piezoelectric stiffening is a weak effect and the value of the stiffening coefficient is, at best, only a fraction of the stiffness coefficient.

The solutions also indicate that a number of different modes of waves can propagate in a piezoelectric solid just as in a non-piezoelectric solid (White, 1970). In a general solid, first there are the longitudinal bulk waves with the direction of propagation and direction of displacement in the same direction. Second, there exist the transverse bulk waves with the direction of propagation and direction of displacement being perpendicular to each other. Then there are the Rayleigh waves propagating in a preferred direction in the vicinity of a free surface (they are rapidly attenuated with depth), with displacements normal to both the direction of propagation and the free surface.

To briefly recapitulate the results obtained in Section 4.4.7, for an elastically isotropic non-piezoelectric solid, the Rayleigh-type surface wave velocity may be explicitly obtained by solving a characteristic equation (equation (4.107)), which may be expressed as

$$\left(2 - v_{rT}^2\right)^2 = 4\sqrt{\left(1 - v_{rT}^2\right)\left(1 - v_{rL}^2\right)},$$

where v_{rT} is the velocity ratio of the surface wave velocity a_R to the transverse or equi-voluminal or shear (S-wave) wave velocity

$$a_2 = \sqrt{\frac{c_{11} - c_{12}}{2\rho}} = \sqrt{\frac{\mu}{\rho}}, \tag{4.99}$$

and v_{rL} is the velocity ratio of the surface wave velocity a_R to the longitudinal or irrotational (P-wave) wave velocity

$$a_1 = \sqrt{\frac{c_{11}}{\rho}} = \sqrt{\frac{\lambda + 2\mu}{\rho}}. \tag{4.98}$$

An approximation to the solution is given by the formula

$$\frac{a_R}{a_2} = \frac{0.87c_{11} + 2c_{12}}{c_{11} + 2c_{12}}. \tag{6.33}$$

A semi-infinite piezoelectric material can also propagate Rayleigh waves in the vicinity of a free surface. Furthermore, Love waves with particle motions entirely transverse to the propagation vector also propagate in a piezoelectric medium provided density variations in the vicinity of the free surface are favourable for the propagation of such waves. The velocity of the Rayleigh waves in an isotropic piezoelectric material, v_p, may be expressed in terms of the electromechanical coefficient as

$$v_p^2 \approx a_R^2\left(1 + K_{66}^2\right). \tag{6.34}$$

This is valid when the 'stiffened' Poisson's ratio for the material is unaffected by the piezoelectric constants. Thus, the fractional decrease in the velocity of the surface wave when the piezoelectric field is shorted out is given by

$$\frac{\Delta v}{a_R} = 1 - \sqrt{1 + K_{66}^2} \cong -\frac{1}{2}K_{66}^2. \tag{6.35}$$

It has become conventional to specify surface wave propagation directions by naming the normal to the free surface and the direction along which propagation occurs. Thus, '*YZ-lithium niobate*', refers to the *lithium niobate* with the normal to the free surface being aligned to the Y crystallographic axis and the direction of propagation along the Z crystallographic axis.

There are in addition a number of other modes of wave propagation as well as effects accompanying a wave as it propagates. Two of these effects are of primary significance. First, as a wave propagates in a piezoelectric solid it is accompanied by a travelling electric field. In fact, the displacement field is directly coupled to it and can be effectively controlled by modifying the electric field and vice versa. The electric field propagates in the same direction as the wave motion. It has a significantly symmetric component on either side of the free surface. Thus, its influence extends to both sides of the free surface and decays away from it. Second, the influence of piezoelectricity is to effectively increase the mechanical stiffness, thereby increasing the wave velocities, which lie in the vicinity of 3488 m/s for YZ-lithium niobate ($K^2 = 0.045$), 3158 m/s for ST-quartz ($K^2 = 0.0011$), 1681 m/s for bismuth germanium oxide ($Bi_{12}GeO_{20}$) ($K^2 = 0.014$), 2840 m/s for gallium arsenide, 2700 m/s for zinc oxide, 1500 m/s for cadmium sulphide, 2100 m/s for lead-zirconate-titanate PZT-4 and 4600 m/s for silicon (compared to a typical terrestrial Rayleigh wave velocity of 3000 m/s). Surface wave velocities in piezoelectric solids are known to range from a value below 1500 m/s in polymeric materials such as PVDF to values above 4700 m/s for polycrystalline materials in the PZT family. By comparison the lowest and highest surface wave velocities in non-piezoelectric solids are 570 m/s in lead and about 11 000 m/s in diamond, respectively. Both wave velocities and wave modes as well as the rate of attenuation with depth or the decay rate are now significantly dependent on dielectric properties of piezoelectric materials.

Different piezoelectric materials are employed in different applications. Thus, zinc oxide is employed to deposit a thin piezoelectric film on non-piezoelectric semiconductor substrates (silicon). Thin films of silicon dioxide are deposited on piezoelectric crystals such as quartz, and Love waves become significant in both these applications while isotropic PZT is often employed to propagate Rayleigh waves.

There is yet another property of certain semiconductor materials known as piezoresistivity, which results in the conductivity of the semiconductor to be directly dependent on stress. In fact, this is another property that plays a significant role in the design of SAW transducers.

Surface wave attenuation is an important effect that is generally minimized in the design of SAW transducers. A typical parameter indicating the relative amplitude attenuation is the attenuation constant αmeasured in decibels per centimetre. The intrinsic component of the attenuation constant α is directly proportional to the square of the frequency and the crystal viscosity coefficient and inversely proportional to the mass density and the cube of the wave velocity. Yet when it can be controlled, for example in a piezoresistive semiconductor such as a doped slice of silicon by the application of external pressure, the amplitude attenuation of surface waves may be exploited to measure the applied pressure. It is this feature that permits the application of a SAW device as a transducer.

Another important application area of piezoelectric crystalline waveguides is communications technology (Matthews, 1977; Morgan, 1985; Campbell, 1998). A key component in optical signal generation and transmission is the modulator, which impresses the high-speed data stream onto the optical signal in time domain multiplexing (TDM) or wavelength domain multiplexing (WDM).

Multiplexing is sending multiple signals or streams of information on a carrier signal at the same time in the form of a single, complex signal and then recovering the separate signals at the receiving end. Analogue signals are commonly multiplexed using frequency-division multiplexing (FDM), in which the carrier bandwidth is divided into sub-channels of different frequency widths, each carrying a signal at the same time in parallel. Digital signals are commonly multiplexed using time-division multiplexing (TDM), in which the multiple signals are carried over the same channel in alternating time slots. Dense wavelength division multiplexing (DWDM) is a fibre-optic transmission technique that employs light wavelengths to transmit data parallel-by-bit or serial-by-character. In some optical fibre networks, multiple signals are carried together as separate wavelengths of light in a multiplexed signal using DWDM. The output from a continuous wave laser optical source, which serves as the carrier, passes through the modulator before transmission.

Over the years lithium niobate ($LiNbO_3$) crystals have been and continue to be the primary materials for optical modulators due to their properties of enabling low-loss optical waveguides, their ability to be integrated with SAW devices and their high electro-optic effect. The electro-optic coupling effect, also known as the Bragg effect, permits the direct modulation of the optical carrier by the acoustic surface wave. Lithium niobate-based travelling wave acousto-optic modulators, transmitting well over 20 Gb/s, and compatible SAW devices have emerged as the most utilized modulators in the fibre-optic communication industry.

6.3.1 Normal Modes of Wave Propagation in Crystalline Media

To be able to synthesize surface wave solutions and estimate the velocities of propagation of specific surface wave modes, it is almost always necessary to first determine the normal modes of wave propagation in an infinite medium of the same material.

Considering the equations of motion, of a general anisotropic elastic material, described by equations (6.3), the stress–strain relationships given by equation (6.4) and the strain–displacement relationships given by equations (4.58a) to (4.58f), and eliminating the stresses and the strains, the equations of wave propagation in an elastic medium are

$$\rho \frac{\partial^2 u_i}{\partial t^2} = \frac{1}{2} c_{ijkl} \frac{\partial}{\partial x_j} \left(\frac{\partial u_l}{\partial x_k} + \frac{\partial u_k}{\partial x_l} \right). \tag{6.36}$$

By symmetry arguments (k and l are interchangeable), we can simplify equation (6.36) to get

$$\rho \frac{\partial^2 u_i}{\partial t^2} = c_{ijkl} \frac{\partial^2 u_l}{\partial x_k \partial x_j}. \tag{6.37}$$

Plane wave solutions u_i, in terms of ζ, the bulk wave number, components of $\mathbf{U}$, the displacement amplitude vector (which defines the polarization) and components of $\mathbf{n}$, the propagation direction unit vector may be expressed as

$$u_i = U_i \exp i \left(\zeta n_k x_k - \omega t \right)$$

and substituting it into equation (6.37) gives

$$\omega^2 U_i = \frac{c_{ijkl}}{\rho} \zeta^2 n_k n_j U_l = \lambda_{ijkl} \zeta^2 n_k n_j U_l, \tag{6.38}$$

where

$$\lambda_{ijkl} = \frac{c_{ijkl}}{\rho}.$$

Introducing the phase velocity

$$v = \frac{\omega}{\zeta}$$

into equation (6.38), this may be expressed as an eigenvalue problem given by

$$\left(v^2 \delta_{il} - \lambda_{ijkl} n_k n_j \right) U_l = 0, \tag{6.39}$$

where $\delta_{il} = 1$ when $i = l$ and equal to 0 otherwise. As the associated characteristic equation may be expressed as a cubic, there are three characteristic solutions representing three distinct plane waves and the associated eigenvectors represent characteristic directions of polarization. The elements of the characteristic determinant are linear functions of the elastic constants. In the case of wave propagation in piezoelectric crystalline media, the elastic constants are replaced by their stiffened counterparts. In an isotropic medium, the eigenvectors are polarized either in the direction of propagation or normal to it. When this is the case in an anisotropic media, the modes are said to be *pure*. Thus, pure modes are defined as being modes that are either normal to or parallel to the direction of propagation.

Two other important concepts associated with waves propagating in piezoelectric media are the *slowness curve* and the *skew curve*. A slowness curve is a plot of the inverse of velocity (units are therefore seconds/metre or equivalent). A slowness curve is generated by choosing a plane in the material of interest and then calculating the different phase velocities for a selection of propagation directions. Slowness is then plotted as a function of propagation direction in a polar plot. For a material that is transversely isotropic in the plane of propagation, the slowness curve is a perfect circle. In the case of a cubic crystalline material with the direction of propagation in one of crystalline principal planes, the slowness curve is an ellipse (e.g. cadmium sulphide). Skew is a measure of how far any particular mode deviates from a pure mode. If the mode is pure, skew will be zero. For other modes, the skew is the angle between the polarization vector and the direction of propagation (for quasi-longitudinal modes) or the normal to the direction of propagation (for quasi-shear modes).

6.3.2 Surface Wave Propagation in Piezoelectric Crystalline Media

To estimate the wave velocities of surface waves in semi-infinite and layered piezoelectric media, the normal modes of wave propagation are first computed. Linear combinations of these modes, which serve as partial wave solutions, may then be employed to construct solutions that meet the boundary conditions at a free surface. The surface velocities are obtained by solving for the roots of the associated characteristic boundary determinant. The procedure is quite similar to the isotropic case, and a number of graded examples are included in the exercises at the end of this chapter. Yet the solutions obtained indicate a number of special features. A variety of SAWs exist in anisotropic solids for arbitrary directions of propagation relative to the crystalline reference axes (White, 1970). In anisotropic solids, the velocity of wave propagation or the phase velocity and nature of displacements depend on the crystal plane used as the free surface and on the direction of SAW propagation (generalized Rayleigh waves). The phase velocity may no longer be parallel to the flow of wave energy, which is generally determined by the group velocity. Directions with collinear phase and group velocity vectors are a special feature of pure modes. In anisotropic crystals, the depth decay factor often has a real part, i.e. it may be complex instead of imaginary, resulting in the appearance of oscillations in the decaying distribution of displacement components with depth.

Altering the boundary conditions may also significantly control the generation and attenuation of SAWs. Boundary conditions may be classified into two groups: mechanical and electrical. Typical mechanical boundary conditions are the stress-free boundary condition at a free surface and the displacement or stress component continuity boundary conditions at an interface. Electrical boundary conditions generally imply that the charge or potential (applied) is specified on portions of the boundary. In particular, when an oscillating potential is applied across a surface (by attaching a metal surface that serves as an electrode) surface waves may be launched. Similarly, surface waves may also be received or measured by tapping an appropriately introduced electrode on a surface of the material.

A solution to acoustic wave propagation in a piezoelectric substrate requires that both Maxwell's and Newton's equations be solved simultaneously. The Maxwell and Newton equations are coupled through the piezoelectric equations. In Section 6.1, the latter were employed to eliminate the magnetic field intensity **H**, the electric displacement **D** and the magnetic induction **B** from the Maxwell and Newton equations. The resulting equations governing the propagation of acoustic waves were given by equations (6.23). To compute the coupled electroacoustic normal modes, the components of both the displacement vector and the electric field are assumed to be of the form

$$u_i = U_i \exp i\,(\zeta n_k x_k - \omega t)$$

and

$$E_i = E_{0i} \exp i\,(\zeta n_k x_k - \omega t)\,.$$

The generalized eigenvalue problem corresponding to equation (6.39) is obtained by substituting the above assumed solutions into equations (6.23a), (6.23b) and (6.23c). The normal modes associated with the corresponding eigenvalue problem serve as partial wave solutions for the synthesis of coupled electroacoustic surface waves, just as in the non-piezoelectric case. In the general case, there are as many as eight normal modes and associate eigenvalues.

A significant feature of both piezoelectric and non-piezoelectric materials is the appearance of coupled electroacoustic surface waves, which are also known as *Bleustein–Gulyaev waves*, with particle motion still normal to the direction of propagation but parallel to the free surface as in the case of Love waves. Associated with this type of wave is an electric field, so it can be coupled electrically just as a Rayleigh wave in a piezoelectric solid. They are quite unlike the Love waves, as they are dependent on certain dielectric and electrical properties of the material rather than on the density variations in the vicinity of the free surface as in the case of classical Love waves. An important feature of these waves is that they can extend far deeper into the solid than ordinary Rayleigh waves and may be amplified by appropriate interaction with an external field generated by drifting carrier. Furthermore, they exist only in certain preferred directions of propagation in a plane parallel to the free surface.

A piezoelectric plate can also propagate an infinite number of 'plate modes' or Lamb modes of the type considered in Section 7.6, and these exhibit cut-off features similar to electromagnetic waveguides. As the 'plate' has a finite thickness, the propagation velocity depends on the frequency and the waves are dispersive. Typical generalizations of plate modes are to waves at the interface of two different piezoelectric media and to layered solids. The basic waves propagating at the interface of two semi-infinite solids are the so-called Stoneley waves that propagate at the common boundary provided the relative properties of the two materials lie in certain restricted ranges. In the piezoelectric case, there exists another extra wave solution, the *Maerfeld–Tournois wave*, that bears the same relationship to the Stoneley wave as the Bleustein–Gulyaev wave does to the Rayleigh wave. It is not only horizontally polarized but also bound to the boundary between two piezoelectric layered half-spaces. Also when a thin (piezoelectric or non-piezoelectric) plate rests on a semi-infinite piezoelectric surface, the resulting wave modes, similar to Love waves, have their particle motion perpendicular to the direction of propagation but parallel to the free surface. In a layered (piezoelectric or non-piezoelectric) solid, there are indications of all the above wave modes. The modes are similar to plate modes, the one like the symmetric mode being like the Rayleigh surface wave and the antisymmetric modes being known as *Sezawa modes*. The particle motions in these modes include components both along and transverse to the propagation vector. Love waves with particle motions entirely transverse to the propagation vector also propagate in layered piezoelectric solids. In addition, when the wave velocity (transverse shear wave velocity) of the upper layer exceeds the corresponding wave velocity in the lower layer, the surface waves tend to 'leak' into the interior and thus loose the vital energy essential for propagation on the surface.

6.3.3 *Influence of Coordinate Transformations on Elastic Constants*

In the analysis of piezoelectric crystalline media, it is often necessary to perform a general rotation of the rectangular coordinates to align them with preferred directions. A general rotation of the rectangular coordinates can be performed by applying successive rotations about different coordinate axis. A standard way of doing this is the following. The coordinates are first rotated clockwise through an angle θ about the z axis, then through a clockwise angle β about the *transformed* y axis and finally rotated clockwise through an angle α about the *transformed* x axis. The net result is a transformation of the reference frame, and the associated transformation matrix $\mathbf{T}$ consists of elements that may all be expressed in terms of trigonometric functions of θ, β and α. The transformation is always orthogonal and satisfies

$$\mathbf{T}^T\mathbf{T} = \mathbf{T}\mathbf{T}^T = \mathbf{I}.$$

Considering a small cuboid of the material, the stresses relate the force and the area vectors, which may be expressed in matrix form as

$$\begin{bmatrix} F_x \\ F_y \\ F_z \end{bmatrix} = \begin{bmatrix} \sigma_{xx} & \sigma_{yx} & \sigma_{zx} \\ \sigma_{xy} & \sigma_{yy} & \sigma_{zy} \\ \sigma_{xz} & \sigma_{yz} & \sigma_{zz} \end{bmatrix} \begin{bmatrix} A_x \\ A_y \\ A_z \end{bmatrix}.$$

In the transformed reference frame, the forces are

$$\begin{bmatrix} F_x^l \\ F_y^l \\ F_z^l \end{bmatrix} = \mathbf{T} \begin{bmatrix} F_x \\ F_y \\ F_z \end{bmatrix} = \mathbf{T} \begin{bmatrix} \sigma_{xx} & \sigma_{yx} & \sigma_{zx} \\ \sigma_{xy} & \sigma_{yy} & \sigma_{zy} \\ \sigma_{xz} & \sigma_{yz} & \sigma_{zz} \end{bmatrix} \begin{bmatrix} A_x \\ A_y \\ A_z \end{bmatrix},$$

which may be expressed in terms of the transformed areas as

$$\begin{bmatrix} F_x^l \\ F_y^l \\ F_z^l \end{bmatrix} = \mathbf{T} \begin{bmatrix} \sigma_{xx} & \sigma_{yx} & \sigma_{zx} \\ \sigma_{xy} & \sigma_{yy} & \sigma_{zy} \\ \sigma_{xz} & \sigma_{yz} & \sigma_{zz} \end{bmatrix} \mathbf{T}^T \mathbf{T} \begin{bmatrix} A_x \\ A_y \\ A_z \end{bmatrix} = \mathbf{T} \begin{bmatrix} \sigma_{xx} & \sigma_{yx} & \sigma_{zx} \\ \sigma_{xy} & \sigma_{yy} & \sigma_{zy} \\ \sigma_{xz} & \sigma_{yz} & \sigma_{zz} \end{bmatrix} \mathbf{T}^T \begin{bmatrix} A_x^l \\ A_y^l \\ A_z^l \end{bmatrix}.$$

Thus, the matrix relationship

$$\begin{bmatrix} \sigma_{xx}^l & \sigma_{yx}^l & \sigma_{zx}^l \\ \sigma_{xy}^l & \sigma_{yy}^l & \sigma_{zy}^l \\ \sigma_{xz}^l & \sigma_{yz}^l & \sigma_{zz}^l \end{bmatrix} = \mathbf{T} \begin{bmatrix} \sigma_{xx} & \sigma_{yx} & \sigma_{zx} \\ \sigma_{xy} & \sigma_{yy} & \sigma_{zy} \\ \sigma_{xz} & \sigma_{yz} & \sigma_{zz} \end{bmatrix} \mathbf{T}^T \tag{6.40}$$

gives the stress matrix in the transformed coordinates.

When the stress matrix is expressed as a vector with the double subscripts replaced by the abbreviated single subscripts, the transformational relationship may be expressed as

$$\begin{bmatrix} \sigma_{xx}^l \\ \sigma_{yy}^l \\ \sigma_{zz}^l \\ \sigma_{yz}^l \\ \sigma_{xz}^l \\ \sigma_{xy}^l \end{bmatrix} = \mathbf{M} \begin{bmatrix} \sigma_{xx} \\ \sigma_{yy} \\ \sigma_{zz} \\ \sigma_{yz} \\ \sigma_{xz} \\ \sigma_{xy} \end{bmatrix} = \mathbf{M} \begin{bmatrix} \sigma_1 \\ \sigma_2 \\ \sigma_3 \\ \sigma_4 \\ \sigma_5 \\ \sigma_6 \end{bmatrix}, \tag{6.41}$$

where $\mathbf{M}$ is the corresponding transformation matrix and is known as the *Bond transformation matrix*. The elements of the $\mathbf{M}$ matrix are quadratic functions of the elements of the $\mathbf{T}$ matrix and are given by the relationships

$$m_{i,j} = t_{i,j}^2, \tag{6.42a}$$

$$m_{i,j+3} = 2t_{i,j+1}t_{i,j+2}, \tag{6.42b}$$

$$m_{i+3,j} = t_{i+1,j}t_{i+2,j} \tag{6.42c}$$

and

$$m_{i+3,j+3} = t_{i+1,j+1}t_{i+2,j+2} + t_{i+1,j+2}t_{i+2,j+1}, \tag{6.42d}$$

where $i, j = 1, 2, 3$. A similar transformation may be employed to transform the strain matrix from the original coordinates to the transformed coordinates. This relationship takes the form

$$\begin{bmatrix} \varepsilon_{xx}^l \\ \varepsilon_{yy}^l \\ \varepsilon_{zz}^l \\ \varepsilon_{yz}^l \\ \varepsilon_{xz}^l \\ \varepsilon_{xy}^l \end{bmatrix} = \left(\mathbf{M}^T\right)^{-1} \begin{bmatrix} \varepsilon_{xx} \\ \varepsilon_{yy} \\ \varepsilon_{zz} \\ \varepsilon_{yz} \\ \varepsilon_{xz} \\ \varepsilon_{xy} \end{bmatrix} = \mathbf{N} \begin{bmatrix} \varepsilon_1 \\ \varepsilon_2 \\ \varepsilon_3 \\ \varepsilon_4 \\ \varepsilon_5 \\ \varepsilon_6 \end{bmatrix}, \tag{6.43}$$

where $\mathbf{N} = \left(\mathbf{M}^T\right)^{-1} \neq \mathbf{M}$. The elements of the $\mathbf{N}$ may be shown to be given by the relationships

$$n_{i,j} = t_{i,j}^2, \tag{6.44a}$$

$$n_{i,j+3} = t_{i,j+1}t_{i,j+2}, \tag{6.44b}$$

$$n_{i+3,j} = 2t_{i+1,j}t_{i+2,j} \tag{6.44c}$$

and

$$n_{i+3,j+3} = t_{i+1,j+1}t_{i+2,j+2} + t_{i+1,j+2}t_{i+2,j+1}, \tag{6.44d}$$

where $i, j = 1, 2, 3$. Thus, the matrix of elastic constants in the transformed coordinates $\mathbf{c}^l$ may be expressed as

$$\mathbf{c}^l = \mathbf{M} \times \mathbf{c} \times \mathbf{M}^T. \tag{6.45}$$

It must be mentioned that it is essential that the piezoelectric stress and permittivity matrices $\mathbf{e}$ and E are also appropriately transformed. It can be shown that, the piezoelectric permittivity matrix is transformed as

$$\mathrm{E}^l = \mathbf{T} \times \mathrm{E} \times \mathbf{T}^T \tag{6.46}$$

and the piezoelectric stress matrix by

$$\mathbf{e}^l = \mathbf{T} \times \mathbf{e} \times \mathbf{M}^T. \tag{6.47}$$

Consider, for example, the clockwise rotation of the coordinates about the z axis by the angle θ. The resulting transformation relating the transformed x, y, z, axes to the original reference axes is given by

$$\begin{bmatrix} x^l \\ y^l \\ z^l \end{bmatrix} = \begin{bmatrix} \cos\theta & \sin\theta & 0 \\ -\sin\theta & \cos\theta & 0 \\ 0 & 0 & 1 \end{bmatrix} \begin{bmatrix} x \\ y \\ z \end{bmatrix} = \mathbf{T} \begin{bmatrix} x \\ y \\ z \end{bmatrix}, \tag{6.48}$$

where the transformation matrix $\mathbf{T}$ is

$$\mathbf{T} = \begin{bmatrix} \cos\theta & \sin\theta & 0 \\ -\sin\theta & \cos\theta & 0 \\ 0 & 0 & 1 \end{bmatrix}. \tag{6.49}$$

The corresponding Bond transformation matrix $\mathbf{M}$ is

$$\mathbf{M} = \begin{bmatrix} \cos^2\theta & \sin^2\theta & 0 & 0 & 0 & \sin 2\theta \\ \sin^2\theta & \cos^2\theta & 0 & 0 & 0 & -\sin 2\theta \\ 0 & 0 & 1 & 0 & 0 & 0 \\ 0 & 0 & 0 & \cos\theta & -\sin\theta & 0 \\ 0 & 0 & 0 & -\sin\theta & \cos\theta & 0 \\ -\dfrac{\sin 2\theta}{2} & \dfrac{\sin 2\theta}{2} & 0 & 0 & 0 & \cos 2\theta \end{bmatrix}. \tag{6.50}$$

In the case of a cubic crystalline material, the generalized Hooke's law is of the form

$$\sigma_{xx} = c_{11}\varepsilon_{xx} + c_{12}\varepsilon_{yy} + c_{12}\varepsilon_{zz}, \tag{6.51a}$$

$$\sigma_{yy} = c_{12}\varepsilon_{xx} + c_{11}\varepsilon_{yy} + c_{12}\varepsilon_{zz}, \tag{6.51b}$$

$$\sigma_{zz} = c_{12}\varepsilon_{xx} + c_{12}\varepsilon_{yy} + c_{11}\varepsilon_{zz}, \tag{6.51c}$$

$$\sigma_{yz} = c_{44}\varepsilon_{yz}, \tag{6.51d}$$

$$\sigma_{zx} = c_{44}\varepsilon_{zx}, \tag{6.51e}$$

$$\sigma_{xy} = c_{44}\varepsilon_{xy}. \tag{6.51f}$$

When the rotational transformation is applied to this material, the transformed stress–strain relationships are

$$\sigma_{xx}^{l} = c_{11}^{l}\varepsilon_{xx}^{l} + c_{12}^{l}\varepsilon_{yy}^{l} + c_{13}^{l}\varepsilon_{zz}^{l} + c_{16}^{l}\varepsilon_{xy}^{l}, \tag{6.52a}$$

$$\sigma_{yy}^{l} = c_{12}^{l}\varepsilon_{xx}^{l} + c_{11}^{l}\varepsilon_{yy}^{l} + c_{13}^{l}\varepsilon_{zz}^{l} - c_{16}^{l}\varepsilon_{xy}^{l}, \tag{6.52b}$$

$$\sigma_{zz}^{l} = c_{13}^{l}\varepsilon_{xx}^{l} + c_{13}^{l}\varepsilon_{yy}^{l} + c_{33}^{l}\varepsilon_{zz}^{l}, \tag{6.52c}$$

$$\sigma_{yz}^{l} = c_{44}^{l}\varepsilon_{yz}^{l}, \tag{6.52d}$$

$$\sigma_{zx}^{l} = c_{44}^{l}\varepsilon_{zx}^{l}, \tag{6.52e}$$

$$\sigma_{xy}^{l} = c_{66}^{l}\varepsilon_{xy}^{l} + c_{16}^{l}\left(\varepsilon_{xx}^{l} - \varepsilon_{yy}^{l}\right), \tag{6.52f}$$

where

$$c_{11}^{l} = c_{11} - \left(\frac{c_{11} - c_{12}}{2} - c_{44}\right)\sin^2(2\theta), \tag{6.53a}$$

$$c_{12}^{l} = c_{12} + \left(\frac{c_{11} - c_{12}}{2} - c_{44}\right)\sin^2(2\theta), \tag{6.53b}$$

$$c_{13}^{l} = c_{12}, \tag{6.53c}$$

$$c_{16}^{l} = -\left(\frac{c_{11} - c_{12}}{2} - c_{44}\right)\sin(2\theta)\cos(2\theta), \tag{6.53d}$$

$$c_{33}^{l} = c_{11}, \tag{6.53e}$$

$$c_{44}^{l} = c_{44} \tag{6.53f}$$

and

$$c_{66}^{l} = c_{44} + \left(\frac{c_{11} - c_{12}}{2} - c_{44}\right)\sin^2(2\theta). \tag{6.53g}$$

The transformational relationships are particularly simple when one considers the conditions of plane stress and plane strain. The first of these is particularly relevant for a thin lamina such as a plate loaded in the plane of the plate, while the latter is relevant for elongated bodies of constant cross-section subjected to uniform loading. The condition of plane stress is based on the assumptions that

$$\sigma_{zx} = \sigma_{zy} = \sigma_{zz} = 0,$$

where the z direction is normal to the plane of the plate and that no stress components vary across the plane of the plates.

Considering the case of plane stress and assuming that the material is transversely isotropic, the stress–strain relationships may be expressed in matrix form as

$$\begin{bmatrix} \sigma_{xx} \\ \sigma_{yy} \\ \sigma_{xy} \end{bmatrix} = \frac{1}{\left(1 - \nu_x\nu_y\right)} \begin{bmatrix} E_x & 0 & 0 \\ 0 & E_y & 0 \\ 0 & 0 & G_{xy} \end{bmatrix} \begin{bmatrix} 1 & \nu_y & 0 \\ \nu_x & 1 & 0 \\ 0 & 0 & \left(1 - \nu_x\nu_y\right) \end{bmatrix} \begin{bmatrix} \varepsilon_{xx} \\ \varepsilon_{yy} \\ \varepsilon_{xy} \end{bmatrix}.$$

The relationships also apply to a whole range of composite materials or multi-layered laminated materials provided the material is considered to be a set of stiffnesses in parallel for the loads in the plane of the lamina and as a set of stiffnesses in series for the loads normal to the lamina. The Bond transformation matrix in this case, **M**, is

$$\mathbf{M} = \begin{bmatrix} \cos^2\theta & \sin^2\theta & \sin 2\theta \\ \sin^2\theta & \cos^2\theta & -\sin 2\theta \\ -\dfrac{\sin 2\theta}{2} & \dfrac{\sin 2\theta}{2} & \cos 2\theta \end{bmatrix},$$

while

$$\mathbf{N} = \begin{bmatrix} \cos^2\theta & \sin^2\theta & -\sin 2\theta \\ \sin^2\theta & \cos^2\theta & \sin 2\theta \\ \dfrac{\sin 2\theta}{2} & -\dfrac{\sin 2\theta}{2} & \cos 2\theta \end{bmatrix}^T.$$

6.3.4 Determination of Piezoelectric Stiffened Coefficients

In this section, we illustrate the general method of deriving the piezoelectric stiffened elastic coefficients. A key assumption in the derivation of the stiffened elastic constants is that the electric field is the gradient of a scalar potential function. This is possible when the electric field vector has just one non-zero component. From Maxwell's electromagnetic equations, for a wave propagating in a plane, all derivatives of the electric displacement and electric field intensity in the normal direction may be set equal to zero. Hence, it follows that the component of the electric displacement in the direction normal to the direction of the wave propagation and in the plane of propagation is also equal to zero. The electric field components may also be set to zero in two of the three directions following the first assumption, and this will result in the quasi-static solution of the governing electromagnetic equations. Because there is only one component of the electric field, the electric displacement is constant in the confines of the piezoelectric material. A further assumption is made that the constant electric displacement, in the direction of the non-zero electric field component, is approximately equal to zero. This permits one to solve for the electric field component and eliminate the same from the generalized Hooke's law, resulting in the stiffened elastic coefficients. The process is illustrated in the following.

Consider a hexagonal (6 mm) crystalline material and assume that the electric field is given by the vector

$$\begin{bmatrix} E_x \\ E_y \\ E_z \end{bmatrix} = \begin{bmatrix} 1 \\ 0 \\ 0 \end{bmatrix} E_x. \tag{6.54}$$

The direction of wave propagation is assumed to be the x axis and in the crystalline XZ plane, but not aligned with the crystalline X axis. Upper case letters are employed to indicate the crystalline axes, especially when they are different from the reference axes employed to describe the wave motion. To align the crystalline X axis with the direction of wave propagation, the crystalline frame is rotated clockwise about the Y axis. Thus, it is essential that the matrices of elastic constants, the piezoelectric stress matrix and the permittivity matrix, are all transformed to the reference frame employed to describe the wave motion. The transformation matrix **T** is

$$\mathbf{T} = \begin{bmatrix} \cos\beta & 0 & -\sin\beta \\ 0 & 1 & 0 \\ \sin\beta & 0 & \cos\beta \end{bmatrix}. \tag{6.55}$$

The corresponding Bond transformation matrix **M** is

$$\mathbf{M} = \begin{bmatrix} \cos^2\beta & 0 & \sin^2\beta & 0 & -\sin 2\beta & 0 \\ 0 & 1 & 0 & 0 & 0 & 0 \\ \sin^2\beta & 0 & \cos^2\beta & 0 & \sin 2\beta & 0 \\ 0 & 0 & 0 & \cos\beta & 0 & \sin\beta \\ \dfrac{\sin 2\beta}{2} & 0 & -\dfrac{\sin 2\beta}{2} & 0 & \cos 2\beta & 0 \\ 0 & 0 & 0 & -\sin\beta & 0 & \cos\beta \end{bmatrix}. \tag{6.56}$$

The transformed permittivity matrix is

$$\mathrm{E}^l = \mathbf{T} \times \mathrm{E} \times \mathbf{T}^T = \begin{bmatrix} \in^l_{xx} & 0 & \in^l_{xz} \\ 0 & \in^l_{yy} & 0 \\ \in^l_{xz} & 0 & \in^l_{zz} \end{bmatrix}, \tag{6.57}$$

where

$$\begin{aligned} &\in^l_{xx} = \in_{xx}\cos^2\beta + \in_{zz}\sin^2\beta, \quad \in^l_{zz} = \in_{xx}\sin^2\beta + \in_{zz}\cos^2\beta, \\ &\in^l_{yy} = \in_{xx}, \quad \in^l_{xz} = (\in_{xx} - \in_{zz})\cos\beta\sin\beta. \end{aligned} \tag{6.58}$$

The transformed piezoelectric stress matrix is

$$\mathbf{e}^l = \mathbf{T} \times \mathbf{e} \times \mathbf{M}^T \tag{6.59}$$

and is given explicitly by 0

$$\begin{bmatrix} e^l_{x1} & e^l_{x2} & e^l_{x3} & 0 & e^l_{x5} & 0 \\ 0 & 0 & 0 & e^l_{y4} & 0 & e^l_{y6} \\ e^l_{z1} & e^l_{z2} & e^l_{z3} & 0 & e^l_{z5} & 0 \end{bmatrix} = \mathbf{T} \times \begin{bmatrix} 0 & 0 & 0 & 0 & e_{x5} & 0 \\ 0 & 0 & 0 & e_{x5} & 0 & 0 \\ e_{z1} & e_{z1} & e_{z3} & 0 & 0 & 0 \end{bmatrix} \times \mathbf{M}^T, \tag{6.60}$$

where

$$e^l_{x1} = -e_{z1}\sin\beta\cos^2\beta - e_{z3}\sin^3\beta - e_{x5}\cos\beta\sin 2\beta, \tag{6.61a}$$

$$e^l_{x2} = -e_{z1}\sin\beta, \tag{6.61b}$$

$$e^l_{x3} = -e_{z1}\sin^3\beta - e_{z3}\cos^2\beta\sin\beta + e_{x5}\cos\beta\sin 2\beta, \tag{6.61c}$$

$$e^l_{x5} = (e_{z3} - e_{z1})\frac{\sin\beta\sin 2\beta}{2} + e_{x5}\cos\beta\cos 2\beta, \tag{6.61d}$$

$$e^l_{y4} = e_{x5}\cos\beta, \tag{6.61e}$$

$$e^l_{y6} = -e_{x5}\sin\beta, \tag{6.61f}$$

$$e^l_{z1} = e_{z1}\cos^3\beta + e_{z3}\cos\beta\sin^2\beta - e_{x5}\sin\beta\sin 2\beta, \tag{6.61g}$$

$$e^l_{z2} = e_{z1}\cos\beta, \tag{6.61h}$$

$$e^l_{z3} = e_{z1}\sin^2\beta\cos\beta + e_{z3}\cos^3\beta + e_{x5}\sin\beta\sin 2\beta, \tag{6.61i}$$

$$e^l_{x5} = (e_{z1} - e_{z3})\frac{\cos\beta\sin 2\beta}{2} + e_{x5}\sin\beta\cos 2\beta. \tag{6.61j}$$

In the case of a hexagonal (6 mm) crystalline material, the generalized Hooke's law, with the piezoelectric terms set to zero, is of the form

$$\sigma_{xx} = c_{11}\varepsilon_{xx} + c_{12}\varepsilon_{yy} + c_{13}\varepsilon_{zz}, \tag{6.62a}$$

$$\sigma_{yy} = c_{12}\varepsilon_{xx} + c_{11}\varepsilon_{yy} + c_{13}\varepsilon_{zz}, \tag{6.62b}$$

$$\sigma_{zz} = c_{13}\varepsilon_{xx} + c_{13}\varepsilon_{yy} + c_{33}\varepsilon_{zz}, \tag{6.62c}$$

$$\sigma_{yz} = c_{44}\varepsilon_{yz}, \tag{6.62d}$$
$$\sigma_{zx} = c_{44}\varepsilon_{zx}, \tag{6.62e}$$
$$\sigma_{xy} = c_{66}\varepsilon_{xy}. \tag{6.62f}$$

When the rotational transformation is applied to this material, the transformed stress–strain relationships, with the piezoelectric terms set to zero, take the form

$$\sigma^l_{xx} = c^l_{11}\varepsilon^l_{xx} + c^l_{12}\varepsilon^l_{yy} + c^l_{13}\varepsilon^l_{zz} + c^l_{15}\varepsilon^l_{zx}, \tag{6.63a}$$
$$\sigma^l_{yy} = c^l_{12}\varepsilon^l_{xx} + c^l_{22}\varepsilon^l_{yy} + c^l_{23}\varepsilon^l_{zz} + c^l_{25}\varepsilon^l_{zx}, \tag{6.63b}$$
$$\sigma^l_{zz} = c^l_{13}\varepsilon^l_{xx} + c^l_{23}\varepsilon^l_{yy} + c^l_{33}\varepsilon^l_{zz} + c^l_{35}\varepsilon^l_{zx}, \tag{6.63c}$$
$$\sigma^l_{yz} = c^l_{44}\varepsilon^l_{yz} + c^l_{46}\varepsilon^l_{xy}, \tag{6.63d}$$
$$\sigma^l_{zx} = c^l_{15}\varepsilon^l_{xx} + c^l_{25}\varepsilon^l_{yy} + c^l_{35}\varepsilon^l_{zz} + c^l_{55}\varepsilon^l_{zx}, \tag{6.63e}$$
$$\sigma^l_{xy} = c^l_{46}\varepsilon^l_{yz} + c^l_{66}\varepsilon^l_{xy}. \tag{6.63f}$$

Since the electric field vector is assumed to be of the form,

$$\begin{bmatrix} E_x \\ E_y \\ E_z \end{bmatrix} = \begin{bmatrix} 1 \\ 0 \\ 0 \end{bmatrix} E_x \equiv \mathbf{L}E_x, \tag{6.64}$$

where the vector $\mathbf{L}$ is given by

$$\mathbf{L} = \begin{bmatrix} 1 \\ 0 \\ 0 \end{bmatrix}, \tag{6.65}$$

the three electric displacement equations, in matrix form, are multiplied by the vector $\mathbf{L}^T$. The resulting equation is

$$D_x = \begin{bmatrix} e^l_{x1} & e^l_{x2} & e^l_{x3} & 0 & e^l_{x5} & 0 \end{bmatrix} \begin{bmatrix} \varepsilon_1 \\ \varepsilon_2 \\ \varepsilon_3 \\ \varepsilon_4 \\ \varepsilon_5 \\ \varepsilon_6 \end{bmatrix} + \epsilon_{xx} E_x. \tag{6.66}$$

Assuming that this component of the electric displacement is almost equal to zero, and solving for the electric field component E_x, we obtain

$$E_x = -\frac{1}{\epsilon_{xx}} \begin{bmatrix} e^l_{x1} & e^l_{x2} & e^l_{x3} & 0 & e^l_{x5} & 0 \end{bmatrix} \begin{bmatrix} \varepsilon_1 \\ \varepsilon_2 \\ \varepsilon_3 \\ \varepsilon_4 \\ \varepsilon_5 \\ \varepsilon_6 \end{bmatrix}. \tag{6.67}$$

Eliminating the electric field vector in the piezoelectric stress–strain relationships, the stiffened elastic coefficient matrix is given by

$$\bar{\mathbf{c}}^l = \mathbf{c}^l - \frac{1}{\in_{xx}} \begin{bmatrix} e^l_{x1} \\ e^l_{x2} \\ e^l_{x3} \\ 0 \\ e^l_{x5} \\ 0 \end{bmatrix} \left[e^l_{x1} e^l_{x2} e^l_{x3} 0 e^l_{x5} 0 \right]. \tag{6.68}$$

6.4 Transmission Line Model

In Section 4.3.5, a transmission line model was obtained for acoustic plane wave propagation in isotropic solids. The basic model must be modified if it is to be suitable for anisotropic crystalline solids. The equivalent transmission line model is first derived for the case of a non-piezoelectric anisotropic crystalline solid.

6.4.1 Transmission Line Model for Wave Propagation in Non-piezoelectric Crystalline Solids

To consider the development of a transmission line model for a crystalline solid, consider the case of a cubic crystalline material. While in the isotropic case the equations of particle motion were found to decouple into three pairs of coupled equations, in this case the equations decouple into two sets: one involving four coupled equations and the other involving a pair of coupled equations. In considering wave propagation in a cubic crystalline material, it is instructive to consider a wave propagating in an arbitrary direction in the x–y plane. One approach is to transform the crystalline reference frame so that the transformed reference x axis is oriented towards the direction of propagation. This generally requires a rotation of the reference frame in the plane of the wave propagation, about the z axis, so as to orient the transformed reference x axis towards the direction of propagation. As in Section 7.4.6, assuming that the waves propagate in the x–y plane, we may set the partial derivatives in the z direction to zero. The equations of motion then reduce to

$$\frac{\partial \sigma_{xx}}{\partial x} + \frac{\partial \sigma_{xy}}{\partial y} = \rho \frac{\partial^2 u}{\partial t^2}. \tag{6.69a}$$

$$\frac{\partial \sigma_{yx}}{\partial x} + \frac{\partial \sigma_{yy}}{\partial y} = \rho \frac{\partial^2 v}{\partial t^2}, \tag{6.69b}$$

$$\frac{\partial \sigma_{zx}}{\partial x} + \frac{\partial \sigma_{zy}}{\partial y} = \rho \frac{\partial^2 w}{\partial t^2}. \tag{6.69c}$$

In terms of the particle velocities v_x, v_y and v_z in the x, y and z directions, the equations may be written as

$$\frac{\partial \sigma_{xx}}{\partial x} + \frac{\partial \sigma_{xy}}{\partial y} = \rho \frac{\partial v_x}{\partial t}, \tag{6.70a}$$

$$\frac{\partial \sigma_{yx}}{\partial x} + \frac{\partial \sigma_{yy}}{\partial y} = \rho \frac{\partial v_y}{\partial t}, \tag{6.70b}$$

$$\frac{\partial \sigma_{zx}}{\partial x} + \frac{\partial \sigma_{zy}}{\partial y} = \rho \frac{\partial v_z}{\partial t}. \tag{6.70c}$$

The stress–strain equations for a cubic material with the reference frame transformed so that the transformed x axis is oriented towards the direction of wave propagation are

$$\sigma_{xx}^l = c_{11}^l \varepsilon_{xx}^l + c_{12}^l \varepsilon_{yy}^l + c_{13}^l \varepsilon_{zz}^l + c_{16}^l \varepsilon_{xy}^l, \tag{6.71a}$$

$$\sigma_{yy}^l = c_{12}^l \varepsilon_{xx}^l + c_{11}^l \varepsilon_{yy}^l + c_{13}^l \varepsilon_{zz}^l - c_{16}^l \varepsilon_{xy}^l, \tag{6.71b}$$

$$\sigma_{zz}^l = c_{13}^l \varepsilon_{xx}^l + c_{13}^l \varepsilon_{yy}^l + c_{33}^l \varepsilon_{zz}^l, \tag{6.71c}$$

$$\sigma_{yz}^l = c_{44}^l \varepsilon_{yz}^l, \tag{6.71d}$$

$$\sigma_{zx}^l = c_{44}^l \varepsilon_{zx}^l, \tag{6.71e}$$

$$\sigma_{xy}^l = c_{66}^l \varepsilon_{xy}^l + c_{16}^l \left(\varepsilon_{xx}^l - \varepsilon_{yy}^l\right). \tag{6.71f}$$

In what follows, the superscripts on the elastic constants are ignored.

The relationships between the strain and particle velocity are

$$\frac{\partial}{dt}\varepsilon_{xx} = \frac{\partial v_x}{\partial x}, \tag{6.72a}$$

$$\frac{\partial}{dt}\varepsilon_{yy} = \frac{\partial v_y}{\partial y}, \tag{6.72b}$$

$$\frac{\partial}{dt}\varepsilon_{yz} = \frac{\partial v_z}{\partial y}, \tag{6.72c}$$

$$\frac{\partial}{\partial t}\varepsilon_{zx} = \frac{\partial v_z}{\partial x}, \tag{6.72d}$$

$$\frac{\partial}{dt}\varepsilon_{xy} = \frac{\partial v_y}{\partial x} + \frac{\partial v_x}{\partial y}. \tag{6.72e}$$

Eliminating the strains from above stress–strain relationships,

$$\frac{\partial}{dt}\sigma_{xx} = c_{11}\frac{\partial v_x}{\partial x} + c_{12}\frac{\partial v_y}{\partial y} + c_{16}\frac{\partial v_y}{\partial x}, \tag{6.73a}$$

$$\frac{\partial}{\partial t}\sigma_{yy} = c_{12}\frac{\partial v_x}{\partial x} + c_{11}\frac{\partial v_y}{\partial y} - c_{16}\frac{\partial v_y}{\partial x}, \tag{6.73b}$$

$$\frac{\partial}{\partial t}\sigma_{zx} = c_{44}\frac{\partial v_z}{\partial x}, \tag{6.73c}$$

$$\frac{\partial}{\partial t}\sigma_{yz} = c_{44}\frac{\partial v_z}{\partial y}, \tag{6.73d}$$

$$\frac{\partial}{\partial t}\sigma_{xy} = c_{66}\left(\frac{\partial v_x}{\partial y} + \frac{\partial v_y}{\partial x}\right) + c_{16}\left(\frac{\partial v_x}{\partial x} - \frac{\partial v_y}{\partial y}\right). \tag{6.73e}$$

Since in the case of a plane wave propagating in the x axis, the derivatives with respect to y are zero, we have, together with the equations of force equilibrium,

$$\frac{\partial}{dt}\sigma_{xx} = c_{11}\frac{\partial v_x}{\partial x} + c_{16}\frac{\partial v_y}{\partial x}, \tag{6.74a}$$

$$\frac{\partial}{\partial t}\sigma_{xy} = c_{66}\frac{\partial v_y}{\partial x} + c_{16}\frac{\partial v_x}{\partial x}, \tag{6.74b}$$

$$\frac{\partial}{\partial t}\sigma_{zx} = c_{44}\frac{\partial v_z}{\partial x}, \tag{6.74c}$$

$$\frac{\partial}{\partial t}\sigma_{yy} = c_{12}\frac{\partial v_x}{\partial x} - c_{16}\frac{\partial v_y}{\partial x}, \tag{6.74d}$$

$$\frac{\partial}{\partial t}\sigma_{yz} = 0, \tag{6.74e}$$

$$\frac{\partial}{\partial t}\sigma_{zz} = c_{13}\frac{\partial v_x}{\partial x} \tag{6.74f}$$

and

$$\frac{\partial \sigma_{xx}}{\partial x} = \rho\frac{\partial v_x}{\partial t}, \tag{6.74g}$$

$$\frac{\partial \sigma_{xy}}{\partial x} = \rho\frac{\partial v_y}{\partial t}, \tag{6.74h}$$

$$\frac{\partial \sigma_{zx}}{\partial x} = \rho\frac{\partial v_z}{\partial t}. \tag{6.74i}$$

Only the first three and the last three are coupled with each other. The coupled equations may be rearranged as

$$\frac{\partial}{dt}\sigma_{xx} = c_{11}\frac{\partial v_x}{\partial x} + c_{16}\frac{\partial v_y}{\partial x}, \tag{6.75a}$$

$$\frac{\partial}{\partial t}\sigma_{xy} = c_{66}\frac{\partial v_y}{\partial x} + c_{16}\frac{\partial v_x}{\partial x}, \tag{6.75b}$$

$$\frac{\partial \sigma_{xx}}{\partial x} = \rho\frac{\partial v_x}{\partial t}, \tag{6.75c}$$

$$\frac{\partial \sigma_{xy}}{\partial x} = \rho\frac{\partial v_y}{\partial t} \tag{6.75d}$$

and

$$\frac{\partial}{\partial t}\sigma_{zx} = c_{44}\frac{\partial v_z}{\partial x}, \tag{6.75e}$$

$$\frac{\partial \sigma_{zx}}{\partial x} = \rho\frac{\partial v_z}{\partial t}. \tag{6.75f}$$

The last two may be represented by a single transmission line, while the first four are equivalent to a pair of coupled transmission lines. Expressing the last two equations in transmission line form

$$\frac{\partial v_z}{\partial x} - \frac{1}{c_{44}}\frac{\partial}{\partial t}\sigma_{zx} = 0, \tag{6.76a}$$

$$-\frac{\partial \sigma_{zx}}{\partial x} - \rho\frac{\partial v_z}{\partial t} = 0. \tag{6.76b}$$

The first four equations may also be expressed in transmission line form as

$$\frac{\partial v_x}{\partial x} - \frac{c_{66}}{c_{11}c_{66} - c_{16}^2}\frac{\partial}{dt}\sigma_{xx} + \frac{c_{16}}{c_{11}c_{66} - c_{16}^2}\frac{\partial}{\partial t}\sigma_{xy} = 0, \tag{6.77a}$$

$$-\frac{\partial \sigma_{xx}}{\partial x} + \rho\frac{\partial v_x}{\partial t} = 0, \tag{6.77b}$$

$$\frac{\partial v_y}{\partial x} - \frac{c_{11}}{c_{11}c_{66} - c_{16}^2}\frac{\partial}{\partial t}\sigma_{xy} + \frac{c_{16}}{c_{11}c_{66} - c_{16}^2}\frac{\partial}{\partial t}\sigma_{xx} = 0, \tag{6.77c}$$

$$-\frac{\partial \sigma_{xy}}{\partial x} + \rho\frac{\partial v_y}{\partial t} = 0. \tag{6.77d}$$

The equivalent transmission line equations are

$$\frac{\partial V_x}{\partial x} + L_0\frac{\partial I_x}{\partial t} = 0, \tag{6.78a}$$

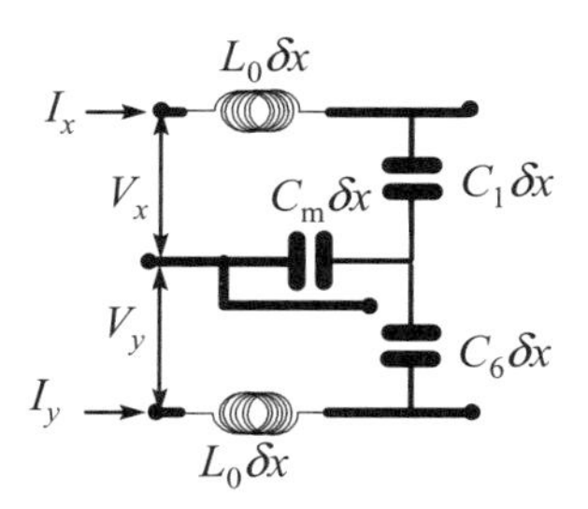

Figure 6.2 Transmission line model for wave propagation in the *x–y* plane of a cubic crystalline material

$$\frac{\partial I_x}{\partial x} + C_1\frac{\partial V_x}{\partial t} - C_m\frac{\partial V_y}{\partial t} = 0, \tag{6.78b}$$

$$\frac{\partial V_y}{\partial x} + L_0\frac{\partial I_y}{\partial t} = 0, \tag{6.78c}$$

$$\frac{\partial I_y}{\partial x} + C_6\frac{\partial V_y}{\partial t} - C_m\frac{\partial V_x}{\partial t} = 0. \tag{6.78d}$$

The equivalent circuit element is illustrated in Figure 6.2.

6.4.2 *Transmission Line Model for Wave Propagation in Piezoelectric Crystalline Solids*

In the case of a piezoelectric cubic crystalline solid with the rotated reference frame considered in Section 6.3.3, the constitutive equations take the form

$$\sigma^l_{xx} = c^l_{11}\varepsilon^l_{xx} + c^l_{12}\varepsilon^l_{yy} + c^l_{13}\varepsilon^l_{zz} + c^l_{16}\varepsilon^l_{xy}, \tag{6.79a}$$

$$\sigma^l_{yy} = c^l_{12}\varepsilon^l_{xx} + c^l_{11}\varepsilon^l_{yy} + c^l_{13}\varepsilon^l_{zz} - c^l_{16}\varepsilon^l_{xy}, \tag{6.79b}$$

$$\sigma^l_{zz} = c^l_{13}\varepsilon^l_{xx} + c^l_{13}\varepsilon^l_{yy} + c^l_{33}\varepsilon^l_{zz}, \tag{6.79c}$$

$$\sigma^l_{yz} = c^l_{44}\varepsilon^l_{yz} - e_{x4}E_x, \tag{6.79d}$$

$$\sigma^l_{zx} = c^l_{44}\varepsilon^l_{zx} - e_{x4}E_y, \tag{6.79e}$$

$$\sigma^l_{xy} = c^l_{66}\varepsilon^l_{xy} + c^l_{16}\left(\varepsilon^l_{xx} - \varepsilon^l_{yy}\right) - e_{x4}E_z. \tag{6.79f}$$

The electric displacement equations are

$$D_x = e^l_{x4}\varepsilon^l_{yz} + \epsilon^l_{x4}\, E^l_x, \tag{6.80a}$$

$$D_y = e^l_{x4}\varepsilon^l_{zx} + \epsilon^l_{x4}\, E^l_y, \tag{6.80b}$$

$$D_z = e^l_{x4}\varepsilon^l_{xy} + \epsilon^l_{x4}\, E^l_z. \tag{6.80c}$$

In what follows the superscripts on the elastic constants are ignored.

From Maxwell's electromagnetic equations, for a wave propagating in the *x–y* plane, all derivatives of the electric displacement and electric field intensity in the *z* direction may be set equal to zero. Hence, it follows that the component of the electric displacement in the direction normal to *z* and the direction of the wave propagation is also equal to zero. The electric field components in the *x* and *z* directions may also be set to zero, and this would result in the quasi-static solution of the governing electromagnetic equations. Because there is only one component of the electric field, the electric displacement is constant in the confines of the piezoelectric material. As a consequence, the electric field may be obtained as the gradient of a scalar potential. From the second of the electric displacement equations,

$$D_y = e^l_{x4}\varepsilon^l_{zx} + \epsilon^l_{x4}\, E^l_y = 0 \tag{6.81}$$

and it follows that

$$E_y = -\frac{e_{x4}}{\epsilon_{x4}}\varepsilon_{zx}. \tag{6.82}$$

Eliminating E_y from the constitutive equation, σ_{zx} takes the form

$$\sigma_{zx} = \left(c_{44} + \frac{e^2_{x4}}{\epsilon_{x4}}\right)\varepsilon_{zx}, \tag{6.83}$$

and introducing the stiffened elastic coefficient

$$\bar{c}_{44} = c_{44} + \frac{e^2_{x4}}{\epsilon_{x4}}. \tag{6.84}$$

The constitutive equations reduce to the same form as in the non-piezoelectric case.

Thus, the same transmission line models, derived in the non-piezoelectric case, are also valid in the piezoelectric case provided the elastic coefficient c_{44} is replaced by the stiffened elastic coefficient $\bar{c}_{44}$.

6.5 Discrete Element Model of Thin Piezoelectric Transducers

Thin piezoelectric transducers, such as disc transducers, are employed extensively in the measurement of vibration. They are also used in vibration actuation, particularly within micro-fluid pumps. The deposited film transducers employed in microwave applications with bandwidths well over 100 MHz, thicknesses in the range of microns and transverse dimensions of the order of millimetres are also typical examples of these transducers. They usually built around hexagonal (6 mm) piezoelectric crystals such as cadmium sulphide (CdS) and zinc oxide (ZnO). The face of the disc is chosen to be normal to the crystalline x axis. The thickness of the disc is chosen to be an order of magnitude less than its lateral dimensions. The transducer disc is usually sandwiched within two electrodes of the same size and shape as the piezoelectric disc itself. The transducer terminals constitute a pair of parallel plates, and the geometrical capacitance between them is given by

$$C_0 = \epsilon_{eff}\,\frac{A}{d}, \tag{6.85}$$

where A is the effective area of the terminal plates, d the separation distance and ϵ_{eff} the effective permittivity. Thus, when the transducer is wired in series with a resistance R_0 and an external voltage applied to it, the potential difference across the transducer disc is

$$V = \frac{\frac{1}{i\omega C_0}}{R_0 + \frac{1}{i\omega C_0}}V_e, \tag{6.86}$$

and there is net uniform field across the disc given by

$$E_x = \frac{V}{d} = \frac{\frac{1}{i\omega C_0}}{R_0 + \frac{1}{i\omega C_0}}\frac{V_e}{d}. \tag{6.87}$$

Because there is only one component of the electric field, the electric displacement component in the direction of the field is constant in the space of the transducer disc. For a hexagonal (6 mm) crystal,

$$D_x = e_{x5}\varepsilon_{xz} + \epsilon_{xx}\, E_x = 0, \tag{6.88}$$

and this results in just one stiffened elastic constant

$$\bar{c}_{44} = c_{44} + \frac{e_{x5}^2}{\epsilon_{xx}}. \tag{6.89}$$

Furthermore, only the transverse shear wave equations are coupled with the electrical input, and we have

$$\frac{\partial}{\partial t}\sigma_{zx} = c_{44}\frac{\partial v_z}{\partial x} + e_{x5}E_x = c_{44}\frac{\partial v_z}{\partial x} + e_{x5}\frac{dV}{dx}, \tag{6.90a}$$

$$\frac{\partial \sigma_{zx}}{\partial x} = \rho\frac{\partial v_z}{\partial t}. \tag{6.90b}$$

Solving these equations, we may express the shear force on the two faces of the transducer F_1 and F_2, in terms of the velocities across the two faces v_{z1} and v_{z2}, the input current I and the voltage V, across the two faces of the transducer as

$$F_1 = Z_{11}v_{z1} + Z_{12}v_{z2} + Z_{13}I, \tag{6.91a}$$

$$F_2 = Z_{21}v_{z1} + Z_{22}v_{z2} + Z_{23}I, \tag{6.91b}$$

$$V = Z_{31}v_{z1} + Z_{32}v_{z2} + Z_{33}I, \tag{6.91c}$$

where

$$v_{z1} = v_z(0), \tag{6.92a}$$

$$v_{z2} = -v_z(d), \tag{6.92b}$$

$$I = i\omega D_x A, \tag{6.92c}$$

$$F_1 = -A\sigma_{zx}(0), \tag{6.92d}$$

$$F_2 = -A\sigma_{zx}(d). \tag{6.92e}$$

To find the impedance elements Z_{ij}, the electric displacement component D_x is assumed to be constant. Furthermore, the velocity component $v_z(x)$ is assumed to be given by a general standing waveform given by

$$v_z(x) = a\sin\left(\frac{\omega x}{a}\right) + b\cos\left(\frac{\omega x}{a}\right), \tag{6.93}$$

where

$$a = \sqrt{\frac{\bar{c}_{44}}{\rho}}.$$

Applying the boundary conditions at $x = 0$ and $x = d$,

$$b = v_z(0) = v_{z1} \tag{6.94a}$$

and

$$a = \frac{v_z(d) - v_z(0)\cos\left(\frac{\omega d}{a}\right)}{\sin\left(\frac{\omega d}{a}\right)} = -\frac{v_{z2} - v_{z1}\cos\left(\frac{\omega d}{a}\right)}{\sin\left(\frac{\omega d}{a}\right)}. \tag{6.94b}$$

The general solution for E_x and $\sigma_{zx}(x)$ are then obtained by substituting in the constitutive equations and given by

$$E_x = \frac{I}{i\omega A\,\epsilon_{xx}} + \frac{ie_{x5}}{a\,\epsilon_{xx}}\left(a\cos\left(\frac{\omega x}{a}\right) - b\sin\left(\frac{\omega x}{a}\right)\right), \tag{6.95a}$$

$$\sigma_{zx} = -\frac{Ie_{x5}}{i\omega A\,\epsilon_{xx}} - \frac{i\bar{c}_{44}}{a}\left(a\cos\left(\frac{\omega x}{a}\right) - b\sin\left(\frac{\omega x}{a}\right)\right). \tag{6.95b}$$

The terminal voltages and the shear forces on the two faces of the transducer may then be evaluated. The resulting impedance equations are (Kikuchi, 1969)

$$F_1 = \frac{A\rho a}{i\sin\left(\frac{\omega d}{a}\right)}\left(v_{z1}\cos\left(\frac{\omega d}{a}\right) + v_{z2}\right) + \frac{e_{x5}}{i\omega \in_{xx}} I, \tag{6.96a}$$

$$F_2 = \frac{A\rho a}{i\sin\left(\frac{\omega d}{a}\right)}\left(v_{z1} + v_{z2}\cos\left(\frac{\omega d}{a}\right)\right) + \frac{e_{x5}}{i\omega \in_{xx}} I, \tag{6.96b}$$

$$V = \frac{e_{x5}}{i\omega \in_{xx}}\left(v_{z1} + v_{z2}\right) + \frac{1}{i\omega C_0} I. \tag{6.96c}$$

In the above equations we observe that

$$Z_0 = A\rho a \tag{6.97a}$$

is the equivalent acoustic impedance,

$$k = \omega/a \tag{6.97b}$$

is the wave number and

$$C_0 = \in_{xx} A/d \tag{6.97c}$$

is the clamped (zero strain) geometrical capacitance.

The three-port equations are particularly useful in evaluating the input impedance and the bandwidth of the transducer. They may be cast in a form that may be interpreted as an equivalent circuit. One of the elements in the equivalent circuit is a non-physical negative capacitance. These equivalents are illustrated in Figure 6.3(a) (Mason, 1948, 1950). In the figure,

$$Z_i = iZ_0\tan\left(\frac{kd}{2}\right),\ Z_p = -iZ_0\csc(kd) \tag{6.98a}$$

and

$$N = C_0\frac{e_{x5}}{\in_{xx}}. \tag{6.98b}$$

In terms of these constants (equations (6.98)), the equations (6.96) are

$$F_1 = \left(Z_i + Z_p\right)v_{z1} + Z_p v_{z2} + \frac{N}{i\omega C_0} I, \tag{6.99a}$$

$$F_2 = Z_p v_{z1} + \left(Z_i + Z_p\right)v_{z2} + \frac{N}{i\omega C_0} I, \tag{6.99b}$$

$$V = \frac{1}{i\omega C_0}\left(N\left(v_{z1} + v_{z2}\right) + I\right). \tag{6.99c}$$

One special case of importance is when $v_{z1} = 0$. This corresponds to the case when there is no input 'current' in the acoustic port. In this case, it reduces to a standard two-port circuit, which is shown in Figure 6.3(b). In most practical applications, the negative capacitance is dropped while the input admittance and the output impedance are modified to account for dielectric and other losses, depending on the nature of the application. On application, the modelling of ultrasonic wave motors (Sashida and Kenjo, 1993) is discussed in Section 6.5.1.

It is also possible to establish an equivalent acoustic transmission line model for the thin piezoelectric transducers considered in this section (Kikuchi, 1969). This is illustrated in Figure 6.3(c). In this model, V and I are the respective voltage and current applied to the piezoelectric crystal, which produce the resulting acoustic forces F_i and particle velocities v_i, at the respective faces of the crystal. The model

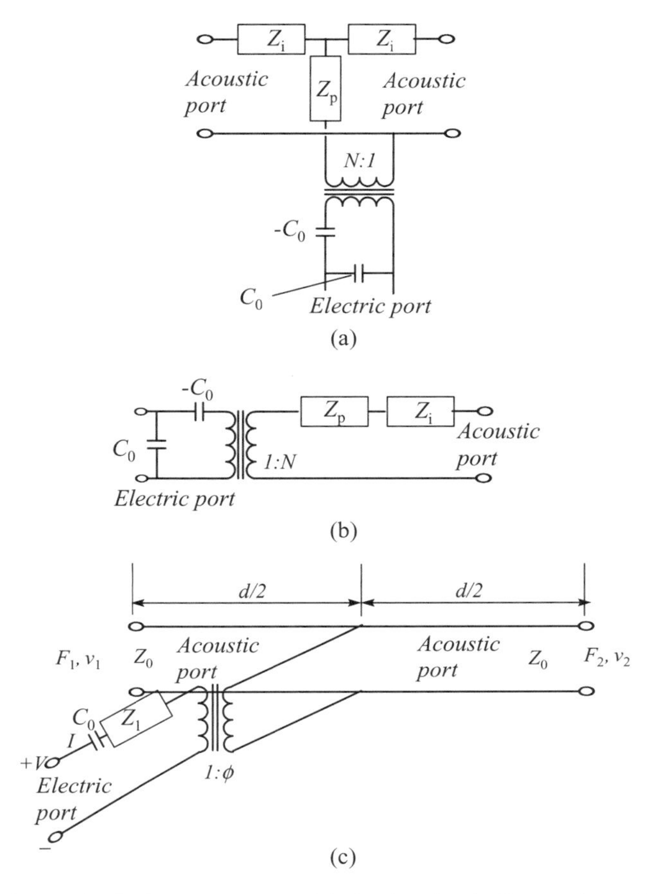

Figure 6.3 (a) Equivalent circuit of the three-port model of a piezoelectric disc transducer, (b) special case when there is no input 'current' in the acoustic port and (c) equivalent three-port transmission line model

parameters include the thickness of the crystal d, the area of the crystal A and the characteristic impedance of the acoustic transmission line (i.e. the radiation impedance) modelling the piezoelectric crystal Z_0. The radiation impedance is the additional mechanical impedance of a transducer due to the presence of the acoustic medium. In order to complete the model, it is also necessary to include a capacitor C_0, a impedance Z_1 and a transformer with the ratio $(1{:}\phi)$ that converts the electrical signal into the appropriate acoustical values. C_0 results from the resonator consisting of a dielectric, the piezoelectric crystal, between two excited conducting surfaces. The values for these parameters are

$$Z_1 = \frac{ih^2}{\omega^2 Z_0}\sin\left(\frac{\omega d}{a}\right), \quad \phi = \frac{\omega Z_0}{2h}\cos ec\left(\frac{\omega}{2a}\right), \quad \text{and} \quad h = \frac{e_{x5}}{\in_{xx}},$$

where $\in_{xx}$ is the permittivity of the piezoelectric under no applied voltage and h is the piezoelectric pressure constant for the crystal. All other parameters are as defined earlier in equation (6.97). The discrete element model and the transmission line model are equivalent to each other when the acoustic inputs are absent $(F_i = 0)$. Unlike the discrete element model, the acoustic transmission line model is made up of real physical elements. However, the discrete element model is often more useful for purposes of designing and analysing a real piezoelectric disc or film transducer.

6.5.1 One-port Modelling of Thin Piezoelectric Transducers

The two-port model presented in the earlier section may be reduced to a one-port model in certain restricted cases. The reduced one-port model permits the interpretation of the losses in the circuit in terms of a radiation impedance. Assuming that one face of the piezoelectric transducer is bonded rigidly to a non-piezoelectric body of the same material, the acoustic wave propagating towards the rigid face of the thin piezoelectric material may be modelled as a non-piezoelectric z-polarized shear wave, and it follows that

$$v_{z2} = v_{z20} \exp(i\omega t - ikx)$$

and

$$\sigma_{zx} = -\rho\sqrt{\frac{c_{44}}{\rho}} \times v_{z20} \exp(i\omega t - ikx).$$

Hence, assuming that face 2 is rigid,

$$F_2 = -A\rho\sqrt{\frac{c_{44}}{\rho}} \times v_{z2} = -Z_{0p} v_{z2}.$$

The other face is assumed to be stress free and so at face 1, $F_1 = 0$. Hence, eliminating v_{z1} and v_{z2} from equation (6.99c), we may express the electrical terminal voltage in terms of the electrical terminal current as

$$V = Z_{IN} I, \quad \frac{1}{Z_{IN}} = i\omega C_0 + \frac{1}{Z_a},$$

where Z_a represents the piezoelectrically induced radiation or motional impedance and Z_{IN} the total electrical input impedance given by

$$Z_{IN} = \frac{1}{i\omega C_0}\left(1 - \frac{N^2}{i\omega C_0} \times \frac{2Z_i + Z_{0p}}{\left(Z_i + Z_p + Z_{0p}\right)\left(Z_i + Z_p\right) - Z_p^2}\right).$$

The inverse of the electrical impedance, the admittance of any transducer, can be expressed as the sum of the motional and damped admittance. The latter is generally the vibration-independent part of the admittance and is only relevant for transducers with a non-vibrational inductance coil wound on a magnetic core. Although not frequency independent, the impedance due to the geometrical capacitance and any resistance in series plays the same role as the damped impedance. The former is entirely vibration or motion dependent, and as a consequence the motional impedance Z_a is generally frequency dependent and negligible in the absence of vibration. The complete frequency dependence of the input impedance may be approximately reproduced by a series *R-L-C* circuit in parallel with the clamped geometrical or clamped electrical capacitance C_0. The equivalent circuit is shown in Figure 6.4.

Considering the circuit in Figure 6.4,

$$Z_p = R_1 + \frac{1}{i\omega C_1} + i\omega L_1.$$

Hence,

$$\frac{1}{Z_{IN}} = i\omega C_0 + \frac{1}{Z_p} = i\omega C_0 + i\omega C_1 \frac{\left(1 - \omega^2 L_1 C_1\right) - i\omega C_1 R_1}{\left(1 - \omega^2 L_1 C_1\right)^2 + (\omega C_1 R_1)^2}.$$

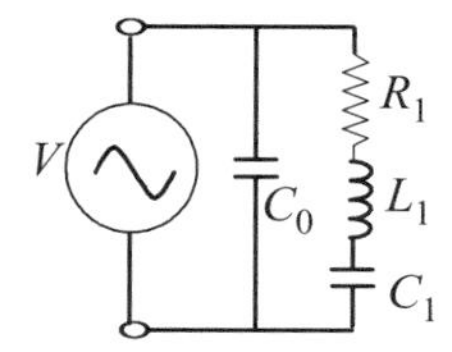

Figure 6.4 Equivalent circuit of the input impedance of a thin piezoelectric transducer

The actual values of the three elements are found by comparing Z_p with the electrical motional impedance or mechanical radiation impedance of the transducer Z_a.

6.5.2 *Two-port Modelling of a Piezoelectric Diaphragm Resting on a Cavity*

Piezoelectric diaphragms, especially fabricated from piezoceramic films, are employed extensively in a variety of transducers, actuators and fluidic pumps. They may be considered as special cases of plate structures and serve as mechanical waveguides (Redwood, 1960; Tiersten, 1969). Broadly these fall into two categories: resonant and non-resonant devices. While the former employ excitation with a narrow band frequency spectrum in the vicinity of the resonance frequency, the latter employ broadband excitation. The general approach for deriving equivalent circuit representations for piezoelectric transducers is based on adopting a normal mode expansion of the governing equations. An equivalent circuit is derived for each of the principal coordinates, and the individual circuits are then coupled by employing ideal transformers with appropriate turns ratios. The modelling theory presented in the preceding sections can also be adapted to a variety of resonant and non-resonant thin piezoelectric devices. The case of a piezoceramic diaphragm is archetypal. In the case of a thin piezoelectric diaphragm, it is reasonable to assume that $v_{z1} = v_{z2}$ in equations (6.99a), (6.99b) and (6.99c). Thus, we have

$$F_1 = \left(Z_i + 2Z_p\right) v_{z1} + \frac{N}{i\omega C_0} I, \quad F_2 = \left(Z_i + 2Z_p\right) v_{z1} + \frac{N}{i\omega C_0} I,$$

$$V = \frac{1}{i\omega C_0} \left(2N v_{z1} + I\right).$$

Since the expressions for F_1 and F_2 are identical, it follows that $F_1 = F_2$. The equations may be represented in an equivalent circuit diagram, as shown in Figure 6.5(a). If in addition the diaphragm is supported over a closed cavity, as shown in Figure 6.5(b), the radiation impedance of the cavity appears in the circuit as a load across the output terminals. The equivalent mechanical radiation impedance of the cavity, closed at both ends, in the vicinity of the resonance frequency, may be shown to be

$$Z_c = \frac{1}{i\omega C_{ac}} = -i\rho_c a_c S \cot\left(\frac{\omega D}{a_c}\right), \tag{6.100}$$

where ρ_c is the density of the fluid in the cavity, a_c the speed of sound in the cavity, D the depth of the cavity and S the cross-sectional area of the cavity.

Together with the cavity, the piezoelectric diaphragm represents a coupled electromechanical-acoustic system with frequency-dependent properties that are determined by the device dimensions and the acoustic and piezoelectric properties of the propagating media. The equivalent circuit of the coupled diaphragm and cavity is shown in Figure 6.5(c). The negative capacitance in these circuits is usually replaced by a motional impedance that must be accounted for.

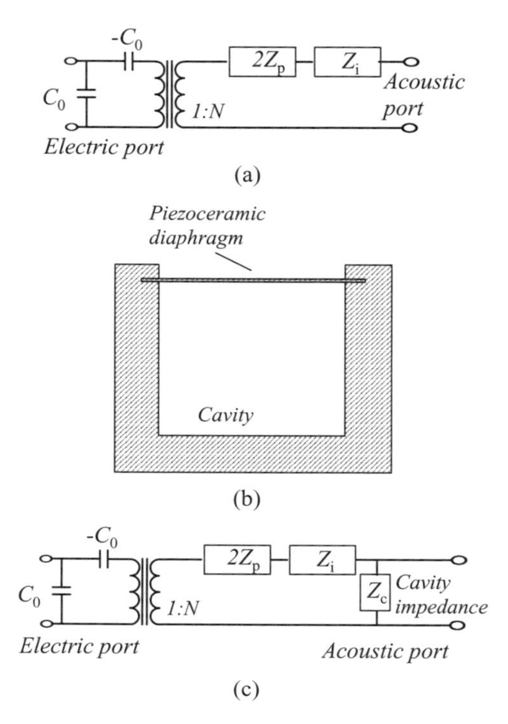

Figure 6.5 (a) Equivalent circuit of the two-port model of a piezoelectric diaphragm, (b) piezoelectric diaphragm supported over a closed cavity and (c) equivalent circuit of a piezoelectric diaphragm supported over a cavity

6.5.3 Modelling of a Helmholtz-type Resonator Driven by a Piezoelectric Disc Transducer

A Helmholtz resonator consists of a large cavity connected with the environment via a small aperture. Typical practical examples of such resonators are a bottle with a long, uniform and narrow neck and the guitar. The lowest resonant frequency of a Helmholtz resonator does not correspond to waves standing between opposite walls of the cavity, as would be expected in the case of a closed cavity. Rather, it is induced by an alternating stream through the aperture compressing and decompressing the air inside the cavity. The phenomenon is known as *Helmholtz resonance*, and the frequency at which it occurs is termed the *Helmholtz resonance frequency*. Both H. von Helmholtz and subsequently Lord Rayleigh predicted this frequency over a hundred years ago by assuming a potential flow through the opening and a uniform pressure inside the cavity, and neglecting radiation losses. They found the Helmholtz resonance frequency to be proportional to the bulk wave speed and to the square-root of the quotient of aperture area and the product of the cavity volume and the length of the neck. The opening to the cavity resonator may be a long duct, as in the case of a bottle, or just a hole in a narrow wall, as in the case of a guitar.

In this section, we consider the modelling of a Helmholtz-type resonator driven by a piezoelectric disc transducer. Our objective is to develop an electrical network model of the system under consideration. To this end we consider the analysis of a typical Helmholtz cavity resonator.

The basic Helmholtz cavity resonator considered is illustrated in Figure 6.6(a). It consists of a one-dimensional cavity of length L_1, cross-sectional are A_1 and internal volume V attached to a long uniform narrow neck of length L_2 and internal cross-sectional area equal to A_2. The displacement, pressure,

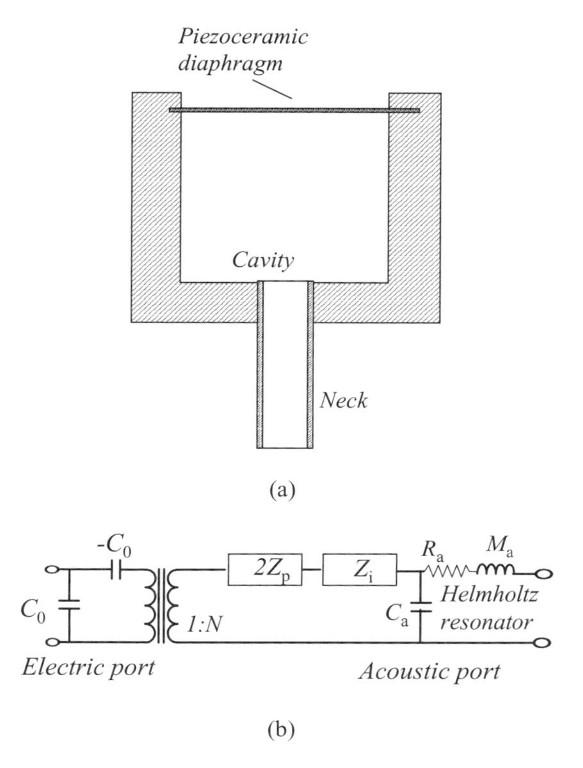

Figure 6.6 (a) Illustration of a Helmholtz cavity resonator driven by a piezoelectric diaphragm, (b) equivalent circuit of a Helmholtz resonator driven by a piezoelectric diaphragm

density, bulk modulus, wave number and wave speed are denoted u_i, p_i, ρ_i, β_i, k_i and a_i, respectively, with $i = 1$ and 2, representing the region within the cavity and the region within the neck, respectively.

Applying Newton's law to a small longitudinal element of mass,

$$\rho_i \frac{\partial^2 u_i}{\partial t^2} = -\frac{\partial p_i}{\partial x},$$

while the pressure is given by

$$p_i = -\beta_i \frac{\partial u_i}{\partial x}.$$

Hence, displacements in the two regions satisfy

$$\frac{1}{a^2}\frac{\partial^2 u_1}{\partial t^2} = \frac{\partial^2 u_1}{\partial x^2}, \quad 0 \leq x \leq L_1 \tag{6.101}$$

and

$$\frac{1}{a^2}\frac{\partial^2 u_2}{\partial t^2} = \frac{\partial^2 u_2}{\partial x^2}, \quad L_1 \leq x \leq L_1 + L_2, \tag{6.102}$$

where it is assumed that $a = a_1 = a_2 = \beta / \rho$. The displacement u_1 is equal to zero at $x = 0$, while the displacement u_2 is equal to zero when $x = L_1 + L_2$. At the interface between the cavity and the neck, i.e.

at $x = L_1$, the flow rate and the pressure are continuous. Hence,

$$A_1\dot{u}_1 = A_2\dot{u}_2 \quad \text{and} \quad \beta_1\frac{\partial u_1}{\partial x} = \beta_2\frac{\partial u_2}{\partial x}.$$

Applying the four boundary conditions to the general solutions of the two wave equations for u_i and assuming that $\beta_1 = \beta_2$, we have the frequency equation

$$\frac{A_2}{A_1}\cot\left(\frac{\omega_n L_1}{a}\right) = \tan\left(\frac{\omega_n L_2}{a}\right).$$

For the fundamental natural frequency, we may assume that both L_1 and L_2 are small compared to the typical wavelength a/ω_n, and it follows that

$$\cot\left(\frac{\omega_n L_1}{a}\right) \approx \frac{a}{\omega_n L_1}, \quad \tan\left(\frac{\omega_n L_2}{a}\right) \approx \frac{\omega_n L_2}{a} \quad \text{and} \quad \frac{A_2}{A_1} = \frac{\omega_n^2 L_1 L_2}{a^2}.$$

Thus,

$$\omega_n = a\sqrt{\frac{A_2}{L_2 V}}. \tag{6.103}$$

Interestingly, we observe that the equation defining the fundamental natural frequency of the resonator could also be derived from considerations of force equilibrium at the interface between the cavity and the neck. Considering the total inertia force of the mass of air in the neck, it must equal the total restoring force. Hence,

$$\rho_2 A_2 L_2 \ddot{w}_2(L_1) = -\frac{A_2 w_2(L_1)}{V}\beta_1 A_2$$

and equation (6.103) follows.

A similar approach may be adopted to derive an equivalent circuit model for the Helmholtz resonator. From considerations of continuity in the cavity, the mass flow rate is given by

$$\frac{\mathrm{d}M}{\mathrm{d}t} = V\frac{\mathrm{d}\rho_1}{\mathrm{d}t} \equiv Q.$$

Assuming the disturbances to be harmonic and isentropic, the pressure in the cavity is

$$p_1 = a^2 \quad \rho = a^2 Q/i\omega V.$$

Hence, the pressure at the entrance to the neck is

$$p_2 = p_1 + i\omega\frac{QL_2}{A_2} = Q\left(\frac{a^2}{i\omega V} + i\omega\frac{L_2}{A_2}\right) = A_2\rho_2\dot{u}_2(L_1)\left(\frac{a^2}{i\omega V} + i\omega\frac{L_2}{A_2}\right).$$

Considering force equilibrium, it follows that

$$A_2 p_2 = M_a\dot{u}_2(L_1) + \frac{1}{C_a}\int u_2(L_1)\,\mathrm{d}t,$$

where M_a is the mechanical equivalent acoustic inductance and C_a is the mechanical equivalent acoustic capacitance. It therefore follows that

$$A_2^2\rho_2\dot{u}_2(L_1)\left(\frac{a^2}{i\omega V} + i\omega\frac{L_2}{A_2}\right) \equiv \dot{u}_2(L_1)\left(\frac{1}{i\omega C_a} + i\omega M_a\right).$$

Hence, by comparing the two sides of the above identity,

$$M_a = L_2 A_2 \rho_2, \quad C_a = \frac{1}{\rho_2 a^2}\frac{V}{A_2^2}.$$

The resistance of the neck has so far been neglected. This could be estimated on the basis of laminar flow theory as

$$R_a = 8\pi \mu L_2.$$

Furthermore, the equivalent mass or inductance is usually modified to include the additional apparent mass of the flow in the cavity. When this is done, mechanical equivalent acoustic inductance M_a is given by

$$M_a = k_a \rho_2 A_2 L_2,$$

where k_a is an apparent mass coefficient and is normally assumed to be $k_a = 4/3$.

Thus, with the equivalent circuit parameters of the Helmholtz resonator defined as

$$M_a = k_a \rho_2 A_2 L_2, \quad R_a = 8\pi \mu L_2, \quad C_a = \frac{1}{\rho_2 a^2} \frac{V}{A_2^2}, \tag{6.104}$$

we may establish the complete equivalent circuit of the thin piezoelectric disc-driven transducer, which is illustrated in Figure 6.6(b).

Up to this point only loss-less piezoelectric transducers have been considered, although it was recognized that all piezoelectric devices have internal energy losses. To account for the influence of the losses, the impedances in Figures 6.5(a), 6.5(c) and 6.6(b) must be replaced by their lossy counterparts. The process is best demonstrated by another archetypal application that is considered in the next section. One further comment is in order. Although the derivation of the model presented above was restricted to the thickness shear wave mode, the general form of the model is valid for most wave modes, including flexural plate modes. Further details on estimating the parameters characterizing the specific models may be found in the work by Mason (1948, 1950) as well as in the text by Kikuchi (1969).

6.5.4 Modelling of Ultrasonic Wave Motors

An ultrasonic motor (USM) is a new actuator that uses mechanical vibrations in the ultrasonic range as its input driver. USMs have important features such as a high stall speed and specific torque, high torque at low speed, compact in size, quiet operation and little or no electromagnetic interference. The torque of a USM is 10–100 times larger than conventional electromagnetic motors of the same size or weight. Because of these features, USMs are presently being employed for several industrial, medical, robotic, space and automotive applications.

The USM is characterized by 'low speed and high torque', contrary to the 'high speed and low torque' of the electromagnetic motors. Two categories of USMs that have been developed are the standing wave and the travelling wave types. The standing wave type is sometimes referred to as a vibratory-coupler type, where a vibratory piece is connected to a piezoelectric driver and the tip portion generates flat elliptical movement. Attached to a rotor or a slider, the vibratory piece provides intermittent rotational torque or thrust, as shown in Figure 6.7.

Although several travelling wave USM types have been designed, the rotary travelling wave USM (TWUSM) is the most commonly used USM type. The TWUSM is driven by a high-frequency two-phase sinusoidal supply with a 90° phase difference between the two phases. The speed of a USM is controlled by the amplitude, frequency and phase difference between the two phase voltages. During the operation of these motors, there is a two-stage conversion of energy. In the first stage, there is the electromechanical energy conversion where the electrical energy is converted into mechanical energy. This is achieved by the excitation of piezoelectric ceramic with ultrasonic range frequency. In the second stage, there is the mechanical energy conversion where the mechanical vibrations are converted into linear or rotary motion largely by the friction force generated in the stator–rotor interface and other non-linear effects. The principal non-linear effect is the phenomenon of *acoustic streaming*, which refers to generation of

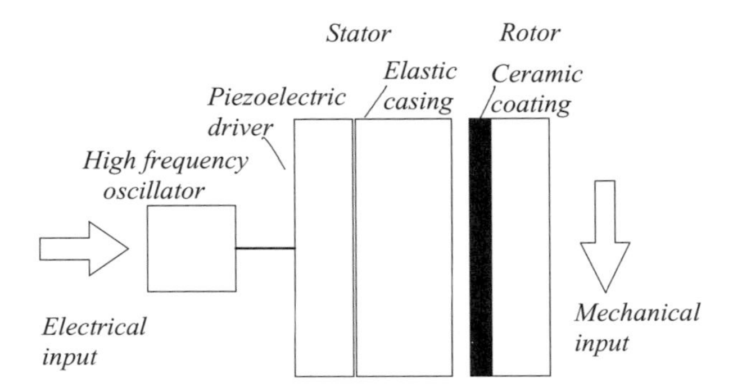

Figure 6.7 Principle of a standing wave USM or vibratory coupler

mean motion due to the propagation of SAWs in the form of travelling waves. Acoustic streaming is also the basis for the design of valve-free fluidic micro pumps and other devices.

The travelling wave-type motor combines two standing waves with a 90° phase difference both in time and space. By means of the travelling elastic wave induced by the thin piezoelectric ring, a ring-type slider in contact with the surface of the elastic body can be driven. Like most electric motors, the USM also consists of a stator and a rotor. The stator consists of the piezoelectric ceramic and the elastic body. The rotor is fabricated from a bronze material pressed against the stator by means of a disc spring. When two-phase voltages are applied to two orthogonal modes of piezoelectric ceramic of USM, elliptical waves occur on the stator surface. The rotor is then driven by the tangential force at the contact surface resulting from the elliptical motion at the wave crests.

A particle at the surface of the stator, in which a flexural travelling wave is guided, moves elliptically. If an object is in contact with the stator, the object will be forced to move by the friction force. The intuitive analogy may be 'a surf board on a surface wave'. By placing a movable object in contact with a flexural travelling surface wave, the movable object, in this case the rotor, will be forced to move along with the velocity of the particles at the crests of transverse travelling flexural waves propagating down the stator. As a result, the rotor surface will move in the direction of the source of the wave. If the direction of the propagation of the wave is changed, the rotor similarly changes its direction.

The equivalent circuit of a two-phase travelling wave motor is illustrated in Figure 6.8. Figure 6.8(a) shows the basic equivalent circuit of just one phase of motor that is based on the standard equivalent circuit of a thin piezoelectric transducer. A number of real effects such as the dielectric loss due to the hysteresis between the applied electric field and the electric displacement in the piezoelectric material must also be included. As a result of this hysteresis between $\boldsymbol{E}$ and $\boldsymbol{D}$, heat is generated and the corresponding dissipation, known as the dielectric loss, is represented by R_d. The combined impedance of C_0 and R_d is known as the blocking impedance. However, R_d may be neglected for the frequency range of most USMs. In the piezoelectric effect for a polarized element, the displacement (or strain) has a slight phase lag behind the electric field, again resulting in a hysteresis loop, which results in losses. Furthermore, various mechanical losses occur in components such as the vibrator's metal block, which is under cyclic deformation. The imperfect bonding between the metal and piezoelectric ceramic element also contributes to the losses. The losses are collectively represented by the by the element R_m. The combined impedance of R_m, C_m and L_m represents the total additional impedance due to the piezoelectrically induced motion or the radiation impedance. Sometimes it is conventional to split the motional impedance into two components, one electrical and the other mechanical and include the two on either side of the ideal transformer. These real effects, as well as the combined load impedance, are included in the equivalent circuit in Figure 6.8(b). Also included at the output side is a diode to model the fact that the principal mechanism of motion transfer to the load is by Coulomb friction. This is an

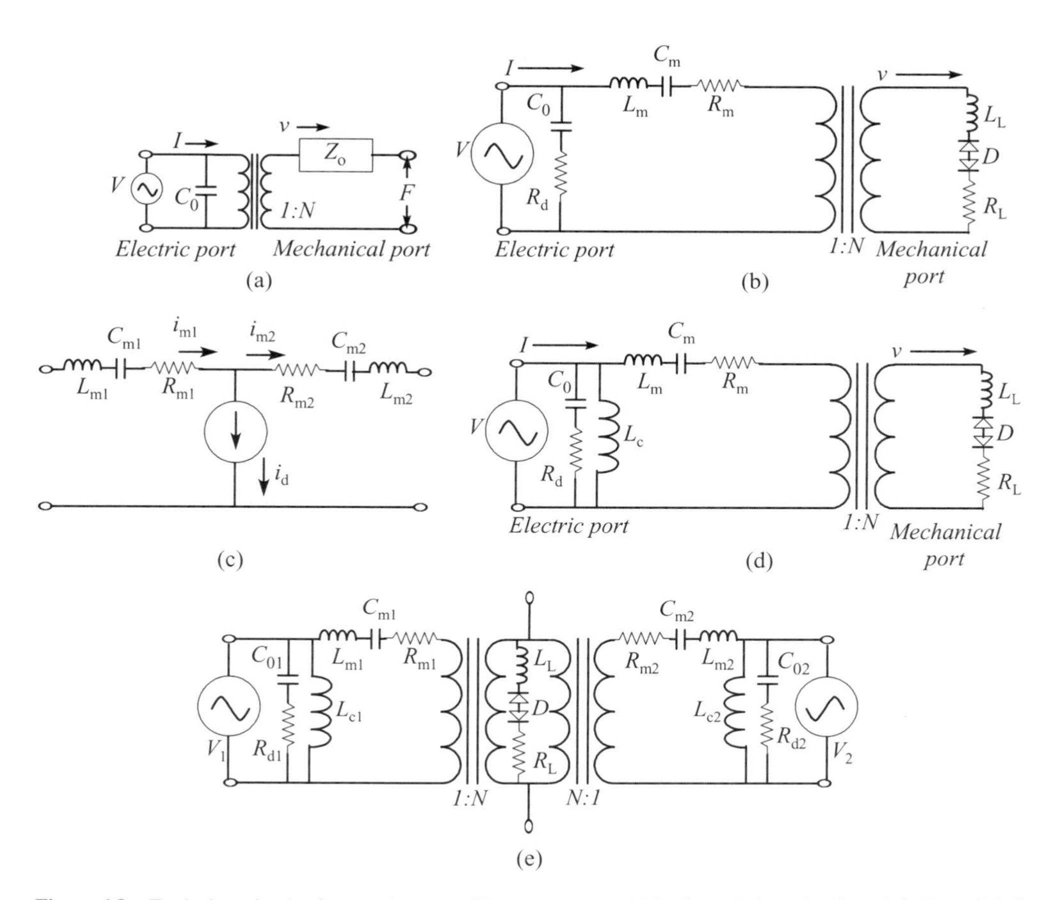

Figure 6.8 Equivalent circuit of a two-phase travelling wave motor: (a) basic equivalent circuit model, (b) modified equivalent circuit for a single phase including effects of dielectric loss, motional impedance, load torque, applied preload pressure and a diode non-linearity to represent the effect of the frictional driving force, (c) detailed model of the non-linear slip-stick behaviour, (d) compensating the effect of the blocking capacitance with an inductor, (e) two-phase input and output model of wave motor, showing the two electrical ports and the mechanical port

idealization of the real behaviour of the motor, which may be characterized as a slip-stick type of non-linearity discussed in Chapter 6. The speed of the unloaded motor depends on the normal forcing, which is responsible for the sliding and the stick-slip behaviour at the contact surface of the driving mechanism (friction). The measured speed is generally less than the ideal speed where the speed drop is due to the normal forcing. The speed drop is almost proportional to the normal forcing and their relationship is linear. The effect of this drop in speed can be modelled by splitting the motional impedance into two components and introducing a negative current generator to form a T circuit, as shown in Figure 6.8(c). The speed of the loaded motor depends on the amount of the torque load at the output shaft of the motor. The variational nature of the torque gives rise to non-linear dynamical changes affecting the sliding and stick-slip behaviour at the contact surface of the driving mechanism (friction). The effect can be modelled, if essential, by adding further T sections of the type illustrated in Figure 6.8(c).

The effect of the blocking capacitance is to reduce the power factor of the circuit. In a real situation, the inertial impedance of the piezoelectric material has a compensating effect. Hence, an inductor is included in parallel with the blocking capacitor in Figure 6.8(d), which then represents the complete equivalent circuit of one phase of the motor. The complete equivalent circuit of the two-phase motor is shown in Figure 6.8(e).

In order to achieve high efficiencies, USMs are driven at or close to the frequency at which the motional impedance is a minimum. This frequency is given by

$$f_m = \frac{1}{2\pi} \frac{1}{\sqrt{L_m C_m}}.$$

The frequency at which the input admittance is a minimum is

$$f_i \approx \frac{1}{2\pi} \frac{1}{\sqrt{L_m \left(\frac{C_m C_0}{C_m + C_0} \right)}}.$$

Thus, the motional impedance parameters are estimated as

$$L_m \approx \frac{1}{4\pi^2 C_0} \left(\frac{1}{f_i^2 - f_m^2} \right), \quad C_m \approx \frac{C_0}{f_m^2} \left(f_i^2 - f_m^2 \right),$$

while R_m is obtained by measuring the total impedance at $f = f_m$. In fact, the motional capacitance may be related to the electromechanical coupling coefficient since it can be shown that

$$K^2 \approx \frac{1}{f_i^2} \left(f_i^2 - f_m^2 \right).$$

These relationships are particularly useful in estimating the parameters in the representative equivalent circuit in designing the motor to a specific requirement.

6.6 The Generation of Acoustic Waves

An acoustic wave is defined as a mechanical disturbance that is transmitted in a medium without any net change in the mass of the transmitting element. Although acoustic waves may be generated mechanically or electrically, the latter is particularly suitable for the design and construction of practical wave generators. The electromechanical generation of acoustic waves may be achieved using both piezoelectric transducers and electromagnetic acoustic transducers (EMATs).

The active element is basically a piece of polarized material (i.e. some parts of the molecule are positively charged, while other parts of the molecule are negatively charged) with electrodes attached to two of its opposite faces. When an electric field is applied across the material, the polarized molecules will align themselves with the electric field, resulting in induced dipoles within the molecular or crystal structure of the material. This alignment of molecules will cause the material to change dimensions. This is the phenomenon of electrostriction. In addition, a permanently polarized material such as quartz (SiO_2) or barium titanate ($BaTiO_3$) will produce an electric field when the material changes dimensions as a result of the strain due to an imposed mechanical force. This phenomenon or the property of piezoelectricity is effectively employed in the generation of acoustic waves.

The active element of most acoustic transducers used today is a piezoelectric ceramic, which can be cut in various ways to produce different wave modes. A large piezoelectric ceramic element can be seen in the image of a sectioned low-frequency transducer. Preceding the advent of piezoelectric ceramic in the early 1950s, piezoelectric crystals made from quartz crystals and magnetostrictive materials were primarily used. When piezoelectric ceramics were introduced, they soon became the dominant material for transducers due to their good piezoelectric properties and their ease of manufacture into a variety of shapes and sizes. They also operate at low voltage and are usable up to about 300°C. The first piezoceramic in general use was barium titanate, and that was followed during the 1960s by lead zirconate titanate compositions, which are now the most commonly employed ceramic for making transducers. New materials such as piezopolymers and composites are also being used in some applications. A wide variety of transducers are now available in the market configured either as contact transducers or as immersion transducers that may be employed as dual element transducers incorporating both a receiver and a transmitter element, delay-line, angle beam and shear wave transducers. Most transducers have a typical constructional architecture.

The thickness of the active element is determined by the desired frequency of the transducer. A thin wafer element vibrates with a wavelength that is twice its thickness. Therefore, piezoelectric crystals are cut to a thickness that is half the desired radiated wavelength. The thickness of the active element is directly proportional to the frequency of operation of the transducer. The primary reason that higher-frequency contact transducers are not produced is because the corresponding elements are very thin and too fragile.

A cut-away view of a typical contact transducer is shown in Figure 6.9(a). To maximize the energy transmitted out of the transducer, an impedance matching wafer is placed between the active element and the face of the transducer. Optimal impedance matching is achieved by sizing the matching layer so that

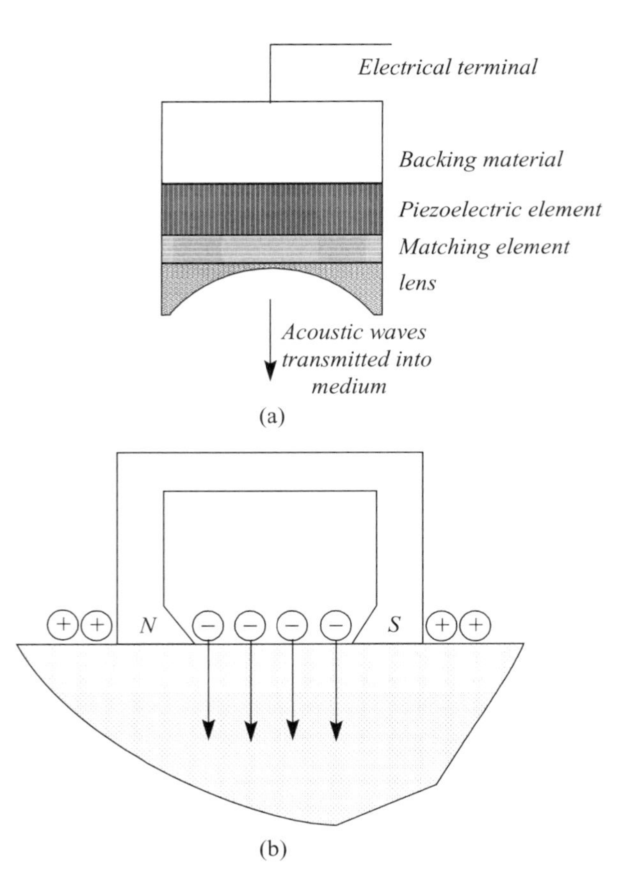

Figure 6.9 Electromechanical transducers for generating acoustic waves: (a) cut-away view of a typical contact transducer and (b) cross-sectional view of an EMAT for exciting polarized longitudinal waves propagating normal to the surface

its thickness is one-quarter of the wavelength. This keeps waves that were reflected within the matching layer in phase when they exit the layer, as illustrated in the image to the right. For contact transducers, the matching layer is made from a material that has an acoustical impedance between the active element and steel. Immersion transducers have a matching layer with an acoustical impedance between the active element and water. Contact transducers also often incorporate a wear plate to protect the matching layer and active element from scratch. In addition, there is also a thick layer of backing material. The entire assembly constitutes a longitudinal wave transmission medium.

The backing material supporting the crystal has a great influence on damping characteristics of a transducer. Using a backing material with an impedance similar to that of the active element will produce the most effective damping. Such a transducer will have a narrow bandwidth, resulting in higher sensitivity. As the mismatch in impedance between the active element and the backing material increases, material penetration is increased but transducer sensitivity is reduced.

The acoustic field that emanates from a piezoelectric transducer does not originate from a point. Rather it originates from most of the surface of the piezoelectric element. Round transducers are often referred to as piston source transducers because the sound field almost resembles a cylindrical mass in front of the transducer. In reality, it is almost always conical with significant percentage of the energy radiating out. This results in beam spreading, which is an undesirable feature that must often be minimized. The acoustic beam is more uniform in the far field, where the beam spreads out in a pattern originating from the centre of the transducer. The transition between the initial non-uniform zone and the far field zone occurs at a distance N and is sometimes referred to as the 'natural focus' of a flat (or unfocused) transducer. It is a function of the transducer diameter. The near/far distance N is significant because amplitude variations that characterize the near field changes to a smoothly declining amplitude at this point. This area just beyond the near field is where the sound wave is well behaved and at its maximum strength. Maximum transducer resolution may be obtained by operating in this region. Beyond this region, the beam spreads out in a spherical fashion in the far field.

In practice, it is also often necessary to artificially focus the acoustic wave field to increase the field intensity. Focusing concentrates the total energy emanating from the relatively large area to the relatively small focused point. A range of acoustic wave focusing lens, acoustic mirrors, specially shaped crystals or electronic focusing methods may be employed to perform one of several intensity-processing operations. Acoustic lenses are generally made from metals such as aluminium or plastics such as polymethylmethacrylate (perspex) and polystyrene. These materials usually have characteristic acoustic impedance of twice that of the acoustic wave-receiving medium. The lenses are directly bonded to the front of the transducer before attaching the impedance-matching element. The lens is usually flat faced on the transducer end and concave on the other side, with a constant radius of curvature R. The focal length of the lens is then determined from the radius of curvature and the refractive index n; it is the ratio of the acoustic wave velocity in the lens to that of the receiving medium and is given by

$$F = \frac{n}{(n-1)} R.$$

The acoustic impedance of the entire lens assembly is then nearly the same as the transducer element, and this is essential to minimize internal reflections.

EMATs act through totally different physical principles and do not need a couplant to match the impedance of the generator to the impedance of the transmitting medium. When a wire is placed near the surface of an electrically conducting object and is driven by a current at the desired acoustic wave frequency, eddy currents will be induced in a near surface region of the object.

If a static magnetic field is also present, these eddy currents will experience Lorentz forces of the form

$$\mathbf{F} = \mathbf{J} \times \mathbf{B},$$

where $\mathbf{F}$ is a body force per unit volume, $\mathbf{J}$ is the induced dynamic current density and $\mathbf{B}$ is the static magnetic induction. The coil and magnet structure can also be designed to excite complex wave patterns and polarizations that would be difficult to realize with fluid-coupled piezoelectric probes. In the inference of material properties from precise velocity or attenuation measurements, use of EMATs can eliminate errors associated with couplant variation, particularly in contact measurements.

Typical practical EMAT configurations are shown in Figure 6.9(b). It consists of a biased magnet structure and a coil. These elements and forces on the surface of the solid are shown in the exploded view. The configuration may be employed to excite acoustic beams propagating normal to the surface of the half-space and produce beams with longitudinal polarizations. Although a great number of variations on this configuration have been conceived and used in practice, consideration of the typical geometry should suffice to introduce the fundamentals. EMATs are suitable for the generation of Lamb waves, shear waves and a variety of surface waves, although they may not be the most efficient method for generating the latter.

Practical EMAT designs are relatively narrow band and require strong magnetic fields and large currents to produce acoustic fields that are often weaker than that produced by piezoelectric transducers. Rare-earth materials such as samarium–cobalt and neodymium–iron–boron are often used to produce sufficiently strong magnetic fields, which may also be generated by pulsed electromagnets. The acoustic waves are then appropriately intensified by the use of focusing techniques.

6.6.1 Launching and Sensing of SAWs in Piezoelectric Media

An important breakthrough in the investigation of SAWs was the invention, around 1950, of the interdigital transducer, which works in the ultrasound range. This device had, and still has, a tremendous impact on the application of piezoelectric materials to the generation of waves and vibrations. The great practical importance of piezoelectric materials stems from the fact that they provide the simplest and most efficient method for exciting high-frequency acoustic waves. One of the most effective ways of utilizing the piezomechanical coupling in a piezoelectric solid is to embed on the surface an array of conducting electrodes that enable one to produce spatially non-uniform time-varying or periodic electric fields on the material surface. The effect of the periodic electric fields is to generate periodic stress fields due to the electromechanical coupling and consequently launch SAWs. The process is particularly effective when the frequencies or wave speeds are close to the natural acoustic wave speeds of the material. The transmission process at these speeds is also reversible, so it can act as a receiver as well as a transmitter.

It is well known that in a typical transmission line the distribution of capacitive and inductive coupling results in the generation of an electromagnetic wave that propagates both forwards and backwards. While in a transmission line the forward and backward propagating wave components are independent of each other, they are also coupled in the case of an array of conducting electrodes due to the spatial variation of the electric fields. Thus, the spatial variation of the electric field is responsible for not only the electromechanical coupling but also the coupling between waves propagating in directions opposite to each other. Using a distribution of electrodes not only increases the efficiency of the acoustic transmission but also limits the bandwidth, and by effective trade-off of these two properties, the device may be tuned to any particular application. Furthermore, the insertion of electrodes could significantly alter the various surface wave modal velocities. This is due to the need to satisfy additional electrical boundary conditions pertaining to the charge distribution along the free surface. Normally, the mechanical boundary conditions at an unelectroded surface relate to a stress-free boundary (normal component of the stress field is zero) or a rigid boundary (normal component of the velocity field is zero), while the electrical boundary conditions relate to a short circuit (the electric potential function is zero) or an open circuit (the normal component of the electric displacement field which relates the current is zero). In the case of an electroded surface, while the mechanical boundary conditions are generally the same, the electrical boundary conditions change because of the need to prescribe the electric potential function as a constant or specify the rate of change normal component of the electric displacement field averaged over the boundary as the latter is proportional to the current flowing across the electroded boundary surface.

Another related breakthrough was the thermoelastic generation of SAWs by pulsed laser radiation in 1968, limited only by the absorption of light. Focusing short laser pulses, as much as possible, to a point or line source can launch broadband elastic surface pulses. Strong non-linear laser excitation, by an ablative mechanism or by an explosive evaporation technique using a highly absorbing carbon containing layer, leads to strongly non-linear SAWs, which develop shock fronts during propagation and may even break covalent solids such as single-crystal silicon. The typical frequencies of broadband ultrasonic SAW pulses are within the 5–500 MHz range, covering a large spectral range of more than one decade simultaneously. This feature is employed for the generation of SWAs in layered and graded non-piezoelectric materials. Frequency spectra within a band with an upper limit approaching 1 GHz were realized recently using near-field optical excitation with fibre-optic devices. A detailed discussion of the thermoelastic generation of SAWs by pulsed laser radiation is well beyond the scope of this chapter.

There are two effective ways of embedding electrodes on the surface of a piezoelectric material:

1. single-phase electrode array;
2. inter-digitated (IDT) electrode arrays.

In the case of single-phase electrode arrays, finger-like electrodes, connected in parallel and all at the same potential with respect to the ground electrode, are embedded on the surface. Two of these electrode arrays, embedded at two different locations along the desired axis of propagation and reception, are employed to generate and launch surface waves.

For the case of a poled ferroelectric ceramic, the electroacoustic surface wave velocities for a bare and electroded surface are given by the expressions

$$\left(\frac{v}{\bar{v}_{44}}\right)^2_{bare} = \left(1 - \left(\frac{K^2}{1+\varepsilon_{11}}\right)^2\right)$$

and

$$\left(\frac{v}{\bar{v}_{44}}\right)^2_{electroded} = \left(1 - K^4\right),$$

where

$$\bar{v}_{44} = \sqrt{\frac{\bar{c}_{44}}{\rho}} \quad \text{and} \quad K^2 = \frac{e_{15}^2}{c_{44}\varepsilon_{11}},$$

with the stiffened elastic coefficient $\bar{c}_{44}$ defined as

$$\bar{c}_{44} = c_{44}\left(1 + K^2\right).$$

The expressions clearly indicate that electroding has a significant effect on the velocity of surface wave propagation.

In the case of the IDT arrays, the two sets of arrays are co-located and IDT, as shown in Figure 6.10. An array of uniformly spaced metallic fingers is deposited on the surface of the piezoelectric substrate, and voltage excitation of opposite polarity is applied to alternate fingers. The electric fields applied to the substrate in this manner produce a spatially periodic distribution of piezoelectric stress, and, when the spacing of the fingers is chosen to match the wavelength of the Rayleigh wave, this stress generates constructive Rayleigh wave radiation in both forward and backward directions. A single IDT array is employed as a transmitter for launching surface waves, while another independent IDT array is employed as a receiver. As a result of employing a single IDT array for launching surface waves by applying an RF signal across its two terminals, the field components change sign from gap to gap. This induces fundamental surface wave generation when the array is driven at the frequency for which the wavelength

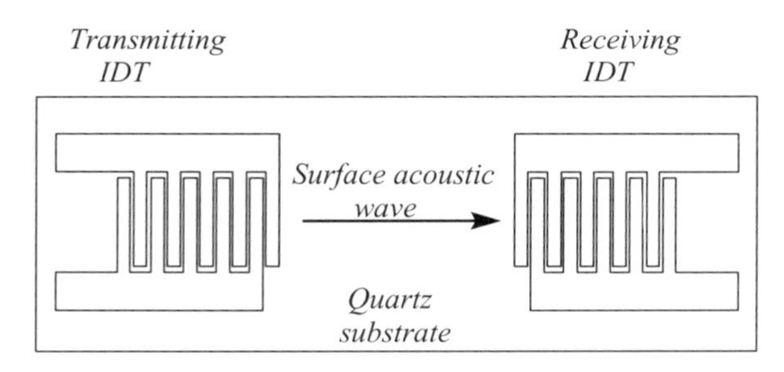

Figure 6.10 Transmitting and receiving IDT electrode arrays

under the array equals the periodic distance L. The IDT configuration is now considered to be 20 times more efficient than single-phase arrays and is the standard mode of generating and receiving SAWs in piezoelectric polycrystalline material-based transducers, unless other design considerations or mitigating circumstances require the use of single-phase arrays. Other important features of IDT arrays are that they may be employed to launch (1) Rayleigh surface waves in both piezoelectric and non-piezoelectric crystals, (2) electroacoustic surface waves in piezoelectric crystals, (3) Love waves in a layered medium with at least one piezoelectric material layer and (4) Lamb waves or waves associated with 'plate modes' in piezoelectric plates with the IDT arrays placed in the medium plane of the plate. Inter-digitated arrays are known to possess a number of other features that can be realized by embedding the electrodes with variable pitch or by introducing multiple electrode arrays. Thus, it is possible to focus, guide and reflect surface waves as is necessary for specific applications. Typical examples of IDT electrode arrays are illustrated in Figure 6.11.

Normally, IDT arrays launch surface waves in both forward and backward directions. However, in several applications, unidirectional launching of surface waves is a primary requirement and this is achieved by employing special IDT arrays. One approach is to employ a pair of identical IDT arrays separated by a distance of

$$d = \left(n + 1/4\right)\lambda,$$

where λ is the wavelength of the surface wave and n is an integer. The pair is driven by two oscillators whose outputs are at a phase difference of 90° from each other. Alternatively, they may be connected by a quarter-wave length electrical transmission line and driven by a single oscillator. Under these

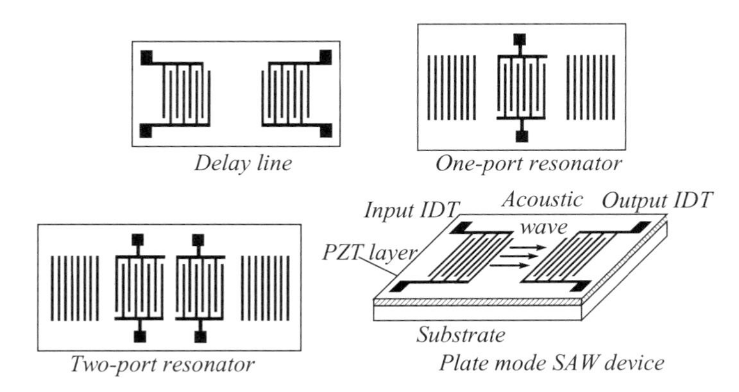

Figure 6.11 Typical examples of IDT electrode arrays

circumstances, surface waves generated by both arrays and travelling to the right add up in phase and waves travelling to the left cancel each other. The second approach also uses two IDT arrays but only the first is driven electrically by the oscillator. The second array is terminated by an inductor. Thus, it is parallel to the array capacitance. The resonance frequency of the inductor capacitor pair is chosen to be equal to the operating frequency so as to cause maximum wave reflection from that array. Although unidirectional operation increases the array's efficiency by 3 dB, since the waves are all launched in one direction only rather than in two, the bandwidth of the array pair is reduced. There are also a number of communication delay-line-related applications for unidirectional surface wave launchers.

6.6.2 Wave Propagation in Periodic Structures

The grating is the principal component of an IDT and essentially a periodic structure, in that the transmission properties in the direction of wave propagation are periodic. To understand the modelling and analysis of IDTs, it is therefore essential to consider the propagation of waves in periodic media. The analysis of wave propagation in periodic media is intimately associated with the theory of differential equations with periodic coefficients and the methodology adopted in the analysis of these equations is the general Floquet approach. Consider the 'inhomogeneous' wave equation

$$\left(\frac{\mathrm{d}^2}{\mathrm{d}x^2}+\frac{\omega^2}{v_0^2}\right)\varphi\left(x\right)=-\zeta\left(x\right)\frac{\omega^2}{v_0^2}\varphi\left(x\right), \tag{6.105}$$

where ω is the angular frequency and $\zeta(x)$ is the periodic perturbation 'load density', that is Λ-periodic in the coordinate x.

$$\zeta\left(x\right)=\zeta_0+\zeta_1\cos\left(\frac{2\pi x}{\Lambda}\right).$$

Parameter ζ_0 measures the perturbation to the uniform loading, while the parameters ζ_1 and Λ determine the magnitude and wavelength of the spatially periodic loading.

The ordinary differential equation (6.105) may be used to describe various types of wave propagation in a periodically perturbed medium, and it is known in mathematical physics as the Mathieu equation; the solutions of the equation are expressed in terms of a family of special functions known as Mathieu functions. For the case of most interest to us, that of SAWs, v_0 denotes the SAW wave velocity in an unloaded substrate and the load density is interpreted to describe the electric and/or mechanical loading due to the presence of the metal electrodes or grooves on the surface. The field $\varphi(x)$ may be interpreted as stress, displacement or electric potential, but usually it is normalized to the power flow. Although not stated explicitly, an implicit time dependence of the form $\exp(i\omega t)$ is assumed, and this results in the ω^2 term in equation (6.105).

The periodicity of the coefficients of governing equation implies that eigenmodes of the wave equation (6.105) satisfy the relationship

$$\varphi\left(x+\Lambda\right)=\exp\left(-i\beta\Lambda\right)\varphi\left(x\right),$$

where β may be interpreted as an unknown wave number of the solution. Floquet's theorem may be applied to the solution, and it can be deduced that

$$\varphi\left(x\right)=\exp\left(-i\beta x\right)\xi\left(x\right),$$

where $\xi(x)$ is a Λ-periodic function:

$$\zeta(x+\Lambda)=\zeta(x).$$

The generalization of the theorem for vector arguments is the famous theorem associated with name of Bloch in solid-state physics. Owing to the Λ-periodicity of the function $\xi(x)$, it has the Fourier series representation

$$\xi(x) = \sum_{n=-\infty}^{n=\infty} \xi_n \exp\left(-i\frac{2\pi nx}{\Lambda}\right)$$

and the field assumes the form

$$\varphi(x) = \sum_{n=-\infty}^{n=\infty} \xi_n \exp\left(-i\left(\frac{2\pi n}{\Lambda} + \beta\right)x\right). \tag{6.106}$$

Thus, the eigenmodes propagating in a periodically perturbed medium consist of an infinite set of discrete harmonics: a fundamental harmonic with the wave number β and an infinite number of higher harmonics with the wave numbers

$$\beta_n = \beta + \frac{2\pi}{\Lambda}n.$$

These harmonics, sometimes referred to as Floquet harmonics, are a fundamental characteristic of wave propagation in periodic structures, independent of the physical nature of the waves and the particular geometry of the structure. Depending on the physical context, any Floquet harmonic may be interpreted as the main or incident wave accompanied by an infinite number of scattered waves.

Substituting the assumed field for $\varphi(x)$ into the wave equation (6.105) yields

$$\sum_{n=-\infty}^{n=+\infty}\left[\left(-\beta_n^2 + (1+\zeta_0)\frac{\omega^2}{v_0^2}\right)\xi_n - \zeta_1\frac{\omega^2}{v_0^2}(\xi_{n+1} + \xi_{n-1})\right]\exp(-i\beta_n x) = 0.$$

Thus, the amplitudes of the harmonics are coupled to each other and must satisfy an infinite set of linear homogeneous equations given by

$$\left(\beta_n^2 - (1+\zeta_0)\frac{\omega^2}{v_0^2}\right)\xi_n + \zeta_1\frac{\omega^2}{v_0^2}(\xi_{n+1} + \xi_{n-1}) = 0$$

for all n considered. The above equation may be formally expressed as a homogenous matrix equation

$$[\mathbf{D}(\beta, \omega, v_0)]\,\Xi = 0,$$

where $\mathbf{D}(\beta, \omega, v_0)$ is an infinite square matrix and Ξ is an infinite vector consisting of the amplitudes ξ_n. For a non-trivial solution to exist, the determinant of the matrix must vanish and this results in the dispersion relation

$$|\mathbf{D}(\beta, \omega, v_0)| = 0$$

and the frequency is now a function of the wave number. The coupling between the harmonics is influenced to a large extent by the actual value of β. The coupling can be shown to be significant when β is a multiple of the first subharmonic of the 'load' variation, i.e. when

$$\beta = \delta + \frac{\pi}{\Lambda}n$$

and δ is small. Thus,

$$\beta_{-n} = \delta - \frac{\pi}{\Lambda}n = 2\delta - \beta.$$

In particular, when $\delta = 0, \beta_{-n} = -\beta = -\beta_0$ and the coupling between ξ_{-n} and ξ_0 is extremely significant. The condition $\beta_{-n} = -\beta = -\beta_0$ is therefore equivalent to the Bragg condition in crystallography and gives rise to strong interference between the β_0 and $-\beta_0$ wave components, leading to complete loss in transmission. The loss of transmission of waves at subharmonic and other related wave numbers is a fundamental feature of the propagation of waves in media with periodic variations in the relevant properties.

In what follows, only the two most strongly resonating harmonics will be retained, yielding the two-mode approximation:

$$\left(\beta_0^2 - (1+\zeta_0)\frac{\omega^2}{v_0^2}\right)\xi_0 + \zeta_1\frac{\omega^2}{v_0^2}\xi_{-1} = 0,$$

$$\left(\beta_{-1}^2 - (1+\zeta_0)\frac{\omega^2}{v_0^2}\right)\xi_{-1} + \zeta_1\frac{\omega^2}{v_0^2}\xi_0 = 0.$$

Retaining only the main harmonics, $n=0$ and $n=-1$ in equation (6.106), the Floquet expansion of the field assumes the form

$$\varphi(x) = \xi_{-1}\exp\left(-i\left(\beta - \frac{2\pi}{\Lambda}\right)x\right) + \xi_0\exp(-i\beta x)$$

and when

$$\beta = \delta + \frac{\pi}{\Lambda}n,$$

$$\varphi(x) = \xi_{-1}\exp\left(-i\left(\delta - \frac{\pi}{\Lambda}\right)x\right) + \xi_0\exp\left(-i\left(\delta + \frac{\pi}{\Lambda}\right)x\right).$$

The spatial variation of the field is governed by strongly oscillating factors $\exp(\pm i\pi x/\Lambda)$, whereas the factor $\exp(-i\delta x)$ describes the slow variations. Therefore, we may define a variable transformation to separate the strong oscillations from the slow variations that relates the field to two new variables,

$$\varphi(x) = W_0^+\exp\left(-i\frac{\pi}{\Lambda}x\right) + W_0^{-1}\exp\left(i\frac{\pi}{\Lambda}x\right). \tag{6.107}$$

This, the two-mode approximation to the field is the central result of the application of Floquet theory to the propagation of waves in periodic media and is the basis for the so-called coupling-of-modes approximation. A discussion of the coupling-of-modes theory is beyond the scope of this chapter, and the reader is directed to the references for further details.

Exercises

1. The piezoelectric constitutive relationships were presented in this chapter in matrix form as

$$\begin{bmatrix}\mathbf{T}\\ \mathbf{D}\end{bmatrix} = \begin{bmatrix}\mathbf{C} & -\mathbf{e}^T\\ \mathbf{e} & \mathrm{E}\end{bmatrix}\begin{bmatrix}\mathbf{S}\\ \mathbf{E}\end{bmatrix}.$$

Alternative formulations are given as

$$\begin{bmatrix}\mathbf{S}\\ \mathbf{D}\end{bmatrix} = \begin{bmatrix}\mathbf{s} & \mathbf{d}\\ \bar{\mathbf{d}} & \mathrm{E}\end{bmatrix}\begin{bmatrix}\mathbf{S}\\ \mathbf{E}\end{bmatrix},$$

$$\begin{bmatrix}\mathbf{T}\\ \mathbf{E}\end{bmatrix} = \begin{bmatrix}\mathbf{C}^D & -\mathbf{h}\\ -\bar{\mathbf{h}} & \mathrm{B}\end{bmatrix}\begin{bmatrix}\mathbf{S}\\ \mathbf{D}\end{bmatrix}$$

and

$$\begin{bmatrix}\mathbf{S}\\ \mathbf{E}\end{bmatrix} = \begin{bmatrix}\mathbf{s}^D & \mathbf{g}\\ \bar{\mathbf{g}} & \bar{\mathrm{B}}\end{bmatrix}\begin{bmatrix}\mathbf{T}\\ \mathbf{D}\end{bmatrix}.$$

Establish a representative set of relationships between the coefficient matrices.

2. A SAW is generated on the surface of a piezoelectric YZ-lithium niobate by means of an AC voltage applied to an IDT at a synchronous frequency of 1 GHz.

(i) Given that the velocity of propagation of the SAW in the material is $a_R = 3488$ m/s, determine the acoustic wavelength.

(ii) Compare this wavelength with that of an electromagnetic wave propagation in free space at the same frequency.

(iii) Hence or otherwise obtain the ratio of the SAW wavelength to that of the *EM* wave.

3. For an ideal transmission line show that

$$\frac{\Delta v}{c_0} \equiv \frac{\Delta c_0}{c_0} = -0.5\frac{\Delta C_0}{C_0},$$

where Δv is the change in velocity of wave propagation c_0, due to a change ΔC_0 in the line capacitance per unit length C_0.

4. For isotropic piezoelectric material, show from first principles that

$$K^2 = -2\frac{\Delta v}{a_R}.$$

State all the relevant assumptions made in the derivation.

5. Cubic-type fused silica is characterized by the following elastic and relative permittivity constants: $c_{11} = 7.85 \times 10^{10}\text{N/m}^2$, $c_{44} = 3.12 \times 10^{10}\text{N/m}^2$, $c_{12} = 6.39 \times 10^{10}\text{N/m}^2$ and $\varepsilon_{11} = 11.7 \times \varepsilon_0$. The density of the material is $\rho = 2332\text{kg/m}^3$. Calculate the Rayleigh SAW velocity. The direction of propagation is the x axis, and the free surface is normal to the z axis. [Hint: You may employ a displacement formulation as in exercise 13 in Chapter 5.]

6. Gallium arsenide is a piezoelectric crystalline material belonging to the cubic group (class $\bar{4}3m$). The elastic constants, piezoelectric constant in Coulombs per square metre and relative permittivity constants are $c_{11} = 11.88 \times 10^{10}\text{N/m}^2$, $c_{44} = 5.94 \times 10^{10}\text{N/m}^2$, $c_{12} = 5.38 \times 10^{10}\text{N/m}^2$, $e_{14} = 0.154\text{C/m}^2$ and $\varepsilon_{11} = 12.5 \times \varepsilon_0$. The density of the material is $\rho = 5307\text{kg/m}^3$. The piezoelectric coupling coefficient and the permittivity matrices are given by

$$\mathbf{e} = \begin{bmatrix} 0 & 0 & 0 & e_{14} & 0 & 0 \\ 0 & 0 & 0 & 0 & e_{14} & 0 \\ 0 & 0 & 0 & 0 & 0 & e_{14} \end{bmatrix} \quad \text{and} \quad \mathrm{E} = \begin{bmatrix} \varepsilon_{11} & 0 & 0 \\ 0 & \varepsilon_{11} & 0 \\ 0 & 0 & \varepsilon_{11} \end{bmatrix}.$$

Show that the governing equations of motion and the constraint on the electric displacement are

$$\begin{aligned}
\rho\ddot{u}_1 &= c_{11}\frac{\partial^2 u_1}{\partial x_1^2} + (c_{12}+c_{44})\frac{\partial^2 u_2}{\partial x_1 \partial x_2} + (c_{12}+c_{44})\frac{\partial^2 u_3}{\partial x_1 \partial x_3} + c_{44}\frac{\partial^2 u_1}{\partial x_2^2} \\
&\quad + c_{44}\frac{\partial^2 u_1}{\partial x_3^2} + 2e_{14}\frac{\partial^2 V}{\partial x_2 \partial x_3}, \\
\rho\ddot{u}_2 &= c_{44}\frac{\partial^2 u_2}{\partial x_1^2} + (c_{12}+c_{44})\frac{\partial^2 u_1}{\partial x_1 \partial x_2} + (c_{12}+c_{44})\frac{\partial^2 u_3}{\partial x_2 \partial x_3} + c_{11}\frac{\partial^2 u_2}{\partial x_2^2} + c_{44}\frac{\partial^2 u_2}{\partial x_3^2} + 2e_{14}\frac{\partial^2 V}{\partial x_1 \partial x_3}, \\
\rho\ddot{u}_3 &= c_{44}\frac{\partial^2 u_3}{\partial x_1^2} + (c_{12}+c_{44})\frac{\partial^2 u_1}{\partial x_1 \partial x_3} + (c_{12}+c_{44})\frac{\partial^2 u_2}{\partial x_2 \partial x_3} + c_{44}\frac{\partial^2 u_3}{\partial x_2^2} \\
&\quad + c_{33}\frac{\partial^2 u_3}{\partial x_3^2} + 2e_{14}\frac{\partial^2 V}{\partial x_1 \partial x_2}, \\
2e_{14}&\left(\frac{\partial^2 u_1}{\partial x_2 \partial x_3} + \frac{\partial^2 u_2}{\partial x_1 \partial x_3} + \frac{\partial^2 u_3}{\partial x_1 \partial x_2}\right) - \varepsilon_{11}\left(\frac{\partial^2 V}{\partial x_1^2} + \frac{\partial^2 V}{\partial x_2^2} + \frac{\partial^2 V}{\partial x_3^2}\right) = 0.
\end{aligned}$$

Hence, show that the general solution is the superposition of four-plane wave solutions, each with a unique decay constant and each satisfying boundary conditions associated with a piezoelectric surface. State the four boundary conditions that the solutions must satisfy and calculate the Rayleigh SAW velocity and the electromechanical coupling coefficient. The direction of propagation is the x_1 axis (*100 axis*) and the free surface is normal to the x_3 axis (*001 axis*). Comment on your results.

7. *Bismuth germanium oxide* is a piezoelectric crystalline material belonging to the cubic group (class 23). The elastic constants, piezoelectric constant in Coulombs per square metre and relative permittivity constants are $c_{11} = 12.8 \times 10^{10}\text{N/m}^2$, $c_{44} = 2.55 \times 10^{10}\text{N/m}^2$, $c_{12} = 3.05 \times 10^{10}\text{N/m}^2$, $e_{14} = 0.99\text{C/m}^2$ and $\varepsilon_{11} = 3 \times \varepsilon_0$. The density of the material is $\rho = 9200\text{kg/m}^3$. The piezoelectric coupling coefficient and the permittivity matrices are identical to the case of a class $\bar{4}3m$ material.

(i) Assume that the material is isotropic and that $c_{44} = (c_{11} - c_{12})/2$, and compute the Rayleigh SAW velocity. The direction of propagation is the x axis and the free surface is normal to the z axis.

(ii) Calculate the Rayleigh SAW velocity and the electromechanical coupling coefficient when the assumption of isotropy is not made. What is the percentage error in the isotropic calculation?

8. Consider the propagation of a surface wave in a hexagonal group (class 6 mm) crystal. The direction of propagation is the z axis, and the free surface is a plane normal to the y axis. These axes are oriented to the crystalline Y and X axes, respectively. By appropriate transformation of matrices, show that that the transformed non-zero elastic constants are given by $c'_{11} = c_{33}$, $c'_{22} = c_{11}$, $c'_{33} = c_{11}$, $c'_{66} = c'_{55} = c_{44}$, $c'_{44} = c_{66}$, $c'_{12} = c_{13}$, $c'_{13} = c_{13}$ and $c'_{23} = c_{12}$, where $2c_{66} = c_{11} - c_{12}$. Hence or otherwise determine the other constitutive matrices corresponding to the directions of propagation and the free surface in the crystal.

9. *Zinc oxide* and *cadmium sulphide* are both piezoelectric crystalline materials belonging to the hexagonal group (class 6 mm). The elastic constants in units of 10^{10}N/m^2, as well as the piezoelectric and relative permittivity constants of *zinc oxide* are $c_{11} = 20.97$, $c_{33} = 21.09$, $c_{44} = 4.247$, $c_{12} = 12.11$, $c_{13} = 10.51$, $e_{15} = -0.4\text{C/m}^2$, $e_{31} = -0.573\text{C/m}^2$, $e_{33} = 1.32\text{C/m}^2$, $\varepsilon_{11} = 8.55 \times \varepsilon_0$ and $\varepsilon_{33} = 10.2 \times \varepsilon_0$. The density of *zinc oxide* is $\rho = 5680\text{kg/m}^3$. The velocity of propagation of the SAW in *zinc oxide* is $a_R = 2680$ m/s. The direction of propagation is the z axis, and the free surface is the plane normal to the y axis.

The elastic constants in units of 10^{10}N/m^2, as well as the piezoelectric and relative permittivity constants of *cadmium sulphide* are $c_{11} = 9.07$, $c_{33} = 9.38$, $c_{44} = 1.504$, $c_{12} = 5.81$, $c_{13} = 5.1$, $e_{15} = -0.21\text{C/m}^2$, $e_{31} = -0.24\text{C/m}^2$, $e_{33} = 0.44\text{C/m}^2$, $\varepsilon_{11} = 9.02 \times \varepsilon_0$ and $\varepsilon_{33} = 9.53 \times \varepsilon_0$. The density of *cadmium sulphide* is $\rho = 4820\text{kg/m}^3$. The direction of propagation remains the z axis, and the free surface continues to be normal to the y axis.

Show by employing similarity arguments that the velocity of propagation of the SAW in cadmium sulphide is approximately equal to $a_R \approx 1730$ m/s.

10. *YZ*-lithium niobate is a crystalline material belonging to the trigonal group (class 3m). The piezoelectric coupling coefficient and the permittivity matrices are given by

$$\mathbf{e} = \begin{bmatrix} 0 & 0 & 0 & 0 & e_{15} & -e_{22} \\ -e_{22} & e_{22} & 0 & e_{15} & 0 & 0 \\ e_{31} & e_{31} & e_{33} & 0 & 0 & 0 \end{bmatrix} \quad \text{and} \quad \mathrm{E} = \begin{bmatrix} \varepsilon_{11} & 0 & 0 \\ 0 & \varepsilon_{11} & 0 \\ 0 & 0 & \varepsilon_{33} \end{bmatrix}.$$

The elastic constants in units of 10^{10}N/m^2, as well as the piezoelectric and relative permittivity constants are $c_{11} = 20.3$, $c_{33} = 24.5$, $c_{44} = 6$, $c_{12} = 5.3$, $c_{13} = 7.5$, $c_{14} = 0.9$, $e_{15} = 3.7\text{C/m}^2$, $e_{22} = 2.5\text{C/m}^2$, $e_{31} = 0.2\text{C/m}^2$, $e_{33} = 1.3\text{C/m}^2$, $\varepsilon_{11} = 44 \times \varepsilon_0$ and $\varepsilon_{33} = 29 \times \varepsilon_0$. The density of the material is $\rho = 4700\text{kg/m}^3$. The velocity of propagation of the SAW in the material is $a_R = 3488$ m/s. The direction of propagation is the z axis, and the free surface is a plane normal to the y axis. Determine the electromechanical coefficient and the stiffened elastic coefficients, given that the applied electric field is in a direction normal to the free surface (2-direction). Hence, compute the increase in the SAW velocity due to the stiffening effect.

11. *St-quartz* is a crystalline material belonging to the trigonal group (class 32). The elastic constants in units of 10^{10}N/m^2, as well as the piezoelectric and relative permittivity constants are $c_{11} = 8.674$, $c_{33} = 10.72$, $c_{44} = 5.794$, $c_{12} = 0.699$, $c_{13} = 1.191$, $c_{14} = -1.791$, $e_{11} = 0.171\text{C/m}^2$, $e_{14} = -0.0436\text{C/m}^2$, $\varepsilon_{11} = 4.5 \times \varepsilon_0$ and $\varepsilon_{33} = 4.6 \times \varepsilon_0$. The density of the material is $\rho = 2651\text{kg/m}^3$. The velocity of propagation of the SAW in the material is $a_R = 3158$ m/s. Determine the electromechanical coefficient and the stiffened elastic coefficients, given that the applied electric field is in a direction normal to the free surface (2-direction). Hence, compute the increase in the SAW velocity due to the stiffening effect.

12. *PZT* is piezoceramic material that is polycrystalline. The elastic constants in units of 10^{10}N/m^2, as well as the piezoelectric and relative permittivity constants of a particular variety of poled *PZT* are $c_{11} = 12.6$, $c_{33} = 11.7$, $c_{44} = 2.3$, $c_{12} = 7.95$, $c_{13} = 0.41$, $e_{15} = 17\text{C/m}^2$, $e_{31} = -6.5\text{C/m}^2$, $e_{33} = 23.3\text{C/m}^2$, $\varepsilon_{11} = 1700 \times \varepsilon_0$ and $\varepsilon_{33} = 1470 \times \varepsilon_0$. The density of the material is $\rho = 7500\text{kg/m}^3$. Estimate the piezoelectric constants d_{31}, d_{33}, g_{31}, g_{31}, k_{55} and k_{33}.

13. The piezoelectric polymer PVDF is characterized by the following piezoelectric constants: $d_{31} = 23 \times 10^{-12}\text{m/V}$, $d_{33} = -33 \times 10^{-12}\text{m/V}$, $g_{31} = 216 \times 10^{-3}\text{V/m/N/m}^2$, $g_{31} = -330 \times 10^{-3}\text{V/m/N/m}^2$, $k_{55} = 0.012$, $k_{33} = 0.014$, and $\varepsilon_{11} = 12 \times \varepsilon_0$. The density of the material is $\rho = 1780\text{kg/m}^3$. The Young's modulus for the material is $E = 0.3 \times 10^{10}\text{N/m}^2$. The 'thin plate' wave velocities in the longitudinal and transverse directions are 2200 and 1500 m/s, respectively. Determine the change in the SAW velocity due to piezoelectric stiffening. State all the relevant assumptions made in the analysis.

14. Derive from first principles, the three-port transmission line model for a thin piezoelectric transducer illustrated in Figure 6.3(c).

15. Consider the admittance matrix of the discrete element model of a typical SWA-IDT device,

$$[y] = \begin{bmatrix} y_{11} & y_{12} & y_{13} \\ y_{12} & y_{11} & -y_{13} \\ y_{13} & -y_{13} & y_{33} \end{bmatrix}.$$

Show that for the case of the in-line field model,

$$y_{11} = -\frac{\sqrt{-1}}{R_0}\cot\left(\frac{\theta}{4}\right)x_1\left(2 - \left(\frac{x_2}{x_1}\right)^2\right),$$

$$y_{12} = \frac{\sqrt{-1}}{R_0}\cot\left(\frac{\theta}{4}\right)\left(\frac{x_2^2}{2\left(2x - \cot\left(\frac{\theta}{4}\right)\right)x_1}\right),$$

$$y_{13} = -\frac{\sqrt{-1}}{R_0}\tan\left(\frac{\theta}{4}\right)\left(\frac{1}{1 - 2x\tan\left(\frac{\theta}{4}\right)}\right),$$

$$y_{33} = \sqrt{-1}\omega C_s\left(\frac{1}{1 - 2x\tan\left(\frac{\theta}{4}\right)}\right),$$

where

$$x_1 = x - \cot\left(\frac{\theta}{2}\right), \quad x_2 = x - \csc\left(\frac{\theta}{4}\right), \quad x = \frac{2}{\omega C_s R_0},$$

θ is the periodic section transit angle, defined as

$$\theta = 2\pi\frac{\omega}{\omega_0},$$

and ω_0 is the grating Bragg frequency.

Hence show that the elements of the transducer admittance matrix

$$[Y] = \begin{bmatrix} Y_{11} & Y_{12} & Y_{13} \\ Y_{12} & Y_{12} & -Y_{13} \\ Y_{13} & -Y_{13} & Y_{33} \end{bmatrix}$$

are given by

$$Y_{11} = Y_{22} = -\frac{S_{11}}{S_{12}}, \quad Y_{12} = Y_{21} = \frac{1}{S_{12}}, \quad Y_{13} = Y_{31} = y_{13}, \quad Y_{23} = Y_{32} = -y_{13}, \quad Y_{33} = Ny_{13},$$

N being the number of periodic sections, and

$$\mathbf{S} = \left(\frac{1}{y_{11}}\right)^N \begin{bmatrix} -y_{11} & 1 \\ y_{11}^2 - y_{12}^2 & -y_{11} \end{bmatrix}^N.$$

16. Consider the admittance matrix of the discrete element model of a typical SWA-IDT device

$$[y] = \begin{bmatrix} y_{11} & y_{12} & y_{13} \\ y_{12} & y_{11} & -y_{13} \\ y_{13} & -y_{13} & y_{33} \end{bmatrix}.$$

Show that for the case of the cross-field model,

$$y_{11} = -\frac{\sqrt{-1}}{R_0}\cot\theta, \quad y_{12} = \frac{\sqrt{-1}}{R_0}\csc\theta, \quad y_{13} = -\frac{\sqrt{-1}}{R_0}\tan\left(\frac{\theta}{4}\right),$$

$$y_{33} = \sqrt{-1}\left(\frac{4}{R_0}\tan\left(\frac{\theta}{4}\right) + \omega C_s\right).$$

Hence show that the elements of the transducer admittance matrix

$$[Y] = \begin{bmatrix} Y_{11} & Y_{12} & Y_{13} \\ Y_{12} & Y_{11} & -Y_{13} \\ Y_{13} & -Y_{13} & Y_{33} \end{bmatrix}$$

are given by

$$Y_{11} = -\frac{\sqrt{-1}}{R_0}\cot N\theta, \quad Y_{12} = \frac{\sqrt{-1}}{R_0}\csc N\theta, \quad Y_{13} = -\frac{\sqrt{-1}}{R_0}\tan\left(\frac{\theta}{4}\right),$$

$$Y_{33} = \sqrt{-1}N\left(\frac{4}{R_0}\tan\left(\frac{\theta}{4}\right) + \omega C_s\right),$$

N being the number of periodic sections.

17. Consider an ideal transmission line of length L. Show that the voltages at the two ends V_1 and V_2 are related to the currents I_1 and I_2 flowing in and out, respectively, at the two ends, by the relationship

$$\begin{bmatrix} V_1 \\ V_2 \end{bmatrix} = \begin{bmatrix} Z_{11} & Z_{12} \\ Z_{21} & Z_{22} \end{bmatrix}\begin{bmatrix} I_1 \\ I_2 \end{bmatrix},$$

where

$$Z_{11} = -iZ_0\cot\left(\frac{\omega L}{v}\right) = Z_{21}\cos\left(\frac{\omega L}{v}\right), \quad Z_{21} = -iZ_0\left(1\Big/\sin\left(\frac{\omega L}{v}\right)\right),$$

$$Z_{12} = iZ_0\left(1\Big/\sin\left(\frac{\omega L}{v}\right)\right) = -Z_{21} \quad \text{and} \quad Z_{22} = iZ_0\cot\left(\frac{\omega L}{v}\right) = -Z_{21}\cos\left(\frac{\omega L}{v}\right).$$

Hence show that the transmission matrix **T**, relating the input wave amplitudes to the output wave amplitudes across a section of the transmission line of length L, defined by the relationship

$$\begin{bmatrix} W_{i-1}^{+} \\ W_{i-1}^{-} \end{bmatrix} = \mathbf{T} \begin{bmatrix} W_i^{+} \\ W_i^{-} \end{bmatrix}$$

is given by

$$\mathbf{T} = \begin{bmatrix} \exp\left(i\frac{\omega L}{v}\right) & 0 \\ 0 & \exp\left(-i\frac{\omega L}{v}\right) \end{bmatrix}.$$

18. Consider a uniformly spaced grating and show that the transmission matrix **T** relating the input wave amplitudes to the output wave amplitudes across a section of the grating of length L, defined by the relationship

$$\begin{bmatrix} W_{i-1}^{+} \\ W_{i-1}^{-} \end{bmatrix} = \mathbf{T} \begin{bmatrix} W_i^{+} \\ W_i^{-} \end{bmatrix}$$

is given by

$$\mathbf{T} = \frac{1}{v_g} \begin{bmatrix} \exp(i\Delta) & i\sqrt{1-v_g^2} \\ -i\sqrt{1-v_g^2} & \exp(-i\Delta) \end{bmatrix},$$

where

$$v_g = \operatorname{sech}(\kappa L).$$

19. Consider a typical SAW-IDT transducer and show that the transmission matrix **T**, relating the input acoustic wave amplitudes and the amplitude of the electrical wave entering the transducer to the output acoustic wave amplitudes and the amplitude of the electrical wave leaving the transducer defined by the relationship

$$\begin{bmatrix} W_{i-1}^{+} \\ W_{i-1}^{-} \\ \beta_i \end{bmatrix} = \mathbf{T} \begin{bmatrix} W_i^{+} \\ W_i^{-} \\ \alpha_i \end{bmatrix},$$

is given by

$$\mathbf{T} = s \begin{bmatrix} (1+t_0)\exp(i\theta_t) & -t_0 & st_{13} \\ t_0 & (1-t_0)\exp(i\theta_t) & st_{13}\exp(-i\theta_t) \\ t_{13} & t_{13}\exp(-i\theta_t) & st_{33} \end{bmatrix},$$

where

$$t_0 = \frac{G_r(R_s+Z_e)}{1+i\theta_e}, \quad s = (-1)^{N_t}, \quad t_{13} = \frac{\sqrt{2G_r Z_e}}{1+i\theta_e}\exp\left(\frac{i\theta_t}{2}\right),$$

$$t_{33} = 1 - \frac{2i\theta_c}{1+i\theta_e}, \quad \theta_t = N_t\Lambda\delta, \quad \theta_c = \omega C_T(R_s+Z_e),$$

$$\theta_e = (\omega C_T + B_r)(R_s+Z_e), \quad C_T = (N_t-1)\frac{C_s}{2},$$

and N_t is the number of electrodes in the transducer, R_s is the effective series electrode resistance, G_r is the transducer radiation conductance, δ is a frequency deviation parameter, B_r is the transducer radiation susceptance and C_s is the static capacitance per electrode pair. For uniform transducers, it is given that

$$G_r = 2G_0(N_t-1)^2\left[\frac{\sin\left(\frac{\theta_t}{2}\right)}{\frac{\theta_t}{2}}\right]^2, \quad G_0 = \frac{\omega}{2\pi}K_c^2 C_s$$

and

$$B_r = 4G_0 (N_t - 1)^2 \frac{\sin\theta_t - \theta_t}{\theta_t^2},$$

where K_c^2 is the electromechanical coupling coefficient.

Hence show that by neglecting the static transducer capacitance and the frequency dependence of the propagation phase shift through the transducer, the transmission matrix $\mathbf{T}$ may be expressed as

$$\mathbf{T} = s \begin{bmatrix} 1 + g_t & -g_t & s\sqrt{2g} \\ g_t & 1 - g_t & s\sqrt{2g} \\ \sqrt{2g} & -\sqrt{2g} & s \end{bmatrix},$$

where

$$g = G_r Z_e, \quad g_t = G_r (R_s + Z_e).$$

References

Auld, B. A. (1973) *Acoustic Fields and Waves in Solids*, Vols. I & II, John Wiley & Sons Inc., New York.

Bhagavantam, S. (1966) *Crystal Symmetry and Physical Properties*, Academic Press, New York, pp. 159–163.

Campbell, C. K. (1998) *Surface Acoustic Wave devices for Mobile and Wireless Communications*, Academic Press Inc., New York, Chapters 1–5, pp. 3–158.

Ikeda, T. (1996) *Fundamentals of Piezoelectricity*, Oxford University Press, Oxford.

Kikuchi, Y. (1969) *Ultrasonic Transducers*, Corona Publishing Co. Ltd, Tokyo.

Kolsky, H. (1963) *Stress Waves in Solids*, Dover Publications Inc., New York.

Mason, W. P. (1948) *Electromechanical Transducers and Wave Filters*, 2nd edn, Van Nostrand Co. Inc., New York.

Mason, W. P. (1950) *Piezoelectric Crystals and Their Application to Ultrasonics*, D. Van Nostrand Co. Inc., Princeton, New Jersey.

Mason, W. P. (1964) *Physical Acoustics*, Harcourt Brace Jovanovich, Academic Press, New York.

Matthews, H. (1977) *Surface Wave Filters: Design, Construction, and Use*, Wiley, New York.

Morgan, D. P. (1985) *Surface-Wave Devices for Signal Processing*, Elsevier, New York.

Redwood, M. (1960) *Mechanical Waveguides*, Pergamon Press Ltd, London.

Sashida, T. and Kenjo, T. (1993) *An Introduction to Ultrasonic Motors*, Clarendon Press, Oxford Science Publishers, Oxford.

Tiersten, H. F. (1969) *Linear Piezoelectric Plate Vibrations*, Plenum Press, New York.

White, R. M. (1970) Surface elastic waves, *Proc. IEEE*, 58 (8), 1238–1276.

7

Mechanics of Electro-actuated Composite Structures

7.1 Mechanics of Composite Laminated Media

Composite structures are broadly classified into three categories: *particulate composite structures*, which consist of particles of various shapes and sizes dispersed randomly and embedded within a matrix, a homogeneous material; *continuous fibre composite structures*, which consist of a matrix, embedded and reinforced with long and continuous fibres; and *short fibre composites*, which contain short discontinuous fibres embedded within a matrix.

The composite laminates generally consist of thin layers, or laminae, that are orthotropic, transverse isotropic or isotropic in behaviour. An arbitrary number of layers of laminae are assumed to be stacked together with arbitrary ply-orientation angles. However, in most composite laminates identical layers or plies are ordered in a particular and regular stacking sequence, which refers to the location and sequence of various plies. The configuration indicating its ply make-up is referred to as its lay-up. A symmetric laminate is a laminate in which, for every layer on one side of the laminate reference surface with a specific thickness, specific material properties and specific fibre orientation, there is another layer at the same distance on the opposite side of the reference surface with the same specific thickness, material properties and fibre orientation.

In classical fibre composites, fibres such as boron, borsic, graphite, Kevlar and fibre-galls are used while the matrix materials are metals such as aluminium, titanium and magnesium. A variety of resin materials are used for bonding the fibres in the matrix or bonding the layers. These include epoxy, bismaleimide, polyimide, polyester, phenolic and a number of thermo-plastics. Piezo-patch-actuated composite laminates and shape memory alloy (SMA)-actuated composite structures include additional layers of piezo-ceramics or SMA wires embedded within layers of resin. These layers are usually bonded on either side of classical fibre composite to provide the structure with effective active control mechanism. Bonding the layers on either side of the composite and applying an unsymmetrical potential difference on both sides provides maximum control forces and moments.

The piezoelectrically and thermally induced stresses developed in symmetric electro-actuated laminates are most often unsymmetric. Thus, the control-configured laminates are generally analysed as laminates that are not symmetric and will cause curvature under the presence of a control loading. The piezoelectric laminates of interest for active control applications are thin unsymmetric laminates.

There are a number of books covering the various aspects of classical laminates such as Daniel and Ishai (2005), Hyer (1998), Jones (1999), Reddy (1996) and Tsai and Hahn (1980). In the following sections, we will briefly consider only aspects that are relevant to the analysis of smart or active laminates. The paper by Lee (1990) is an early classic reference covering some aspects of the theory of laminated piezoelectric plates.

7.1.1 Classical Lamination Theory

While exact solutions exist only for certain rectangular bidirectional composites and sandwich plates (Pagano, 1970), in most practical situations a number of simplifying assumptions must be made in the analysis of laminated plates. Classical lamination theory (CLT) is a plate theory for laminated composite materials, in which properties are smeared through the thickness and defined at a reference surface, and in-plane loads are defined in terms of force and moment resultants acting on the reference surface. Assuming that the composite laminates are thin, one neglects the strains and stresses through the thickness of the laminate and formulates the stiffness and strength characteristics for the middle surface directions only. This is the basic premise of CLT. CLT uses similar premises to the Kirchhoff–Love hypothesis, which states that the strains in the transverse direction are relatively small, to represent the stiffness properties of a three-dimensional body in terms of middle surface quantities only. The objective is to establish a set of constitutive relationships for the composite structure from the constitutive relations for the materials and individual lamina. In CLT, a number of simplifying assumptions are made, including the assumptions that

1. the Kirchhoff hypothesis is valid;
2. the plane stress assumption (though the thickness stresses are much smaller than in-plane stresses) is valid;
3. the material behaviour is linearly elastic;
4. the strain–displacement equations are linear.

As can be seen, CLT is an extension of classical linear elastic Kirchhoff theory for homogeneous plates to laminated composite plates. While CLT cannot always accurately predict the shapes of unsymmetric laminates with control loading, it provides an initial starting reference for developing several other theories that can be used for the control of smart structures.

The application of CLT to model and analyse unsymmetric laminates involves several standard steps. Two coordinate systems are used with CLT, the first a structural or global coordinate system in x, y and z, with the z direction as the thickness direction, and a material coordinate system in the '1', '2' and '3' directions, with the fibres in the '1' direction and the thickness in the '3' direction. To start with, one considers the constitutive relation of several typical laminae and transforms them into global coordinates.

7.1.2 Orthotropic, Transverse Isotropic and Isotropic Elastic Laminae

The constitutive equations of a piezoelectric material in the absence of thermal effects can be expressed as

$$\sigma_{ij} = C_{ijkl}\varepsilon_{kl} - e_{mij}E_{mm}, \quad D_i = e_{ijk}\varepsilon_{jk} + \bar{\epsilon}_{im}E_{mm},$$

while the inverse constitutive equation for the strain tensor may be expressed as

$$\varepsilon_{ij} = S_{ijkl}\sigma_{kl} + d_{mij}E_{mm}, \quad D_i = e_{ijk}\sigma_{jk} + \epsilon_{im}\ E_{mm},$$

where σ_{ij} and ε_{ij} are the stress and strain tensor components, E_{mm} is the electric field components, S_{ijkl} and C_{ijkl} are the elastic compliance and stiffness tensors, e_{mij} and d_{mij} are the piezoelectric coefficients,

D_i is the electric displacement and $\bar{\in}_{im}$ and $\in_{im}$ are the dielectric coefficients, which are assumed to be non-zero only for $i \neq m$.

The generalized Hooke's law for a 'special' orthotropic plate may be expressed as

$$\begin{Bmatrix} \sigma_{11} \\ \sigma_{22} \\ \sigma_{33} \\ \sigma_{23} \\ \sigma_{31} \\ \sigma_{12} \end{Bmatrix} = \begin{bmatrix} C_{11} & C_{12} & C_{13} & 0 & 0 & 0 \\ C_{12} & C_{22} & C_{23} & 0 & 0 & 0 \\ C_{13} & C_{23} & C_{33} & 0 & 0 & 0 \\ 0 & 0 & 0 & C_{44} & 0 & 0 \\ 0 & 0 & 0 & 0 & C_{55} & 0 \\ 0 & 0 & 0 & 0 & 0 & C_{66} \end{bmatrix} \begin{Bmatrix} \varepsilon_{11} \\ \varepsilon_{22} \\ \varepsilon_{33} \\ \varepsilon_{23} \\ \varepsilon_{31} \\ \varepsilon_{12} \end{Bmatrix} + \mathbf{E} \begin{Bmatrix} E_{11} \\ E_{22} \\ E_{33} \end{Bmatrix} \tag{7.1}$$

where σ_{ij} are the stresses, ε_{ij} are the strains, C_{ij} are the elastic constants relating the stresses to the strains, e_{ij} are the piezoelectric constants and E_{ii} are components of the applied electric field vector. For piezoelectric crystals with a hexagonal structure,

$$\mathbf{E} = \begin{bmatrix} 0 & 0 & e_{31} \\ 0 & 0 & e_{32} \\ 0 & 0 & e_{33} \\ 0 & e_{24} & 0 \\ e_{15} & 0 & 0 \\ 0 & 0 & 0 \end{bmatrix}, \tag{7.2}$$

which is also representative of a poled PZT patch actuator.

In the case of an anisotropic material, the elastic constants $C_{ij} \neq 0$, for all i and j. A transversely isotropic material is characterized by five independent constants:

$$C_{11}, \quad C_{22} = C_{33}, \quad C_{12} = C_{13}, \quad C_{55} = C_{66} \quad \text{and} \quad C_{44} = (C_{22} - C_{23})/2.$$

In the case where the material is isotropic,

$$C_{11} = C_{22} = C_{33} = \frac{E(1-\nu)}{(1+\nu)(1-2\nu)}, \qquad C_{12} = C_{13} = C_{23} = \frac{E\nu}{(1+\nu)(1-2\nu)}$$

and

$$C_{44} = C_{55} = C_{66} = \frac{C_{11} - C_{12}}{2} = \frac{E}{2(1+\nu)}. \tag{7.3}$$

Here E is the Young modulus and ν is the Poisson ratio.

When the normal stress σ_{33} is assumed to be **small** in comparison with other normal stresses and it is neglected. With this assumption, equation (7.1) (Whitney and Pagano, 1970) reduces to the following contracted form:

$$\begin{Bmatrix} \sigma_{11} \\ \sigma_{22} \\ \sigma_{12} \\ \sigma_{23} \\ \sigma_{13} \end{Bmatrix} = \begin{bmatrix} \bar{C}_{11} & \bar{C}_{12} & 0 & 0 & 0 \\ \bar{C}_{12} & \bar{C}_{22} & 0 & 0 & 0 \\ 0 & 0 & \bar{C}_{66} & 0 & 0 \\ 0 & 0 & 0 & \bar{C}_{44} & 0 \\ 0 & 0 & 0 & 0 & \bar{C}_{55} \end{bmatrix} \begin{Bmatrix} \varepsilon_{11} \\ \varepsilon_{22} \\ \varepsilon_{12} \\ \varepsilon_{23} \\ \varepsilon_{13} \end{Bmatrix} + \bar{\mathbf{E}} \begin{Bmatrix} E_{11} \\ E_{22} \\ E_{33} \end{Bmatrix}, \tag{7.4}$$

where

$$\bar{C}_{ij} = C_{ij} - \frac{C_{i3}C_{3j}}{C_{33}}, \qquad \bar{e}_{ji} = e_{ji} - \frac{C_{i3}e_{j3}}{C_{33}}, \qquad \bar{\mathbf{E}} = \begin{bmatrix} 0 & 0 & \bar{e}_{31} \\ 0 & 0 & \bar{e}_{32} \\ 0 & 0 & 0 \\ 0 & \bar{e}_{24} & 0 \\ \bar{e}_{15} & 0 & 0 \end{bmatrix}.$$

In the case of the plane stress approximation, which applies to a thin plate, all out-of-plane stresses are assumed to be zero, i.e.

$$\sigma_{13} = \sigma_{23} = \sigma_{33} = 0.$$

For a 'special' orthotropic plate, the elastic compliance matrix may be obtained by inverting the stiffness matrix:

$$\mathbf{S} = \mathbf{C}^{-1} = \begin{bmatrix} C_{11} & C_{12} & C_{13} & 0 & 0 & 0 \\ C_{12} & C_{22} & C_{23} & 0 & 0 & 0 \\ C_{13} & C_{23} & C_{33} & 0 & 0 & 0 \\ 0 & 0 & 0 & C_{44} & 0 & 0 \\ 0 & 0 & 0 & 0 & C_{55} & 0 \\ 0 & 0 & 0 & 0 & 0 & C_{66} \end{bmatrix}^{-1}. \tag{7.5}$$

The compliance matrix may also be expressed in terms of the engineering elastic constants, the Young's moduli, shear moduli and Poisson's ratios as

$$\mathbf{S} = \begin{bmatrix} \frac{1}{E_1} & -\frac{\nu_{21}}{E_2} & -\frac{\nu_{31}}{E_3} & 0 & 0 & 0 \\ -\frac{\nu_{12}}{E_1} & \frac{1}{E_2} & -\frac{\nu_{32}}{E_3} & 0 & 0 & 0 \\ -\frac{\nu_{13}}{E_1} & -\frac{\nu_{23}}{E_2} & \frac{1}{E_3} & 0 & 0 & 0 \\ 0 & 0 & 0 & \frac{1}{G_{23}} & 0 & 0 \\ 0 & 0 & 0 & 0 & \frac{1}{G_{13}} & 0 \\ 0 & 0 & 0 & 0 & 0 & \frac{1}{G_{12}} \end{bmatrix}. \tag{7.6}$$

The compliance matrix $\mathbf{S}$ is symmetric.

In the case of plane stress,

$$\begin{Bmatrix} \varepsilon_{11} \\ \varepsilon_{22} \\ \varepsilon_{12} \end{Bmatrix} = \begin{bmatrix} S_{11} & S_{12} & 0 \\ S_{12} & S_{22} & 0 \\ 0 & 0 & S_{33} \end{bmatrix} \begin{Bmatrix} \sigma_{11} \\ \sigma_{22} \\ \sigma_{12} \end{Bmatrix} + \begin{bmatrix} S_{11}e_{31} + S_{12}e_{32} + S_{13}e_{33} \\ S_{12}e_{31} + S_{22}e_{32} + S_{23}e_{33} \\ 0 \end{bmatrix} E_{33}, \tag{7.7}$$

where S_{ij} are the elements of the compliance matrix $\mathbf{S}$.

The inverse relationship is

$$\begin{Bmatrix} \sigma_{11} \\ \sigma_{22} \\ \sigma_{12} \end{Bmatrix} = \frac{1}{(1-\nu_{12}\nu_{21})} \begin{bmatrix} E_1 & \nu_{12}E_2 & 0 \\ \nu_{21}E_1 & E_2 & 0 \\ 0 & 0 & G_{12} \end{bmatrix} \begin{Bmatrix} \varepsilon_{11} \\ \varepsilon_{22} \\ \varepsilon_{12} \end{Bmatrix} + \begin{bmatrix} \bar{e}_{31}^p \\ \bar{e}_{32}^p \\ 0 \end{bmatrix} E_{33}, \tag{7.8}$$

where the electric field to stress distribution vector is

$$\begin{bmatrix} \bar{e}_{31}^p \\ \bar{e}_{32}^p \\ 0 \end{bmatrix} = \left(\begin{bmatrix} e_{31} \\ e_{32} \\ 0 \end{bmatrix} - \frac{1}{(1-\nu_{12}\nu_{21})} \begin{bmatrix} \nu_{31}E_1 + \nu_{32}\nu_{12}E_2 \\ \nu_{31}\nu_{21}E_1 + \nu_{32}E_2 \\ 0 \end{bmatrix} \frac{e_{33}}{E_3} \right). \tag{7.9}$$

7.1.3 Axis Transformations

In order to be able combine laminar stress and strain, it is vital that all the properties are with reference to the global frame. Thus, axis transformations play a key role in establishing the constitutive relationships for composite plates.

Consider the resolution of any stress component into the normal stress in a different direction. Since stress is the ratio of the components to two vectors, the resolved normal stress component σ_{nn} may be expressed as $\sigma_{nn} = n_i n_j \sigma_{ij}$ for all i and j, where $n_i = \cos\theta_{ni}$ is the direction cosine of the axis i with reference to the axis n. Similarly, the resolved shear stress component σ_{nm}, $n \neq m$, may be expressed as $\sigma_{nm} = \left(n_i m_j + m_i n_j\right)\sigma_{ij}$, for all i and j. Thus, the resolved stress component may be expressed in matrix form as

$$\begin{Bmatrix} \sigma_{ll} \\ \sigma_{mm} \\ \sigma_{nn} \\ \sigma_{mn} \\ \sigma_{nl} \\ \sigma_{lm} \end{Bmatrix} = \mathbf{T}_{\sigma\sigma} \begin{Bmatrix} \sigma_{11} \\ \sigma_{22} \\ \sigma_{33} \\ \sigma_{23} \\ \sigma_{31} \\ \sigma_{12} \end{Bmatrix}, \tag{7.10}$$

$$\mathbf{T}_{\sigma\sigma} = \begin{bmatrix} l_1^2 & m_1^2 & n_1^2 & 2m_1n_1 & 2n_1l_1 & 2l_1m_1 \\ l_2^2 & m_2^2 & n_2^2 & 2m_2n_2 & 2n_2l_2 & 2l_2m_2 \\ l_3^2 & m_3^2 & n_3^2 & 2m_3n_3 & 2n_3l_3 & 2l_3m_3 \\ l_2l_3 & m_2m_3 & n_2n_3 & m_2n_3+m_3n_2 & n_2l_3+n_3l_2 & l_2m_3+l_3m_2 \\ l_3l_1 & m_3m_1 & n_3n_1 & m_3n_1+m_1n_3 & n_3l_1+n_1l_3 & l_3m_1+l_1m_3 \\ l_1l_2 & m_1m_2 & n_1n_2 & m_1n_2+m_2n_1 & n_1l_2+n_2l_1 & l_1m_2+l_2m_1 \end{bmatrix}.$$

In the case of the strain components, because of the definition of the strains and the strain–displacement relationships, it follows that

$$\begin{Bmatrix} \varepsilon_{ll} \\ \varepsilon_{mm} \\ \varepsilon_{nn} \\ 0.5\varepsilon_{mn} \\ 0.5\varepsilon_{nl} \\ 0.5\varepsilon_{lm} \end{Bmatrix} = \mathbf{T}_{\sigma\sigma}^{-1} \begin{Bmatrix} \varepsilon_{11} \\ \varepsilon_{22} \\ \varepsilon_{33} \\ 0.5\varepsilon_{23} \\ 0.5\varepsilon_{31} \\ 0.5\varepsilon_{12} \end{Bmatrix}. \tag{7.11}$$

Thus, one may write

$$\begin{Bmatrix} \varepsilon_{ll} \\ \varepsilon_{mm} \\ \varepsilon_{nn} \\ \varepsilon_{mn} \\ \varepsilon_{nl} \\ \varepsilon_{lm} \end{Bmatrix} = \begin{bmatrix} \mathbf{I} & \mathbf{0} \\ \mathbf{0} & 2\mathbf{I} \end{bmatrix} \mathbf{T}_{\sigma\sigma}^{-1} \begin{bmatrix} \mathbf{I} & \mathbf{0} \\ \mathbf{0} & 0.5\mathbf{I} \end{bmatrix} \begin{Bmatrix} \varepsilon_{11} \\ \varepsilon_{22} \\ \varepsilon_{33} \\ \varepsilon_{23} \\ \varepsilon_{31} \\ \varepsilon_{12} \end{Bmatrix} = \mathbf{T}_{\varepsilon\varepsilon} \begin{Bmatrix} \varepsilon_{11} \\ \varepsilon_{22} \\ \varepsilon_{33} \\ \varepsilon_{23} \\ \varepsilon_{31} \\ \varepsilon_{12} \end{Bmatrix}. \tag{7.12}$$

Hence,

$$\mathbf{T}_{\varepsilon\varepsilon} = \begin{bmatrix} \mathbf{I} & \mathbf{0} \\ \mathbf{0} & 2\mathbf{I} \end{bmatrix} \mathbf{T}_{\sigma\sigma}^{-1} \begin{bmatrix} \mathbf{I} & \mathbf{0} \\ \mathbf{0} & 0.5\mathbf{I} \end{bmatrix}. \tag{7.13a}$$

In most practical situations, one is interested in a two-dimensional rotation about the third axis. In this case, if one denotes l_1 as l and m_1 as m, then $l_2 = -m$ and $m_2 = l$ and $\mathbf{T}_{\sigma\sigma}$ is given by

$$\mathbf{T}_{\sigma\sigma} = \begin{bmatrix} l^2 & m^2 & 0 & 0 & 0 & 2lm \\ m^2 & l^2 & 0 & 0 & 0 & -2lm \\ 0 & 0 & 1 & 0 & 0 & 0 \\ 0 & 0 & 0 & l & -m & 0 \\ 0 & 0 & 0 & m & l & 0 \\ -lm & ml & 0 & 0 & 0 & l^2 - m^2 \end{bmatrix}. \tag{7.13b}$$

The inverse of $\mathbf{T}_{\sigma\sigma}$ is found by reversing the direction of rotation about the third axis. Hence,

$$\mathbf{T}_{\sigma\sigma}^{-1} = \begin{bmatrix} l^2 & m^2 & 0 & 0 & 0 & -2lm \\ m^2 & l^2 & 0 & 0 & 0 & 2lm \\ 0 & 0 & 1 & 0 & 0 & 0 \\ 0 & 0 & 0 & l & m & 0 \\ 0 & 0 & 0 & -m & l & 0 \\ lm & -ml & 0 & 0 & 0 & l^2 - m^2 \end{bmatrix} \tag{7.13c}$$

and

$$\mathbf{T}_{\varepsilon\varepsilon} = \begin{bmatrix} l^2 & m^2 & 0 & 0 & 0 & -lm \\ m^2 & l^2 & 0 & 0 & 0 & lm \\ 0 & 0 & 1 & 0 & 0 & 0 \\ 0 & 0 & 0 & l & m & 0 \\ 0 & 0 & 0 & -m & l & 0 \\ 2lm & -2ml & 0 & 0 & 0 & l^2 - m^2 \end{bmatrix}. \tag{7.13d}$$

Hence, the transformed constitutive relationships are

$$\begin{Bmatrix} \sigma_{ll} \\ \sigma_{mm} \\ \sigma_{nn} \\ \sigma_{mn} \\ \sigma_{nl} \\ \sigma_{lm} \end{Bmatrix} = \mathbf{T}_{\sigma\sigma} \begin{bmatrix} C_{11} & C_{12} & C_{13} & 0 & 0 & 0 \\ C_{21} & C_{22} & C_{23} & 0 & 0 & 0 \\ C_{31} & C_{32} & C_{33} & 0 & 0 & 0 \\ 0 & 0 & 0 & C_{44} & 0 & 0 \\ 0 & 0 & 0 & 0 & C_{55} & 0 \\ 0 & 0 & 0 & 0 & 0 & C_{66} \end{bmatrix} \mathbf{T}_{\varepsilon\varepsilon}^{-1} \begin{Bmatrix} \varepsilon_{ll} \\ \varepsilon_{mm} \\ \varepsilon_{nn} \\ \varepsilon_{mn} \\ \varepsilon_{nl} \\ \varepsilon_{lm} \end{Bmatrix} + \mathbf{T}_{\sigma\sigma}\mathbf{E} \begin{Bmatrix} E_{11} \\ E_{22} \\ E_{33} \end{Bmatrix} \tag{7.14a}$$

with

$$\mathbf{T}_{\varepsilon\varepsilon}^{-1} = \begin{bmatrix} l^2 & m^2 & 0 & 0 & 0 & lm \\ m^2 & l^2 & 0 & 0 & 0 & -lm \\ 0 & 0 & 1 & 0 & 0 & 0 \\ 0 & 0 & 0 & l & -m & 0 \\ 0 & 0 & 0 & m & l & 0 \\ -2lm & 2ml & 0 & 0 & 0 & l^2 - m^2 \end{bmatrix}. \tag{7.14b}$$

7.1.4 Laminate Constitutive Relationships

In the CLT, the constitutive relationships for laminates are generally expressed in terms of the strains and curvatures at the middle plane. Thus, given the middle plane displacements in the three reference directions as $u_0(x, y)$, $v_0(x, y)$ and $w(x, y)$, the in-plane displacement components at any point away from the middle plane may be expressed as

$$u\,(x, y, z) = u_0(x, y) - z\theta_x, \quad v\,(x, y, z) = v_0(x, y) - z\theta_y, \tag{7.15}$$

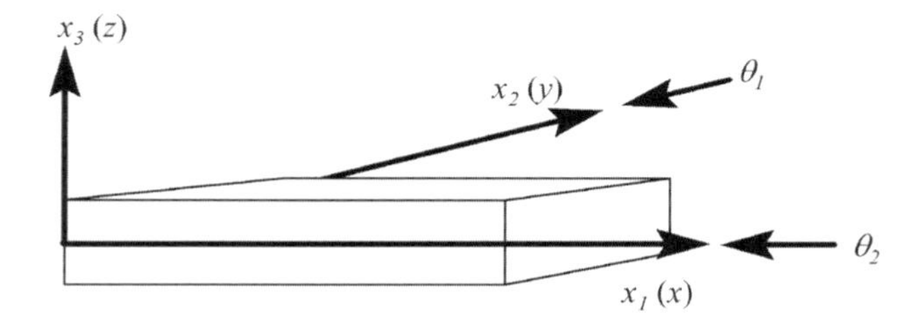

Figure 7.1 Reference axes and rotations

where z is the distance of the point from the middle plane, and θ_x and θ_y are the rotations of the normals to the x and y axes in the middle surface plane, as illustrated in Figure 7.1.

The strain–displacement relationship in the Cartesian coordinate system is obtained by substituting equations (7.14) into the linear strain–displacement relationship. The in-plane strains are given by

$$\begin{bmatrix} \varepsilon_{xx} \\ \varepsilon_{yy} \\ \varepsilon_{xy} \end{bmatrix} = \begin{bmatrix} \varepsilon_{xx}^0 \\ \varepsilon_{yy}^0 \\ \varepsilon_{xy}^0 \end{bmatrix} - z \begin{bmatrix} \kappa_{xx} \\ \kappa_{yy} \\ \kappa_{xy} \end{bmatrix}, \tag{7.16}$$

where the strains and curvatures at the middle surface are, respectively, given by

$$\begin{bmatrix} \varepsilon_{xx}^0 \\ \varepsilon_{yy}^0 \\ \varepsilon_{xy}^0 \end{bmatrix} = \begin{bmatrix} \partial u_0/\partial x \\ \partial v_0/\partial y \\ (\partial u_0/\partial y) + (\partial v_0/\partial x) \end{bmatrix} \tag{7.17a}$$

and

$$\begin{bmatrix} \kappa_{xx} \\ \kappa_{yy} \\ \kappa_{xy} \end{bmatrix} = \begin{bmatrix} \partial \theta_x/\partial x \\ \partial \theta_y/\partial y \\ (\partial \theta_x/\partial y) + (\partial \theta_y/\partial x) \end{bmatrix}. \tag{7.17b}$$

The in-plane strains are given by

$$\begin{bmatrix} \varepsilon_{yz} \\ \varepsilon_{xz} \end{bmatrix} = \begin{bmatrix} (\partial v/\partial z) + (\partial w/\partial y) \\ (\partial u/\partial z) + (\partial w/\partial x) \end{bmatrix} = \begin{bmatrix} (\partial w/\partial y) - \theta_y \\ (\partial w/\partial x) - \theta_x \end{bmatrix}. \tag{7.18b}$$

As a consequence of the Euler–Bernoulli hypothesis, θ_x and θ_y are approximated by

$$\theta_x = \frac{w(x, y)}{dx}, \quad \theta_y = \frac{w(x, y)}{dy}, \quad \begin{bmatrix} \varepsilon_{yz} \\ \varepsilon_{xz} \end{bmatrix} = \begin{bmatrix} 0 \\ 0 \end{bmatrix}, \tag{7.19b}$$

i.e. the transverse shear strains are zero.

The constitute relationships for the in-plane stresses in the ith layer and in the transformed coordinates are expressed as

$$\begin{Bmatrix} \sigma_{xx} \\ \sigma_{yy} \\ \sigma_{xy} \end{Bmatrix}^i = \begin{bmatrix} Q_{11}^i & Q_{12}^i & Q_{13}^i \\ Q_{21}^i & Q_{22}^i & Q_{23}^i \\ Q_{31}^i & Q_{32}^i & Q_{33}^i \end{bmatrix} \begin{Bmatrix} \varepsilon_{xx} \\ \varepsilon_{yy} \\ \varepsilon_{xy} \end{Bmatrix}^i + \begin{bmatrix} Q_1^{pi} \\ Q_2^{pi} \\ Q_3^{pi} \end{bmatrix} E_{33}. \tag{7.20}$$

For the out-of-plane stresses, although the strains are small, one has

$$\begin{bmatrix} \sigma_{yz} \\ \sigma_{xz} \end{bmatrix}^i = \begin{bmatrix} Q^i_{44} & Q^i_{45} \\ Q^i_{54} & Q^i_{55} \end{bmatrix} \begin{bmatrix} \varepsilon_{yz} \\ \varepsilon_{xz} \end{bmatrix}^i + \begin{bmatrix} Q^{in-pi}_{11} & Q^{in-pi}_{12} \\ Q^{in-pi}_{21} & Q^{in-pi}_{22} \end{bmatrix} \begin{bmatrix} E_{11} & 0 \\ 0 & E_{22} \end{bmatrix}. \tag{7.21}$$

Substituting for the strains in the above expressions for the in-plane stresses,

$$\begin{Bmatrix} \sigma_{xx} \\ \sigma_{yy} \\ \sigma_{xy} \end{Bmatrix}^i = \begin{bmatrix} Q^i_{11} & Q^i_{12} & Q^i_{13} \\ Q^i_{21} & Q^i_{22} & Q^i_{23} \\ Q^i_{31} & Q^i_{32} & Q^i_{33} \end{bmatrix} \left\{ \begin{bmatrix} \varepsilon^0_{xx} \\ \varepsilon^0_{yy} \\ \varepsilon^0_{xy} \end{bmatrix} - z \begin{bmatrix} \kappa_{xx} \\ \kappa_{yy} \\ \kappa_{xy} \end{bmatrix} \right\}^i + \begin{bmatrix} Q^{pi}_1 \\ Q^{pi}_2 \\ Q^{pi}_3 \end{bmatrix} E_{33}. \tag{7.22}$$

From the above expressions, it is seen that while the strains vary linearly through the laminate, the stresses do not. By integrating the stresses and their moments across the thickness of the laminate h, from the bottom to the top surface, one may define the in-plane force, shear force and moment resultants from the stress components as

$$\begin{Bmatrix} N_{xx} \\ N_{yy} \\ N_{xy} \\ Q_{yz} \\ Q_{xz} \end{Bmatrix} \equiv \int_{-\frac{h}{2}}^{\frac{h}{2}} \begin{Bmatrix} \sigma_{xx} \\ \sigma_{yy} \\ \sigma_{xy} \\ \sigma_{yz} \\ \sigma_{xz} \end{Bmatrix} \mathrm{d}z = \sum_{i=1}^{n} \int_{z_{i-1}}^{z_i} \begin{Bmatrix} \sigma_{xx} \\ \sigma_{yy} \\ \sigma_{xy} \\ \sigma_{yz} \\ \sigma_{xz} \end{Bmatrix}^i \mathrm{d}z, \tag{7.23}$$

$$\begin{Bmatrix} M_{xx} \\ M_{yy} \\ M_{xy} \end{Bmatrix} \equiv -\int_{-\frac{h}{2}}^{\frac{h}{2}} \begin{Bmatrix} \sigma_{xx} \\ \sigma_{yy} \\ \sigma_{xy} \end{Bmatrix} z\mathrm{d}z = -\sum_{i=1}^{n} \int_{z_{i-1}}^{z_i} \begin{Bmatrix} \sigma_{xx} \\ \sigma_{yy} \\ \sigma_{xy} \end{Bmatrix}^i z\mathrm{d}z. \tag{7.24}$$

Denote the integrals

$$\Delta z_i \equiv \int_{z_{i-1}}^{z_i} \mathrm{d}z, \quad \bar{z}_i \Delta z_i \equiv \int_{z_{i-1}}^{z_i} z\mathrm{d}z \quad \text{and} \quad \bar{\bar{z}}_i^2 \Delta z_i \equiv \int_{z_{i-1}}^{z_i} z^2 \mathrm{d}z. \tag{7.25}$$

It may be noted that, for a uniform isotropic plate, the rotations to the normals to the x and y axes in the middle surface plane θ_x and θ_y, respectively, may be expressed as

$$\begin{bmatrix} \theta_x \\ \theta_y \end{bmatrix} = -\frac{12}{h^3} \int_{-\frac{h}{2}}^{\frac{h}{2}} \begin{bmatrix} u \\ v \end{bmatrix} z\mathrm{d}z. \tag{7.26}$$

Recognizing that the middle plane strains and curvatures are the same for all layers and eliminating the stresses in the terms of the middle surface strains and curvatures, one obtains the expressions for the force and moment resultants as

$$\begin{aligned} \begin{Bmatrix} N_{xx} \\ N_{yy} \\ N_{xy} \end{Bmatrix} &= \left[\sum_{i=1}^{n} \begin{bmatrix} Q^i_{11} & Q^i_{12} & Q^i_{13} \\ Q^i_{21} & Q^i_{22} & Q^i_{23} \\ Q^i_{31} & Q^i_{32} & Q^i_{33} \end{bmatrix} \Delta z_i \right] \begin{bmatrix} \varepsilon^0_{xx} \\ \varepsilon^0_{yy} \\ \varepsilon^0_{xy} \end{bmatrix} \\ &+ \left[-\sum_{i=1}^{n} \begin{bmatrix} Q^i_{11} & Q^i_{12} & Q^i_{13} \\ Q^i_{21} & Q^i_{22} & Q^i_{23} \\ Q^i_{31} & Q^i_{32} & Q^i_{33} \end{bmatrix} \bar{z}_i \Delta z_i \right] \begin{bmatrix} \kappa_{xx} \\ \kappa_{yy} \\ \kappa_{xy} \end{bmatrix} + \left[\sum_{i=1}^{n} \begin{bmatrix} Q^{pi}_1 \\ Q^{pi}_2 \\ Q^{pi}_3 \end{bmatrix} \Delta z_i \right] E_{33}, \end{aligned} \tag{7.27}$$

$$\begin{Bmatrix} M_{xx} \\ M_{yy} \\ M_{xy} \end{Bmatrix} = \left[-\sum_{i=1}^{n} \begin{bmatrix} Q_{11}^{i} & Q_{12}^{i} & Q_{13}^{i} \\ Q_{21}^{i} & Q_{22}^{i} & Q_{23}^{i} \\ Q_{31}^{i} & Q_{32}^{i} & Q_{33}^{i} \end{bmatrix} \bar{z}_i \Delta z_i \right] \begin{bmatrix} \varepsilon_{xx}^{0} \\ \varepsilon_{yy}^{0} \\ \varepsilon_{xy}^{0} \end{bmatrix}$$

$$+ \left[\sum_{i=1}^{n} \begin{bmatrix} Q_{11}^{i} & Q_{12}^{i} & Q_{13}^{i} \\ Q_{21}^{i} & Q_{22}^{i} & Q_{23}^{i} \\ Q_{31}^{i} & Q_{32}^{i} & Q_{33}^{i} \end{bmatrix} \bar{z}_i^2 \Delta z_i \right] \begin{bmatrix} \kappa_{xx} \\ \kappa_{yy} \\ \kappa_{xy} \end{bmatrix} + \left[-\sum_{i=1}^{n} \begin{bmatrix} Q_{1}^{pi} \\ Q_{2}^{pi} \\ Q_{3}^{pi} \end{bmatrix} \bar{z}_i \Delta z_i \right] E_{33}. \tag{7.28}$$

The above are the coupled force–strain and moment–curvature relationships for a laminate. It is customary to write these in a single matrix given by

$$\begin{bmatrix} \mathbf{N} \\ \mathbf{M} \end{bmatrix} = \begin{bmatrix} \mathbf{A} & \mathbf{B} \\ \mathbf{C} & \mathbf{D} \end{bmatrix} \begin{bmatrix} \boldsymbol{\varepsilon} \\ \boldsymbol{\kappa} \end{bmatrix} + \begin{bmatrix} \mathbf{A}_e \\ \mathbf{B}_e \end{bmatrix} E_{33}, \tag{7.29}$$

where

$$\mathbf{N} = \begin{Bmatrix} N_{xx} \\ N_{yy} \\ N_{xy} \end{Bmatrix}, \quad \mathbf{M} = \begin{Bmatrix} M_{xx} \\ M_{yy} \\ M_{xy} \end{Bmatrix}, \quad \boldsymbol{\varepsilon} = \begin{bmatrix} \varepsilon_{xx}^{0} \\ \varepsilon_{yy}^{0} \\ \varepsilon_{xy}^{0} \end{bmatrix}, \quad \boldsymbol{\kappa} = \begin{bmatrix} \kappa_{xx} \\ \kappa_{yy} \\ \kappa_{xy} \end{bmatrix}. \tag{7.30}$$

These expressions are particularly useful in developing expressions for the force and moment resultants developed due to the presence of piezoelectric excitation due to a single-poled PZT patch actuator. To do this, it is convenient to split the above matrix relationship for the force and stress resultants and the strains, curvatures and excitation into two components: one representing the structure without the patch and the other with the PZT patch in place but without the structure.

The structure being non-piezoelectric, one may write for the structure alone

$$\begin{bmatrix} \mathbf{N} \\ \mathbf{M} \end{bmatrix}^{s} = \begin{bmatrix} \mathbf{A} & \mathbf{B} \\ \mathbf{C} & \mathbf{D} \end{bmatrix}^{s} \begin{bmatrix} \boldsymbol{\varepsilon} \\ \boldsymbol{\kappa} \end{bmatrix}. \tag{7.31}$$

For the patch, one has

$$\begin{bmatrix} \mathbf{N} \\ \mathbf{M} \end{bmatrix}^{p} = \begin{bmatrix} \mathbf{A} & \mathbf{B} \\ \mathbf{C} & \mathbf{D} \end{bmatrix}^{p} \begin{bmatrix} \boldsymbol{\varepsilon} \\ \boldsymbol{\kappa} \end{bmatrix} + \begin{bmatrix} \mathbf{A}_e \\ \mathbf{B}_e \end{bmatrix} E_{33}. \tag{7.32}$$

Since the structure is in equilibrium, it follows that

$$\begin{bmatrix} \mathbf{N} \\ \mathbf{M} \end{bmatrix} = \begin{bmatrix} \mathbf{N} \\ \mathbf{M} \end{bmatrix}^{s} + \begin{bmatrix} \mathbf{N} \\ \mathbf{M} \end{bmatrix}^{s} = \begin{bmatrix} \mathbf{0} \\ \mathbf{0} \end{bmatrix}. \tag{7.33}$$

Hence,

$$\begin{bmatrix} \mathbf{A} & \mathbf{B} \\ \mathbf{C} & \mathbf{D} \end{bmatrix}^{s} \begin{bmatrix} \boldsymbol{\varepsilon} \\ \boldsymbol{\kappa} \end{bmatrix} + \begin{bmatrix} \mathbf{A} & \mathbf{B} \\ \mathbf{C} & \mathbf{D} \end{bmatrix}^{p} \begin{bmatrix} \boldsymbol{\varepsilon} \\ \boldsymbol{\kappa} \end{bmatrix} + \begin{bmatrix} \mathbf{A}_e \\ \mathbf{B}_e \end{bmatrix} E_{33} = \begin{bmatrix} \mathbf{0} \\ \mathbf{0} \end{bmatrix} \tag{7.34}$$

and it follows that

$$\begin{bmatrix} \mathbf{A} & \mathbf{B} \\ \mathbf{C} & \mathbf{D} \end{bmatrix} \begin{bmatrix} \boldsymbol{\varepsilon} \\ \boldsymbol{\kappa} \end{bmatrix} = -\begin{bmatrix} \mathbf{A}_e \\ \mathbf{B}_e \end{bmatrix} E_{33}. \tag{7.35}$$

Solving for the strains and curvatures,

$$\begin{bmatrix} \boldsymbol{\varepsilon} \\ \boldsymbol{\kappa} \end{bmatrix} = -\begin{bmatrix} \mathbf{A} & \mathbf{B} \\ \mathbf{C} & \mathbf{D} \end{bmatrix}^{-1} \begin{bmatrix} \mathbf{A}_e \\ \mathbf{B}_e \end{bmatrix} E_{33}. \tag{7.36}$$

The force and stress resultants in the structure are then given as

$$\begin{bmatrix} \mathbf{N} \\ \mathbf{M} \end{bmatrix}^s = \begin{bmatrix} \mathbf{A} & \mathbf{B} \\ \mathbf{C} & \mathbf{D} \end{bmatrix}^s \begin{bmatrix} \varepsilon \\ \kappa \end{bmatrix} = -\begin{bmatrix} \mathbf{A} & \mathbf{B} \\ \mathbf{C} & \mathbf{D} \end{bmatrix}^s \begin{bmatrix} \mathbf{A} & \mathbf{B} \\ \mathbf{C} & \mathbf{D} \end{bmatrix}^{-1} \begin{bmatrix} \mathbf{A}_e \\ \mathbf{B}_e \end{bmatrix} E_{33}. \tag{7.37}$$

For independently excited by p patches, one has for the total strains and curvatures

$$\begin{bmatrix} \varepsilon \\ \kappa \end{bmatrix} = -\begin{bmatrix} \mathbf{A} & \mathbf{B} \\ \mathbf{C} & \mathbf{D} \end{bmatrix}^{-1} \sum_{j=1}^{p} \begin{bmatrix} \mathbf{A}_e \\ \mathbf{B}_e \end{bmatrix}^j E_{33}^j. \tag{7.38}$$

The force and stress resultants in the structure due to the independent excited p patches are then given as

$$\begin{bmatrix} \mathbf{N} \\ \mathbf{M} \end{bmatrix}^s = \begin{bmatrix} \mathbf{A} & \mathbf{B} \\ \mathbf{C} & \mathbf{D} \end{bmatrix}^s \begin{bmatrix} \varepsilon \\ \kappa \end{bmatrix} = -\begin{bmatrix} \mathbf{A} & \mathbf{B} \\ \mathbf{C} & \mathbf{D} \end{bmatrix}^s \begin{bmatrix} \mathbf{A} & \mathbf{B} \\ \mathbf{C} & \mathbf{D} \end{bmatrix}^{-1} \sum_{j=1}^{p} \begin{bmatrix} \mathbf{A}_e \\ \mathbf{B}_e \end{bmatrix}^j E_{33}^j. \tag{7.39}$$

7.1.5 Dynamics of Laminated Structures

The displacement–stress equations of motion may be expressed as

$$\frac{\partial \sigma_{xx}}{\partial x} + \frac{\partial \sigma_{yx}}{\partial y} + \frac{\partial \sigma_{zx}}{\partial z} = \rho \frac{\partial^2 u}{\partial t^2}, \tag{7.40a}$$

$$\frac{\partial \sigma_{xy}}{\partial x} + \frac{\partial \sigma_{yy}}{\partial y} + \frac{\partial \sigma_{zy}}{\partial z} = \rho \frac{\partial^2 v}{\partial t^2}, \tag{7.40b}$$

$$\frac{\partial \sigma_{xz}}{\partial x} + \frac{\partial \sigma_{yz}}{\partial y} + \frac{\partial \sigma_{zz}}{\partial z} = \rho \frac{\partial^2 w}{\partial t^2}. \tag{7.40c}$$

Assuming the laminate to be symmetric about the middle surface and integrating these equations over the total thickness of the laminate h_t,

$$\begin{aligned} &\frac{\partial N_{xx}}{\partial x} + \frac{\partial N_{xy}}{\partial y} + \Delta\sigma_{zx} = \bar{\rho} h_t \frac{\partial^2 u_0}{\partial t^2}, \quad \frac{\partial N_{xy}}{\partial x} + \frac{\partial N_{yy}}{\partial y} + \Delta\sigma_{zy} = \bar{\rho} h_t \frac{\partial^2 v_0}{\partial t^2} \\ &\frac{\partial Q_{xz}}{\partial x} + \frac{\partial Q_{yz}}{\partial y} + \Delta\sigma_{zz} = \bar{\rho} h_t \frac{\partial^2 w}{\partial t^2}, \quad \bar{\rho} = \int_{-h_t/2}^{h_t/2} \rho \mathrm{d}z \bigg/ h_t, \end{aligned} \tag{7.41c}$$

where $\Delta\sigma_{pq}$ is the difference in the stresses σ_{pq} between the upper (z positive) and lower surfaces of the laminate. Taking moments of the laminate about the middle surface and integrating the moments over the thickness of the laminate,

$$\frac{\partial M_{xx}}{\partial x} + \frac{\partial M_{xy}}{\partial y} + Q_{xz} - h_t \bar{\sigma}_{zx} = \bar{\rho} \frac{\bar{h}_t^3}{12} \frac{\partial^2 \theta_x}{\partial t^2}, \tag{7.42a}$$

$$\frac{\partial M_{xy}}{\partial x} + \frac{\partial M_{yy}}{\partial y} + Q_{zy} - h_t \bar{\sigma}_{zy} = \bar{\rho} \frac{\bar{h}_t^3}{12} \frac{\partial^2 \theta_y}{\partial t^2}, \tag{7.42b}$$

$$\bar{h}_t^3 = 12 \int_{-h_t/2}^{h_t/2} \rho z^2 \mathrm{d}z \bigg/ \bar{\rho}, \tag{7.42c}$$

where $\bar{\sigma}_{pq}$ is the mean shear stresses on the upper and lower surfaces of the laminate, and Q_{pq} are the shear stress resultants. Equations (7.40) represent the most general form of the equilibrium equations for the displacements and sectional rotations.

7.1.6 Equations of Motion of an Orthotropic Thin Plate

Consider a typical rectangular orthotropic laminated plate with a uniform thickness h and actuated by a single piezoelectric patch on either side of the plate each of thickness t_p. The laminate is assumed to be composed of perfectly bonded orthotropic layers. The orientation of the axes of orthotropy is assumed to be parallel to geometric orthotropic principle axes of the plate. A coordinate system is adopted such that the X–Y plane coincides with mid-plane of the plate and the Z axis is perpendicular to the plane. In the orthotropic case, the constitutive relations may be expressed in form presented earlier. It is assumed that the plate is not piezoelectric but symmetrically actuated by piezoelectric material or other similar patches, located on both the top and the bottom of the plate. In symmetric laminates, there is no coupling between the in-plane loading and out-of-plane curvature and twist, and between the bending and twisting moments and the in-plane deformations. It is also assumed that the applied transverse component of the electric field is in opposite directions in the upper and lower patches, while the in-plane components are in the same direction. This will ensure the moment resultants due to both patches, which are proportional to the distance between the middle surfaces of the two patches, are cumulative and do not cancel one another. Thus, the relationships are expressed as

$$\begin{Bmatrix} M_{xx} \\ M_{yy} \\ M_{xy} \end{Bmatrix} = \mathbf{D} \begin{Bmatrix} \kappa_{xx} \\ \kappa_{yy} \\ \kappa_{xy} \end{Bmatrix} - t_p\left(h + t_p\right)\Psi(x, y) \begin{Bmatrix} M_{xx}^P \\ M_{yy}^P \\ M_{xy}^P \end{Bmatrix} E_{33}, \tag{7.43a}$$

$$\begin{Bmatrix} Q_{zy} \\ Q_{zx} \end{Bmatrix} = \mathbf{E}_z \begin{Bmatrix} \varepsilon_{zy} \\ \varepsilon_{zx} \end{Bmatrix} + 2\kappa\Psi(x, y)t_p \begin{bmatrix} l & m \\ -m & l \end{bmatrix} \begin{bmatrix} 0 & \bar{e}_{24} \\ \bar{e}_{15} & 0 \end{bmatrix} \begin{bmatrix} E_{11} \\ E_{22} \end{bmatrix}, \tag{7.43b}$$

where

$$\mathbf{D} = \mathbf{D}^s + \begin{bmatrix} l^2 & m^2 & -lm \\ m^2 & l^2 & lm \\ 2lm & -2lm & l^2 - m^2 \end{bmatrix} \mathbf{D}^p \begin{bmatrix} l^2 & m^2 & 2lm \\ m^2 & l^2 & -2lm \\ -lm & lm & l^2 - m^2 \end{bmatrix},$$

$$\mathbf{D}^s = \begin{bmatrix} D_{11}^s & \nu_{21}^s D_{11}^s & 0 \\ \nu_{12}^s D_{22}^s & D_{22}^s & 0 \\ 0 & 0 & D_{33}^s(1 - \nu_{33}^s)/2 \end{bmatrix},$$

$$\mathbf{D}^p = \Psi(x, y) \begin{bmatrix} D_{11}^p & \nu_{21}^p D_{11}^p & 0 \\ \nu_{12}^p D_{22}^p & D_{22}^p & 0 \\ 0 & 0 & D_{33}^p(1 - \nu_{33}^p)/2 \end{bmatrix},$$

$$\begin{Bmatrix} M_{xx}^P \\ M_{yy}^P \\ M_{xy}^P \end{Bmatrix} = \begin{bmatrix} l^2 & m^2 & -lm \\ m^2 & l^2 & lm \\ 2lm & -2lm & l^2 - m^2 \end{bmatrix} \begin{bmatrix} \bar{e}_{31}^p \\ \bar{e}_{32}^p \\ 0 \end{bmatrix},$$

$$\mathbf{E}_z = \kappa \left\{ h \begin{bmatrix} \mu_y^s & 0 \\ 0 & \mu_x^s \end{bmatrix} + 2\Psi t_p \begin{bmatrix} l & m \\ -m & l \end{bmatrix} \begin{bmatrix} \mu_y^p & 0 \\ 0 & \mu_x^p \end{bmatrix} \begin{bmatrix} l & -m \\ m & l \end{bmatrix} \right\}.$$

The superscripts refer either to the structure (superscript s) or to the actuator patch (superscript p). $\Psi = \Psi(x, y)$ is a function that is equal to 1 where the patches are bonded to the plate and to 0 where the plate is patch free, ν_{ij} are generalizations of Poisson's ratio, κ is the Reissner–Mindlin shear force correction factor and μ is a Lamé parameter in the x or y directions.

The vectors $\left\{ M_{xx}^{P} \; M_{yy}^{P} \; M_{xy}^{P} \right\}^{T}$ are the moment resultants that appear due to the piezoelectric contributions to the stresses in the constitutive relations relating to the actuator patches. The parameter D_{33}^{x} is related to the shear modulus G_{12}^{x} and is estimated from E_{12}^{x}, where the superscript $x = s, p$. E_{12}^{x} and the parameters D_{ij}^{x} are then consistently defined by

$$\begin{aligned} &E_{11}^{x} \equiv E_{1}^{x}, \quad E_{22}^{x} \equiv E_{2}^{x}, \quad E_{12}^{x} = 2G_{12}^{x}\left(1 - \nu_{21}^{x}\nu_{12}^{x}\right) + \nu_{21}^{x}E_{1}^{x} \equiv E_{33}^{x}, \\ &D_{ij}^{x} = E_{ij}^{x}I^{x} \Big/ \left(1 - \nu_{21}^{x}\nu_{12}^{x}\right), \end{aligned} \tag{7.44}$$

where the superscript $x = s, p$, and

$$I^{s} = h^{3}/12, \quad I^{p} = \left(\left(h + 2t_{p}\right)^{3} - h^{3}\right)\Big/12. \tag{7.45}$$

The parameters ν_{21}^{x} and ν_{12}^{x} are the usual generalizations of Poisson's ratio, which satisfy the relationship $\nu_{12}^{x} = \nu_{21}^{x}E_{11}^{x}/E_{22}^{x}$, while the parameter ν_{33}^{x} is defined by $\nu_{33}^{x} = \nu_{21}^{x}E_{11}^{x}/E_{22}^{x}$.

The special case of the *Kirchhoff* theory is considered first. The principal directions of the poled PZT patch actuators are assumed to be aligned with principal directions of the plate. Hence, $l = 1$ and $m = 1$. In this case, the curvature–displacement relationships are

$$\begin{bmatrix} \kappa_{xx} & \kappa_{yy} & \kappa_{xy} \end{bmatrix} = \begin{bmatrix} \dfrac{\partial^2}{\partial x^2} & \dfrac{\partial^2}{\partial y^2} & 2\dfrac{\partial^2}{\partial x \partial y} \end{bmatrix} w(x, y). \tag{7.46}$$

Substituting in the moment–curvature relationship and ignoring the external moment contributions,

$$\begin{Bmatrix} M_{xx} \\ M_{yy} \\ M_{xy} \end{Bmatrix} = \left(\mathbf{D}^{s} + \mathbf{D}^{p}\right) \begin{Bmatrix} \partial^2/\partial x^2 \\ \partial^2/\partial y^2 \\ 2\partial^2/\partial x \partial y \end{Bmatrix} w - t_{p}\left(h + t_{p}\right)\Psi \begin{bmatrix} \bar{e}_{31}^{p} \\ \bar{e}_{32}^{p} \\ 0 \end{bmatrix} E_{33}. \tag{7.47}$$

The orthotropic plate is assumed to be of infinite extent and to be subjected to an external reactive pressure $q(x, y)$. In the absence of rotary inertia and the out-of-plane shear stresses, the moment equilibrium equations reduce to

$$\frac{\partial^2 M_{xx}}{\partial x^2} + 2\frac{\partial^2 M_{xy}}{\partial x \partial y} + \frac{\partial^2 M_{yy}}{\partial y^2} = q(x, y) + \bar{\rho}h_{t}\omega^2 w(x, y), \tag{7.48}$$

where $h_{t} = h + 2t_{p}\Psi(x, y)$.

Consider an orthotropic thin plate of infinite extent and assume that the plate is subjected to an external reactive pressure $q(x, y)$. The *Kirchhoff* plate equation for this case is

$$\left(a\frac{\partial^4}{\partial x^4} + 2b\frac{\partial^2}{\partial x^2}\frac{\partial^2}{\partial y^2} + c\frac{\partial^4}{\partial y^4} - \bar{\rho}h_{t}\omega^2\right) w(x, y) = q_{e}(x, y), \tag{7.49}$$

where the coefficients in the equation are defined as

$$\begin{aligned} a &= D_{11}^{s} + \Psi(x, y)D_{11}^{p}, \\ c &= D_{22}^{s} + \Psi(x, y)D_{22}^{p}, \\ b &= b^{s} + b^{p}, \quad b^{s} = D_{33}^{s} + 0.5\left(\nu_{21}^{s}D_{11}^{s} + \nu_{12}^{s}D_{22}^{s}\right) - \nu_{33}^{s}D_{33}^{s}, \\ b^{p} &= D_{33}^{p} + 0.5\left(\nu_{21}^{p}D_{11}^{p} + \nu_{12}^{p}D_{22}^{s}\right) - \nu_{33}^{p}D_{33}^{p}, \\ q_{e}(x, y) &= q(x, y) + t_{p}\left(h + t_{p}\right)\left(\frac{\partial^2 \bar{e}_{31}^{p}\Psi(x, y)}{\partial x^2} + \frac{\partial^2 \bar{e}_{32}^{p}\Psi(x, y)}{\partial y^2}\right) E_{33}, \end{aligned}$$

with the appropriate radiation boundary condition at infinity. When the piezoelectric constants $\bar{e}^p_{31}$ and $\bar{e}^p_{32}$ are constants,

$$q_e(x, y) = q(x, y) + t_p \left(h + 2t_p\right) \left(\bar{e}^p_{31} \frac{\partial^2 \Psi(x, y)}{\partial x^2} + \bar{e}^p_{32} \frac{\partial^2 \Psi(x, y)}{\partial y^2}\right) E_{33}.$$

Equation (7.49) may be written as

$$\left(b\nabla^4 + (a-b)\frac{\partial^4}{\partial x^4} + (c-b)\frac{\partial^4}{\partial y^4} - \bar{\rho} h_t \omega^2\right) w(x, y)$$
$$= q(x, y) + t_p \left(h + t_p\right) \left(\bar{e}^p_{31} \frac{\partial^2 \Psi(x, y)}{\partial x^2} + \bar{e}^p_{32} \frac{\partial^2 \Psi(x, y)}{\partial y^2}\right) E_{33}. \quad (7.50)$$

It may be noted that in principle for a uniform plate, with no excitation, a, b, c and h are constants and the solution may be expressed by employing the Fourier transform as

$$w(x, y) = \frac{1}{4\pi 2} \int_{-\infty} \infty \int_{-\infty} \infty W(\alpha, \beta)\, \mathrm{d}\alpha \mathrm{d}\beta,$$

where

$$W(\alpha, \beta) = \frac{\int_{-\infty}^{\infty}\int_{-\infty}^{\infty} q_e(\xi, \eta) \exp\left(-i\left(\alpha(x-\xi) + \beta(y-\eta)\right)\right) \mathrm{d}\xi \mathrm{d}\eta}{\Delta(\alpha, \beta)} \Delta(\alpha, \beta)$$
$$= a\alpha^4 + 2b\alpha^2\beta^2 + c\beta^4 - \rho h \omega^2. \quad (7.51)$$

However, in the above equation the coefficients a, b, c and h are functions of the spatial coordinates on the surface of the plate.

The governing equation for vibration of an isotropic thin plate excited by an in-plane isotropic $\left(\bar{e}^p_{31} = \bar{e}^p_{32}\right)$ PZT patch actuator is obtained by setting $a = b = c = D^s + \Psi(x, y)D^p$ and is written in terms of the bi-Laplace and Laplace operators as follows:

$$((D^s + \Psi(x, y)D^p)\nabla^4 - \bar{\rho}(h + 2t_p\Psi(x, y))\omega^2)w(x, y) = q(x, y) + t_p(h + t_p)\bar{e}^p_{31}\nabla^2\Psi(x, y)E_{33}. \quad (7.52)$$

Equations (7.49) and (7.52) are partial differential equations defined over the domain of the middle surface of the plate and the time domain. From equation (7.52), we may conclude that when a plate is covered with a piezoelectric patch covering a certain area over which $\Psi(x, y)$ is non-zero, the equivalent piezoelectric load consists of a distributed load proportional to $\nabla^2\Psi(x, y)$. In general, when using these relationships one generally integrates twice by parts to avoid differentiating the function $\Psi(x, y)$. Similarly, we may show that when an isotropic piezoelectric patch is used as a sensor in a bimorph configuration, the charge generated is given by

$$Q = e_{31} t_p \left(h + t_p\right) E_{33} \int_S \Omega(x, y)\nabla^2 w(x, y) \mathrm{d}x \mathrm{d}y. \quad (7.53)$$

These relationships will be employed in Chapter 9, where a feedback controller is synthesized for optimal control of a laminated structure using piezoelectric actuators and piezoelectric sensors.

Although there are basically three methods for checking the stability of a linear partial differential equation, they are generally quite restrictive and difficult to apply. One can calculate its spectrum, construct a Lyapunov function or show that for all s (Laplace transform variable) with real part larger than zero the resolvent of the associated infinitesimal generator is uniformly bounded. These methods could be extended for the purpose of synthesizing stabilizing controllers. Yet the synthesis of a control system is relatively easier if the partial differential equation is reduced to a set of ordinary differential equations for representative degrees of freedom evolving in the time domain. Typically, such methods involve modal superposition and residual minimization, leading to such well-known techniques as Galerkin's method and the method of collocation. Such methods will be discussed in a latter section.

7.1.7 First-order Shear Deformation Theory

The first-order shear deformation theory (FSDT) is the result of dropping the Euler–Bernoulli hypothesis. In this case, θ_x and θ_y are not approximated by the slopes of the transverse displacement and

$$\theta_x \neq \frac{w(x,y)}{dx}, \quad \theta_y \neq \frac{w(x,y)}{dy}, \quad \begin{bmatrix} \varepsilon_{yz} \\ \varepsilon_{xz} \end{bmatrix} \neq \begin{bmatrix} 0 \\ 0 \end{bmatrix}; \tag{7.54}$$

i.e. the transverse shear strains are not ignored. Inserting the constitutive relationships into the equilibrium equations, the governing equations for the Reissner–Mindlin isotropic plate (Reissner, 1945; Mindlin, 1951) to first order can be expressed in terms of the transverse displacement w and the section bending rotations θ_x and θ_y as

$$\begin{aligned}
&\left(D^s\frac{(1-\nu^s)}{2} + \Psi D^p\frac{(1-\nu^p)}{2}\right)\nabla^2\theta_x + \left(D^s\frac{(1+\nu^s)}{2} + \Psi D^p\frac{(1+\nu^p)}{2}\right)\left(\theta_{x,xx} + \theta_{y,yx}\right)\\
&\quad + \kappa\left(h\mu^s + 2t_p\Psi(x,y)\mu^p\right)(w_{,x} - \theta_x) = \bar{\rho}\frac{\bar{h}_t^3}{12}\frac{\partial^2\theta_x}{\partial t^2}\\
&\quad + t_p\left(h + 2t_p\right)\bar{e}_{31}^p E_{33}\frac{\partial}{\partial x}\Psi(x,y) - 2\kappa\Psi(x,y)t_p\bar{e}_{15}^p E_{11},
\end{aligned} \tag{7.55a}$$

$$\begin{aligned}
&\left(D^s\frac{(1-\nu^s)}{2} + \Psi D^p\frac{(1-\nu^p)}{2}\right)\nabla^2\theta_y + \left(D^s\frac{(1+\nu^s)}{2} + \Psi D^p\frac{(1+\nu^p)}{2}\right)\left(\theta_{x,xy} + \theta_{y,yy}\right)\\
&\quad + \kappa\left(h\mu^s + 2t_p\Psi(x,y)\mu^p\right)(w_{,y} - \theta_y) = \bar{\rho}\frac{\bar{h}_t^3}{12}\frac{\partial^2\theta_y}{\partial t^2}\\
&\quad + t_p\left(h + 2t_p\right)\bar{e}_{32}^p E_{33}\frac{\partial}{\partial x}\Psi(x,y) - 2\kappa\Psi(x,y)t_p\bar{e}_{24}^p E_{22},
\end{aligned} \tag{7.55b}$$

$$\begin{aligned}
&\kappa\left(h\mu^s + 2t_p\Psi(x,y)\mu^p\right)\left(w_{,xx} + w_{,yy} - \theta_{x,x} - \theta_{y,y}\right) + q\\
&\quad = \bar{\rho}h_t\frac{\partial^2 w}{\partial t^2} - 2\kappa t_p\left(\bar{e}_{15}^p E_{11}\frac{\partial}{\partial x} + \bar{e}_{24}^p E_{22}\frac{\partial}{\partial x}\right)\Psi(x,y).
\end{aligned} \tag{7.55c}$$

The applied moment and shear force resultants are functions of the distance of the actuator patch from the plate middle surface, the shape function of the active area of the actuator patch and certain actuator constants that could be derived from the constitutive relationships. These have been included, so we could derive the general equations governing an orthotropic plate externally driven by embedded piezoelectric actuator patches. The governing displacement and rotation equations are

$$\begin{aligned}
&\left(D_{33}^s\frac{(1-\nu_{33}^s)}{2} + \Psi D_{33}^p\frac{(1-\nu_{33}^p)}{2}\right)\nabla^2\theta_x\\
&\left(D_{11}^s - D_{33}^s\frac{(1-\nu_{33}^s)}{2} + \Psi\left(D_{11}^p - D_{33}^p\frac{(1-\nu_{33}^p)}{2}\right)\right)\theta_{x,xx}\\
&\quad + \left(\nu_{21}^s D_{11}^s + D_{33}^s\frac{(1-\nu_{33}^s)}{2} + \Psi\left(\nu_{21}^p D_{11}^p + D_{33}^p\frac{(1-\nu_{33}^p)}{2}\right)\right)\theta_{y,yx}\\
&\quad + \kappa\left(h\mu_1^s + 2t_p\Psi(x,y)\mu_1^p\right)(w_{,x} - \theta_x) = \bar{\rho}\frac{\bar{h}_t^3}{12}\frac{\partial^2\theta_x}{\partial t^2}\\
&\quad + t_p\left(h + 2t_p\right)\bar{e}_{31}^p E_{33}\frac{\partial}{\partial x}\Psi(x,y) - 2\kappa\Psi(x,y)t_p\bar{e}_{15}^p E_{11},
\end{aligned} \tag{7.56a}$$

$$\left(D_{33}^{s}\frac{(1-\nu_{33}^{s})}{2}+\Psi D_{33}^{p}\frac{(1-\nu_{33}^{p})}{2}\right)\nabla^{2}\theta_{y}$$
$$+\left(D_{22}^{s}-D_{33}^{s}\frac{(1-\nu_{33}^{s})}{2}+\Psi\left(D_{22}^{p}-D_{33}^{p}\frac{(1-\nu_{33}^{p})}{2}\right)\right)\theta_{y,yy}$$
$$+\left(\nu_{12}^{s}D_{22}^{s}+D_{33}^{s}\frac{(1-\nu_{33}^{s})}{2}+\Psi\left(\nu_{12}^{p}D_{22}^{p}+D_{33}^{p}\frac{(1-\nu_{33}^{p})}{2}\right)\right)\theta_{x,xy}$$
$$+\kappa\left(h\mu_{2}^{s}+2t_{p}\Psi(x,y)\mu_{2}^{p}\right)\left(w_{,y}-\theta_{y}\right)=\bar{\rho}\frac{\bar{h}_{t}^{3}}{12}\frac{\partial^{2}\theta_{y}}{\partial t^{2}}$$
$$+t_{p}\left(h+2t_{p}\right)\bar{e}_{32}^{p}E_{33}\frac{\partial}{\partial x}\Psi(x,y)-2\kappa\Psi(x,y)t_{p}\bar{e}_{24}^{p}E_{22}, \tag{7.56b}$$
$$\kappa\left(h\mu_{1}^{s}+2t_{p}\Psi(x,y)\mu_{1}^{p}\right)\left(w_{,xx}-\theta_{x,x}\right)+\kappa\left(h\mu_{2}^{s}+2t_{p}\Psi(x,y)\mu_{2}^{p}\right)\left(w_{,yy}-\theta_{y,y}\right)+q$$
$$=\bar{\rho}h_{t}\frac{\partial^{2}w}{\partial t^{2}}-2\kappa t_{p}\left(\bar{e}_{15}^{p}E_{11}\frac{\partial}{\partial x}+\bar{e}_{24}^{p}E_{22}\frac{\partial}{\partial x}\right)\Psi(x,y). \tag{7.56c}$$

These are a set of three coupled partial differential equations for the out-of-plane displacement and in-plane rotations. They must be reduced to a set of ordinary differential equations for representative degrees of freedom evolving in the time domain.

7.1.8 *Composite Laminated Plates: First-order Zig-zag Theory*

In the theories involving the modelling of the displacements of thin and moderately thick plates discussed so far, the displacement distribution in thickness direction was approximated by continuously differentiable linear functions. Theories based on such functions are named first-order C_z^1 function theories. The standard models of this type are the CLT and the FSDT presented in earlier sections. Both these theories can be extended to laminated composites, although obtaining representative shear correction factors for laminates is an involved process. To avoid this process, several higher-order theories have been proposed.

Many higher-order shear deformation theories (HSDT) have been proposed to partially remedy these deficiencies and short comings in the CLT. Higher-order theories have also been proposed to model piezoelectric plate structures (Batra and Vidoli, 2002). Primarily, these higher-order plate theories include additional terms in Taylor series expansion for the displacements in the three Cartesian directions in terms of generalized middle plane displacements and rotations. In the case of the simplest of these higher-order theories, this approach results in at least a cubic variation of in plane strains (εx, εy, γxy), quadratic variation of transverse shear strains (γxz, γyz) and linear variation of transverse normal strain (εz) through the thickness of the plate. The HSDT are able to model both in-plane and out-of-plane modes of deformation.

It is also important to note that in the modelling and analysis discussed here, the role of non-linearities has been ignored. Non-linearities will have to be considered, particularly when the post bucking response is of importance (Stein et al., 1990). Non-linearities are also important when considering interaction phenomenon such as flutter (Shiau and Wu, 2001). In these applications, geometrical non-linearities are significant and must be represented in the analysis or synthesis processes. Examples of plate structures with moderate rotations are discussed in the exercises.

Three of the higher-order theories are quite novel and deserve special mention. The first, due to Murthy (1981), is based introducing two second-order rotations based on a least squares approximation, which influence both the shear deformations and the rotations. The second, due to Reddy (1984) and Phan and Reddy (1985), is based on the condition of vanishing transverse shear strains at the upper and lower surface, which is used to replace higher-order effects without introducing any additional degrees of freedom. Several investigators in the early 1970s noted that for slender composite laminates, the in-plane displacements show a pronounced layer-wise zig-zagging distribution due to an abrupt change of material properties across the layer interfaces. This led to the development of the zig-zag plate theories based on C_z^0 approximations.

Introduced by Murakami (1986), the zig-zag plate theories employ a zig-zag distribution of the in-plane displacements along the thickness coordinate with compatibility enforced at the interfaces. The layer-wise first-order zig-zag approximations due to Carrera (2003, 2004) and Demasi (2007) give a better prediction of the transverse stresses and the in-plane displacements. The zig-zag theories are particularly suitable for use with laminates, where one or more of the lamina is engineered with functionally graded properties. In such circumstances, the individual physical lamina may have to be modelled as a multi-layered structure using the zig-zag representations of not just the stresses and trains but also of the material properties of each sub-layer within the physical lamina, i.e. the elastic or piezoelectric parameters are themselves represented by zig-zag functions and not assumed to be constants across each sub-layer. However, the complexity of the analysis, and in particular the number of independent variables, increases rapidly even in the case of simple multi-layered structures, and resorting to a numerical approach is unavoidable. Here the application of a first-order zig-zag plate theory to an electro-actuated isotropic plate is considered to illustrate the process of the derivation of the equations for dynamic equilibrium.

A zig-zag in-plane displacement variation across the thickness of the plate, whose amplitude is expressed by $\psi_i(x, y)$, is included in addition to the standard linear variation. Thus, given the middle plane displacements in the three reference directions as $u_0(x, y)$, $v_0(x, y)$ and $w(x, y)$, the in-plane displacement components at any point away from the middle plane may be expressed as

$$u(x, y, z) = u_0(x, y) - z\theta_x - \psi_{1,x} S_k^1, \tag{7.57a}$$

$$v(x, y, z) = v_0(x, y) - z\theta_y - \psi_{1,y} S_k^1, \tag{7.57b}$$

where z is the distance of the point from the middle plane, θ_x and θ_y are the rotations of the normals to the x and y axes in the middle surface plane, and the layer-wise zig-zag function S_k^1 is defined in each of the $(2m + 1)$ layers as

$$S_k^i = 2(-1)^k (z - z_{0k})^i / h_k, \quad k = -m, -(m-1), \ldots, -1, 0, 1, \ldots, (m-1), m, \tag{7.58}$$

z_{0k} is the z coordinate of the middle surface of the kth layer and h_k is the corresponding thickness of the kth layer. The zig-zag function is piecewise linear with values of (-1) and 1 alternately at the different interfaces. The zig-zag function accounts for the slope discontinuity of in-plane displacements at the interfaces of laminate without substantially increasing the number of degrees of freedom.

The layer-wise in-plane strains may then be expressed as

$$\begin{bmatrix} \varepsilon_{xx} \\ \varepsilon_{yy} \\ \varepsilon_{xy} \end{bmatrix} = \begin{bmatrix} \varepsilon_{xx}^0 \\ \varepsilon_{yy}^0 \\ \varepsilon_{xy}^0 \end{bmatrix} - z \begin{bmatrix} \kappa_{xx} \\ \kappa_{yy} \\ \kappa_{xy} \end{bmatrix} - S_k^1 \begin{bmatrix} \kappa_{xx}^1 \\ \kappa_{yy}^1 \\ \kappa_{xy}^1 \end{bmatrix}, \tag{7.59}$$

where the strains, curvatures and curvature distortions at the middle surface are, respectively, given by

$$\begin{bmatrix} \varepsilon_{xx}^0 \\ \varepsilon_{yy}^0 \\ \varepsilon_{xy}^0 \end{bmatrix} = \begin{bmatrix} \partial u_0/\partial x \\ \partial v_0/\partial y \\ (\partial u_0/\partial y) + (\partial v_0/\partial x) \end{bmatrix}, \quad \begin{bmatrix} \kappa_{xx} \\ \kappa_{yy} \\ \kappa_{xy} \end{bmatrix} = \begin{bmatrix} \partial \theta_x/\partial x \\ \partial \theta_y/\partial y \\ (\partial \theta_x/\partial y) + (\partial \theta_y/\partial x) \end{bmatrix},$$
$$\begin{bmatrix} \kappa_{xx}^1 \\ \kappa_{yy}^1 \\ \kappa_{xy}^1 \end{bmatrix} = \begin{bmatrix} \partial \psi_{1,x}/\partial x \\ \partial \psi_{1,y}/\partial y \\ (\partial \psi_{1,x}/\partial y) + (\partial \psi_{1,y}/\partial x) \end{bmatrix}. \tag{7.60}$$

The layer-wise in-plane strains are given by

$$\begin{bmatrix} \varepsilon_{yz} \\ \varepsilon_{xz} \end{bmatrix} = \begin{bmatrix} (\partial v/\partial z) + (\partial w/\partial y) \\ (\partial u/\partial z) + (\partial w/\partial x) \end{bmatrix} = \begin{bmatrix} (\partial w/\partial y) - \theta_y - 2(-1)^k \psi_{1,y}/h_k \\ (\partial w/\partial x) - \theta_x - 2(-1)^k \psi_{1,x}/h_k \end{bmatrix}. \tag{7.61}$$

The constitutive relationship for the in-plane stresses in the ith layer and in the transformed coordinates is expressed as

$$\begin{Bmatrix} \sigma_{xx} \\ \sigma_{yy} \\ \sigma_{xy} \end{Bmatrix}^i = \begin{bmatrix} Q_{11}^i & Q_{12}^i & Q_{13}^i \\ Q_{21}^i & Q_{22}^i & Q_{23}^i \\ Q_{31}^i & Q_{32}^i & Q_{33}^i \end{bmatrix} \left\{ \begin{bmatrix} \varepsilon_{xx}^0 \\ \varepsilon_{yy}^0 \\ \varepsilon_{xy}^0 \end{bmatrix} - z \begin{bmatrix} \kappa_{xx} \\ \kappa_{yy} \\ \kappa_{xy} \end{bmatrix} - S_i^1 \begin{bmatrix} \kappa_{xx}^1 \\ \kappa_{yy}^1 \\ \kappa_{xy}^1 \end{bmatrix} \right\}^i + \begin{bmatrix} Q_1^{pi} \\ Q_2^{pi} \\ Q_3^{pi} \end{bmatrix} E_{33} \tag{7.62}$$

and for the out-of-plane stresses one has

$$\begin{bmatrix} \sigma_{yz} \\ \sigma_{xz} \end{bmatrix}^i = \begin{bmatrix} Q_{44}^i & Q_{45}^i \\ Q_{54}^i & Q_{55}^i \end{bmatrix} \begin{bmatrix} (\partial w/\partial y) - \theta_y - 2(-1)^i \psi_{1,y}/h_i \\ (\partial w/\partial x) - \theta_x - 2(-1)^i \psi_{1,x}/h_i \end{bmatrix} + \begin{bmatrix} Q_{11}^{in-pi} & Q_{12}^{in-pi} \\ Q_{21}^{in-pi} & Q_{22}^{in-pi} \end{bmatrix} \begin{bmatrix} E_{11} & 0 \\ 0 & E_{22} \end{bmatrix}. \tag{7.63}$$

For simplicity, the in-plane electric field components will be assumed to be zero, i.e.

$$\begin{bmatrix} E_{11} & E_{22} \end{bmatrix} = \begin{bmatrix} 0 & 0 \end{bmatrix}. \tag{7.64}$$

By integrating the stresses and their moments across the thickness of the laminate h, from the bottom to the top surface, one may define the in-plane force, shear force, moment and generalized moment resultants from the stress components as

$$\begin{bmatrix} N_{xx} & M_{xx} & L_{xx} \\ N_{yy} & M_{yy} & L_{yy} \\ N_{xy} & M_{xy} & L_{xy} \end{bmatrix} \equiv \sum_{i=1}^{n} \int_{z_{i-1}}^{z_i} \begin{Bmatrix} \sigma_{xx} \\ \sigma_{yy} \\ \sigma_{xy} \end{Bmatrix}^i \begin{bmatrix} 1 & -z & -S_i^1 \end{bmatrix} \mathrm{d}z, \tag{7.65a}$$

$$\begin{bmatrix} Q_{yz} & R_{yz} \\ Q_{xz} & R_{xz} \end{bmatrix} \equiv \sum_{i=1}^{n} \left(\int_{z_{i-1}}^{z_i} \begin{Bmatrix} \sigma_{yz} \\ \sigma_{xz} \end{Bmatrix}^i \mathrm{d}z \right) \begin{bmatrix} 1 & 2(-1)^i/h_i \end{bmatrix}. \tag{7.65b}$$

The laminate constitutive relationships may be expressed in terms of the integrals

$$h_i \equiv \int_{z_{i-1}}^{z_i} \mathrm{d}z, \quad \bar{z}_{0i} h_i \equiv \int_{z_{i-1}}^{z_i} z\mathrm{d}z, \quad \bar{\bar{z}}_i^2 h_i \equiv \int_{z_{i-1}}^{z_i} z^2 \mathrm{d}z = \frac{h_i}{3}\left(z_i^2 + 2z_i z_{i-1} + z_{i-1}^2\right),$$

$$\bar{S}_{0i} h_i \equiv \int_{z_{i-1}}^{z_i} S_i^1 \mathrm{d}z = \frac{2(-1)^i}{h_i} \int_{z_{i-1}}^{z_i} (z - z_{0i})\,\mathrm{d}z = \frac{2(-1)^i}{h_i} \int_{-h_i/2}^{h_i/2} z\mathrm{d}z = 0,$$

$$\bar{S}_{1i} h_i \equiv \int_{z_{i-1}}^{z_i} S_i^1 z\mathrm{d}z = \frac{2(-1)^i}{h_i} \int_{-h_i/2}^{h_i/2} z^2 \mathrm{d}z = (-1)^i \frac{h_i^2}{6}$$

and

$$\bar{\bar{S}}_i h_i \equiv \int_{z_{i-1}}^{z_i} \left(S_i^1\right)^2 \mathrm{d}z = \frac{4}{h_i^2} \int_{z_{i-1}}^{z_i} (z - z_{0i})^2\,\mathrm{d}z = \frac{h_i}{3}. \tag{7.66b}$$

$$\begin{bmatrix} N_{xx} & M_{xx} & L_{xx} \\ N_{yy} & M_{yy} & L_{yy} \\ N_{xy} & M_{xy} & L_{xy} \end{bmatrix} = \left[\sum_{i=-m}^{m} \begin{bmatrix} Q_{11}^i & Q_{12}^i & Q_{13}^i \\ Q_{21}^i & Q_{22}^i & Q_{23}^i \\ Q_{31}^i & Q_{32}^i & Q_{33}^i \end{bmatrix} \begin{bmatrix} \varepsilon_{xx}^0 \\ \varepsilon_{yy}^0 \\ \varepsilon_{xy}^0 \end{bmatrix} \begin{bmatrix} 1 & -z_{0i} & 0 \end{bmatrix} h_i \right]$$

$$+\left[\sum_{i=-m}^{m}\begin{bmatrix} Q_{11}^{i} & Q_{12}^{i} & Q_{13}^{i} \\ Q_{21}^{i} & Q_{22}^{i} & Q_{23}^{i} \\ Q_{31}^{i} & Q_{32}^{i} & Q_{33}^{i} \end{bmatrix}\begin{bmatrix} \kappa_{xx} \\ \kappa_{yy} \\ \kappa_{xy} \end{bmatrix}\begin{bmatrix} -z_{0i} & \bar{\bar{z}}_{i}^{2} & \bar{S}_{1i} \end{bmatrix} h_i\right]$$
$$+\left[\sum_{i=-m}^{m}\begin{bmatrix} Q_{11}^{i} & Q_{12}^{i} & Q_{13}^{i} \\ Q_{21}^{i} & Q_{22}^{i} & Q_{23}^{i} \\ Q_{31}^{i} & Q_{32}^{i} & Q_{33}^{i} \end{bmatrix}\begin{bmatrix} \kappa_{xx}^{1} \\ \kappa_{yy}^{1} \\ \kappa_{xy}^{1} \end{bmatrix}\begin{bmatrix} 0 & \bar{S}_{1i} & \bar{S}_{i} \end{bmatrix} h_i\right]+\left[\sum_{i=1}^{n}\begin{bmatrix} Q_{1}^{pi} \\ Q_{2}^{pi} \\ Q_{3}^{pi} \end{bmatrix}\begin{bmatrix} 1 & -\bar{z}_{0i} & 0 \end{bmatrix} h_i\right] E_{33}$$
(7.67a)

and

$$\begin{bmatrix} Q_{yz} & R_{yz} \\ Q_{xz} & R_{xz} \end{bmatrix} \equiv \sum_{i=1}^{n}\begin{bmatrix} Q_{44}^{i} & Q_{45}^{i} \\ Q_{54}^{i} & Q_{55}^{i} \end{bmatrix}\begin{bmatrix} (\partial w/\partial y) - \theta_y - 2(-1)^i\,\psi_{1,y}/h_i \\ (\partial w/\partial x) - \theta_x - 2(-1)^i\,\psi_{1,x}/h_i \end{bmatrix}\begin{bmatrix} 1 \\ 2(-1)^i/h_i \end{bmatrix}^T h_i. \quad (7.67\text{b})$$

For symmetric laminates with an anti-symmetric out-of-plane electric field,

$$\begin{bmatrix} N_{xx} & M_{xx} & L_{xx} \\ N_{yy} & M_{yy} & L_{yy} \\ N_{xy} & M_{xy} & L_{xy} \end{bmatrix} = \left[\sum_{i=-m}^{m}\begin{bmatrix} Q_{11}^{i} & Q_{12}^{i} & Q_{13}^{i} \\ Q_{21}^{i} & Q_{22}^{i} & Q_{23}^{i} \\ Q_{31}^{i} & Q_{32}^{i} & Q_{33}^{i} \end{bmatrix}\begin{bmatrix} \varepsilon_{xx}^{0} \\ \varepsilon_{yy}^{0} \\ \varepsilon_{xy}^{0} \end{bmatrix}\begin{bmatrix} 1 & 0 & 0 \end{bmatrix} h_i\right]$$
$$+\left[\sum_{i=-m}^{m}\begin{bmatrix} Q_{11}^{i} & Q_{12}^{i} & Q_{13}^{i} \\ Q_{21}^{i} & Q_{22}^{i} & Q_{23}^{i} \\ Q_{31}^{i} & Q_{32}^{i} & Q_{33}^{i} \end{bmatrix}\begin{bmatrix} \kappa_{xx} \\ \kappa_{yy} \\ \kappa_{xy} \end{bmatrix}\begin{bmatrix} 0 & \bar{z}_{i}^{2} & \bar{S}_{1i} \end{bmatrix} h_i\right]$$
$$+\left[\sum_{i=-m}^{m}\begin{bmatrix} Q_{11}^{i} & Q_{12}^{i} & Q_{13}^{i} \\ Q_{21}^{i} & Q_{22}^{i} & Q_{23}^{i} \\ Q_{31}^{i} & Q_{32}^{i} & Q_{33}^{i} \end{bmatrix}\begin{bmatrix} \kappa_{xx}^{1} \\ \kappa_{yy}^{1} \\ \kappa_{xy}^{1} \end{bmatrix}\begin{bmatrix} 0 & \bar{S}_{1i} & \bar{S}_{i} \end{bmatrix} h_i\right]+\left[\sum_{i=1}^{n}\begin{bmatrix} Q_{1}^{pi} \\ Q_{2}^{pi} \\ Q_{3}^{pi} \end{bmatrix}\begin{bmatrix} 0 & -\bar{z}_{0i} & 0 \end{bmatrix} h_i\right] E_{33}.$$
(7.68b)

To solve for the two additional distributed degrees of freedom, two more equilibrium equations are required. Assuming the laminate to be symmetric about the middle surface, multiplying the stresses by the zig-zag functions and integrating the products over the thickness of the laminate,

$$\frac{\partial L_{xx}}{\partial x}+\frac{\partial L_{xy}}{\partial y}=\bar{\rho}h_1\left(\frac{\sum_i \rho_i h_i \bar{S}_i}{\sum \rho_i h_i}\right)\frac{\partial^2\psi_{1,x}}{\partial t^2}, \quad (7.69\text{a})$$

$$\frac{\partial L_{yx}}{\partial x}+\frac{\partial L_{yy}}{\partial y}=\bar{\rho}h_t\left(\frac{\sum_i \rho_i h_i \bar{\bar{S}}_i}{\sum_i \rho_i h_i}\right)\frac{\partial^2\psi_{1,y}}{\partial t^2}, \quad (7.69\text{b})$$

where ρ_i is the density of the ith layer.

The above pair may be combined into a single equation for $\psi_1(x, y)$, which is

$$\frac{\partial^2 L_{xx}}{\partial x^2}+2\frac{\partial^2 L_{xy}}{\partial x\partial y}+\frac{\partial^2 L_{yy}}{\partial y^2}=\bar{\rho}h_t\left(\frac{\sum_i \rho_i h_i \bar{S}_i}{\sum_i \rho_i h_i}\right)\frac{\partial^2}{\partial t^2}\nabla^2\psi_1(x, y). \quad (7.70)$$

Considering an isotropic plate of thickness h with an isotropic PZT piezoelectric patch actuator of thickness t_p, on either side of the plate,

$$\bar{z}_{0i} = -(h+t_p)/2 \quad i=-1, \quad \bar{z}_{00}=0, \quad \bar{z}_{0i}=(h+t_p)/2, \quad i=1, \quad \bar{\bar{z}}_0^2 = \frac{h^2}{12},$$
$$\bar{\bar{z}}_1^2 = \bar{\bar{z}}_{-1}^2 = \frac{t_p^2}{3} + \frac{h\,(h+2t_p)}{4} = \frac{4t_p^2+6t_p h+3h^2}{12}, \quad \bar{S}_{10}=\frac{h}{6}, \quad \bar{\bar{S}}_0=\frac{1}{3}, \quad \bar{S}_{1i}=(-1)^i\frac{t_p}{6}, \quad \bar{S}_i=\frac{1}{3},$$
$$i=\pm 1. \tag{7.71}$$

Hence,

$$\begin{Bmatrix} M_{xx} \\ M_{yy} \\ M_{xy} \end{Bmatrix} = \mathbf{D}^s \left\{ \begin{bmatrix} \kappa_{xx} \\ \kappa_{yy} \\ \kappa_{xy} \end{bmatrix} + \frac{2}{h} \begin{bmatrix} \kappa_{xx}^1 \\ \kappa_{yy}^1 \\ \kappa_{xy}^1 \end{bmatrix} \right\} + \mathbf{D}^p \left\{ \begin{bmatrix} \kappa_{xx} \\ \kappa_{yy} \\ \kappa_{xy} \end{bmatrix} - \frac{t_p}{6\bar{\bar{z}}_i^2} \begin{bmatrix} \kappa_{xx}^1 \\ \kappa_{yy}^1 \\ \kappa_{xy}^1 \end{bmatrix} \right\}$$
$$-t_p\,(h+t_p)\,\Psi(x,y) \begin{Bmatrix} M_{xx}^P \\ M_{yy}^P \\ M_{xy}^P \end{Bmatrix} E_{33} \tag{7.72a}$$

and

$$\begin{Bmatrix} L_{xx} \\ L_{yy} \\ L_{xy} \end{Bmatrix} = \frac{2\mathbf{D}^s}{h} \left\{ \begin{bmatrix} \kappa_{xx} \\ \kappa_{yy} \\ \kappa_{xy} \end{bmatrix} + \frac{2}{h} \begin{bmatrix} \kappa_{xx}^1 \\ \kappa_{yy}^1 \\ \kappa_{xy}^1 \end{bmatrix} \right\} - \frac{\mathbf{D}^p}{6\bar{\bar{z}}_i^2} \left\{ t_p \begin{bmatrix} \kappa_{xx} \\ \kappa_{yy} \\ \kappa_{xy} \end{bmatrix} - 2 \begin{bmatrix} \kappa_{xx}^1 \\ \kappa_{yy}^1 \\ \kappa_{xy}^1 \end{bmatrix} \right\}. \tag{7.72b}$$

Finally, the shear forces are given by

$$\begin{Bmatrix} Q_{zy} \\ Q_{zx} \end{Bmatrix} = \left\{ \kappa h \begin{bmatrix} \mu_y^s & 0 \\ 0 & \mu_x^s \end{bmatrix} + \kappa \Psi t_p \begin{bmatrix} \mu_y^p & 0 \\ 0 & \mu_x^p \end{bmatrix} \right\} \begin{bmatrix} (\partial w/\partial y) - \theta_y \\ (\partial w/\partial x) - \theta_x \end{bmatrix}$$
$$-2\kappa \left\{ \begin{bmatrix} \mu_y^s & 0 \\ 0 & \mu_x^s \end{bmatrix} - \Psi \begin{bmatrix} \mu_y^p & 0 \\ 0 & \mu_x^p \end{bmatrix} \right\} \begin{bmatrix} \psi_{1,y} \\ \psi_{1,x} \end{bmatrix}. \tag{7.72c}$$

Hence, using the expressions for the moment, shear forces and generalized moment resultant, one may obtain the equations for dynamic equilibrium in terms of the transverse displacement $w(x, y)$, rotations θ_x and θ_y, and the zig-zag distribution function $\psi_1(x, y)$. The solution of the equations may then be obtained by imposing an appropriate set of boundary conditions. This aspect will be considered in a later section.

7.1.9 Elastic Constants Along Principal Directions

So far in the analysis of lamina and laminates, the micromechanics of the lamina have not been given due consideration. A lamina is usually made of fibre composites, including such fibres as SMA wires. Thus, the longitudinal properties associated with loading in the fibre direction are dominated by the fibres that are stronger and stiffer, and are characterized by lower ultimate strains. Assuming perfect bonding between the matrix, in which the fibres are embedded, and the fibres, the longitudinal strains may be assumed to be uniform throughout the lamina and equal for both the matrix and the fibres. Thus, one may formulate the rule of mixtures, based on the parallel model for the longitudinal Young's modulus, as

$$E_L = V_f E_{lf} + V_m E_m, \tag{7.73}$$

where E_{lf} and E_m are the fibre's longitudinal and the matrix Young's modulus, respectively, and V_f and V_m are the volume fraction ratios, respectively. In formulating the rule of mixtures, it is implicitly assumed that the fibre can be anisotropic while the matrix is isotropic.

For the transverse moduli, it is usually assumed that the state of stress is dominated by the matrix. A uni-direction lamina is modelled as alternate strips of fibre and matrix for purposes of modelling the transverse moduli and consequently the series model is most appropriate, which is

$$\frac{1}{E_T} = \frac{V_f}{E_{tf}} + \frac{V_m}{E_m}. \tag{7.74}$$

A similar approach is adopted for the longitudinal and transverse Poisson's ratio. Thus for the longitudinal Poisson's ratio, one obtains

$$\nu_{12} = V_f \nu_{12f} + V_m \nu_{12m}. \tag{7.75}$$

The shear modulus may be estimated from the Young's modulus and the Poisson's ratio. Alternatively, shear in a composite may be assumed to be dominated by the matrix and consequently modelled as

$$G_{12} = \frac{V_f}{G_{12f}} + \frac{V_m}{G_{12m}}. \tag{7.76}$$

A number of empirical formulae have been proposed that combine the parallel and series models, given above for the Young's modulus and other elastic constants. The Halpin–Tsai relationship is one such formula, which gives

$$E_{HT} = \frac{E_m \left(1 + \xi \eta V_f\right)}{1 - \eta V_f}, \qquad \eta = \frac{E_f - E_m}{E_f + \xi E_m}, \tag{7.77}$$

ξ being the reinforcing efficiency.

Hence,

$$E_{HT} = E_m \frac{\left(E_f \left(1 + \xi V_f\right) + \xi E_m \left(1 - V_f\right)\right)}{E_f \left(1 - V_f\right) + E_m \left(\xi + V_f\right)}. \tag{7.78}$$

In particular when $\xi \to \infty$, one obtains the parallel model, while when $\xi \to 0$, one obtains the series model. The appropriate value of the parameter ξ is determined experimentally. Another formula that is well representative of real composite lamina is the Shaffer formula, which is based on the premise that a fraction of the matrix acts in parallel with a series element while the remaining portion of matrix is in series with the fibres, constituting the series element. For the portion of the matrix in series with the fibres,

$$\frac{1}{E_{2s}} = \frac{V_{fs}}{E_{2f}} + \frac{V_{ms}}{E_m}. \tag{7.79}$$

Hence,

$$E_{2s} = \frac{E_{2f} E_m}{E_m V_{fs} + E_{2f} V_{ms}}. \tag{7.80}$$

Thus, the overall modulus is given by

$$E_S = \left(1 - V_{mp}\right) E_{2s} + V_{mp} E_m. \tag{7.81}$$

Thus,

$$E_S = E_m \left(\frac{\left(1 - V_{mp} \left(1 - V_{ms}\right)\right) E_{2f} + E_m V_{fs} V_{mp}}{E_{2f} V_{ms} + E_m V_{fs}} \right). \tag{7.82}$$

Finally, similar empirical formulae have been formulated for short fibre composites.

The application of these empirical formulae is to estimate the elastic coefficients of a lamina with embedded SMA wires. Control of the elastic properties of such lamina is affected by controlling the fraction of the martensite to austenite in the SMA wires. The fraction of the martensite to austenite in the SMA wires is therefore the control input for controlling laminated composites with embedded SMA wires. Thus, by combining the theories of modelling laminated composites with the empirical formulae for the elastic moduli in terms of the volume fraction ratios, one could effectively model flexible laminated composite structures with embedded SMA wires.

It is now possible to assemble all the formulae essential to model a lamina with embedded SMA wires. They are given by

$$E_1 = E_{1m}V_m + E_{1f}V_f, \quad G_{23} = G_{23m}V_m + G_{23f}V_f, \quad E_2 = \frac{E_{2m}E_{2f}}{E_{2m}V_f + E_{2f}V_m},$$
$$G_{12} = \frac{G_{12m}G_{12f}}{G_{12m}V_f + G_{12f}V_m}, \tag{7.83}$$

$$\nu_{ij} = \nu_{ijm}V_m + \nu_{ijf}V_f, \quad i,j = 1,2,3, \quad \rho = \rho_m V_m + \rho_f V_f, \quad \alpha_1 = \frac{E_{1m}V_m\alpha_{1m} + E_{1f}V_f\alpha_{1f}}{E_1},$$
$$\alpha_2 = V_m\alpha_{2m} + V_f\alpha_{2f}, \tag{7.84}$$

where the subscripts 'm' and 'f'' refer to the composite matrix and fibre, respectively. The Young's modulus, the shear modulus, Poisson's ratio, the material density and the thermal expansion coefficient, respectively, are denoted by E, G, ν, ρ and α in the directions indicated by the subscripts. In addition, V_m and V_f are the volume fractions of the composite matrix and fibres, respectively. The thermal expansion coefficients are defined and discussed in Chapter 8.

7.2 Failure of Fibre Composites

The emergence of modern composites as standard structural materials began with the development of glass fibre–polyester composites in the 1940s. However, recent developments in the understanding of failure mechanisms in these and other composite materials has been the primary reason for their acceptance by structural engineers for important applications in aircraft and automobiles. An understanding of the mechanisms of failure in a composite laminate is extremely important if one is interested in introducing active controls. Control inputs generate electrically and thermally induced stress, and it is therefore important to recognize at the design stage that limits must be imposed on the magnitudes of the control inputs.

While methods of failure analysis and corresponding failure theories for commonplace materials have been well understood, the application of these theories to composite structures requires an additional dimension to be considered. For this reason, the modern approach to understanding the mechanisms of failure in a composite structure begins with an understanding of the modes of failure that one can expect with such structures. It is therefore useful to begin by enunciating the primary failure modes in composite structure. Failure modes in structures may broadly be classified as

1. fibre-related failure modes, which refer to predominantly fibre-related failures involving fibre breaking in tension or fibre buckling in compression;
2. matrix-related failure modes, which refer to matrix fracture due to tension in the transverse direction or compression failure or crazing due to compression in the transverse direction;
3. debonding and delamination resulting from either a failure in the fibre–matrix bond or separation between layers in laminates;
4. other laminate failure modes due to variety of causes related to inter-laminar shear and the presence of voids and other imperfections within the laminates.

At the outset, it is important to recognize that a fibre-related failure would lead to the failure of the complete composite while matrix-related failures may not.

On the other hand, number damage modes are precursors to the initiation of failure modes. There are many different damage modes that may develop and interact before a final failure occurs. The damage modes are dependent on the type of loading, the stacking sequence and the geometry of the structure or lamina. Knowledge of these damage modes makes it possible to design components from composite structures that are relatively more damage tolerant. To exploit the ability to design for damage tolerance, one must be able to predict the stiffness, strength and life of these materials in variable environmental conditions. Damage development may include matrix cracking, fibre fracture, fibre–matrix separation, delamination and environmental degradation. For most composite structural lamina, the tensile strength is controlled by the stress distributions surrounding fibre fractures. In particular, the stress concentrations in fibres in the vicinity of the fractured ones and the domain over which the perturbed stress field acts are required for tensile strength predictions.

Several researchers have considered the stress fields around typically shaped cracks under the influence of normal and shear loads, and have established criteria for the growth of the cracks, based on the energy dissipated during the growth of crack per unit of the developed cracked surface area. This is known as the strain energy release rate per unit area of developed crack and is the basis for the Griffith's energy criterion for predicting the critical stresses in the vicinity of a crack that must be exceeded for failure to occur.

The stress in the vicinity of a crack may be shown to be inversely proportional to the distance from the crack tip and a non-linear function of the orientation of the plane of the crack relative to the point under consideration. The normalized proportionality constants are known as stress intensity factors, and several numerical methods have evolved for the prediction of these factors. When combined with Griffith's energy criterion, the criterion for failure may be expressed in terms of the relevant stress intensity factors that should not be exceeded at a particular point, for a hypothetical crack to propagate.

Delamination is an extremely complex process caused by the ply-by-ply mismatch in interlamina shear stresses. Interlamina shear stresses can grow following defects during manufacture. Delamination is one of the best examples of 'self-similar' crack growth in composites. Delamination grows in a stable manner until a critical delamination size is reached. After this state is reached, the growth tends to become unstable. The complex state of stress at the delamination crack tip makes the strain energy release rate approach easier to implement without the need to calculate the stress intensity factors.

The next step in the understanding of failures in composites is to establish typical failure criteria corresponding to each of the failure modes and for multi-mode failures. An attempt to do this has led to the development of

1. failure criteria for single plies in laminates;
2. single-mode failure criteria;
3. interactive multi-mode failure criteria;
4. fibre–matrix failure criteria.

Most of the failure criteria are based on the state of stress within a lamina. The failure criteria for single plies are probably the simplest. Usually, these take the form of the maximum stress (in the principal lamina direction) being required to be less than a certain critical allowable value. When the critical allowable value is exceeded in a given layer, the material is no longer elastic and consequently the elastic constants influenced by the particular mode of failure are no longer representative of that layer. This generally leads to a reduction of stiffness that only hastens the failure.

There are two approaches to single-mode failure analysis; the first based on the state of stress that is responsible for the failure and the second based on strain criteria. The strain criterion is a more representative measure, since it is directly obtained from measurements or by calculations that can be validated. On the other hand, the state of stress is difficult to determine both experimentally or by prediction. The problem is that the constitutive relationships may not be known precisely at the actual strain level. Since damage accumulation is cumulative, and failure is usually progressive rather than catastrophic and instantaneous, errors introduced by assuming the material constants can accumulate. If these errors are responsible for the damage accumulation, the failure will not be predicted. Single-mode failure theories do not account for multi-mode interaction.

Two interactive multi-mode failure criteria have emerged for composite structures:

1. the Tsai–Hill criterion;
2. the Tsai–Wu criterion.

Hill adapted the classic von Mises criterion for failure based on the shear strain energy for isotropic materials. Hill's adaptation extends the principle to anisotropic materials. Tsai made further contributions to this criterion by expressing it in a general quadratic form as

$$\frac{\sigma_{11}\left(\sigma_{11}-\sigma_{22}\right)}{\left(\sigma_{11}^{f}\right)^{2}}+\left(\frac{\sigma_{22}}{\sigma_{22}^{f}}\right)^{2}+\left(\frac{\sigma_{12}}{\sigma_{12}^{f}}\right)^{2}<1, \tag{7.85}$$

where σ_{ij}^{f} are the maximum allowable stresses before failure can occur in the principal (longitudinal and transverse) directions and in shear. For orthotropic materials, the criterion may be expressed as

$$\frac{(\sigma_{11})^2}{(F_1)^2}-\frac{(\sigma_{11}\sigma_{22})}{(F_1)^2}+\frac{(\sigma_{22})^2}{(F_2)^2}+\frac{(\sigma_{12})^2}{(F_6)^2}+\frac{(\sigma_{23})^2}{(F_4)^2}+\frac{(\sigma_{31})^2}{(F_5)^2}<1. \tag{7.86}$$

The criterion is suitable for an orthotropic lamina that is equally strong in tension and compression. To generalize the criterion to lamina that is not equally strong in tension and compression, the Tsai–Wu strength criterion was proposed, which assumes the existence of a failure surface in the stress space. The criterion is of the following form:

$$F_1\sigma_{11}+F_2\sigma_{22}+F_6\sigma_{12}+F_{11}\sigma_{11}^2+F_{22}\sigma_{22}^2+F_{66}\sigma_{12}^2+2F_{12}\sigma_{11}\sigma_{22}=1, \tag{7.87}$$

where

$$F_i=\frac{1}{\sigma_{ii}^{ft}}-\frac{1}{\sigma_{ii}^{fc}},\quad F_{ii}=\frac{1}{\sigma_{ii}^{ft}}\times\frac{1}{\sigma_{ii}^{fc}},\quad i=1,2,\quad F_6=\frac{1}{\sigma_{12}^{ft}}-\frac{1}{\sigma_{12}^{fc}},\quad F_{66}=1\Big/\left(\sigma_{12}^{ft}\times\sigma_{12}^{fc}\right),$$
$$F_{12}=-0.5\sqrt{F_{11}F_{22}},$$

and σ_{ij}^{ft} and σ_{ij}^{fc} are the maximum allowable stresses in tension and compression, respectively, before failure can occur in the principal (longitudinal and transverse) directions and in shear. Because of the presence of six independent constants, the Tsai–Wu criterion is capable of dealing with six failure modes.

The Hoffman criterion for orthotropic brittle materials is a slight modification of the Tsai–Wu criterion in that the coefficient F_{12} is modified as

$$F_{12}=-\frac{1}{2}\frac{1}{\sigma_{11}^{ft}}\times\frac{1}{\sigma_{11}^{fc}}. \tag{7.88}$$

There have been several other generalizations of the Tsai–Hill and Tsai–Wu failure criteria. Several of these may be expressed in quadratic forms as

$$\begin{bmatrix}\sigma_{11}\\\sigma_{22}\\\sigma_{12}\\\sigma_{31}\\\sigma_{32}\end{bmatrix}^T\begin{bmatrix}f_{11}&f_{12}&0&f_{14}&f_{15}\\f_{12}&f_{22}&0&f_{24}&f_{25}\\0&0&0&f_{34}&f_{35}\\f_{14}&f_{24}&f_{34}&0&f_{45}\\f_{15}&f_{25}&f_{35}&f_{45}&0\end{bmatrix}\begin{bmatrix}\sigma_{11}\\\sigma_{22}\\\sigma_{12}\\\sigma_{31}\\\sigma_{32}\end{bmatrix}$$
$$+\begin{bmatrix}f_1&f_2&0&f_4&f_5\end{bmatrix}\begin{bmatrix}\sigma_{11}&\sigma_{22}&\sigma_{12}&\sigma_{31}&\sigma_{32}\end{bmatrix}^T<1. \tag{7.89}$$

Such constraints are particularly easy to handle when synthesizing control laws for electro-actuated composite structures. These belong to a class of matrix inequalities that could be met while designing optimal control laws. Advanced techniques for designing control laws based on linear and quadratic matrix inequalities are discussed in Chapter 9.

A number of fibre–matrix failure criteria have been developed. Most of these use a quadratic for fibre failures and a different quadratic form for the failure of the matrix. The criteria are applied with different coefficients in the longitudinal and transverse directions, and for debonding and delamination. Delamination quadratic criteria are based on the maximum out-of-plane stress σ_{31}, σ_{32} and σ_{33}, both in tension and in compression and take the form

$$\frac{(\sigma_{31})^2}{(F_{31})^2} + \frac{(\sigma_{32})^2}{(F_{32})^2} + \frac{(\sigma_{33}^c)^2}{(F_{33}^c)^2} + \frac{(\sigma_{33}^t)^2}{(F_{33}^t)^2} < 1. \tag{7.90}$$

The coefficients in the failure criteria are determined by extensive experimentation and simulated failures in each of the modes. To design a laminate for a particular application, several failure criteria should be considered to ensure that the design does not meet any of them.

7.3 Flexural Vibrations in Laminated Composite Plates

Prior to any synthesis of control laws for flexible actuated composite structure, it is always essential to obtain the free vibration natural frequencies and the corresponding mode shapes. To do this one often makes simplifying assumptions. Thus, in obtaining the natural modes the presence piezoelectric patches could be ignored. Alternately, one could obtain the natural frequencies and mode shapes of an equivalent orthotropic thin structure (where the Kirchhoff assumptions are assumed to be valid).

The equations for free vibration of a plate ignoring shear, rotary inertia and external loading can be shown to be

$$\frac{\partial N_{xx}}{\partial x} + \frac{\partial N_{xy}}{\partial y} = \bar{\rho} h_t \frac{\partial^2 u_0}{\partial t^2}, \quad \frac{\partial N_{xy}}{\partial x} + \frac{\partial N_{yy}}{\partial y} = \bar{\rho} h_t \frac{\partial^2 v_0}{\partial t^2}, \tag{7.91a}$$

$$\frac{\partial^2 M_{xx}}{\partial x^2} + 2\frac{\partial^2 M_{xy}}{\partial x \partial y} + \frac{\partial^2 M_{yy}}{\partial y^2} + \bar{\rho} h_t \frac{\partial^2 w}{\partial t^2} = 0. \tag{7.91b}$$

The equations are valid for isotropic, orthotropic and composite plates. One could substitute for the stress and moment resultants in terms of the middle surface strains and curvatures, and then eliminate them in favour of the middle surface displacement. For general anisotropic plate, these are a set of three couple partial differential equations in $u_0(x, y)$, $v_0(x, y)$ and $w(x, y)$. For an orthotropic plate, these equations decouple and for the transverse displacement alone one obtains

$$\frac{\partial^2}{\partial x^2} D_{11} \frac{\partial^2 w}{\partial x^2} + 2\frac{\partial^2}{\partial x \partial y}(D_{12} + 2D_{33})\frac{\partial^2 w}{\partial x \partial y} + \frac{\partial^2}{\partial y^2} D_{22} \frac{\partial^2 w}{\partial y^2} + \bar{\rho} h_t \frac{\partial^2 w}{\partial t^2} = 0. \tag{7.92b}$$

For a rectangular plate with sides a and b, which is simply supported on all edges, the mode shapes take the form

$$w_{mn}(x, y) = w_{mn}^0 \sin\left(\frac{m\pi x}{a}\right) \sin\left(\frac{n\pi x}{b}\right), \tag{7.93b}$$

and the corresponding natural frequencies are

$$\omega_{mn} = \lambda_{mn} \sqrt{\frac{\pi^4 D_{22}}{\bar{\rho} b^4}}, \tag{7.94a}$$

$$\lambda_{mn} = \sqrt{\frac{D_{11}}{D_{22}}\left(\frac{b}{a}\right)^4 + \frac{2(D_{12} + 2D_{33})}{D_{22}}\left(\frac{b}{a}\right)^2 + 1}. \tag{7.94b}$$

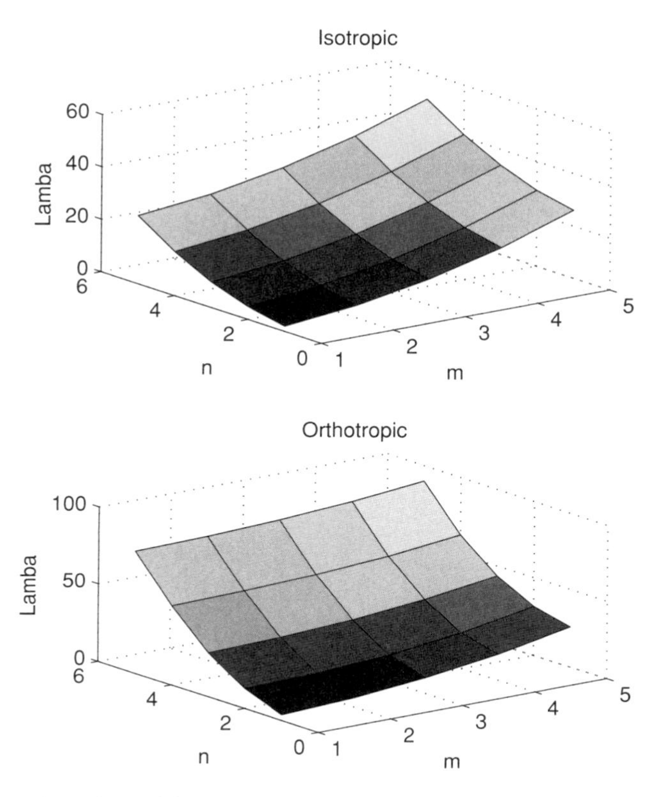

Figure 7.2 Comparison of natural frequencies of an orthotropic square plate with those of an isotropic plate

The typical values for λ_{mn} for orthotropic plate $(D_{11}/D_{22} = 10)$ are compared with those for an isotropic plate $(D_{11}/D_{22} = 1)$ in Figure 7.2. In both cases, $(D_{12} + 2D_{33})/D_{22} = 1$. The figure demonstrates the substantial differences in the natural frequencies between isotropic and orthotropic plates.

7.3.1 Equations of Motion of Continuous Systems in Principal Coordinates: The Energy Method

Considering the kinetic energy of an element of the structure vibrating transversely and integrating over the whole of the structure, it can be shown that the total kinetic of the structure is related to the operator M by the relationship

$$T = \frac{1}{2}\int_{\Omega} \dot{w}(x,t)\, M(\dot{w}(x,t))\,\mathrm{d}x. \tag{7.95}$$

By representing the general solution $w(x,t)$ in terms of the free vibration *eigenfunctions* or the *normal modes* of vibration as

$$w(x,t) = \sum_{n=1}^{\infty} \eta_n(x)\, P_n(t) \tag{7.96}$$

and recognizing that the operator M is a linear operator, it follows that

$$M\left(\dot{w}\left(x,t\right)\right)=\sum_{n=1}^{\infty}M\left(\eta_{n}\left(x\right)\right)\dot{P}_{n}\left(t\right). \tag{7.97}$$

As a result of the orthogonality condition,

$$\int_{\Omega}\eta_{j}\left(x\right)M\left(\dot{w}\left(x,t\right)\right)\mathrm{d}x=\sum_{i=1}^{\infty}\dot{P}_{i}\left(t\right)\int_{\Omega}\eta_{j}\left(x\right)M\left(\eta_{i}\left(x\right)\right)\mathrm{d}x=\dot{P}_{j}\left(t\right)\alpha_{jj} \tag{7.98}$$

By a similar argument,

$$T=\frac{1}{2}\int_{\Omega}\dot{w}\left(x,t\right)M\left(\dot{w}\left(x,t\right)\right)\mathrm{d}x=\frac{1}{2}\sum_{j=1}^{\infty}\dot{P}_{j}\left(t\right)\int_{\Omega}\eta_{j}\left(x\right)M\left(\dot{w}\left(x,t\right)\right)\mathrm{d}x \tag{7.99}$$

and hence it follows that

$$T=\frac{1}{2}\sum_{j=1}^{\infty}\dot{P}_{j}^{2}\left(t\right)\alpha_{jj}. \tag{7.100}$$

By adopting a similar approach, we may also show that

$$\frac{1}{2}\int_{\Omega}w\left(x,t\right)L\left(w\left(x,t\right)\right)\mathrm{d}x=\frac{1}{2}\sum_{j=1}^{\infty}P_{j}^{2}\left(t\right)\gamma_{jj}. \tag{7.101}$$

We can also show that the integral quantity

$$V\equiv\frac{1}{2}\sum_{j=1}^{\infty}P_{j}^{2}\left(t\right)\gamma_{jj} \tag{7.102}$$

is the potential energy of the structure. This is illustrated in the special case of the transverse vibrations of a beam.

Consider the integral

$$V=\frac{1}{2}\int_{\Omega}w\left(x,t\right)L\left(w\left(x,t\right)\right)\mathrm{d}x. \tag{7.103}$$

In the special case of transverse vibrations of a beam, we have

$$V=\frac{1}{2}\int_{\Omega}w\left(x,t\right)\frac{\partial^{2}}{\partial x^{2}}\left(EI_{zz}\frac{d^{2}w\left(x,t\right)}{d^{2}x}\right)\mathrm{d}x=\frac{1}{2}\int_{0}^{L}w\left(x,t\right)\frac{\partial^{2}}{\partial x^{2}}\left(EI_{zz}\frac{d^{2}w\left(x,t\right)}{d^{2}x}\right)\mathrm{d}x. \tag{7.104}$$

Integrating the preceding equation twice by parts and assuming that the beam satisfies either certain geometric or natural boundary conditions at each of its two ends,

$$V=\frac{1}{2}\int_{0}^{L}EI_{zz}\left(\frac{d^{2}w\left(x,t\right)}{d^{2}x}\right)^{2}\mathrm{d}x. \tag{7.105}$$

Employing the fact that according to Bernoulli–Euler theory of bending, the bending moment is given by

$$M_{z}=-EI_{zz}\frac{\partial\theta}{\partial x} \tag{7.106}$$

and since

$$\theta = \frac{\partial w(x,t)}{\partial x}, \tag{7.107}$$

the quantity V may be written as

$$V = -\frac{1}{2}\int_0^L M_z \mathrm{d}\theta = -W, \tag{7.108}$$

where W is the work done by the applied external moment. Hence, we may conclude that V is the potential energy due to bending.

Finally, the work done by an applied external transverse force distribution $q_y(x,t)$ is given by

$$W_e = \int_0^L q_y(x,t)\, w(x,t)\, \mathrm{d}x, \tag{7.109}$$

which may expressed as

$$W_e = \sum_{j=1}^{\infty} Q_j(t)\, P_j(t), \tag{7.110}$$

where

$$Q_n(t) = \int_0^L q_y(x,t)\, \eta_n(x)\, \mathrm{d}x. \tag{7.111}$$

Hence, it follows that the virtual work done by a set of virtual displacements $\delta P_j(t)$ may be expressed as

$$\delta W_e = \sum_{j=1}^{\infty} Q_j(t)\, \delta P_j(t). \tag{7.112}$$

The equations of motion for the transverse vibrations of a beam with arbitrary distributed loading may then be obtained by employing the Euler–Lagrange equations,

$$\frac{\mathrm{d}}{\mathrm{d}t}\frac{\partial T}{\partial \dot{P}_n} - \frac{\partial T}{\partial P_n} + \frac{\partial V}{\partial P_n} = Q_n, \tag{7.113}$$

which reduces to

$$\alpha_{nn}\frac{\mathrm{d}^2 P_n(t)}{\mathrm{d}t^2} + \alpha_{nn}\omega_n^2 P_n(t) = Q_n(t). \tag{7.114}$$

These are the same equations (3.115) and (3.127) derived in Chapter 3 for the undamped forced lateral vibrations of thin beams. However, they have been derived here by employing an equivalent energy approach.

Although by the application of the fundamental orthogonality relations, one may reduce the governing equations of motion of a structure with arbitrary loading to the elegant principal coordinate form given above, it is in fact not the preferred approach in practice. For most engineering structures, it is now customary to model the structure by an equivalent finite degree of freedom model and apply the techniques of analysis suitable for systems with multiple but finite degrees of freedom. The principal technique for this is the so-called *finite element method*. A complete presentation of the method is well beyond the scope of this text.

7.3.2 *Energy Methods Applied to Composite Plates*

Often in the case real practical structures it is almost impossible to find the exact solution and one must resort to an approximate method. There are several approximate methods for the calculation of the natural frequencies and mode shapes of vibration of structures. These are a class of methods that are used to establish upper bounds to the natural frequencies. The most popular method in this class is Rayleigh's method.

Rayleigh's method is most useful when applied to structural systems. It is based on first estimating Rayleigh's quotient from the kinetic and potential energy functions using an approximation for the deflection function in the product form:

$$w(x,t) = \eta_1(x) P(t), \tag{7.115}$$

where $\eta_1(x)$ must satisfy all the geometric constraints that manifest themselves in the form of the geometric boundary conditions. The function $P(t)$ represents the time-dependent modal amplitude function and no longer represents a principal coordinate unless $\eta_1(x)$ represents the first eigenfunction or normal mode of vibration. However, in order to estimate Rayleigh's quotient, one must assume a deflection shape for the first mode; if this is the exact eigenfunction then the resulting natural frequency is exact. Any other assumed deflection shape that is appropriate will result in an upper bound for the natural frequency.

There are also a number of extensions to Rayleigh's method such as the Rayleigh–Ritz and Galerkin methods. The Rayleigh–Ritz method is based on an energy formulation of the equations of motion. The first step in the application of the Rayleigh–Ritz method is the formulation of the potential and kinetic energies of the plate.

The strain energy stored in a moderately thin plate of thickness h is given by the integral

$$U_{strain} = \frac{1}{2E}\int_{-\frac{h}{2}}^{\frac{h}{2}}\int_{S} \mathsf{U}\mathrm{d}S\mathrm{d}z, \tag{7.116}$$

where the strain energy density is given by

$$\mathsf{U} = \sigma_{xx}\varepsilon_{xx} + \sigma_{yy}\varepsilon_{yy} + \sigma_{xy}\varepsilon_{xy} + \sigma_{xz}\varepsilon_{xz} + \sigma_{yz}\varepsilon_{yz}. \tag{7.117}$$

Substituting for the strains in terms of the middle surface strains and curvatures and performing the integrations across the thickness of the plate gives

$$\bar{\mathsf{U}} = \frac{1}{2}\int_{-\frac{h}{2}}^{\frac{h}{2}} \mathsf{U}\,\mathrm{d}z = \left(N_{xx}\varepsilon_{xx}^{0} + N_{xy}\varepsilon_{xy}^{0} + N_{yy}\varepsilon_{yy}^{0} + Q_{xz}\varepsilon_{xz} + Q_{yz}\varepsilon_{yz}\right) + \left(M_{xx}\kappa_{xx} + M_{xy}\kappa_{xy} + M_{yy}\kappa_{yy}\right). \tag{7.118}$$

For an isotropic thin plate, the strain energy density reduces to

$$\mathsf{U} = \sigma_{xx}^2 + \sigma_{yy}^2 - 2\mu\sigma_{xx}\sigma_{yy} + 2(1+\mu)\sigma_{xy}^2. \tag{7.119}$$

For a rectangular uniform plate, the strain energy reduces to

$$U_{strain} = \frac{D}{2}\int_{0}^{L}\int_{-b}^{b} \bar{\mathsf{U}}\mathrm{d}x\mathrm{d}y,$$
$$\bar{\mathsf{U}} = \left(\frac{\partial^2 w}{\partial x^2}\right)^2 + \left(\frac{\partial^2 w}{\partial y^2}\right)^2 + 2\mu\frac{\partial^2 w}{\partial x^2}\frac{\partial^2 w}{\partial y^2} + 2(1-\mu)\left(\frac{\partial^2 w}{\partial x \partial y}\right)^2 \tag{7.120}$$
$$\text{and} \quad D = \frac{Eh^3}{12\left(1-\nu^2\right)}.$$

The kinetic energy of a thin plate of thickness h is given by the integral

$$T = \frac{1}{2}\int_{0}^{L}\int_{-b}^{b} m\left(\mathrm{d}w/\mathrm{d}t\right)^2 \mathrm{d}x\mathrm{d}y. \tag{7.121}$$

After formulating the total potential energy of the plate structure, approximate natural frequencies and mode shapes of the plate can be obtained by employing the Rayleigh–Ritz approach. This is done by approximating the reference surface displacement fields in the forms of linear combinations of known functions, which satisfy the geometric boundary conditions, multiplied by unknown coefficients and evaluating the total potential and kinetic energies with respect to these coefficients. The coefficients are treated as generalized coordinates, and the equations of motion are obtained by applying the Euler–Lagrange equations. Thus, the assumed lateral deflection function of the plate has the form

$$w(x,t) = \sum_{i=1}^{n} q_i(t)\,\eta_i(x). \tag{7.122}$$

The potential and kinetic energy functions are then evaluated in terms of the unknown functions of time $q_i(t)$ as

$$V = \frac{1}{2}\sum_{i=1}^{n}\sum_{j=1}^{n} k_{ij}q_iq_j \quad \text{and} \quad T = \frac{1}{2}\sum_{i=1}^{n}\sum_{j=1}^{n} k_{ij}\dot{q}_i\dot{q}_j. \tag{7.123}$$

A Lagrangian is defined by the equation

$$L = T - V, \tag{7.124}$$

and the equations of motion are then obtained by using Lagrange's method. These are

$$\frac{\mathrm{d}}{\mathrm{d}t}\frac{\partial L}{\partial \dot{q}} - \frac{\partial L}{\partial q} = 0. \tag{7.125}$$

The equations are then cast in matrix form, which are

$$\mathbf{M}\ddot{\mathbf{q}} + \mathbf{K}\mathbf{q} = \mathbf{0}, \tag{7.126}$$

where $\mathbf{M} = \left[m_{ij}\right]$ and $\mathbf{K} = \left[k_{ij}\right]$. Assuming simple harmonic motion, the associated eigenvalue problem is defined by

$$\left[\mathbf{K} - \omega^2\mathbf{M}\right]\mathbf{q}_0 = \mathbf{0}, \tag{7.127}$$

which may then be solved to obtain estimates of the natural frequencies and the corresponding approximations in terms of $\eta_i(x)$, to the normal modes.

7.4 Dynamic Modelling of Flexible Structures

The dynamic modelling of flexible structures using one of a variety of matrix methods is a common strategy adopted for modelling flexible structures for a number of advanced applications. In modelling smart structures for various applications, there is a need to consider a number of pertinent issues related to the synthesis of controllers (see, for example, Suleman *et al.*, 1995; Franco-Correia *et al.*, 1998; Sze and Yao, 2000). The matrix methods employ matrix algebra to set up the problem in a convenient form for numerical computation, the emphasis being on large matrix operations. While the finite element method is the most popular of these, methods based on finite difference approximations of differential equations, methods based on influence coefficients, methods using circuit theory analogies and Myklestad's method based on *transfer matrices* are also in widespread use.

All these methods are highly developed with regard to the numerical computational algorithms and their implementation on personal computers. In particular, the finite element method is an extremely powerful computational tool for the practical analysis and there are now available a number of commercial computational packages for running on a personal computer.

7.4.1 The Finite Element Method

The finite element method provides approximate solutions for boundary value problems. In the finite element method, a given domain over which the problem is solved is partitioned into a large number of smaller sub-regions. These sub-regions or elements are the building blocks of the solution to the same problem but over a much larger domain. Over each of the elements a generic solution to the governing differential equation is approximated, usually by a variational method.

Finite element-based structural analysis consists of four main steps:

1. Element choice and interpolation function generation.
2. Discretization of the solution to the boundary value problem and the calculation of element properties such as the mass and stiffness matrices through a variational method such as the Rayleigh–Ritz method.
3. Transformation of the element mass, stiffness and other characteristic matrices from an element-based reference frame to a global reference frame.
4. Assembly of element equations into matrix format and obtaining a solution. The final step involves the solution of a large set of simultaneous algebraic equations.

The element, in finite element analysis, represents a local domain in which the solution to boundary value problem is approximated. Breaking up the domain into smaller elements permits the representation of the general solutions by simple polynomials over each of the elements. Each element is distinguished by a set of nodes, which are discrete points on the element boundary. Elements are connected together at their common boundaries, or sides, which are defined by the nodes. At a node, the solution must be continuous with respect to the neighbouring element's nodal solution and possibly its derivatives. The order of continuity required is dependent on the order of the polynomials, or interpolation functions selected to represent the solution over the domain of the element. The collection of all of these finite elements and nodes is referred to as the mesh.

Most commercially available finite element packages, such as ABACUS, provide a variety of elements and element shapes such as triangular, quadrilateral and rectangular elements. The element choice for any mesh depends primarily on the domain that is being represented, the accuracy of the desired approximate solution relative to the exact solution as well as the order and size of the resulting set of simultaneous algebraic equation. Although the description of the method has been presented succinctly here, the method is quite extensive, requiring a specialized text devoted only to it. For further details on implementing the finite element method in a MATLAB® environment, the reader is referred to Kwon and Bang (2000).

7.4.2 Equivalent Circuit Modelling

Equivalent circuit representations that describe the quasi-static electromechanical behaviour of piezoelectric crystal resonators and transducer stacks driven in simple thickness modes of vibration have been developed and used extensively. These model are based not only on the normal passive circuit analogies relating a spring, a dash-pot and an inertia to a capacitor, a resistance and an inductance but also on transmission line-type analogies. These analogies were briefly discussed and used in Chapter 6.

Piezoelectric ceramic and polymeric actuators are generally modelled by equivalent circuit methods or by the finite element approach. Ballato (2001) provides in excellent detail the circuit modelling methods applicable to these materials. The circuit models are particularly useful in estimating optimum power requirements via the maximum power transfer theorem. Sequentially interconnected block models may be established for these actuators. Thus, the control problems associated with piezo-ceramic actuators are, to a large extent, quite generic, although the specific methods employed would differ based on the modelling uncertainties.

7.5 Active Composite Laminated Structures

The primary focus of this chapter is electro-actuated active composite laminated structures. A key feature of these structures is that they can be controlled by the application of external potential difference across the faces of an embedded actuator. While methods of modelling based on the variety of techniques discussed earlier are extremely useful in the construction of dynamic models, which could be used for the design of feedback control laws in the time domain, they usually result in relative high order and must be post-processed to reduce the order of the model by methods such as principal orthogonal decomposition and other related methods. However, control engineers have been developing a number of custom methods, which on one hand try to preserve some of the advantages of the finite element method or circuit theory-based analogical modelling while trying to also substantially reduce the order of the model, without having to perform any model order reduction. This is done by modelling directly in the frequency domain. In the next sub-section, one such method that facilitates direct design of the controller in the frequency domain is briefly discussed. The method is not an alternative to time domain modelling but is complementary, as it facilitates the validation of a controller in the frequency domain.

7.5.1 Frequency Domain Modelling for Control

A method of analysing structural dynamic problems is by exploiting the fact that most structures can be modelled by a set of coupled Helmholtz equations representing an equivalent plate model. The concept of using an equivalent plate model was proposed by Kapania and Liu (2000). The Helmholtz equation is a basic equation for wave propagation. Vepa (2008) has shown how the governing equations for the Reissner–Mindlin first-order shear deformation model of a non-isotropic plate may be cast as a set of coupled Helmholtz equations, which are then partially decoupled by the application of a transformation. The equations are then expressed as integral equations where the Green's functions are expressed in terms of the traditional Hankel functions of the first kind. The integral equations are then discretized over the surface area of the wing, which is divided into a finite number of trapezoidal elemental areas or panels. The degrees of freedom are assumed to be constant over the face of each panel. The discrete integral equations are then employed to determine the static and dynamic influence coefficients, which are inverted to generate the stiffness and dynamic matrices. The method shares a number features with non-conforming finite elements.

To facilitate their solution we rewrite the governing equations of motion, for the Reissner–Mindlin first-order shear deformation model, for an orthotropic plate in a slightly different form. As our intention is to solve the above governing equations of motion, we first differentiate the first two equations with respect to x_1 and x_2, respectively, and add and subtract the two resulting equations, respectively. Eliminating the w terms from the first of these equations, they are then written in matrix form as

$$\begin{bmatrix} D_0 & D_1 & 0 \\ D_2 & D_4 & C_1 \\ 0 & 0 & C_0 \end{bmatrix} \nabla^2 \begin{Bmatrix} \theta_{1,1}+\theta_{2,2} \\ \theta_{1,1}-\theta_{2,2} \\ w \end{Bmatrix} + \left(\begin{bmatrix} D_2 & D_3 & 0 \\ D_5 & D_1 & C_0 \\ 0 & 0 & C_1 \end{bmatrix} \left(\frac{\partial^2}{\partial x_1^2} - \frac{\partial^2}{\partial x_2^2} \right) - \begin{bmatrix} 0 & 0 & 0 \\ C_1 & C_0 & 0 \\ C_0 & C_1 & 0 \end{bmatrix} \right)$$

$$\begin{Bmatrix} \theta_{1,1}+\theta_{2,2} \\ \theta_{1,1}-\theta_{2,2} \\ w \end{Bmatrix} + \begin{Bmatrix} \dfrac{\partial^2 M_{11}^P}{\partial x_1^2} + 2\dfrac{\partial^2 M_{12}^P}{\partial x_1 \partial x_2} + \dfrac{\partial^2 M_{22}^P}{\partial x_2^2} - \dfrac{\partial Q_1^P}{\partial x_1} - \dfrac{\partial Q_2^P}{\partial x_2} \\ \dfrac{\partial^2 M_{11}^P}{\partial x_1^2} - \dfrac{\partial^2 M_{22}^P}{\partial x_2^2} \\ \dfrac{\partial Q_1^P}{\partial x_1} + \dfrac{\partial Q_2^P}{\partial x_2} \end{Bmatrix}$$

$$+ \begin{Bmatrix} -1 \\ 0 \\ 1 \end{Bmatrix} q = m \begin{bmatrix} k_g^2 & 0 & -1 \\ 0 & k_g^2 & 0 \\ 0 & 0 & 1 \end{bmatrix} \frac{\partial^2}{\partial t^2} \begin{Bmatrix} \theta_{1,1}+\theta_{2,2} \\ \theta_{1,1}-\theta_{2,2} \\ w \end{Bmatrix}, \tag{7.128}$$

where the constants D_i, $i=0,1,2,\ldots 5$ and C_i, $i=0$ and 1 are defined in the notation. For an isotropic plate,

$$D_i = 0, \quad i = 1, \quad 2 \quad \text{and} \quad 3, \quad D_0 = D, \quad D_4 = D(1-\nu)/2, \quad D_5 = D(1+\nu)/2$$

and

$$C_1 = 0, \quad C_0 = 0.5\kappa h\,(\mu_1 + \mu_2),$$

and it follows that the second of the three governing equations in equation (7.128) is uncoupled from the other two.

Multiplying the equation by

$$\begin{bmatrix} D_0 & D_1 & 0 \\ D_2 & D_4 & C_1 \\ 0 & 0 & C_0 \end{bmatrix}^{-1} = \frac{1}{\Delta \times C_0} \begin{bmatrix} C_0 D_4 & -C_0 D_1 & C_1 D_1 \\ -C_0 D_2 & C_0 D_0 & -C_1 D_0 \\ 0 & 0 & \Delta \end{bmatrix},$$

where $\Delta = D_0 D_4 - D_1 D_2$, the matrix equation reduces to

$$\left(\nabla^2 + \mathbf{A} \left(\frac{\partial^2}{\partial x^2} - \frac{\partial^2}{\partial y^2} \right) \right) \begin{Bmatrix} \theta_{1,1}+\theta_{2,2} \\ \theta_{1,1}-\theta_{2,2} \\ w \end{Bmatrix},$$

$$+\mathbf{M_0} \begin{Bmatrix} \theta_{1,1}+\theta_{2,2} \\ \theta_{1,1}-\theta_{2,2} \\ w \end{Bmatrix} + \mathbf{g} + \mathbf{f} q = \mathbf{M_1} \frac{m}{C_0} \frac{\partial^2}{\partial t^2} \begin{Bmatrix} \theta_{1,1}+\theta_{2,2} \\ \theta_{1,1}-\theta_{2,2} \\ w \end{Bmatrix}, \tag{7.129}$$

where

$$\mathbf{A} = \begin{bmatrix} D_0 & D_1 & 0 \\ D_2 & D_4 & C_1 \\ 0 & 0 & C_0 \end{bmatrix}^{-1} \begin{bmatrix} D_2 & D_3 & 0 \\ D_5 & D_1 & C_0 \\ 0 & 0 & C_1 \end{bmatrix}, \quad \mathbf{M}_0 = -\begin{bmatrix} D_0 & D_1 & 0 \\ D_2 & D_4 & C_1 \\ 0 & 0 & C_0 \end{bmatrix}^{-1} \begin{bmatrix} 0 & 0 & 0 \\ C_1 & C_0 & 0 \\ C_0 & C_1 & 0 \end{bmatrix},$$

$$\mathbf{g} = \begin{bmatrix} D_0 & D_1 & 0 \\ D_2 & D_4 & C_1 \\ 0 & 0 & C_0 \end{bmatrix}^{-1} \begin{Bmatrix} \dfrac{\partial^2 M_{11}^P}{\partial x_1^2} + 2\dfrac{\partial^2 M_{12}^P}{\partial x_1 \partial x_2} + \dfrac{\partial^2 M_{22}^P}{\partial x_2^2} - \dfrac{\partial Q_1^P}{\partial x_1} - \dfrac{\partial Q_2^P}{\partial x_2} \\ \dfrac{\partial^2 M_{11}^P}{\partial x_1^2} - \dfrac{\partial^2 M_{22}^P}{\partial x_2^2} \\ \dfrac{\partial Q_1^P}{\partial x_1} + \dfrac{\partial Q_2^P}{\partial x_2} \end{Bmatrix},$$

$$\mathbf{f} = \begin{bmatrix} D_0 & D_1 & 0 \\ D_2 & D_4 & C_1 \\ 0 & 0 & C_0 \end{bmatrix}^{-1} \begin{Bmatrix} -1 \\ 0 \\ 1 \end{Bmatrix},$$

$$\text{and} \quad \mathbf{M}_1 = \begin{bmatrix} D_0 & D_1 & 0 \\ D_2 & D_4 & C_1 \\ 0 & 0 & C_0 \end{bmatrix}^{-1} \begin{bmatrix} C_0 k_g^2 & 0 & -C_0 \\ 0 & C_0 k_g^2 & 0 \\ 0 & 0 & C_0 \end{bmatrix}.$$

At this stage we may assume that the motion in the time domain is simple harmonic, and this assumption results in one of the equation coefficients being frequency dependent. We now introduce a transformation of the independent variables defined by

$$\begin{Bmatrix} \theta_{1,1} + \theta_{2,2} \\ \theta_{1,1} - \theta_{2,2} \\ w \end{Bmatrix} = \mathbf{T} \begin{Bmatrix} \phi_1 \\ \phi_2 \\ \phi_3 \end{Bmatrix}, \tag{7.130}$$

where the matrix $\mathbf{T}$ is the set of eigenvectors obtained by solving the eigenvalue problem

$$\left[\mathbf{M_0} + \frac{m\omega^2}{C_0} \mathbf{M_1} \right] \mathbf{t}_i = \lambda_i \mathbf{t}_i. \tag{7.131}$$

The matrix $\mathbf{T}$ is assumed to be non-singular. If $\mathbf{T}$ is singular, it is assumed to be the Jordan canonical form or an identity matrix. In the case when $\omega = 0$, it may be almost impossible to reduce $\mathbf{M_0}$ to a diagonal form due to existence of multiple equal eigenvalues. Thus, the matrix equation governing the plate motion (7.129) may be expressed as

$$\nabla^2 \begin{Bmatrix} \phi_1 \\ \phi_2 \\ \phi_3 \end{Bmatrix} + \bar{\mathbf{M}} \begin{Bmatrix} \phi_1 \\ \phi_2 \\ \phi_3 \end{Bmatrix} + \mathbf{T}^{-1}\mathbf{A}\mathbf{T} \left(\frac{\partial^2}{\partial x^2} - \frac{\partial^2}{\partial y^2} \right) \begin{Bmatrix} \phi_1 \\ \phi_2 \\ \phi_3 \end{Bmatrix} + \mathbf{T}^{-1}\mathbf{g} + \mathbf{T}^{-1}\mathbf{f}q = 0, \tag{7.132}$$

where $(x, y, z) = (x_1, x_2, x_3)$ and

$$\bar{\mathbf{M}} = \mathbf{T}^{-1} \left[\mathbf{M_0} + \frac{m\omega^2}{C_0} \mathbf{M_1} \right] \mathbf{T}. \tag{7.133}$$

When the last three terms on the left-hand side of the above matrix equation are ignored, it reduces to a set of three coupled two-dimensional Helmholtz equations.

To employ the Green's function for the two-dimensional Helmholtz equation, which can be expressed in terms of the Hankel function of the first kind and zero order, we may write the solutions for $\{ \phi_1 \ \phi_2 \ \phi_3 \}^T$ in equation (7.132) as integral equations over the surface area of the plate. Thus in terms of the operator ∇^2, the integral equations are

$$\begin{Bmatrix} \phi_1 \\ \phi_2 \\ \phi_3 \end{Bmatrix} = \left(\frac{1}{\bar{\mathbf{M}} + \nabla^2} \right) \left(\mathbf{T}^{-1}\mathbf{A}\mathbf{T} \left(\frac{\partial^2}{\partial x^2} - \frac{\partial^2}{\partial y^2} \right) \begin{Bmatrix} \phi_1 \\ \phi_2 \\ \phi_3 \end{Bmatrix} + \mathbf{T}^{-1}\mathbf{g} + \mathbf{T}^{-1}\mathbf{f}q \right). \tag{7.134}$$

Let the eigenvalues (poles) of $\bar{\mathbf{M}}$ and the corresponding right and left eigenvectors be given by the triplets $(\lambda_i, x_i v_i)$ and let the right and left eigenvectors (column vectors) be scaled so that $v_i^T x_i = 1$. Note that $v_j^T x_k = 0$ for $j \neq k$. The dyadic expansion of the integral operator can be expressed as a sum of residue matrices $\mathbf{R}_i$ over first-order poles each of multiplicity m_i:

$$\mathbf{R}\left(\nabla^2\right) = \left(\frac{1}{\bar{\mathbf{M}} + \nabla^2} \right) = \sum_{i=1}^{3} \frac{\mathbf{R}_i}{\left(\nabla^2 + \lambda_i\right)^{m_i}}, \tag{7.135}$$

where the residues $\mathbf{R}_i$ are $\mathbf{R}_i = x_i v_i^T$.

Thus, the integral equation (7.134) may be expressed as

$$\begin{Bmatrix} \phi_1(x,y) \\ \phi_2(x,y) \\ \phi_3(x,y) \end{Bmatrix} = \sum_{i=1}^{3} \mathbf{R}_i \iint_S \mathbf{G}_i \times \left[\left(\mathbf{T}^{-1}\mathbf{A}\mathbf{T} \left(\frac{\partial^2}{\partial \xi^2} - \frac{\partial^2}{\partial \eta^2} \right) \right) \begin{Bmatrix} \phi_1(\xi,\eta) \\ \phi_2(\xi,\eta) \\ \phi_3(\xi,\eta) \end{Bmatrix} \right]$$

$$\mathbf{dS} + \sum_{i=1}^{3} \mathbf{R}_i \iint_S \mathbf{G}_i \times [\mathbf{B}(\xi,\eta)\,\mathbf{u}(\xi,\eta)]\,\mathbf{dS}, \tag{7.136}$$

where $\mathbf{G}_i$ is a diagonal matrix with elements

$$g_i = \frac{\sqrt{-1}}{4} H_0^{(1)} \left(\sqrt{\lambda_i} \times r \right), \tag{7.137}$$

when the pole multiplicity $m_i = 1$ and

$$g_i = \frac{\sqrt{-1}}{4} \frac{d}{d\lambda_i} H_0^{(1)} \left(\sqrt{\lambda_i} \times r \right), \tag{7.138}$$

when the pole multiplicity $m_i = 2$,

$$\mathbf{Bu} = \mathbf{T}^{-1} (\mathbf{g}(\xi,\eta) + \mathbf{f}q(\xi,\eta)). \tag{7.139}$$

$H_0^{(1)} \left(\sqrt{\lambda_i} \times r \right)$ is the Hankel function of the first kind and zero order, (ξ, η) are the integration variables and r is defined by $r = \sqrt{(x-\xi)^2 + (y-\eta)^2}$.

When $\lambda_i = 0$, the Green's functions reduce to those corresponding to the Laplace operator $(-\log(r)/2\pi)$ and bi-Laplace operators $(r^2 \log(r)/8\pi)$. It must be said that one could in principle add a constant to the Laplace operator's Green's function and a quadratic in r to the bi-Laplace operator's Green's function. In fact for small values of $z = kr$, the Hankel function and its derivative, ignoring higher-order terms, may, respectively, be expressed as

$$H_0^{(1)}(z) = \frac{2i}{\pi} \log(r) \left(1 - \frac{z^2}{4} \right) + \left(1 + \frac{2i}{\pi} \left(\log\left(\frac{k}{2}\right) + \gamma \right) \right) - \frac{z^2}{4} \left(1 + \frac{2i}{\pi} \left(\log\left(\frac{k}{2}\right) + \gamma - 1 \right) \right), \tag{7.140}$$

$$2\lambda_i \frac{\mathrm{d}}{\mathrm{d}\lambda_i} H_0^{(1)} \left(\sqrt{\lambda_i} \times r \right) = z \frac{\mathrm{d}}{\mathrm{d}z} H_0^{(1)}(z) = -\frac{i}{\pi} z^2 \log r + \frac{2i}{\pi} - \frac{z^2}{2} \left(1 + \frac{2i}{\pi} \left(\log\left(\frac{k}{2}\right) + \gamma - 0.5 \right) \right). \tag{7.141}$$

This anomaly between the Green's functions in the steady case and in the limiting cases does cause the convergence of the method to be frequency dependent.

To evaluate the integral in equation (7.136), a number of methods are available. For instance, when $\sqrt{\lambda_i}$ is sufficiently large, the integration can be performed efficiently by employing the fast multi-pole method or the Hankel transform based on the expansion of the Hankel function in terms of its derivatives and the Bessel functions of the first kind. In fact, the Hankel transform is equivalent to the two-dimensional Fourier transform when the input is rotationally symmetric. However, such expansions are restricted in their usefulness to far field approximations, and our concern is more in the near field. Moreover, the set of integral equation in equation (7.136) are not merely a set of coupled Helmholtz equations.

To solve the integral equation (7.136), we express it as the sum given by

$$\begin{Bmatrix} \phi_1(x,y) \\ \phi_2(x,y) \\ \phi_3(x,y) \end{Bmatrix} = \sum_{i=1}^{3} \mathbf{R}_i \iint_S \mathbf{G}_i \mathbf{T}^{-1}\mathbf{A}\mathbf{T} \left(\frac{\partial^2}{\partial \xi^2} - \frac{\partial^2}{\partial \eta^2} \right) \begin{Bmatrix} \phi_1(\xi,\eta) \\ \phi_2(\xi,\eta) \\ \phi_3(\xi,\eta) \end{Bmatrix} \mathbf{dS} + \sum_{i=1}^{3} \mathbf{R}_i \iint_S \mathbf{G}_i \mathbf{B}(\xi,\eta)\,\mathbf{u}(\xi,\eta)\,\mathbf{dS}, \tag{7.142}$$

and integrating the first of the two summation terms twice by parts we have

$$\begin{Bmatrix}\phi_1(x,y)\\ \phi_2(x,y)\\ \phi_3(x,y)\end{Bmatrix}=\sum_{i=1}^{3}\mathbf{R}_i\iint_S\left(\frac{\partial^2}{\partial x^2}-\frac{\partial^2}{\partial y^2}\right)\mathbf{G}_i\mathbf{T}^{-1}\mathbf{A}\mathbf{T}\begin{Bmatrix}\phi_1(\xi,\eta)\\ \phi_2(\xi,\eta)\\ \phi_3(\xi,\eta)\end{Bmatrix}\mathrm{d}\mathbf{S}+\sum_{i=1}^{3}\mathbf{R}_i\iint_S\mathbf{G}_i\mathbf{B}(\xi,\eta)\,\mathbf{u}(\xi,\eta)\,\mathrm{d}\mathbf{S}. \tag{7.143}$$

The integral equation (7.143) may be solved numerically by dividing the area of integration into panels and evaluating the integrals over the surface of integration.

It is worth noting that in the case of an isotropic plate only ϕ_1 and ϕ_3 are required, and with the appropriate choice of the matrix **T,** the integral equations reduce to the two uncoupled integral equations, which may be expressed as

$$\begin{Bmatrix}\phi_1(x,y)\\ \phi_3(x,y)\end{Bmatrix}=\sum_{i=1}^{3}\mathbf{P}\mathbf{R}_i\iint_S\mathbf{G}_i\times\mathbf{B}(\xi,\eta)\,\mathbf{u}(\xi,\eta)\,\mathrm{d}\mathbf{S}, \tag{7.144}$$

where

$$\mathbf{P}=\begin{bmatrix}1&0&0\\0&0&1\end{bmatrix}. \tag{7.145}$$

It represents the exact solution to the problem in integral form. The solution for the transverse displacement is then recovered from equation (7.130).

The special case of the *Kirchhoff* theory is quite useful for purposes of comparison. In this case, the curvature–displacement relationships are

$$\begin{bmatrix}\kappa_{xx}&\kappa_{yy}&\kappa_{xy}\end{bmatrix}=\begin{bmatrix}\dfrac{\partial^2}{\partial x^2}&\dfrac{\partial^2}{\partial y^2}&\dfrac{\partial^2}{\partial x\partial y}\end{bmatrix}w(x,y). \tag{7.146}$$

Substituting in the moment–curvature relationship and ignoring the external moment contributions,

$$\begin{Bmatrix}M_{xx}\\ M_{yy}\\ M_{xy}\end{Bmatrix}=\begin{bmatrix}D_{11}&\nu_{21}D_{11}&0\\ \nu_{12}D_{22}&D_{22}&0\\ 0&0&D_{33}(1-\nu_{33})/2\end{bmatrix}\begin{Bmatrix}\partial^2/\partial x^2\\ \partial^2/\partial y^2\\ 2\partial^2/\partial x\partial y\end{Bmatrix}w(x,y). \tag{7.147}$$

In the absence of rotary inertia, the moment equilibrium equations reduce to

$$\frac{\partial^2 M_{xx}}{\partial x^2}+2\frac{\partial^2 M_{xy}}{\partial x\partial y}+\frac{\partial^2 M_{yy}}{\partial y^2}=q(x,y)+\rho h\omega^2 w(x,y). \tag{7.148}$$

We consider an orthotropic thin plate of infinite extent and assume that the external flow exerts a reactive pressure. The *Kirchhoff* plate equation for this case is

$$\left(a\frac{\partial^4}{\partial x^4}+2b\frac{\partial^2}{\partial x^2}\frac{\partial^2}{\partial y^2}+c\frac{\partial^4}{\partial y^4}-\rho h\omega^2\right)w(x,y)=q(x,y), \tag{7.149}$$

where

$$a=D_{11},\quad c=D_{22},\quad b=D_{33}+0.5\left(\nu_{21}D_{11}+\nu_{12}D_{22}\right)-\nu_{33}D_{33},$$

with the appropriate radiation boundary condition at infinity. Equation (7.149) may be written as

$$\left(b\nabla^4+(a-b)\frac{\partial^4}{\partial x^4}+(c-b)\frac{\partial^4}{\partial y^4}-\rho h\omega^2\right)w(x,y)=q(x,y). \tag{7.150}$$

The governing equation for vibration of an isotropic thin plate is obtained by setting $a=b=c=D$ and is written in terms of the bi-Laplace operator as follows:

$$\left(D\nabla^4-\rho h\omega^2\right)w(x,y)=q(x,y). \tag{7.151}$$

The solution may be expressed in operator form as

$$w(x, y) = \frac{1}{\left(\nabla^4 - \lambda^2\right)} \frac{q(x, y)}{D} = \frac{1}{2\lambda}\left(\frac{1}{\nabla^2 - \lambda} - \frac{1}{\nabla^2 + \lambda}\right)\frac{q(x, y)}{D}, \tag{7.152}$$

where $\lambda^2 = \rho h \varpi^2 / D$. Equation (7.134) with (7.135) is a generalization of equation (7.152).

The Green's function for the isotropic *Kirchhoff* plate may be expressed as

$$\mathbf{G} = -\frac{1}{2\lambda D} \times \frac{i}{4}\left(H_0^1\left(i\sqrt{\lambda}r\right) - H_0^1\left(\sqrt{\lambda}r\right)\right). \tag{7.153}$$

In the case when $\omega = 0$, both the roots $\pm\lambda$ reduce to zero, and we have a situation corresponding to the multiplicity of the root being equal to 2.

Thus, when $\omega \neq 0$, the solution may be obtained as

$$w(x, y) = -\frac{i}{8\lambda D} \int\int_{\mathbf{S}} \left(H_0^1\left(i\sqrt{\lambda}r\right) - H_0^1\left(\sqrt{\lambda}r\right)\right) q\,(\xi, \eta)\,\mathrm{d}\mathbf{S}. \tag{7.154}$$

The application of interest is the vibration analysis of an anisotropic thick plate. The plate surface is divided into trapezoidal panels. The integral equation (7.152) is collocated, and the kernel functions are evaluated at a set of collocation or receiving points and at a set of integration or sending points. The solution variables in the integral equation (7.152) for the plate structural dynamics are assumed to be constants over the panels. This greatly simplifies the solution of the integral equation, although it is now not possible to numerically differentiate the discretized variables. The integral equation (7.152) is thus reduced to an algebraic equation of the form

$$\mathbf{\Phi} = \mathbf{A}^0\mathbf{\Phi} + \mathbf{B}^0 q, \tag{7.155}$$

where $\mathbf{\Phi}$ is the vector of discrete values of $\left\{\phi_1\ \phi_2\ \phi_3\right\}^T$ with the discrete values of ϕ_1 at each of the panels appearing first, followed by the discrete values of ϕ_2 and then by those of ϕ_3, and $\mathbf{A}^0$ and $\mathbf{B}^0$ are influence coefficient matrices obtained by the process of discretizing the governing matrix integral equation (7.152). In the evaluation of the influence coefficient matrices $\mathbf{A}^0$ and $\mathbf{B}^0$, two cases must be considered corresponding to the two multiplicities of the eigenvalues. Furthermore, the steady case, the low-frequency case and the general frequency case are separately evaluated. As the coefficients are obtained by integrating the corresponding Green's functions over the area of each panel, the process is equivalent to averaging the appropriate Green's functions on the area of each panel. As each of the Green's functions is a only a function of a single variable, the averaging process is carried out equivalently by integrating with respect to this single variable of representing limits for each panel. For simplicity, it is assumed that there is no piezoelectric control loading. The solution for $\mathbf{\Phi}$ is obtained as

$$\mathbf{\Phi} = \left(\mathbf{I} - \mathbf{A}^0\right)^{-1}\mathbf{B}^0 q. \tag{7.156}$$

The dynamic influence coefficient matrix relating the discrete values of the transverse displacement, w to q, is obtained by applying the appropriate transformation to the solution given in equation (7.156). The static influence coefficients may also be obtained in the same way by considering the case when the frequency of oscillation is set to zero. Thus, it is possible to include the effects of structural damping by introducing a suitable structural damping coefficient. Furthermore, assuming a set of response modes and applying a similarity transformation, one could construct the reduced-order modal dynamic matrix per unit area of the plate, $\mathbf{D}(\boldsymbol{\omega})$, which plays the same role as $\left[\mathbf{K}(1 + i\boldsymbol{\omega}g) - \boldsymbol{\omega}^2\mathbf{M}\right]$ in a classical vibrating system, where $\mathbf{K}$ and $\mathbf{M}$ are the classical stiffness and mass matrices and g is the structural damping coefficient. The damped natural frequencies, damping ratios and normal modes of the plate may be verified by solving the eigenvalue problem defined by $[\mathbf{D}(\boldsymbol{\omega})]\,\mathbf{w_i} = \lambda\mathbf{w_i}$, i.e. by determining when the eigenvalues of $\mathbf{D}(\boldsymbol{\omega})$ are closest to the origin.

The method presented above can be used to model actively controlled structural dynamic problems directly in the frequency domain. Thus, given a typical control law, one could directly obtain the input–output Nyquist plot and estimate the frequency domain stability margins.

7.5.2 *Design for Controllability*

In the design of electro-actuated composites with embedded actuators and sensors, the issue of actuator and sensor location is of vital importance. If the application of such structures involves active control of shape or vibration, it is essential that the locations of sensors and actuators are selected for maximum controllability. This means that the modes that are intended to be controlled are excited with minimum energy while the modes that are not intended to be controlled are not affected at all. Displacement sensors must necessarily be located either in the vicinity of the zeros of the displacement modes that are not intended to be controlled or at points where the displacements of the modes that are intended to be controlled are a maximum.

Force actuators similarly must be located where force is most effective in influencing the modes that are intended to be controlled. Similarly, moment actuators must be located where they would be most effective in generating a rotation in modes that are intended to be controlled. Bending stress sensors likewise must be located where the curvature is maximum or minimum for modes, depending on whether they are intended to be controlled or not controlled.

The selected locations for the transducer therefore depend on a host condition: the nature of the transducer, the nature of the boundary conditions the structure is subjected to and the nature of the application.

Exercises

1. Consider a uniform beam of length L, the mass per unit length being m and J the rotary inertia per unit length of the beam about an axis perpendicular to the transverse plane of deflection. The transverse deflection of the beam along the beam axis is assumed to be $w(x,t)$. Let the slope of the element due to the applied moment M_z be θ and let K be the shear stress correction factor defined as the ratio of the average shear stress across the section to the shear stress at the centre of the section, A the area of cross-section and G the shear modulus.

(i) Show that

$$m\frac{\partial^2 w(x,t)}{\partial t^2} = KAG\frac{\partial}{\partial x}\left(\frac{\partial w(x,t)}{\partial x} - \theta\right),$$

$$J\frac{\partial^2 \theta(x,t)}{\partial t^2} = -\frac{\partial M_z}{\partial x} + KAG\left(\frac{\partial w(x,t)}{\partial x} - \theta\right).$$

From the Bernoulli–Euler theory of bending, the moment due to flexure is proportional to the flexural rigidity EI_{zz} and in the absence of other external forces is given by

$$M_z = -EI_{zz}\frac{\partial \theta}{\partial x}.$$

Eliminate θ to show that the equation for translational motion is

$$\frac{\partial^2}{\partial x^2}\left(EI\frac{\partial^2 w(x,t)}{\partial x^2} - \left(J + \frac{mEI}{KAG}\right)\frac{\partial^2 w(x,t)}{\partial t^2}\right) + m\frac{\partial^2 w(x,t)}{\partial t^2} + \frac{mJ}{KAG}\frac{\partial^4 w(x,t)}{\partial t^4} = 0,$$

where the subscripts have been dropped for clarity. Hence show that the equation of transverse displacement in the presence of a distributed load $q(x,t)$, acting in the transverse direction, is

$$\frac{\partial^2}{\partial x^2}\left(EI\frac{\partial^2 w(x,t)}{\partial x^2} - J\frac{\partial^2 w(x,t)}{\partial t^2} - \frac{EI}{KAG}\left(m\frac{\partial^2 w(x,t)}{\partial t^2} - q\right)\right)$$
$$+m\frac{\partial^2 w(x,t)}{\partial t^2} - q + \frac{J}{KAG}\frac{\partial^2}{\partial t^2}\left(m\frac{\partial^2 w(x,t)}{\partial t^2} - q\right) = 0.$$

(ii) For a simply supported beam show that the equation for the natural frequency satisfies the constraint

$$EI\frac{n^4\pi^4}{mL^4} = \omega_n^2\left(1 + n^2\pi^2\left(\frac{J}{mL^2} + \frac{E}{KG}\frac{I}{AL^2}\right) - \frac{J}{mL^2}\frac{E}{KG}\frac{I}{AL^2}\frac{mL^4}{EI}\omega_n^2\right).$$

Hence show that an approximate solution for ω_n^2 is given by

$$\omega_n^2 = n^4\pi^4\left(\frac{EI}{mL^4}\right)\left(1 - n^2\pi^2\left(\frac{J}{mL^2} + \frac{E}{KG}\frac{I}{AL^2}\right) + n^4\pi^4\frac{J}{mL^2}\frac{E}{KG}\frac{I}{AL^2}\right).$$

Thus, for a uniform homogeneous steel beam 0.3 m deep by 0.15 m wide by 3 long, $E/G = 1 + 2\nu = 1.6$; assume that $K = 0.85$ and show that the natural frequency in the first mode is reduced by about 1.8% due to the combined effects of shear deformation and rotary inertia.

(iii) Make suitable assumptions and generalize this equation for a laminated beam where some of the lamina could be made from a piezoelectric ceramic.

2.

(i) A finite beam element for modelling the transverse bending vibration is required to be developed. Both the effects of rotary inertia and shear must be included in the element model. The finite element is assumed to be one dimensional of length L with the element coordinate system centred at one end of it. The two nodes are defined in terms of non-dimensional coordinates at $\xi = 0, 1$.

The displacement and slope at each node are the element degrees of freedom. The displacement is approximated in terms of the shape functions $\mathbf{N}_i$ as

$$w(\xi) = \sum_{i=1}^{2}\mathbf{N}_i\begin{bmatrix} w_i & L\theta_i \end{bmatrix}^T,$$

$$\mathbf{N}_1 = \begin{bmatrix} N_{11} \\ N_{12} \end{bmatrix}^T = \begin{bmatrix} 1 - 3\xi^2 + 2\xi^3 \\ -\xi(1-\xi)^2 \end{bmatrix}^T, \quad \mathbf{N}_2 = \begin{bmatrix} N_{21} \\ N_{22} \end{bmatrix}^T = \begin{bmatrix} 3\xi^2 - 2\xi^3 \\ \xi^2(1-\xi) \end{bmatrix}^T.$$

What boundary conditions do the shape functions satisfy?

(ii) Show that the bending strain can be expressed as

$$\varepsilon = \mathbf{Bw}, \quad \mathbf{B} = \frac{1}{L^2}\begin{bmatrix} (6-12\xi) & (4-6\xi) & (12\xi-6) & (2-6\xi) \end{bmatrix}.$$

Hence, given that the flexural rigidity of the beam is EI, show that the stiffness matrix is given by

$$\mathbf{K} = \int_0^L \mathbf{B}^T EI\mathbf{B}d\xi = \frac{EI}{L^3}\begin{bmatrix} 12 & 6 & -12 & 6 \\ 6 & 4 & -6 & 2 \\ -12 & -6 & 12 & -6 \\ 6 & 2 & -6 & 4 \end{bmatrix}.$$

(iii) Apply the Galerkin method by integrating by parts wherever necessary and show that the element's 4×4 translational inertia and the rotary inertia matrices are, respectively, given by

$$\mathbf{M} = m\int_0^L \mathbf{N}^T\mathbf{N}d\xi + \left(J + \frac{mEI}{KAG}\right)\int_0^L \mathbf{B}_1^T\mathbf{B}d\xi, \quad \mathbf{J} = \frac{mJ}{KAG}\int_0^L \mathbf{N}^T\mathbf{N}d\xi,$$

where $\mathbf{N} = \left[\mathbf{N}_1 \ \mathbf{N}_2 \right]$,

$$\mathbf{B}_1 = \frac{1}{L} \left[\left(6\xi - 12\xi^2\right) \left(4\xi - 1 - 3\xi^2\right) \left(6\xi^2 - 6\xi\right) \left(2\xi - 3\xi^2\right) \right].$$

Hence or otherwise derive the load distribution matrix due to a point load P acting at the point $\xi = \xi_P$.

Observe a few features one can exploit to evaluate element matrices and set up the equations of motion for the synthesis of feedback controls in MATLAB. Products in polynomials in the ξ can be easily done by convolution in MATLAB using the *conv.m* *m*-file. The polynomials may each be integrated in one dimension using the numerical quadrature *m*-file, *quad.m*. Thus, one should first establish all the coefficients of the element stiffness matrix prior to performing the integration as polynomials in descending powers.

(iv) Make suitable assumptions and generalize the element matrices for a beam excited by piezoelectric patches.

(v) Explain why it might be useful to consider rotary inertia and shear effects when exciting the beam with piezoelectric patches. Develop a finite element model of a cantilever beam using 20 self-similar elements and 10 patches. Hence or otherwise develop a model for exciting only the first three modes.

3. In several practical applications of smart structures, it often becomes necessary to control the curvature of beam-like elastic structure. The finite element method may be employed to model, analyse and simulate the closed-loop dynamics of such structure. However, in order to model such a structure, it often becomes necessary to employ a curved beam element. The beam element developed in exercise 7.2 may be modified to establish a finite element for a curved beam by employing the method of Kassegne and Reddy (1997).

The in-plane and transverse displacements (v, w) at a generic point (y, z) in the curved beam element are assumed to be of the form:

$$v\left(y, z\right) = \sum_{i=1}^{N+1} v_i\left(y\right) \mathbf{\Phi}^i\left(z\right), \quad w\left(y, z\right) = \sum_{i=1}^{N+1} w_i\left(y\right) \Phi^i\left(z\right),$$

where N is the number of layers in the curved beam element, v_i and w_i are the nodal functions of the in-plane coordinate, y representing the in-plane and transverse displacements of each interface, and $\mathbf{\Phi}^i\left(z\right)$ is a Lagrangian interpolation function through the thickness. Typical examples of $\mathbf{\Phi}^i\left(z\right)$ for layered beam element are

$$\mathbf{\Phi}^i\left(z\right) = \begin{cases} \phi_2^{i-1} = \dfrac{z - z_{i-1}}{h_{i-1}}, \ z_{i-1} \le z \le z_i \\ \phi_1^i = \dfrac{z_i - z}{h_i}, \ z_i \le z \le z_{i+1} \end{cases}, \quad i = 1, 2 \ldots N,$$

where h_i is the thickness of the ith layer and ϕ_j^i is the local Lagrangian interpolation function associated with the jth node in the ith layer. The strain–displacement relationships are

$$\begin{bmatrix} \varepsilon_{yy} \\ \varepsilon_{yz} \\ \varepsilon_{zz} \end{bmatrix} = \begin{bmatrix} \dfrac{\partial v}{\partial y} + \dfrac{w}{R} + \dfrac{1}{2}\left(\dfrac{\partial w}{\partial y}\right)^2 \\ \dfrac{\partial v}{\partial z} + \dfrac{\partial w}{\partial y} - \dfrac{v}{R} \\ \dfrac{\partial w}{\partial z} \end{bmatrix} = \begin{bmatrix} \left(\dfrac{\partial v_i}{\partial y} + \dfrac{w_i}{R}\right)\mathbf{\Phi}^i + \dfrac{1}{2}\left(\dfrac{\partial w_i}{\partial y}\dfrac{\partial w_j}{\partial y}\right)\mathbf{\Phi}^i\mathbf{\Phi}^j \\ v_i\dfrac{\partial \mathbf{\Phi}^i}{\partial z} + \left(\dfrac{\partial w_i}{\partial y} - \dfrac{v_i}{R}\right)\mathbf{\Phi}^i \\ w_i\dfrac{\partial \mathbf{\Phi}^i}{\partial z} \end{bmatrix},$$

where repeated indices are summed over all layers and R is the radius of curvature of the element. The strain energy stored in a curved beam of thickness h is given by the integral

$$U_{strain} = \frac{1}{2E} \int_{-\frac{h}{2}}^{\frac{h}{2}} \int_{S} \mathsf{U} \mathrm{d}S \mathrm{d}z,$$

where the strain energy density is given by

$$\mathsf{U} = \sigma_{yy}\varepsilon_{yy} + \sigma_{zz}\varepsilon_{zz} + \sigma_{yz}\varepsilon_{yz}.$$

In the expression for the strain energy density, U, σ_{yy}, σ_{yz}, σ_{zz} are the in-plane, transverse shear and transverse normal stresses, respectively. The integral of the strain energy density across the thickness may be expressed in terms of the force and moment resultants and assumed displacement functions as

$$\bar{\mathsf{U}} = N_{yy}^{i}\left(\frac{\partial v_i}{\partial y} + \frac{w_i}{R}\right) + N_{yz}^{i}\left(\frac{\partial w_i}{\partial y} - \frac{v_i}{R}\right) + \frac{M_{yy}^{ij}}{2}\left(\frac{\partial w_i}{\partial y}\frac{\partial w_j}{\partial y}\right) + Q_{zz}^{i} w_i + Q_{yz}^{i} v_i.$$

The force and moment resultants are defined by the integrals

$$N_{yy}^{i} = \int_{-\frac{h}{2}}^{\frac{h}{2}} \sigma_{yy}\Phi^{i}\mathrm{d}z, \quad N_{yz}^{i} = \int_{-\frac{h}{2}}^{\frac{h}{2}} \sigma_{yz}\Phi^{i}\mathrm{d}z, \quad M_{yy}^{ij} = \int_{-\frac{h}{2}}^{\frac{h}{2}} \sigma_{yy}\Phi^{i}\Phi^{j}\mathrm{d}z,$$

$$Q_{yz}^{i} = \int_{-\frac{h}{2}}^{\frac{h}{2}} \sigma_{yz}\frac{\mathrm{d}\Phi^{i}}{\mathrm{d}z}\mathrm{d}z \quad \text{and} \quad Q_{zz}^{i} = \int_{-\frac{h}{2}}^{\frac{h}{2}} \sigma_{zz}\frac{\mathrm{d}\Phi^{i}}{\mathrm{d}z}\mathrm{d}z.$$

The material constitutive relationships are defined as

$$\begin{Bmatrix} \sigma_{yy} \\ \sigma_{zz} \\ \sigma_{yz} \end{Bmatrix}^{i} = \begin{bmatrix} C_{11}^{i} & C_{13}^{i} & 0 \\ C_{13}^{i} & C_{33}^{i} & 0 \\ 0 & 0 & C_{44}^{i} \end{bmatrix} \begin{Bmatrix} \varepsilon_{yy} \\ \varepsilon_{zz} \\ \varepsilon_{yz} \end{Bmatrix}^{i},$$

where the reduced elastic constants C_{kl}^{i} are derived from equation (7.1).

(i) Making suitable assumptions, obtain expressions for the reduced elastic constants C_{kl}^{i} in terms of the elastic constants defined in equation (7.1).

(ii) Substitute the reduced constitutive relationships and employ the strain–displacement relationships to write the force and moment resultants in the form

$$\begin{aligned}
N_{yy}^{i} &= A_{11}^{ij}\left(\frac{\partial v_j}{\partial y} + \frac{w_j}{R}\right) + \bar{A}_{13}^{ij} w_j + \frac{D_{11}^{ijk}}{2}\left(\frac{\partial w_j}{\partial y}\frac{\partial w_k}{\partial y}\right), \\
N_{yz}^{i} &= A_{44}^{ji}\left(\frac{\partial w_j}{\partial y} - \frac{v_j}{R}\right) + \bar{A}_{44}^{ji} v_j, \\
M_{yy}^{ij} &= D_{11}^{ijk}\left(\frac{\partial v_k}{\partial y} + \frac{w_k}{R}\right) + \frac{F_{11}^{ijkl}}{2}\left(\frac{\partial w_k}{\partial y}\frac{\partial w_l}{\partial y}\right), \\
Q_{yz}^{i} &= \bar{A}_{44}^{ji}\left(\frac{\partial w_j}{\partial y} - \frac{v_j}{R}\right) + \bar{\bar{A}}_{44}^{ji} v_j, \\
Q_{zz}^{i} &= \bar{A}_{13}^{ij}\left(\frac{\partial v_j}{\partial y} + \frac{w_j}{R}\right) + \bar{\bar{A}}_{33}^{ij} w_j + \frac{\bar{D}_{13}^{jki}}{2}\left(\frac{\partial w_j}{\partial y}\frac{\partial w_k}{\partial y}\right),
\end{aligned}$$

where repeated indices are summed over all layers and with the coefficients A_{kl}^{ij}, $\bar{A}_{kl}^{ij}$, $\bar{A}_{kl}^{ji}$, D_{11}^{ijk}, $\bar{D}_{mn}^{ijk}$ and F_{11}^{ijkl} defined as

$$A_{mn}^{ij} = \int_{-\frac{h}{2}}^{\frac{h}{2}} C_{mn}^{r} \delta_{ri} \Phi^{i} \Phi^{j} \mathrm{d}z, \quad \bar{A}_{mn}^{ij} = \int_{-\frac{h}{2}}^{\frac{h}{2}} C_{mn}^{r} \delta_{ri} \Phi^{i} \frac{\mathrm{d}\Phi^{j}}{\mathrm{d}z} \mathrm{d}z,$$

$$\bar{\bar{A}}_{mn}^{ij} = \int_{-\frac{h}{2}}^{\frac{h}{2}} C_{mn}^{r} \delta_{ri} \frac{\mathrm{d}\Phi^{i}}{\mathrm{d}z} \frac{\mathrm{d}\Phi^{j}}{\mathrm{d}z} \mathrm{d}z, \quad D_{mn}^{ijk} = \int_{-\frac{h}{2}}^{\frac{h}{2}} C_{mn}^{r} \delta_{ri} \Phi^{i} \Phi^{j} \Phi^{k} \mathrm{d}z,$$

$$\bar{D}_{mn}^{ijk} = \int_{-\frac{h}{2}}^{\frac{h}{2}} C_{mn}^{r} \delta_{ri} \Phi^{i} \Phi^{j} \frac{\mathrm{d}\Phi^{k}}{\mathrm{d}z} \mathrm{d}z \quad \text{and} \quad F_{mn}^{ijkl} = \int_{-\frac{h}{2}}^{\frac{h}{2}} C_{mn}^{r} \delta_{ri} \Phi^{i} \Phi^{j} \Phi^{k} \Phi^{l} \mathrm{d}z.$$

The quantity δ_{ri} is the Kronecker delta, which is equal to 1 when $r = i$ and zero otherwise.
(iii) Assume suitable shape functions $\mathbf{N}_j$ for the nodal functions of the in-plane coordinate y, representing the in-plane and transverse displacements of each interface v_i and w_i,

$$\left[v_i(y) \quad w_i(y) \right] = \sum_{j=1}^{2} \mathbf{N}_j \left[v_{ij} \; w_{ij} \; L\theta_{ij} \right]^T$$

and hence write the strain–displacement relationships in the form $\varepsilon = \mathbf{B}(\mathbf{w})\mathbf{w}$. Obtain expressions for the elements of the matrix $\mathbf{B}(\mathbf{w})$.
(iv) Show that the stiffness matrix for a curved beam element may be expressed as

$$\mathbf{K} = \int_{S} \mathbf{B}^{\mathrm{T}} \mathbf{D} \mathbf{B} \mathrm{d}S.$$

Obtain expressions for the elements of the matrix $\mathbf{D}$.
Hence or otherwise derive the corresponding mass matrix. (The rotary inertia may be ignored.)
(v) Verify the expressions for a flat beam element by setting the radius of curvature of the element R to infinity and the number of layers to unity.
(vi) Make suitable assumptions and generalize the element matrices for a beam excited by layers of piezoelectric patches.

4. Define the force and moment resultants across a uniform, homogeneous plate performing moderate rotations under the action of a uniformly distributed force q, as

$$N_{ij} = \int_{-h/2}^{h/2} \sigma_{ij} \mathrm{d}z, \quad M_{ij} = \int_{-h/2}^{h/2} \sigma_{ij} z \mathrm{d}z,$$

where σ_{ij} are components of the Piola–Kirchhoff stress tensor.
(i) Show that the force equilibrium equations may be expressed as

$$\frac{\partial N_{11}}{\partial x} + \frac{\partial N_{21}}{\partial y} = 0, \quad \frac{\partial N_{12}}{\partial x} + \frac{\partial N_{22}}{\partial y} = 0, \quad \frac{\partial N_{13}}{\partial x} + \frac{\partial N_{23}}{\partial y} + q = 0.$$

(ii) Show that the moment equilibrium equations may be expressed as

$$\frac{\partial M_{11}}{\partial x}+\frac{\partial M_{21}}{\partial y}=N_{31},\quad \frac{\partial M_{12}}{\partial x}+\frac{\partial M_{22}}{\partial y}=N_{32}.$$

(iii) Observe that the force resultants N_{31} and N_{32} are not equal to N_{13} and N_{23}, respectively. From strain energy considerations and from expressions for the deformation tensor components, it can be shown that

$$N_{13}=N_{31}+N_{11}\frac{\partial w}{\partial x}+N_{21}\frac{\partial w}{\partial y},\quad N_{23}=N_{32}+N_{12}\frac{\partial w}{\partial x}+N_{22}\frac{\partial w}{\partial y}.$$

Hence show that

$$\frac{\partial^2 M_{11}}{\partial x^2}+\frac{\partial^2}{\partial x\partial y}(M_{12}+M_{21})+\frac{\partial^2 M_{22}}{\partial y^2}+N_{11}\frac{\partial^2 w}{\partial x^2}+2N_{12}\frac{\partial^2 w}{\partial x\partial y}+N_{22}\frac{\partial^2 w}{\partial y^2}=0.$$

5. Consider a uniform, homogeneous plate performing moderate rotations under the action of a uniformly distributed force q and show that the strain energy in bending and extension may be expressed as

$$\begin{aligned}P=&\frac{D}{2}\int_S\left(\left(\nabla^2 w\right)^2-2(1-\nu)\left(\frac{\partial^2 w}{\partial x^2}\frac{\partial^2 w}{\partial y^2}-\left(\frac{\partial^2 w}{\partial x\partial y}\right)^2\right)\right)\mathrm{d}S\\&+\frac{1}{2Eh}\int_S\left(\left(\nabla^2\Phi\right)^2+Eh\left(\frac{\partial^2 w}{\partial x^2}\frac{\partial^2 w}{\partial y^2}-\left(\frac{\partial^2 w}{\partial x\partial y}\right)^2\right)\Phi\right)\mathrm{d}S,\end{aligned}$$

where $\Phi(x, y)$ is the Airy stress function.

6. Use the results of exercises 4 and 5 to show that the coupled differential equation governing the transverse displacement $w(x, y)$ of a uniform isotropic thin plate with moderate in-plane stresses and the Airy stress function $\Phi(x, y)$ in rectangular Cartesian coordinates may be expressed as

$$\nabla^4\Phi+Eh\left(\frac{\partial^2 w}{\partial x^2}\frac{\partial^2 w}{\partial y^2}-\left(\frac{\partial^2 w}{\partial x\partial y}\right)^2\right)=0,$$

$$D\nabla^4 w-\frac{\partial^2\Phi}{\partial y^2}\frac{\partial^2 w}{\partial x^2}-\frac{\partial^2\Phi}{\partial x^2}\frac{\partial^2 w}{\partial y^2}+2\frac{\partial^2\Phi}{\partial x\partial y}\frac{\partial^2 w}{\partial x\partial y}+q=0,$$

where D is the plate's flexural rigidity, E the Young's modulus, q the distributed loading and h the plate thickness. The in-plane stress resultants are

$$N_{xx}=\frac{\partial^2\Phi}{\partial y^2},\quad N_{yy}=\frac{\partial^2\Phi}{\partial x^2},\quad N_{xy}=-\frac{\partial^2\Phi}{\partial x\partial y}.$$

7.

(i) Consider a rectangular thin, uniform, homogeneous plate, with sides of length a and b, simply supported along two of the opposite sides and fixed along the remaining two opposite sides.

Show that the displacement of the plate may be expressed as

$$w(x, y) = \sum_{m=1}^{\infty} A_m g\left(m\pi y/a\right) \sin\left(m\pi x/a\right),$$

where, $g(\bar{y}) = (\alpha_m \coth \alpha_m + 1)\cosh(\bar{y}) + \bar{y}\sinh(\bar{y})$, $\alpha_m = m\pi b/2a$.

(ii) Consider a rectangular thin, uniform, homogeneous plate, simply supported along two of the opposite sides and free along the remaining two opposite sides.

Show that the boundary conditions along these two edges are of the form

$$\frac{\partial^2 w}{\partial x^2} + \nu \frac{\partial^2 w}{\partial y^2} = 0, \quad D\frac{\partial}{\partial x}\left(\frac{\partial^2 w}{\partial x^2} + (2-\nu)\frac{\partial^2 w}{\partial y^2}\right) = 0.$$

(iii) Consider a rectangular thin, uniform, homogeneous plate, simply supported along two of the opposite sides and subject to a compressive in-plane conservative load P along the remaining two opposite sides.

Show that the boundary conditions along these two edges are of the form

$$\frac{\partial^2 w}{\partial x^2} + \nu \frac{\partial^2 w}{\partial y^2} = 0, \quad D\frac{\partial}{\partial x}\left(\frac{\partial^2 w}{\partial x^2} + (2-\nu)\frac{\partial^2 w}{\partial y^2}\right) + P\frac{\partial w}{\partial x} = 0.$$

(iv) Consider a rectangular thin, uniform, homogeneous plate, simply supported along two of the opposite sides and subject to a compressive in-plane non-conservative follower load along the remaining two opposite sides.

Show that the boundary conditions along these two edges are of the form

$$\frac{\partial^2 w}{\partial x^2} + \nu \frac{\partial^2 w}{\partial y^2} = 0, \quad D\frac{\partial}{\partial x}\left(\frac{\partial^2 w}{\partial x^2} + (2-\nu)\frac{\partial^2 w}{\partial y^2}\right) = 0.$$

(v) Consider a rectangular thin, uniform, homogeneous plate, with sides of length a and b, simply supported along two of the opposite sides and rotationally restrained along the remaining two opposite sides by identical distributed springs of stiffness K_r.

Show that one of the boundary conditions along each of these two edges is of the form

$$D\left(\frac{\partial^2 w}{\partial x^2} + \nu \frac{\partial^2 w}{\partial y^2}\right) = K_r\frac{\partial w}{\partial x} \text{ along } x = 0, \quad D\left(\frac{\partial^2 w}{\partial x^2} + \nu \frac{\partial^2 w}{\partial y^2}\right) = -K_r\frac{\partial w}{\partial x} \text{ along } x = a.$$

State the other boundary condition along each of these two edges.

8. Consider a rectangular thin, uniform, homogeneous plate, simply supported along two of the opposite sides and fixed along the remaining two opposite sides. The plate is loaded by electrostatic forces generated by maintaining the plate at a fixed potential with reference to the ground plane.

(i) Show that the electrostatic force is directly proportional to the first derivative of the capacitance with respect to the transverse displacement.

(ii) Also show that the loading on the plate is directly proportional to the plate's transverse displacement.

(iii) Hence obtain the natural frequencies and normal mode shapes of the plate.

9.

(i) It is desired to develop a finite element for the transverse vibration of a thin plate described by Kirchhoff's equation of motion. The finite element is assumed to be rectangular in shape with the element coordinate system centred at the mid-point of the plate. The four nodes are defined in terms of non-dimensional coordinates at $\xi = \pm 1$ and $\eta = \pm 1$.

The displacement, slopes and twist at each node are the element degrees of freedom. The displacement is approximated in terms of the shape functions $\mathbf{N}_i$ as

$$w(x,y)=\sum_{i=1}^{4}\mathbf{N}_i\left[w_i\quad(\partial w/\partial x)_i\quad(\partial w/\partial y)_i\quad(\partial^2 w/\partial x\partial y)_i\right]^T$$

or

$$w=\sum_{i=1}^{4}w_iN_{i1}+(\partial w/\partial x)_iN_{i2}+(\partial w/\partial y)_iN_{i3}+(\partial^2 w/\partial x\partial y)_iN_{i4},$$

$$\mathbf{N}_i=\begin{bmatrix}N_{i1}\\N_{i2}\\N_{i3}\\N_{i4}\end{bmatrix}^T=\frac{1}{16}\begin{bmatrix}(2\pm 3\xi\mp\xi^3)(2\pm 3\eta\mp\eta^3)\\(\mp 1-\xi\pm\xi^2+\xi^3)(2\pm 3\eta\mp\eta^3)\\(2\pm 3\xi\mp\xi^3)(\mp 1-\eta\pm\eta^2+\eta^3)\\(\mp 1-\xi\pm\xi^2+\xi^3)(\mp 1-\eta\pm\eta^2+\eta^3)\end{bmatrix}^T,$$

where the upper signs apply within each of the bracketed expressions for the node at $\xi=1$ or at $\eta=1$ and the lower signs for the node at $\xi=-1$ or $\eta=-1$.

Show that the element stiffness matrix is given by

$$\mathbf{K}=\int_S\mathbf{B}^T\mathbf{D}\mathbf{B}\,dS,$$

where

$$\mathbf{D}=D\begin{bmatrix}1&\nu&0\\\nu&1&0\\0&0&-(1-\nu)/2\end{bmatrix},\quad\mathbf{B}=\begin{bmatrix}-\dfrac{\partial^2\mathbf{N}}{\partial x^2}\\-\dfrac{\partial^2\mathbf{N}}{\partial y^2}\\-2\dfrac{\partial^2\mathbf{N}}{\partial x\partial y}\end{bmatrix},$$

$$\frac{\partial}{\partial x}=\frac{1}{a}\times\frac{\partial}{\partial\xi},\quad\frac{\partial}{\partial y}=\frac{1}{b}\times\frac{\partial}{\partial\eta},\quad\mathbf{N}=\left[\mathbf{N}_1\ \mathbf{N}_2\ \mathbf{N}_3\ \mathbf{N}_4\right].$$

Observe a few features one can exploit to write a FEM code for the control of plate-like structures in MATLAB.

(a) The shape functions can all be written in terms of just two independent polynomials known as Hermite polynomials.

(b) Products in polynomials in the ξ and in η can be easily done by convolution in MATLAB using the *conv.m* *m*-file. The polynomials may each be integrated in one dimension using the numerical quadrature *m*-file, *quad.m*. Thus, one should first establish all the coefficients of the element stiffness matrix prior to performing the integration as polynomials in descending powers.

(c) If the structure is modelled in terms of a number of self-similar elements, the element matrices need to be computed just once.

(d) If the entire problem could be modelled in two dimensions, there is no need to introduce the third dimension.

(ii) Derive the representative finite element stiffness and inertia matrices.

(iii) Now assume the material to be orthotropic and repeat the derivations in (i) and (ii).

(iv) Assume that the plate is piezoelectric and determine the corresponding forcing vector for a transverse electric field excitation.

(v) Assume a three-layer laminate with the middle layer non-piezoelectric, and derive the finite element stiffness, mass and forcing vector distribution matrices when the transverse excitation is for a bimorph configuration.

(vi) Using MATLAB/Simulink, model a typical rectangular plate by 10×10 elements and establish the equations of motion for the plate in a global reference frame. Obtain the natural frequencies and mode shapes of the plate. (You will probably need to organize your code into five or six m-files. The first one defines the various shape functions and the derivatives. The second could define the element matrix densities prior to integration. The third could evaluate the element matrices by numerical quadrature. The fourth could assemble and transform them into global coordinates. The fifth and final one is used to remove the restrained degrees of freedom and to analyse the global matrices. Save your m-files for use with the latter exercises.)

10.

(i) It is desired to develop a finite element for the transverse vibration of a moderately thick plate described by the Mindlin–Reissner equation of motion. The finite element is assumed to be rectangular in shape with the element coordinate system centred at the mid-point of the plate. The four nodes are defined in terms of non-dimensional coordinates at $\xi = \pm 1$ and $\eta = \pm 1$.

The displacement and rotations at each node are the element degrees of freedom. The displacement is approximated as

$$\begin{bmatrix} w \\ \theta_x \\ \theta_y \end{bmatrix} = \sum_{i=1}^{4} \begin{bmatrix} w_i N_i \\ \theta_{xi} N_i \\ \theta_{yi} N_i \end{bmatrix},$$

where the shape functions

$$N_i = (1 \pm \xi)(1 \pm \eta)/4.$$

The upper signs apply within each of the bracketed expressions for the node at $\xi = 1$ or at $\eta = 1$ and the lower signs for the node at $\xi = -1$ or $\eta = -1$.

Show that the element stiffness matrix is given by

$$\mathbf{K} = \int_S \mathbf{B}_1^T \mathbf{D}_1 \mathbf{B}_1 \mathrm{d}S + \int_S \mathbf{B}_2^T \mathbf{D}_2 \mathbf{B}_2 \mathrm{d}S,$$

where

$$\mathbf{D}_1 = D \begin{bmatrix} 1 & \nu & 0 \\ \nu & 1 & 0 \\ 0 & 0 & -(1-\nu)/2 \end{bmatrix}, \quad \mathbf{B}_1 = -\begin{bmatrix} \dfrac{\partial \mathbf{N}_2}{\partial x} \\ \dfrac{\partial \mathbf{N}_3}{\partial y} \\ \dfrac{\partial \mathbf{N}_2}{\partial y} + \dfrac{\partial \mathbf{N}_3}{\partial x} \end{bmatrix},$$

$$\frac{\partial}{\partial x} = \frac{1}{a} \times \frac{\partial}{\partial \xi}, \quad \frac{\partial}{\partial y} = \frac{1}{b} \times \frac{\partial}{\partial \eta}, \quad \mathbf{D_2} = \frac{\kappa h E}{2(1+\nu)} \begin{bmatrix} 1 & 0 \\ 0 & 1 \end{bmatrix},$$

$$\mathbf{B}_2 = \begin{bmatrix} \dfrac{\partial \mathbf{N}_1}{\partial x} - \mathbf{N}_2 \\ \dfrac{\partial \mathbf{N}_1}{\partial y} - \mathbf{N_3} \end{bmatrix}, \quad \begin{bmatrix} \mathbf{N}_1 \\ \mathbf{N}_2 \\ \mathbf{N}_3 \end{bmatrix} = \begin{bmatrix} N_1 \mathbf{I}_{3\times 3} & N_2 \mathbf{I}_{3\times 3} & N_3 \mathbf{I}_{3\times 3} & N_4 \mathbf{I}_{3\times 3} \end{bmatrix}.$$

(ii) Derive the representative finite element stiffness and inertia matrices. (While performing the integrations use either exact or a high-order numerical integration to avoid the problems of 'shear locking'.)
(iii) Now assume the material to be orthotropic and repeat the derivations in (i) and (ii).

(iv) Assume that the plate is piezoelectric and determine the corresponding forcing vector for a transverse electric field excitation.
(v) Assume a three-layer laminate with the middle layer non-piezoelectric and derive the finite element stiffness, mass and forcing vector distribution matrices when the transverse excitation is for a bimorph configuration. All layers are assumed to be of the same thickness.
(vi) Using MATLAB/Simulink, model a typical rectangular plate by 10×10 elements and establish the equations of motion for the plate in a global reference frame. Obtain the natural frequencies and mode shapes of the plate. (Save your *m*-files for use with later exercises.)

11.

(i) It is desired to develop a finite element for the transverse vibration of a moderately thick plate described by the first-order zig-zag theory. The finite element is assumed to be rectangular ($2a \times 2b$) in shape with the element coordinate system centred at the mid-point of the plate. The four nodes are defined in terms of non-dimensional coordinates at $\xi = \pm 1$ and $\eta = \pm 1$.

Assume a three-layer laminate with the middle layer non-piezoelectric. All layers are assumed to be of the same thickness. The displacement and rotations at each node are the element degrees of freedom. The displacement is approximated as

$$\begin{bmatrix} w \\ \theta_x \\ \theta_y \\ \psi_x \\ \psi_y \end{bmatrix} = \sum_{i=1}^{4} \begin{bmatrix} w_i N_i \\ \theta_{xi} N_i \\ \theta_{yi} N_i \\ \psi_{xi} N_i \\ \psi_{yi} N_i \end{bmatrix},$$

where the shape functions

$$N_i = (1 \pm \xi)(1 \pm \eta)/4$$

and the upper signs apply within each of the bracketed expressions for the node at $\xi = 1$ or at $\eta = 1$ and the lower signs for the node at $\xi = -1$ or $\eta = -1$.

Show that the element stiffness matrix is given by

$$\mathbf{K} = \int_S \mathbf{B}_1^T \mathbf{D}_1 \mathbf{B}_1 \mathrm{d}S + \int_S \mathbf{B}_2^T \mathbf{D}_2 \mathbf{B}_2 \mathrm{d}S,$$

where

$$\mathbf{D}_1 = D \begin{bmatrix} 1 & \nu & 0 \\ \nu & 1 & 0 \\ 0 & 0 & -(1-\nu)/2 \end{bmatrix}, \quad \mathbf{B}_1 = -\begin{bmatrix} \dfrac{\partial \mathbf{N}_2}{\partial x} + (-1)^k \dfrac{\partial \mathbf{N}_4}{\partial x} \\ \dfrac{\partial \mathbf{N}_3}{\partial y} + (-1)^k \dfrac{\partial^2 \mathbf{N}_5}{\partial y} \\ \dfrac{\partial \mathbf{N}_2}{\partial y} + (-1)^k \dfrac{\partial \mathbf{N}_4}{\partial y} + \dfrac{\partial \mathbf{N}_3}{\partial x} + (-1)^k \dfrac{\partial \mathbf{N}_5}{\partial x} \end{bmatrix},$$

$$\mathbf{D}_2 = \frac{\kappa h E}{2(1+\nu)} \begin{bmatrix} 1 & 0 \\ 0 & 1 \end{bmatrix},$$

$$\mathbf{B}_2 = \begin{bmatrix} \dfrac{\partial \mathbf{N}_1}{\partial x} - \mathbf{N}_2 - (-1)^k \mathbf{N}_4 \\ \dfrac{\partial \mathbf{N}_1}{\partial y} - \mathbf{N}_3 - (-1)^k \mathbf{N}_5 \end{bmatrix}, \quad \frac{\partial}{\partial x} = \frac{1}{a} \times \frac{\partial}{\partial \xi}, \quad \frac{\partial}{\partial y} = \frac{1}{b} \times \frac{\partial}{\partial \eta}, \quad \begin{bmatrix} \mathbf{N}_1 \\ \mathbf{N}_2 \\ \mathbf{N}_3 \\ \mathbf{N}_4 \\ \mathbf{N}_5 \end{bmatrix}$$

$$= \begin{bmatrix} N_1 \mathbf{I}_{5\times5} & N_2 \mathbf{I}_{5\times5} & N_3 \mathbf{I}_{5\times5} & N_4 \mathbf{I}_{5\times5} \end{bmatrix},$$

and k is the layer number. It is zero for the middle layer, 1 for the top layer and -1 for the bottom layer.
(ii) Derive the finite element stiffness, mass and forcing vector distribution matrices when the transverse excitation is for a bimorph configuration.
(iii) Using MATLAB/Simulink, model a typical rectangular plate by 10×10 elements and establish the equations of motion for the plate in a global reference frame. Obtain the natural frequencies and mode shapes of the plate. (Save your m-files for use with later exercises.)
(iv) Compare your results with the results obtained in the previous exercise.

12.

(i) The general theory of elastic shells is based on the following three assumptions:
(a) The displacement vector is linear and normal to the coordinate curves in the middle surface.
(b) The distances between points on a normal to the undeformed middle surface do not change during deformation.
(c) The stresses are replaced by a system of stress and moment resultants.
Show that moment and stress and moment resultants for shallow spherical shell satisfy the relationships

$$\frac{\partial N_{12}}{\partial x} + \frac{\partial N_{22}}{\partial y} + \frac{N_{32}}{R} = 0, \qquad \frac{\partial N_{13}}{\partial x} + \frac{\partial N_{23}}{\partial y} - \frac{N_{11} + N_{22}}{R} + q = 0,$$
$$\frac{\partial M_{11}}{\partial x} + \frac{\partial M_{21}}{\partial y} = N_{31}, \qquad \frac{\partial M_{12}}{\partial x} + \frac{\partial M_{22}}{\partial y} = N_{32},$$
$$N_{12} - N_{21} + \frac{M_{12} - M_{21}}{R} = 0.$$

(ii) Starting from the general expressions for the gradient and divergence operators in spherical coordinates, obtain expressions for the strain–displacement relationships in spherical coordinates and show that the strain–displacement relationships for a shallow spherical shell are given by

$$\varepsilon_{xx} = u_{x,x} + \frac{u_z}{R}, \qquad \varepsilon_{yy} = u_{y,y} + \frac{u_z}{R}, \qquad \varepsilon_{xy} = u_{x,y} + u_{y,x},$$
$$\varepsilon_{xz} = u_{z,x} + \beta_x - \frac{u_x}{R}, \qquad \varepsilon_{yz} = u_{z,y} + \beta_y - \frac{u_y}{R}.$$

The angles β_x and β_y are the angles through which the normal vector to the middle surface rotates in the direction of the x and y coordinate lines, respectively. Show that the curvature relationships may be expressed in terms of β_x and $\beta_y$as

$$\kappa_{,xx} = \beta_{x,x}, \qquad \kappa_{,yy} = \beta_{y,y} \quad \text{and} \quad \kappa_{,xy} = \beta_{x,y} + \beta_{y,x}.$$

Hence obtain the displacement equilibrium equations.
(iii) The results of exercise 4 form the basis of shallow shell theory for shells that moderately curved. To develop a shallow shell theory, we assume that the squares and products of the derivates of the independent variables are negligible in comparison with unity. The transverse shear stress resultants are ignored in the in-plane equilibrium equations. The tangential displacements are ignored in the expressions for the transverse shear strains. Hence show that the equations for the transverse displacement may be expressed as

$$\nabla^4\Phi + \frac{Eh}{R}\nabla^2 w = 0,$$
$$D\nabla^4 w - \frac{1}{R}\nabla^2\Phi + q = 0,$$

where $\Phi(x, y)$ is the Airy stress function.

13. In cylindrical coordinates x, φ, z, assume the displacements of the middle surface of a cylindrical shell of radius R to be u, v, w, and rotations of the normal at point in the middle surface about the x, φ

coordinates be ψ_x and ψ_φ. The strain– and curvature–displacement relationships are

$$\varepsilon_{xx} = u_{,x} + z\psi_{x,x}, \quad \varepsilon_{\varphi\varphi} = \frac{v_{,\varphi}}{R} + \frac{z}{R}\psi_{\varphi,\varphi} + \frac{w}{R},$$

$$\varepsilon_{x\varphi} = v_{,x} + \frac{u_{,\varphi}}{R} + z\left(\frac{\psi_{x,\varphi}}{R} + \psi_{\varphi,x}\right), \quad \varepsilon_{xz} = w_{,x} + \psi_x,$$

$$\varepsilon_{yz} = \frac{w_{,\varphi}}{R} + \psi_\varphi - \frac{v}{R}, \quad \kappa_{,xx} = \psi_{x,x}, \quad \kappa_{,\varphi\varphi} = \psi_{\varphi,\varphi}$$

and

$$\kappa_{,x\varphi} = \frac{\psi_{x,\varphi}}{R} + \psi_{\varphi,x}.$$

(i) Show that the kinetic energy of an element of the material of density ρ is given by

$$\mathsf{T} = \frac{1}{2}\mathbf{v}^T\mathbf{J}\mathbf{v}, \quad \mathbf{J} = \rho\begin{bmatrix} 1 & 0 & 0 & \beta & 0 \\ 0 & 1 & 0 & 0 & \beta \\ 0 & 0 & 1 & 0 & 0 \\ \beta & 0 & 0 & \beta R & 0 \\ 0 & \beta & 0 & 0 & \beta R \end{bmatrix}, \quad \beta = h^2/12R \text{ and } \mathbf{v}^T = \frac{\partial}{\partial t}\begin{bmatrix} u & v & w & \psi_x & \psi_\varphi \end{bmatrix}.$$

(ii) Obtain expressions for the stress, moment and shear force resultants.
(iii) Assume that the shell is transversely loaded by a distribution of forces $q(x,t)$ and derive the displacement equations of motion of cylindrical shell. (These equations are often referred to as the Mirsky–Herrmann theory.)
(iv) If the distribution of force actuators is replaced by a distribution of PZT patch actuators, derive the displacement equations of motion of cylindrical shell.

References

Ballato, A. (2001) Modeling Piezoelectric and piezomagnetic devices and structures via equivalent networks, Invited Paper, *IEEE Transactions on Ultrasonics, Ferroelectrics, and Frequency Control* 48 (5), 1189–1240.

Batra, R. C. and Vidoli, S. (2002) Higher order piezoelectric plate theory derived from a three-dimensional variational principle, *AIAA Journal* 40(1), 91–104.

Carrera, E. (2003) Historical review of zig-zag theories for multilayered plates and shells, *Applied Mechanics Reviews* 56(3), 287–308.

Carrera, E. (2004) On the use of Murakami's zig-zag functions in the modeling of layered plates and shells. *Computers and Structures* 82, 541–554.

Daniel, I. M. and Ishai, O. (2005) *Engineering Mechanics of Composite Materials*, 2nd edn, Oxford University Press, Oxford.

Demasi, L. (2007) Plate theories for thick and thin plates: the generalized unified formulation, *Composite Structures*, doi: 10.1016/j.compstruct.2007.08.004.

Franco-Correia, V., Mota-Soares, C. M. and Mota-Soares, C. A. (1998) Modeling and design of adaptive composite structures. Proceedings NATO Advanced Study Institute on Mechanics of Composite Materials and Structures, Troia, Portugal, Vol. II, 59–92.

Hyer, M. W. (1998) *Stress Analysis of Fiber-Reinforced Composite Materials*, WCB/McGraw-Hill Inc., New York.

Jones, R. M. (1999) *Mechanics of Composite Materials*, 2nd edn, Taylor & Francis, Philadelphia.

Kapania, R. K. and Liu, Y. (2000) Static and vibration analyzes of general wing structures using equivalent-plate models, *AIAA Journal* 38(7), 1269–1277.

Kassegne, S. K. and Reddy, J. N. (1997) A layerwise shell stiffener and stand-alone curved beam element, *Asian Journal of Structural Engineering* 2(1 and 2), 1–14.
Kwon, Y. W. and Bang, H. (2000) *The Finite Element Method Using MATLAB*, 2nd edn, Dekker Mechanical Engineering Series, CRC Press, Boca Raton.
Lee, C. K. (1990) Theory of laminated piezoelectric plates for the design of distributed sensors/actuators. Part I: governing equations and reciprocal relationships, *Journal of Acoustical Society of America* 89, 1144–1158.
Mindlin, R. D. (1951) Influence of rotatory inertia and shear on flexural motions of isotropic, elastic plates, *Journal of Applied Mechanics* 18, 31–38.
Murakami, H. (1986) Laminated composite plate theory with improved in-plane response, *Journal of Applied Mechanics* 53, 661–666.
Murthy, M. V. V. (1981) An Improved Transverse Shear Deformation Theory for Laminated Anisotropic Plates, NASA Technical Paper 1903.
Pagano, N. J. (1970) Exact solutions for rectangular bidirectional composites and sandwich plates, *Journal of Composite Materials* 4, 20–34.
Phan, N. D. and Reddy, J. N. (1985) Analysis of laminated composite plates using a higher-order shear deformation theory, *International Journal for Numerical Methods in Engineering* 21, 2201–2219.
Reddy, J. N. (1984) A simple higher-order theory for laminated composite plates, *Journal of Applied Mechanics* 51, 745–752.
Reddy, J. N. (2004) *Mechanics of Laminated Composite Plates and Shells*, 2nd edn, CRC Press, Boca Raton.
Reissner, E. (1945) The effect of transverse shear deformation on the bending of elastic plates, *Journal of Applied Mechanics* 12 (1), A-69–A-77.
Shiau, L.-C. and Wu, T.-Y. (2001) Nonlinear flutter of laminated plates with in-plane force and transverse shear effects, *Mechanics Based Design of Structures and Machines* 29(1), 121–142.
Stein, M., Sydow, P. D. and Librescu, L. (1990) Postbuckling Response Of Long Thick Plates Loaded In Compression Including Higher Order Transverse Shearing Effects, NASA Technical Memorandum 102663.
Suleman, A., Modi, V.J. and Venkayya, V. B. (1995) Structural modeling issues in flexible systems, *AIAA Journal* 33, 919–923.
Sze, K. Y. and Yao, L. Q. (2000) Modeling smart structures with segmented piezoelectric sensors and actuators, *Journal of Sound and Vibration* 235, 495–520.
Tsai, S. W. and Hahn, H. T. (1980) *Introduction to Composite Materials*, Technomic Publishing Co., Inc., Lancaster, PA., USA.
Vepa, R. (2008) Aeroelastic Analysis of Wing Structures Using Equivalent Plate Models, *AIAA Journal* 46(5), 1216–1223.
Whitney, J. M. and Pagano, N. J. (1970) Shear deformation in heterogeneous anisotropic plates, *Journal of Applied Mechanics* 37(4), 1031–1036.

8

Dynamics of Thermoelastic Media: Shape Memory Alloys

8.1 Fundamentals of Thermoelasticity

Layered elastic structures are usually subject to a variety of dynamic loads, including changes in temperature. Because the layers may possess varying thermo-mechanical properties, temperature changes would induce non-uniform stresses within the structure. Thus, the study of the thermoelasticity of such structures is of paramount importance. It is not enough to understand the influence of temperature changes on the state of stress in the structure, it is also essential to understand how the changes in the state of stress influence the temperature distributions within the structure. To be able to systematically model the constitutive relationships, it is essential to revisit the basic thermodynamic concepts (Vedantam, 2000) to establish the foundations of linear thermoelasticity.

8.1.1 Basic Thermodynamic Concepts

The first law of thermodynamics is essentially the statement of the principle of conservation of energy applied to thermodynamic systems. Thus, the variation in the internal energy of a thermodynamic system due to transformation of the state of the system ΔU can be expressed in terms of the work ΔW performed during the transformation and the amount of energy Q received by the system in forms other than work, as $\Delta U = Q - \Delta W$.

The first law arose primarily to state the impossibility of constructing a machine that could create energy. It places no limitations on the possibility of transforming heat into work or work into heat, as they are different forms of energy. However, the second law explicitly places certain limitations on how much heat can be transformed to useful work. Thus, the second law of thermodynamics states that 'A transformation whose only final result is to transform into work, heat extracted from a source which is the same temperature throughout is impossible'. Alternatively, the second law may be stated as 'A transformation whose only final result is to transfer heat from a body at a given temperature to a body at a higher temperature is impossible'. This implies that only part of the heat that is absorbed by the system from the source at a higher temperature can be transformed to useful work and that the rest of the heat must be surrendered to the source at lower temperature. This leads naturally to the concept of entropy, and the second law could be stated as 'For any transformation occurring in an isolated system the entropy of the final state can never be less than that of the initial state'. Thus, if an isolated system is

Dynamics of Smart Structures Dr. Ranjan Vepa

in a state of maximum entropy, when this state corresponds to a certain energy state, it is the most stable state as it cannot undergo any further transformations. The second law may also be stated in terms of the temperature T and the entropy S as

$$\int T\mathrm{d}S \geq Q, \tag{8.1}$$

where the equality sign holds for a reversible transformation.

In thermodynamics, it is customary to define three other thermodynamic potentials in addition to the internal energy. The first of these is the enthalpy H, which is defined to account for the fact that it is generally a lot more convenient to treat the pressure as an input variable and is defined as

$$H = U + PV. \tag{8.2}$$

Assuming that the pressure is prescribed (or held constant),

$$\mathrm{d}H = \mathrm{d}U + P\mathrm{d}V = \mathrm{d}Q, \tag{8.3}$$

which is a succinct statement of the first law. However, both U and H are natural functions of entropy. Yet it is not at all convenient to have entropy as an input variable. For this reason, one defines the Helmholtz free energy as

$$F = U - TS. \tag{8.4}$$

For a constant temperature transformation (or when the temperature is prescribed),

$$\mathrm{d}F = \mathrm{d}U - T\mathrm{d}S = \mathrm{d}W + \mathrm{d}W_{other}. \tag{8.5}$$

Thus, for a constant temperature process, the Helmholtz free energy gives all the work done, including both the useful and the other work done. Consequently, if a system is in thermal contact with the environment at a temperature T, and it is dynamically isolated in such a way that external work can be performed and absorbed by the system, the change in the free energy is equal to the change in the total work done. When the free energy is a minimum, the system is in state of stable equilibrium. Thus, the free energy is analogous to potential energy in a mechanical system. However, the free energy, like the internal energy, is a function of the volume, which is not an input variable that is easily controllable. For this reason, thermodynamicists define yet another potential, the Gibbs free energy,

$$G = H - TS. \tag{8.6}$$

However, it is the specific Helmholtz free energy which is the Helmholtz free energy per unit mass that plays the primary role in determining the dynamics of the system. If the specific Helmholtz free energy is defined by the function ψ_F, it is a function of the deformation gradient and the temperature T. At a particular equilibrium point, the elastic coefficients, evaluated at constant temperature, may be expressed in terms of the linearized elastic coefficients as

$$C_{ijkl} = \frac{\partial^2 \psi_F}{\partial \varepsilon_{ij} \partial \varepsilon_{kl}}. \tag{8.7}$$

Thus, the stress components are related to the strain components as

$$\sigma_{ij} = \rho \frac{\partial \psi_F}{\partial \varepsilon_{ij}} = C_{ijkl} \left(\varepsilon_{kl} - \alpha_k \left(T - T_0 \right) \delta_{kl} \right), \tag{8.8}$$

where α_i are the coefficients of thermal expansion.

The specific entropy η, evaluated at constant strain, is obtained from the specific Helmholtz free energy as

$$\eta = -\frac{\partial \psi_F}{\partial T} = c_v \frac{(T - T_0)}{T_0} + \left(\beta_1 \varepsilon_{11} + \beta_2 \varepsilon_{22} + \beta_3 \varepsilon_{33} \right), \tag{8.9}$$

where β_i are the coefficients of thermal stressing. The specific heat capacity c_v is defined at constant strain and is different from the specific heat capacity c_p, defined at constant stress but related to it.

The generalized constitutive equations for a 'special' non-piezoelectric, thermoelastic orthotropic plate at a given constant temperature may be expressed as

$$\begin{Bmatrix} \sigma_{11} \\ \sigma_{22} \\ \sigma_{33} \\ \sigma_{23} \\ \sigma_{31} \\ \sigma_{12} \end{Bmatrix} = \begin{bmatrix} C_{11} & C_{12} & C_{13} & 0 & 0 & 0 \\ C_{12} & C_{22} & C_{23} & 0 & 0 & 0 \\ C_{13} & C_{23} & C_{33} & 0 & 0 & 0 \\ 0 & 0 & 0 & C_{44} & 0 & 0 \\ 0 & 0 & 0 & 0 & C_{55} & 0 \\ 0 & 0 & 0 & 0 & 0 & C_{66} \end{bmatrix} \begin{Bmatrix} \varepsilon_{11} \\ \varepsilon_{22} \\ \varepsilon_{33} \\ \varepsilon_{23} \\ \varepsilon_{31} \\ \varepsilon_{12} \end{Bmatrix} + (T - T_0) \begin{bmatrix} \beta_1 \\ \beta_2 \\ \beta_3 \\ 0 \\ 0 \\ 0 \end{bmatrix}. \tag{8.10}$$

They may also expressed in terms of the coefficients of thermal expansion α_i as

$$\begin{Bmatrix} \sigma_{11} \\ \sigma_{22} \\ \sigma_{33} \\ \sigma_{23} \\ \sigma_{31} \\ \sigma_{12} \end{Bmatrix} = \begin{bmatrix} C_{11} & C_{12} & C_{13} & 0 & 0 & 0 \\ C_{12} & C_{22} & C_{23} & 0 & 0 & 0 \\ C_{13} & C_{23} & C_{33} & 0 & 0 & 0 \\ 0 & 0 & 0 & C_{44} & 0 & 0 \\ 0 & 0 & 0 & 0 & C_{55} & 0 \\ 0 & 0 & 0 & 0 & 0 & C_{66} \end{bmatrix} \left(\begin{Bmatrix} \varepsilon_{11} \\ \varepsilon_{22} \\ \varepsilon_{33} \\ \varepsilon_{23} \\ \varepsilon_{31} \\ \varepsilon_{12} \end{Bmatrix} - (T - T_0) \begin{bmatrix} \alpha_1 \\ \alpha_2 \\ \alpha_3 \\ 0 \\ 0 \\ 0 \end{bmatrix} \right). \tag{8.11}$$

Thus, the coefficients β_i are related to the coefficients α_i by

$$\begin{bmatrix} \beta_1 \\ \beta_2 \\ \beta_3 \end{bmatrix} = - \begin{bmatrix} C_{11} & C_{12} & C_{13} \\ C_{12} & C_{22} & C_{23} \\ C_{13} & C_{23} & C_{33} \end{bmatrix} \begin{bmatrix} \alpha_1 \\ \alpha_2 \\ \alpha_3 \end{bmatrix}. \tag{8.12}$$

The inverse constitutive relationships may be expressed as

$$\begin{Bmatrix} \varepsilon_{11} \\ \varepsilon_{22} \\ \varepsilon_{33} \\ \varepsilon_{23} \\ \varepsilon_{31} \\ \varepsilon_{12} \end{Bmatrix} = \begin{bmatrix} S_{11} & S_{12} & S_{13} & 0 & 0 & 0 \\ S_{12} & S_{22} & S_{23} & 0 & 0 & 0 \\ S_{13} & S_{23} & S_{33} & 0 & 0 & 0 \\ 0 & 0 & 0 & S_{44} & 0 & 0 \\ 0 & 0 & 0 & 0 & S_{55} & 0 \\ 0 & 0 & 0 & 0 & 0 & S_{66} \end{bmatrix} \begin{Bmatrix} \sigma_{11} \\ \sigma_{22} \\ \sigma_{33} \\ \sigma_{23} \\ \sigma_{31} \\ \sigma_{12} \end{Bmatrix} + (T - T_0) \begin{bmatrix} \alpha_1 \\ \alpha_2 \\ \alpha_3 \\ 0 \\ 0 \\ 0 \end{bmatrix}. \tag{8.13}$$

In the above relationship S_{ij} are elements of the elastic compliance matrix.

In the case of plane stress,

$$\begin{Bmatrix} \varepsilon_{11} \\ \varepsilon_{22} \\ \varepsilon_{12} \end{Bmatrix} = \begin{bmatrix} S_{11} & S_{12} & 0 \\ S_{12} & S_{22} & 0 \\ 0 & 0 & S_{33} \end{bmatrix} \begin{Bmatrix} \sigma_{11} \\ \sigma_{22} \\ \sigma_{12} \end{Bmatrix} - (T - T_0) \begin{bmatrix} \alpha_1 \\ \alpha_2 \\ 0 \end{bmatrix}, \tag{8.14}$$

where S_{ij} are the elements of the compliance matrix $\mathbf{S}$.

The inverse relationship is

$$\begin{Bmatrix} \sigma_{11} \\ \sigma_{22} \\ \sigma_{12} \end{Bmatrix} = \frac{1}{(1 - \nu_{12}\nu_{21})} \begin{bmatrix} E_1 & \nu_{12}E_2 & 0 \\ \nu_{21}E_1 & E_2 & 0 \\ 0 & 0 & G_{12} \end{bmatrix} \begin{Bmatrix} \varepsilon_{11} \\ \varepsilon_{22} \\ \varepsilon_{12} \end{Bmatrix} - \frac{T - T_0}{(1 - \nu_{12}\nu_{21})} \begin{bmatrix} E_1\alpha_1 + \nu_{12}E_2\alpha_2 \\ \nu_{21}E_1\alpha_1 + E_2\alpha_2 \\ 0 \end{bmatrix}. \tag{8.15}$$

8.2 The Shape Memory Effect: The Phase-transformation Kinetics

Shape memory alloys (SMAs) have already been introduced in Chapter 2. These alloys have the remarkable ability to recover from apparently large permanent deformations. The recovery from such large deformations is facilitated by deformation twinning and solid–solid phase transformations that occur due to changes in temperature or due to mechanical loading. Materials that exhibit the shape memory property are characterized by several different symmetric and partially symmetric crystallographic structures such as cubic, tetragonal, orthorhombic and monoclinic. At high temperatures in the austenitic phase they are predominantly cubic and at lower temperatures they are one or more of the remaining partially symmetric structures that are collectively referred to as the martensitic phase. The different variants of the martensitic phase are separated by specific twin boundaries. Below a certain critical temperature the material is fully martensitic, while above another critical temperature it is fully austenitic.

The thermoelastic martensitic transformation occurs when the alloy in the austenitic parent phase is cooled through the transformation temperature range from M_s to M_f, producing martensite. The reverse transformation occurs when the alloy is heated through the reverse transformation temperature range from A_s to A_f and the material structure returns to that of austenite. The martensite phase being only in a state of partial symmetry when compared with the austenite phase, the crystalline structure change in the forward transformation produces several self-accommodating or twinned variants of martensite. Because of the self-accommodation property, there is no net macroscopic deformation.

8.2.1 Pseudo-elasticity

The transformations from one phase to the other are induced by temperature changes and are not known to be homogeneous. These phase transformations involve changes in crystallographic structure and do not involve diffusive processes. The phase domain boundaries are capable of propagating within the material, especially under the influence of external loads. Thus, the transformation temperatures are sensitive to the applied loading. The property of load-induced phase transformations from austenite to martensite at constant temperature is known as the pseudo-elastic effect.

When a stress is applied to the twinned martensite, the variant boundary expands in the sense of the stress at the expense of the not-so-self-accommodating or de-twinned variants. The shape memory property refers to the ability of the material that is deformed in a twinned martensite state to recover its shape on heating above the austenite transformation temperature. Under the action of external loads, the twinned variant of martensite is transformed into a de-twinned variant and in this state it is easily deformable. When the load is removed it remains in the de-twinned state. On heating beyond the austenite transformation temperature it transforms into the austenitic phase, and on subsequent cooling returns to the twinned martensitic state. Thus, the shape memory property is due to a combination of both temperature-induced transformation and pseudo-elastic properties.

While SMAs undergo a change in crystalline structure from a parent cubic austenitic phase to a number of martensitic variants on cooling, the reverse phase change also occurs but is accompanied with considerable hysteresis on increasing the temperature above A_f. These phase changes are accompanied by significant deformations and, when suitably constrained, can be harnessed to produce large actuation forces. To understand the ability of SMAs to induce martensitic phase transformations by pure mechanical loading, it is useful to consider the highly simplified two-dimensional representation of the material's crystalline arrangement, shown in Figure 8.1.

In Figure 8.1 each grain of material with its corresponding grain boundaries is represented by a box. The grains form a heavily 'twinned' structure. Consequently, the grains are oriented symmetrically across grain boundaries.

The twinned structure constrains the interface with adjacent grains while permitting the internal lattice of individual grains to change under load. As a result, an SMA can experience large macroscopic deformations while maintaining a certain order within its microscopic structure. If an SMA alloy starts as austenite (Figure 8.1a), which has a cubic internal atomic lattice structure, the grains remain as square boxes as the grain boundaries are more or less at right angles to each other. If the alloy is cooled below the phase transition temperature, the crystalline structure changes to martensite (Figure 8.1b). The grain boundaries are distorted and the squares now change to a rhombic pattern. The grains lean in different

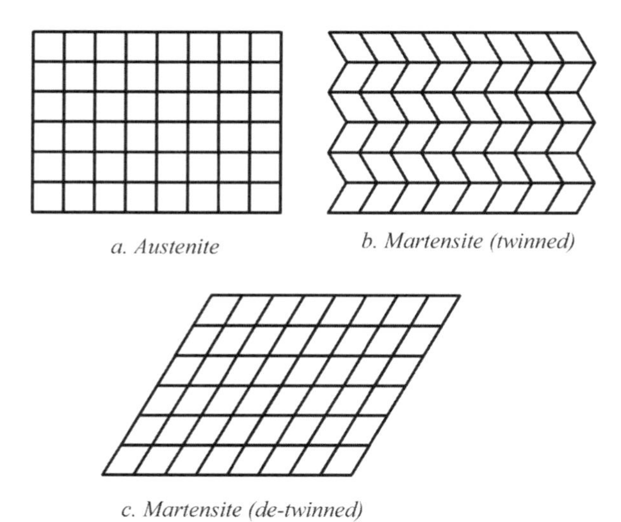

Figure 8.1 Crystalline arrangements in a shape memory alloy

directions in different layers. If the alloy is loaded sufficiently, the applied stress alters the martensitic structure represented in Figure 8.1b, as the grains will start to yield and 'de-twin' as they re-orient and align in the same direction (Figure 8.1c). This behaviour is represented by a typical stress–strain curve for the martensite phase (Figure 8.2), which differs from the same curve in the austenite phase.

Initially, for small stresses, the structure (Figure 8.1b) behaves elastically (region 0 to 1). As the applied stress reaches that at point 1, the material yields and de-twinning occurs between 1 and 2. Beyond point 2, the martensitic structure is entirely de-twinned and the elastic behaviour is again linear, although it is

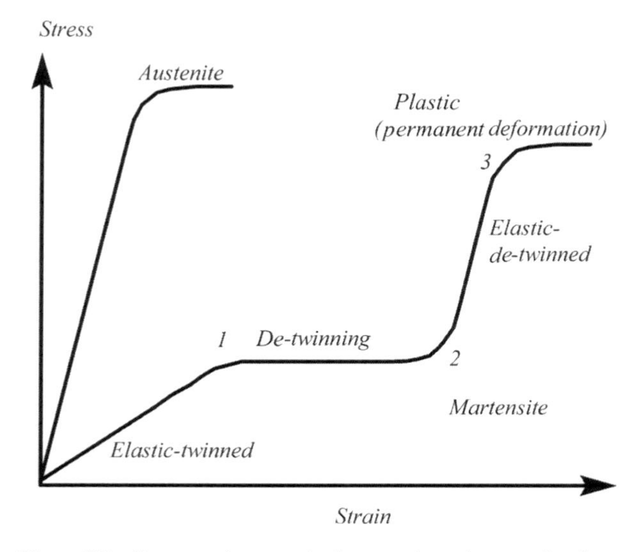

Figure 8.2 Stress–strain curves in the austenite and martensite phases

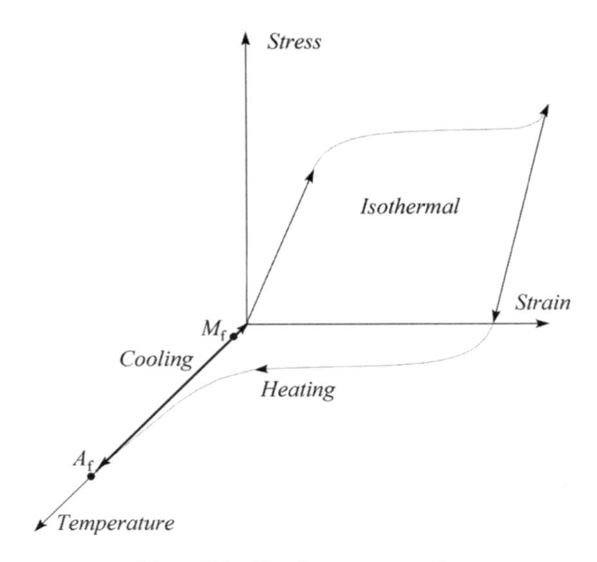

Figure 8.3 The shape memory effect

now a lot stiffer (Figure 8.1c). Thus, there is a second elastic region between points 2 and 3. At point 3, permanent plastic deformation begins that is not recoverable unloading.

8.2.2 The Shape Memory Effect

The shape memory effect is best illustrated by a diagram. In Figure 8.3, an austenite sample is cooled through the martensitic transformation to a temperature below the martensite finish temperature M_f. At this point, the sample is deformed to within ~8–10% strain.

After unloading, the sample is heated through the martensite to austenite transformation. The martensite reverts to austenite along the original paths, such that the deformation previously applied disappears as the temperature is increased. When the temperature exceeds A_f the strain is completely recovered, and the material has the same original shape as before the deformation portion of the cycle. This is a result of the material's thermoelastic, low-symmetry martensite crystal structure and the role of twinning in the martensite phase. The point is that the strain cannot be recovered unless the material is heated beyond the austenite finish temperature. The austenite phase transformation is complete only when the material is heated beyond the austenite finish temperature.

8.2.3 One-way and Two-way Shape Memory Effects

Consider an SMA specimen that is first deformed, via appropriate loading, in the martensitic phase at temperatures lower than austenite start temperature A_s. The deformation results in a re-orientation or 'de-twinning' of the multiple variants of martensite into a single variant and is subsequently unloaded. The specimen will return to its original shape in the austenitic phase if the temperature is raised above the austenite finish temperature A_f. This process is the primary or one-way shape memory effect (SME). In this situation, following a process of free recovery, no shape change is produced in an SMA when it is cooled again below M_f and the specimen must be strained again to repeat the SME.

If the above process is repeated (that is, if a specimen is deformed in martensite, heated to austenite and cooled to martensite), the specimen will develop a secondary or two-way shape memory, which can only be acquired by the process of cyclic loading. In the case of two-way shape memory, the alloy 'remembers' both a low-temperature martensitic configuration and a high-temperature austenitic configuration. An SMA can be trained to exhibit stable and predictable one- or two-way shape memory behaviour by a number of means.

One-way training can be accomplished by imparting a specific amount of plastic (de-twinning) deformation to the SMA in the low-temperature martensitic condition, heating the material to induce the austenite phase transformation and allowing the alloy to recover the initial strain, and cooling the material to induce the transformation to martensite. This procedure is repeated until the hysteresis becomes stable. Two-way shape memory training can be performed by restraining the SMA element, after being plastically deformed at low temperature, and subjecting it to a number of thermal cycles, alternating between specific low and high temperatures. The benefit of this training method is that the shape memory element 'memorizes' two stable shapes (a low-temperature shape and a high-temperature shape) at the expense of some degradation in performance.

8.2.4 Superelasticity

The martensite phase in an SMA can be induced by loading the material if load is applied at a temperature above A_f (austenite finish temperature). The total energy required to stress-induce and deform the material in the martensite phase is less than that required to deform the material in the austenite phase. After the induction of martensite is complete, up to 8% strain can be typically accommodated. Since austenite is the stable phase at this temperature under no-load conditions, the material returns back to its original shape when the stress is removed. This is the property of superelasticity. It finds applications in products such as catheters that would generally not be deformed beyond certain limits. However, when they are, they will undergo a stress-induced transformation, which will facilitate the recovery from deformation.

Superelastic behaviour is illustrated in Figure 8.4. The elastic deformation in the first stage shows a great stress increase over a small strain range $\Delta\varepsilon_1$. Following this initial deformation, the curve flattens with almost no change in stress over a larger strain range $\Delta\varepsilon_2$. During this stage, the martensite grains nucleate and grow into the preferred martensite variant. After the martensitic transformation is complete, elastic deformation continues over the strain range $\Delta\varepsilon_3$. This increase in stress is not as great as in the first stage and continues until yielding of the martensite or unloading. During unloading, the stress is decreased and the reverse process is observed. After elastic recovery over the range $\Delta\varepsilon_3$ (and an additional small strain increment before the lower flat region is reached), the martensite grains that

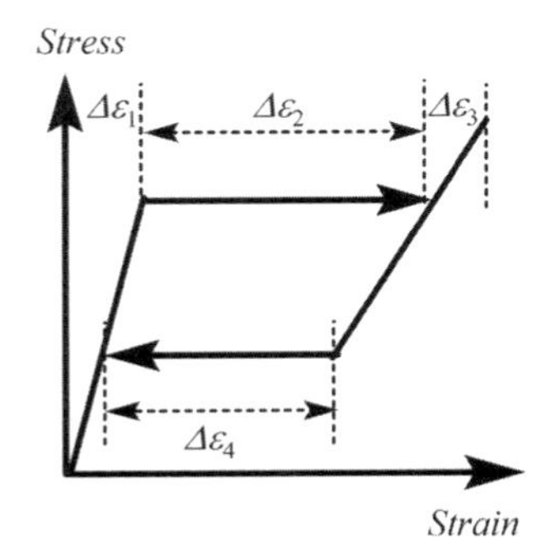

Figure 8.4 Illustration of the superelastic effect

formed during loading over the range $\Delta\varepsilon_2$ revert along the previous crystallographic path, and hence the volume fraction of martensite decreases over the strain range $\Delta\varepsilon_4$. This is the process by which the austenite phase is recovered, so the original undeformed material is restored.

8.3 Non-linear Constitutive Relationships

To be able to integrate SMA materials within lamina, it is important to develop the correct behavioural models for the constitutive equations. Because of the presence of multiple equilibrium states, it could be postulated that the Helmholtz free energy is characterized by multiple minima and multi-well structure. The crystallographic structure of the different phases could be examined and by using the properties of crystal symmetry and thermodynamic properties, the local behaviour of the Helmholtz free energy near each of these minima could be well approximated.

To establish the constitutive equations, one must consider the recovery stress. The recovery stress during constrained recovery is a function of initial strain and temperature. The recovery stress will approach, but not exceed, the austenitic yield stress due to plastic slip or production of stress-induced martensite. The austenitic yield stress is approximately an order of magnitude greater than that of the corresponding martensitic phase. The Young's modulus also increases by a factor of 3–4 under the same conditions.

As a first step one could consider the uniaxial state of stress within the embedded SMA material along the principal axis. Thus, the stress–stain relationship is expressed as

$$\sigma_{ms} = E_{ms}\varepsilon_1 + \sigma_r, \quad T > A_s, \quad \sigma_{ms} = E_{ms}\left(\varepsilon_1 - \alpha_{ms}\Delta T\right), \quad T < A_s, \tag{8.16}$$

where E_{ms} is the Young's modulus in the martensite phase, ε_1 is the uniaxial strain in the principal direction, σ_r is the recovery stress, α_{ms} is the thermal expansion coefficient and $\Delta T = T - T_0$. Thus, for the case of plane stress distribution, the stress is expressed as

$$\begin{Bmatrix} \sigma_{11} \\ \sigma_{22} \\ \sigma_{12} \end{Bmatrix} = \frac{1}{(1-\nu_{12}\nu_{21})}\begin{bmatrix} E_1 & \nu_{12}E_2 & 0 \\ \nu_{21}E_1 & E_2 & 0 \\ 0 & 0 & G_{12} \end{bmatrix}\begin{Bmatrix} \varepsilon_{11} \\ \varepsilon_{22} \\ \varepsilon_{12} \end{Bmatrix} + \begin{bmatrix} \sigma_r \\ 0 \\ 0 \end{bmatrix}$$

$$- \frac{\Delta T}{(1-\nu_{12}\nu_{21})}\begin{bmatrix} E_1\alpha_1(T) + \nu_{12}E_2\alpha_2(T) \\ \nu_{21}E_1\alpha_1(T) + E_2\alpha_2(T) \\ 0 \end{bmatrix}. \tag{8.17}$$

Thus, once a proper recovery stress model is established, it could be incorporated into the complete plane stress–strain relationships or even into the full three-dimensional stress–strain relationships. Establishing the stress–strain relationships in a uniaxial state of stress is of primary importance.

A suitable constitutive equation model based on the one-dimensional stress–strain relationship is the Brinson (1993) model, which has been studied by Prahlad and Chopra (2001). This model makes a clear distinction between the stress-induced and temperature-induced martensite fractions.

The martensite phase can itself exist in two forms, as illustrated in the stress–temperature plot in Figure 8.5. The region where the martensite transition from one state to another occurs, quite independent of the temperature, is labelled '*S*' in the stress–temperature plot shown in Figure 8.5. This describes the Brinson model.

In the Brinson model, when the material is stressed between two critical limits and the temperature is reduced, the alloy transitions from a temperature-induced martensite state to a stress-induced martensite state. In the latter state, the deformations are said to be pseudo-plastic while the strains are macroscopic. While in the temperature-induced martensite state multiple randomly oriented martensite variants can coexist, in the stress-induced case only the variant that corresponds to the direction of applied stress is in equilibrium. Thus, the two-way energy output increases rapidly with applied stress. When the stress is again relieved, however, the martensite remains in the stress-induced state and maintains its shape. Strain

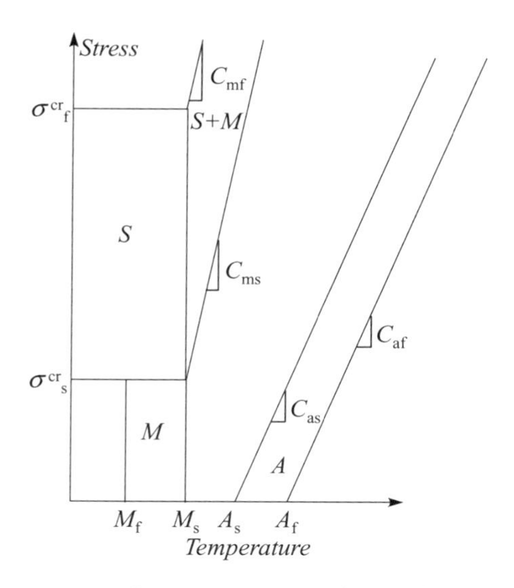

Figure 8.5 Stress–temperature diagram based on the original Brinson model (Brinson, 1993)

restoration can only be achieved by heating to the austenite state. Regions where there is a state transition to the austenite state are labelled '*A*' while regions where there is a state transition to the martensite state are labelled '*M*' and have an approximate linear dependence with applied stress.

Variations in stress above A_f but below the permanent martensite deformation temperature M_d result in complete strain recovery, known as the pseudo-elastic effect. The pseudo-elastic behaviour of SMAs is associated with recovery of the transformation strain on unloading. (On the other hand, superelastic behaviour is observed during loading and unloading above A_s and is associated with stress-induced martensite and reversal to austenite on unloading.)

The Brinson model experiences discontinuities when the material is stressed between two critical limits and the temperature is reduced to soothe alloy transitions from a temperature-induced martensite state to a stress-induced martensite state, so we adopt the modified Brinson model (1993) as enunciated by DeCastro *et al.* (2005, 2006). In this modified state, the stress-temperature diagram is as shown in Figure 8.6. The modified model not only permits *M* to be a linear function of the stress but is also justified by the close agreement between the model and experimental data. The isothermal stress–strain characteristics are depicted in Figure 8.7. Even in the modified Brinson model, the regions *A* and *M* are distinct and a hysteresis loop results when the stress and temperature are dynamically varied.

8.3.1 The Shape Memory Alloy Constitutive Relationships

To model the quasi-static thermo-mechanical evolution, the Brinson model first separates the total martensite fraction ξ into two components, the stress-induced martensite fraction component $\xi_S(\sigma)$ and the temperature-induced martensite fraction component $\xi_T(T)$, which are both hysteretic functions of their respective arguments and satisfy the relationship

$$\xi = \xi_S + \xi_T. \tag{8.18}$$

The hysteresis in the martensite fraction components is discussed in a later subsection. The constitutive equation relating the two martensite states and the strain ε, stress σ and temperature T is

$$\sigma - \sigma_0 = E(\xi)\varepsilon - E(\xi_0)\varepsilon_0 + \Omega(\xi)\xi_S - \Omega(\xi_0)\xi_{S0} + \alpha(\xi)(T - T_0), \tag{8.19}$$

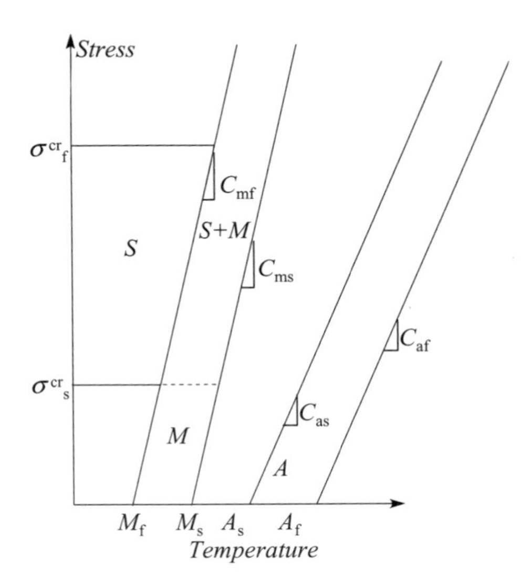

Figure 8.6 Stress–temperature diagram based on the modified Brinson model (De Castro et al., 2005, 2006)

where the subscript 0 denotes the initial states, while $\alpha(\xi)$ is the coefficient of thermal expansion. The corresponding term in the constitutive equation is usually neglected on grounds of smallness. Furthermore, the modulus of elasticity $E(\xi)$ and the coefficient of thermal expansion $\alpha(\xi)$ are defined by the linear relationships

$$E(\xi) = E_A + \xi(E_M - E_A), \quad \alpha(\xi) = \alpha_A + \xi(\alpha_M - \alpha_A), \tag{8.20}$$

with subscripts A and M referring to austenite and martensite, respectively. The phase transformation coefficient in the constitutive relationship, $\Omega(\xi)$, is defined as

$$\Omega(\xi) = -\varepsilon_L E(\xi), \tag{8.21}$$

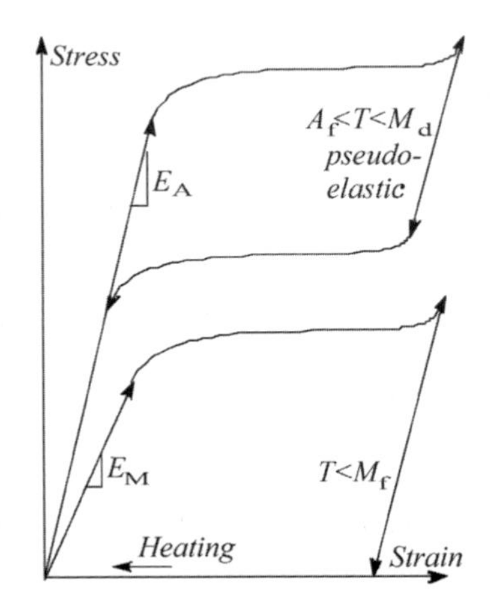

Figure 8.7 Isothermal stress–strain characteristics of shape memory alloys

where ε_L is the maximum recoverable strain. Thus for a restrained actuator,

$$\sigma - \sigma_0 = E(\xi)(\varepsilon - \varepsilon_L \xi_S) - E(\xi_0)(\varepsilon_0 - \varepsilon_L \xi_{S0}) + \alpha(\xi)(T - T_0) \tag{8.22}$$

and for a steadily loaded actuator, with the stresses prescribed, it follows that resulting strain is given by

$$\varepsilon = \varepsilon_L \xi_S + (E(\xi))^{-1}\left[\sigma - \sigma_0 + E(\xi_0)(\varepsilon_0 - \varepsilon_L \xi_{S0}) - \alpha(\xi)(T - T_0)\right]. \tag{8.23}$$

The boundaries of the stress-induced martensite region S are defined by the starting critical stress σ_s^{cr} and the finishing critical stress σ_f^{cr}. When the transition temperatures are known, the bounding stresses for the austenite region A may be defined in terms of the stress influence coefficients C_{as} and C_{af}, as

$$\sigma_{As}^{cr} = C_{as}(T - A_s), \quad \sigma_{Af}^{cr} = C_{af}(T - A_f). \tag{8.24}$$

Similarly, for the martensite region M, the bounding stresses may be defined in terms of the stress influence coefficients C_{ms} and C_{mf}, as

$$\sigma_{Ms}^{cr} = C_{ms}(T - M_s), \quad \sigma_{Mf}^{cr} = C_{mf}(T - M_f). \tag{8.25}$$

In the above relationships, the austenite start temperatureA_s, austenite finish temperature A_f, martensite start temperature M_s and martensite finish temperature M_f are evaluated at zero stress. The stress influence coefficients C_{as}, C_{af}, C_{ms} and C_{mf} are the corresponding slopes of the transition temperature–stress boundaries. The austenite start temperature $A_{s\sigma}$, austenite finish temperature $A_{f\sigma}$, martensite start temperature $M_{s\sigma}$ and martensite finish temperature $M_{f\sigma}$, at a given stress σ, are defined as

$$A_{s\sigma} = A_s + \sigma / C_{as} \quad A_{f\sigma} = A_f + \sigma / C_{af} \quad M_{s\sigma} = M_s + \sigma / C_{ms} \text{ and } M_{f\sigma} = M_f + \sigma / C_{mf}. \tag{8.26}$$

8.4 Thermal Control of Shape Memory Alloys

Heat transfer to and from the SMA wire is largely driven by convection, during both heating and cooling. A provision is also made for heating by passing a current in the wire. Spatial diffusion of temperature by conduction is also included but may be ignored. The relationship between the specific entropy and the heat supplied has been presented earlier and may be expressed as

$$T_0 \frac{\partial \eta}{\partial t} = c_v \frac{\partial T}{\partial t} + T_0 \frac{\partial}{\partial t}(\beta_1 \varepsilon_{11} + \beta_2 \varepsilon_{22} + \beta_3 \varepsilon_{33}) \approx c_v \frac{\partial T}{\partial t}. \tag{8.27}$$

Thus, the heating and cooling model is governed by

$$c_v \frac{\partial T}{\partial t} = k \frac{\partial^2 T}{\partial x^2} - \frac{4h_c(t)}{d}(T - T_\infty) + \frac{4h_e(t)}{d}(T_e - T) + \left(\frac{4}{\pi d^2}\right)^2 \rho_e I(t)^2, \tag{8.28}$$

where c_v is the specific heat capacity, k is the thermal conductivity, $h_c(t)$ is the convection coefficient during air cooling, $h_e(t)$ is the convection coefficient during exhaust heating, d is the diameter of the SMA wire, ρ_e is the electrical resistivity and $I(t)$ is the current in the wire. Ignoring the conductivity, one obtains

$$\frac{c_v d}{4(h_c(t) + h_e(t))} \frac{\partial T}{\partial t} + T = \frac{(h_c(t)T_\infty + h_e(t)T_e)}{(h_c(t) + h_e(t))} + \left(\frac{4}{\pi d^2}\right)^2 \frac{\rho_e d}{4(h_c(t) + h_e(t))} I(t)^2. \tag{8.29}$$

The above heat transfer equation may be expressed as a one-dimensional lag equation given by

$$\tau \frac{\partial T}{\partial t} + T = T_c, \tag{8.30}$$

where

$$\tau = \frac{c_v d}{4(h_c(t) + h_e(t))}$$

and

$$T_c = \frac{(h_c(t)T_\infty + h_e(t)T_e)}{(h_c(t) + h_e(t))} + \left(\frac{4}{\pi d^2}\right)^2 \frac{\rho_e d}{4\left(h_c(t) + h_e(t)\right)} I(t)^2.$$

In the above we have neglected radiation losses, which could be included in terms of the Thring number, frequently referred to in the literature as the Sparrow's number, representing the ratio of the heat transferred by radiation from the SMA surface and the heat transferred by conduction. The time constant can be defined in terms of the Biot number hd/k and the Fourier number $k\tau/\gamma c_v d$, where γ is the ratio of specific heat capacities. However, since we have ignored all spatial diffusion, the thermal conductivity may be eliminated from the equations.

8.5 The Analysis and Modelling of Hysteresis

The hysteresis in the components of the martensite fraction can now be considered. However, before this is done we shall briefly discuss the phenomenon of hysteresis and the many models that can be used to represent it.

8.5.1 The Nature of Hysteresis

Hysteresis in SMAs and in piezoelectric materials is generally rate dependent. There are a number of techniques available to capture and model the hysteresis behaviour in smart materials. Although the hysteretic behaviour of piezoelectric materials shares a number of features with the classical hysteresis found in magnetism and other SMAs, there are some significant differences, in that it is much more rate dependent. Moreover the dependence is dynamic characterized by complex multiple looping structure. While the theory of hysteresis behaviour and its complex looping structure has been extensively studied, literature on its rate dependence is limited.

In a restricted frequency range, it is possible to consider that hysteresis is rate independent and acts as an additive disturbance on the linear dynamics of the system. Hysteresis could be seen as a parallel connection of a linear dynamical system with a rate-independent hysteresis with memory. Most static hysteresis models are based on elementary rate-independent operators and are not suitable for modelling actuator behaviour across a wide frequency band.

Hysteresis non-linearities can be classified into two categories:

1. non-linearities with local memory where the future output depends only on the future input and the present output;
2. non-linearities with non-local memory where the future output depends not only on the current output and the future input but also on the past extreme values of the input. This type of effect can be observed in shape memory and piezoelectric materials.

The modelling of the latter type of hysteresis is generally very difficult as it is necessary to store a number of behavioural parameters over the time history of operation.

The main technique for dealing with hysteresis is to formulate a mathematical model of hysteresis and use an appropriate filter to compensate for the effects of hysteresis. The compensation could be in the form of an inverse, additive or other model to cancel the effects of the hysteresis. Thus, the modelling of the hysteresis is vital for several reasons.

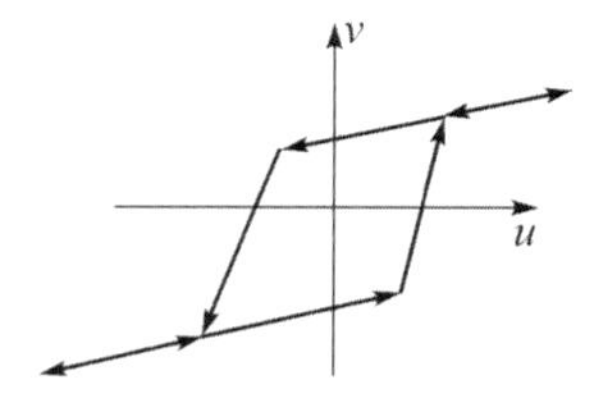

Figure 8.8 A generic hysteron (Pare, 2000)

8.5.2 Hysteresis and Creep

Creep is a phenomenon that appears in the input–output characteristics of smart materials that tend to have the same time-dependent features of viscoelastic materials (Croft *et al.*, 2001). As a consequence of the property of creep in hysteretic elements, there is a drift of the hysteresis loops with cyclic loading during partial transformation in SMAs. Such behaviour has been observed experimentally in *NiTi* materials (Rediniotis *et al.*, 2002b). Thus, the presence of creep complicates the modelling of hysteresis as the phenomenon is no longer time independent; however, it could be modelled by a series connection of parallel spring damper elements collectively called the generalized Kelvin–Voigt model.

8.5.3 Hysteresis Modelling: The Hysteron

In general, the memory and looping characteristics of systems with hysteresis can be quite complicated and adequate representation of these can be achieved by a composition of several hysteretic elements called hysterons. A typical hysteron with a counter-clockwise input–output 'circulation' is depicted in Figure 8.8 (Pare, 2000). The counter-clockwise circulation generally guarantees energy dissipation. In order to simplify the development, a set of properties is required, which will limit the choice to a particular class of hysterons.

The ability to simulate hysteresis in a typical SMA actuator is of some importance. The hysteresis in SMAs is characterized with dynamic time-dependent features and cannot be considered as static. It can be shown that the Duhem model is the one that is most suitable for implementing a dynamic simulation of the hysteresis. The simulation of the Preisach model, which is essentially a superposition of relay operators, and the Prandtl–Ishlinskii model is also important for the modelling of hysteresis. The latter is based on the superposition of play-type or stop-type linear operators and explains why they are inappropriate for modelling hysteresis with creep.

8.5.3.1 The Preisach Model

The classical Preisach model results from the combination of relay operators and defines the hysteresis curve by a parallel superposition of the response of a finite number of individual 'hysterons', as illustrated in Figures 8.9 and 8.10 (Kozek and Gross, 2005). This corresponds to a linear combination of delayed relays with hysteresis at different thresholds. The hysterons are defined as simple two-state relays with hysteresis, where the upper switching point is $\alpha = \alpha_i$, the lower switching point is $\beta = \beta_i$ and the output is defined by

$$\phi(t) = \begin{cases} -1 : u(t) \leq \beta \\ +1 : u(t) \geq \alpha \\ \pm 1 : \beta \langle u(t) \langle \alpha \end{cases} . \tag{8.31}$$

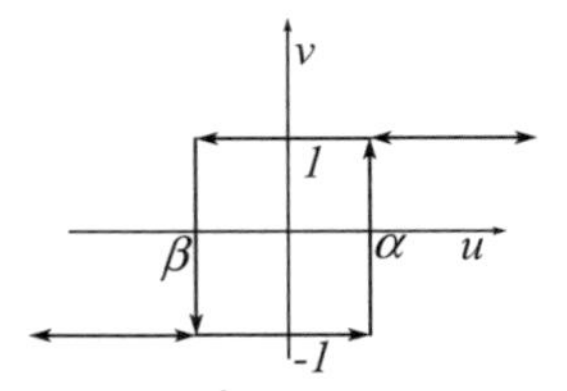

Figure 8.9 The Preisach hysteron (Kozek and Gross, 2005)

The parameters $\alpha$and β are assumed to be constants, i.e. the relays are assumed to be autonomous. Only hysterons with $\alpha \geq \beta$ are physically meaningful, because of their energy-dissipating property. In the case where $\alpha = \beta$, the hysterons correspond to a simple relay without hysteresis. Moreover, the number of hysterons that belong to the set of all possible hysterons P defined by

$$P = \left\{(\alpha, \beta) \in R^2 \,|\alpha \geq \beta \right\}.$$

is assumed to be finite. Without further details, it is noted that the efficient computation is achieved by means of the Everett function or the Everett integral of the Preisach distribution, $\mu_i(t) = \mu\,(\alpha_i, \beta_i)$ (Tan and Baras, 2004).

The Preisach hysteron has a nice interpretation and can be viewed as the output of a feedback system, as illustrated in Figure 8.11. This allows it to be completely invertible, i.e. under a certain restricted class of assumptions the inverse of the Preisach model may also be expressed in terms of the Preisach model. Moreover, by applying suitable transformations to the general hysteron in Figure 8.8, it can be reduced to the Preisach hysteron. Furthermore, by applying transformations and feedback to the Preisach hysteron,

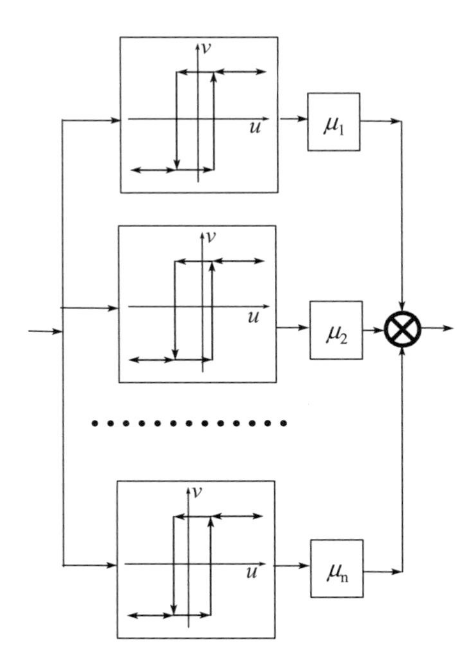

Figure 8.10 The superposition of hysterons (Kozek and Gross, 2005)

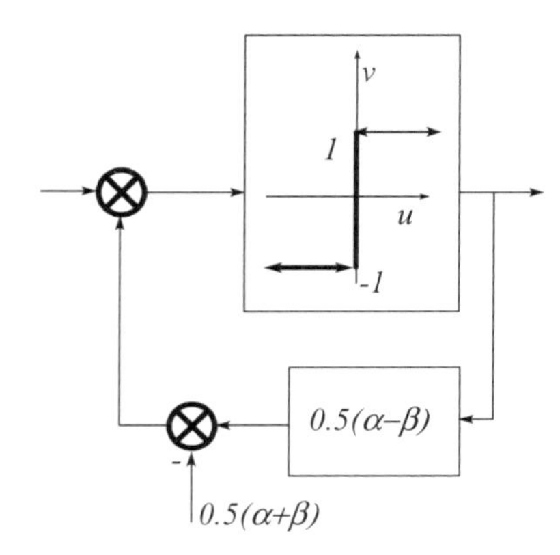

Figure 8.11 Feedback representation of the Preisach hysteron (Cruz-Hernández and Hayward, 2005)

other hysterons based on backlash and play/stop-type non-linearities may be derived, as illustrated in Figures 8.12 and 8.13.

The Preisach operator has *non-local* memory, and it 'remembers' the dominant maximum and minimum values of the past input. In α, β coordinates, each point of the half plane $\alpha \geq \beta$ is identified with only one particular operator $\phi_i(t)$, whose on and off switching values are, respectively, found at the corresponding α, β coordinates of a point. The weighted output of a particular relay is identified by

$$v_i(t) = \mu_i(t)\phi_i(t). \tag{8.32}$$

In practice, the support for $\mu_i(t) = \mu(\alpha_i, \beta_i)$ is finite. This means that $\mu_i(t) = 0$ outside the triangle $\alpha = u_2$, $\beta = u_1$ and $\alpha = \beta$. This can be thought of as restricting the domain of $\mu_i(t)$ to a triangle in the α, β plane. Furthermore, for physical hysteresis it is reasonable to assume that $\mu_i(t)$ must be non-negative, otherwise an increase in the input $\phi_i(t)$ could cause a finite decrease in the output $v_i(t)$, since the major hysteresis loop is closed in the input–output plane.

Hysteresis is a non-linearity with memory, which implies that the output may be affected by the entire history of past inputs. Thus in the Preisach model, the same input applied to a hysteretic system at $(t + t_0) > 0$, will yield same outputs, provided the input history over the period $[t, t + t_0]$ is also

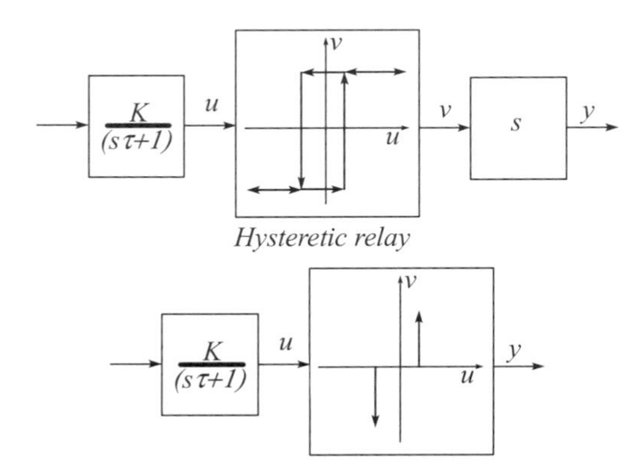

Figure 8.12 The effect of a transformation on the hysteretic relay

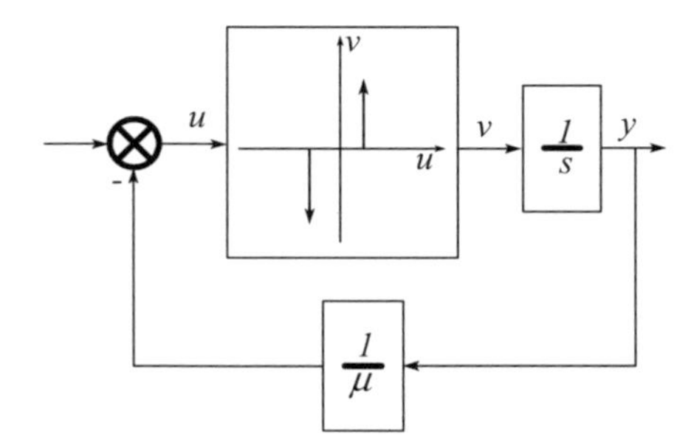

Figure 8.13 Derivation of the backlash non-linearity by applying feedback (Pare, 2000)

the same for any t, when t_0 is assumed to sufficiently large but finite. Thus, only past states within a finite past will influence the current response. However, when creep is present the entire history of past inputs over an infinite time span will influence the current response. There is no provision in the Preisach model to include the effects of the entire time history on the current response within the framework of the assumptions of a finite number of relays and constancy of the parameters α and β. It is important to remember that that the Preisach model is phenomenological and does not explain physical causes of hysteresis. Although the Preisach models have been employed extensively to represent the hysteresis in SMAs, they are unable to represent drift of hysteresis loops with cyclic loading during partial transformation in SMAs.

8.5.3.2 The Prandtl–Ishlinskii Model

The Prandtl–Ishlinskii model was the first attempt to establish a physically meaningful hysteresis model. The model is based on a generalization of the Maxwell and Kelvin–Voight models for the constitutive relationships of viscoelastic materials, with the viscous dampers in these models replaced by Coulomb friction dampers. Two of the simplest types of elementary hysteresis operators that may employed to synthesize the Prandtl–Ishlinskii model by a superposition of hysterons are the linear-play operator (LPO) and the linear-stop operator (LSO) (Brokate and Sprekels, 1996). $u(t)$ is the input signal and $v(t)$ is the output signal of the elementary hysteron. Figures 8.14 and 8.15 show the rate-independent transfer characteristic of the LPO and the LSO in a $v(u)$ plane.

Both operators are characterized by a threshold parameter. For the precise modelling of real hysteresis phenomena, several elementary hysterons with different threshold values ought to be superimposed. This parallel connection of elementary LPO hysterons leads to the discrete Prandtl–Ishlinskii operator of play type (PIOP) while parallel connection of elementary LSO hysterons leads to the discrete Prandtl–Ishlinskii operator of stop type (PIOS).

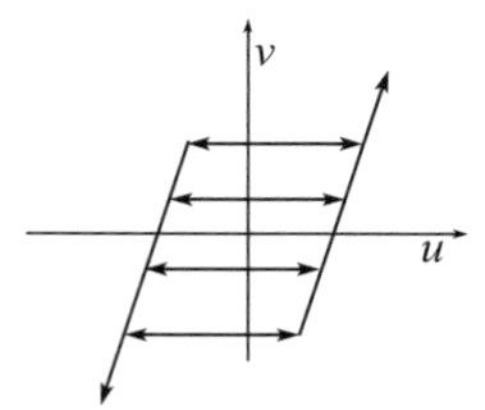

Figure 8.14 Rate-independent transfer characteristic of an LPO (Brokate and Sprekels, 1996)

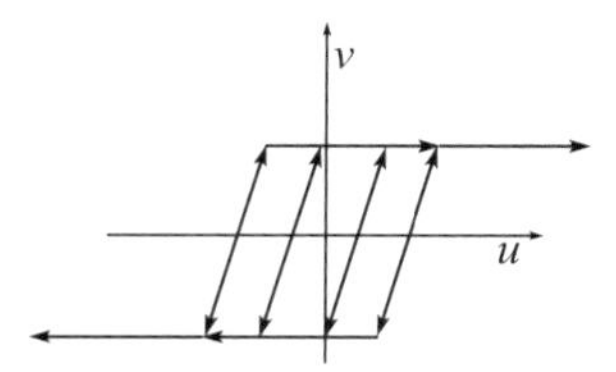

Figure 8.15 Rate-independent transfer characteristic of an LSO (Brokate and Sprekels, 1996)

Many of the characteristics of the Prandtl–Ishlinskii models are no different from the Preisach models. Like the Preisach models, Prandtl–Ishlinskii models are unable to represent drift of hysteresis loops with cyclic loading during partial transformation in SMAs.

8.5.3.3 The Duhem Model

The Duhem model (Visintin, 1994) enables the description of active hysteresis by solutions to either a differential equation or an integral equation. The Duhem model explicitly represents the fact that the output of a hysteretic system can only change its character when the input changes direction. This model uses a phenomenological approach, postulating an integral operator or differential equation to model active hysteretic behaviour, and is not restricted to dissipative hysteresis. Amongst all models of hysteresis, the Duhem model is the one model that can also be derived from physical considerations.

Duhem models can capture the drift of hysteresis loops with cyclic loading, during partial transformation in SMAs such as Nitinol. Bekker and Brinson (1997) have indicated that Duhem models can be separated into those derived from the complete mechanics of the physical process and those derived from a general, simplified physical model of the transformation process. In either case, they may be interpreted physically as the material constitutive relationships or their equivalents in most cases, and the behaviour can be mapped to the actual physical behaviour of the system. It is for this reason that the Duhem model is considered the most appropriate model available for describing hysteresis with creep.

Consider an input (independent) variable $u(t)$, an output (dependent) variable $v(t)$ and a hysteresis relationship on the $u(t)$–$v(t)$ input–output plane. The Duhem differential model follows intuitively from the fact that if $u(t)$ is an increasing function then $v(t)$ increases along one path, and if $u(t)$ is a decreasing function then $v(t)$ decreases along another path, the slopes of the paths being given by continuous functions g_+ and g_-, respectively, known as slope functions. Mathematically, the Duhem model is given by

$$\dot{v}(t) = g_+ (u(t), v(t)) (\dot{u})^+ - g_- (u(t), v(t)) (\dot{u})^- , \quad v(0) = v_0, \tag{8.33}$$

where

$$(\dot{u})^{\pm} = \frac{|\dot{u}| \pm (\dot{u})}{2}.$$

Typical choices for the continuous functions g_+ and g_- have been proposed by Hodgdon (1988a,b), and Coleman and Hodgdon (1986). The Duhem model has been shown to be suitable for representing hysteresis in piezoelectric and ferromagnetic materials and in SMAs.

8.5.3.4 The Bouc–Wen Model

One model that is easy to implement numerically and has been used extensively for modelling hysteretic systems is the Bouc–Wen model. The Bouc–Wen model has also proved to be quite versatile in that it has been used to simulate a variety of hysteretic behaviours. This model is illustrated in Figure 8.16.

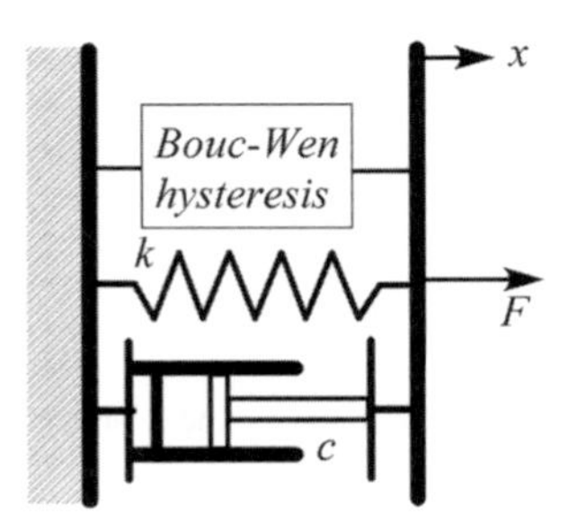

Figure 8.16 Bouc–Wen model

The Bouc–Wen model is based on a parallel combination of two linear elements and one non-linear element, and the force is represented as

$$F = c_{BW}\dot{x} + k_{BW}(x - x_0) + \kappa z. \tag{8.34}$$

The additional state z representing the third element is governed by a dynamic relationship given by

$$\dot{z} = -\gamma\,|\dot{x}| \times z \times |z|^{n-1} - \beta\dot{x} \times |z|^{n} + A\dot{x}. \tag{8.35}$$

In this dynamic relationship, γ, β and A are parameters that could be tuned so as to match the characteristics of the element to any type of hysteresis behaviour. There are a number of ways in which this hysteretic element may be interfaced to other systems.

8.5.3.5 Other Models: The Tan–Balas–Krishnaprasad Model

This is a dynamic model (Tan and Balas, 2004) based on the Preisach hysteron and has been shown to be capable of modelling piezoelectric, ferromagnetic and SMA hysteresis. The general form of the model, which can be adapted to model all of the above-mentioned hysteresis phenomenon, is

$$\tau_x\dot{x}(t) + \tau_z\dot{z}(t) = u(t) - dx(t), \quad z(t) = \varepsilon x(t) + \mathsf{H}(z), \tag{8.36}$$

where $z(t)$ is the model output, $u(t)$ is the model input, $x(t)$ is an internal state, τ_x and τ_z are time constants associated with the dynamic model, $\mathsf{H}(z)$ is a generic input–output representation of a hysteron, and d and ε are constants associated with the static model. A further set of constants may be needed to define a particular hysteron such as the Preisach hysteron. The model has been shown to be suitable for purposes of controller design and simulation.

8.5.4 Modelling the Martensite Fraction–temperature Hysteresis

The hysteresis model describing the relationship between temperature and martensite fraction within the critical stress regions, and employed extensively in the literature, is based on the cosine transformation kinetic approximant proposed by Liang and Rogers (1990) and later modified by Brinson (1993) to accommodate stress-induced martensite. It is employed to describe both the major and the minor loop behaviours of total and stress-induced martensite fractions. Following DeCastro *et al.* (2006), we adopt the modified Brinson model and introduce the normalized stress martensite fraction

$$\xi_{nS} = \xi_S/\xi. \tag{8.37}$$

Thus, we describe the hysteresis in terms of the two variables ξ and ξ_{nS} and the austenite start temperature $A_{s\sigma}$, austenite finish temperature $A_{f\sigma}$, martensite start temperature $M_{s\sigma}$ and martensite finish temperature

$M_{f\sigma}$, at a given stress σ. The hysteresis loops in terms of the clipped cosine functions are defined by the transformation from austenite to martensite, which is represented by the following:

$$\xi^{A\to M}(\sigma,T)=\xi_0(\sigma_{init},T_{init})+\frac{(1-\xi_0(\sigma_{init},T_{init}))}{2}\left(1+\cos\left(\pi\frac{T-M_{f\sigma}}{M_{s\sigma}-M_{f\sigma}}\right)\right), \tag{8.38a}$$

$$\xi_{nS}^{A\to S}(\sigma,T)=\xi_{nS0}(\sigma_{init},T_{init})+\frac{(1-\xi_{nS0}(\sigma_{init},T_{init}))}{2}\left(1+\cos\left(\pi\frac{\sigma-\sigma_{\mathrm{f}}^{\mathrm{cr}}}{\sigma_{\mathrm{s}}^{\mathrm{cr}}-\sigma_{\mathrm{f}}^{\mathrm{cr}}}\right)\right) \tag{8.38b}$$

and the transformation from martensite to austenite, which is represented by the following:

$$\xi^{M\to A}(\sigma,T)=(1-\xi_0(\sigma_{init},T_{init}))+\frac{\xi_0(\sigma_{init},T_{init})}{2}\left(1+\cos\left(\pi\frac{T-A_{s\sigma}}{A_{f\sigma}-A_{s\sigma}}\right)\right) \tag{8.38c}$$

$$\xi_{nS}^{S\to A}(\sigma,T)=\xi_{nSF}\xi^{M\to A}(\sigma,T), \tag{8.38d}$$

where the superscripts designate the direction of transformation, i.e. $A\to M$ is the austenite to martensite and $A\to S$ is the austenite to stress-induced martensite and vice versa, $\xi_0(\sigma_{init},T_{init})$ is the initial martensite fraction corresponding to an initial stress σ_{init} and initial temperature T_{init} at the start of the transformations, $\xi_{nS0}(\sigma_{init},T_{init})$ is the initial value of ξ_{nS} at the start of the transformations, and ξ_{nSF} represents the value of ξ_{nS} at the instant when $\mathrm{d}\xi_{nS}/\mathrm{d}t$ changes sign. The arguments of the cosine functions are limited to domain $(0,\pi)$.

A typical simulation of the martensite fraction R_{m} versus temperature T phase transformation hysteresis is illustrated in Figure 8.17. The model is based on a adaptation of the Jiles–Atherton model (Jiles and Atherton, 1984) for of ferromagnetic hysteresis. It can be shown that after appropriate scaling the curve can be made to match experimentally determined data at certain key points. In particular, the drift of the hysteresis loops with cyclic loading during partial transformation is clearly apparent. Unfortunately, although the Duhem model is invertible, one of the practical problems that arises is the need to identify the parameters in the Duhem model before the model can be inverted. To circumvent this need, we adopt a slightly different approach to decompose the hysteresis behaviour in practice.

8.5.5 Decomposition of Hysteretic Systems

The decomposition of the hysteresis is done by employing the notion of the *phaser*. The phaser is an ideal operator in the frequency domain that shifts the phase of a signal by a given amount but leaves the magnitude unchanged and has been shown experimentally to be very effective at reducing the hysteresis present in strain-based actuators (Cruz-Hernández and Hayward, 2005). An ideal phaser cannot be implemented but can be approximated by a causal filter. The phaser operator can be seen as being complementary to the proportional controller for which the phase is zero, but the magnitude is adjusted according to design specifications. The phaser keeps the magnitude constant but applies a given phase shift at all frequencies according to design specifications. As found by Cruz-Hernández and Hayward (2005), the effect of hysteresis is seen as a phase shift between the input to the system and its output, therefore only periodic signals are considered. An elementary operator termed the phaser ($\boldsymbol{L}_{\mathrm{pa}}$) shifts its periodic input signal by a constant angle θ, with a constant magnitude of 1, independent of the frequency or the magnitude of the input signal.

The response of a system with hysteresis is decomposed to a *phaser* operation followed by a nonlinear operator as shown in Figure 8.18. The *phaser* is obtained by the application of this decomposition process to a typical SMA hysteresis response and is illustrated in Figure 8.19. The output of the phase is now used to compensate the input to the hysteretic system, and the corresponding input–output behaviour of the non-linear block is illustrated in Figure 8.20.

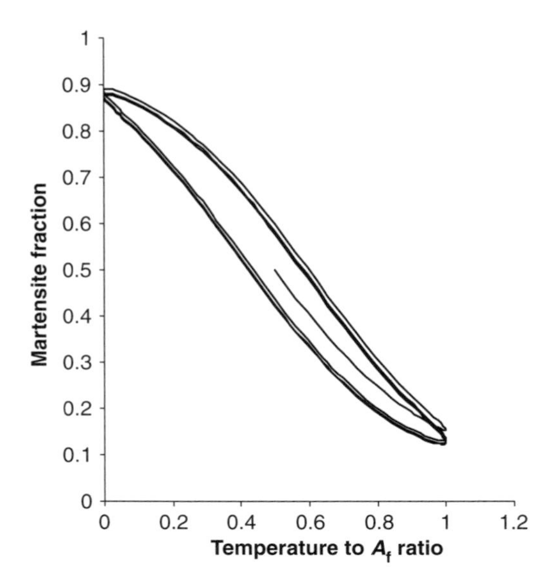

Figure 8.17 Martensite fraction R_m versus temperature T phase transformation hysteresis

The inverse model of this behaviour may be easily approximated by a polynomial curve, while the inverse model of the phaser is constructed by introducing lead compensation. An ideal inverse model of the phaser cannot be implemented but can be approximated by a causal lead compensator. Thus, it is possible in principle to construct an approximate inverse of this system, which is extremely valuable when one is interested in closed-loop control of the hysteretic system.

8.6 Constitutive Relationships for Non-linear and Hysteretic Media

The approach adopted throughout this text in modelling smart structures is to start with the constitutive relationships. If a similar approach is adopted in modelling hysteresis, it is essential that the constitutive relationships are generalized to represent a host of observed non-linear behaviours, including hysteresis.

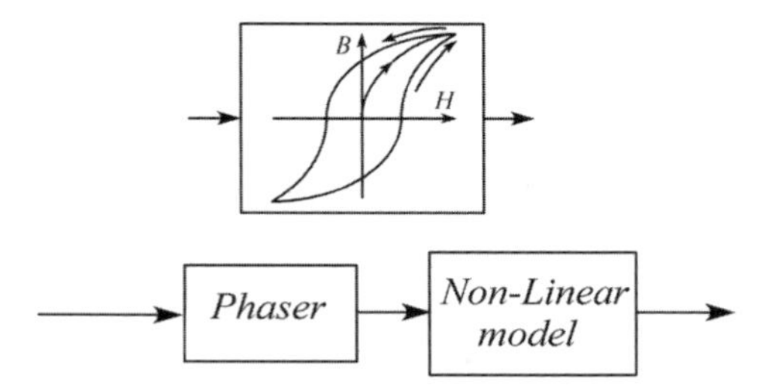

Figure 8.18 Illustration of the decomposition of the hysteretic system

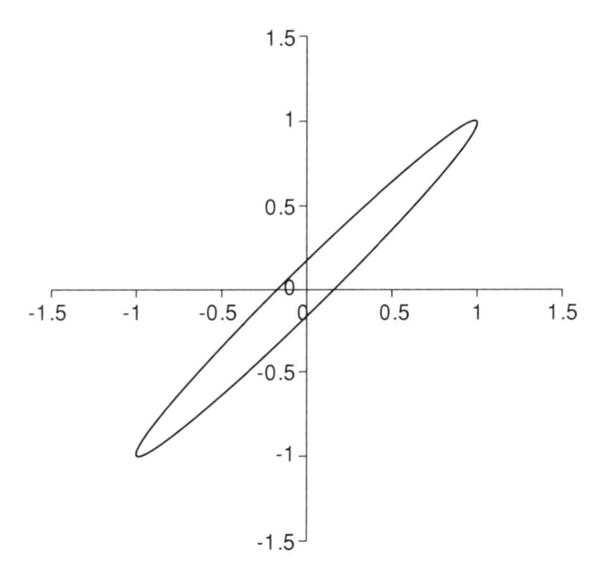

Figure 8.19 Input–output model of the *phaser*

One such generalization has been proposed by Meurer *et al.* (2002) and will be briefly discussed here. The generalized non-linear constitutive relationship may be expressed as

$$\sigma(\varepsilon, \dot{\varepsilon}) = f(\varepsilon) + \{\sigma(\varepsilon_0, 0) - f(\varepsilon_0)\} \exp(\alpha\, sign(\dot{\varepsilon})(\varepsilon_0 - \varepsilon)) + I(\varepsilon, \varepsilon_0), \tag{8.39}$$

$$I(\varepsilon, \varepsilon_0) = \int_{\varepsilon_0}^{\varepsilon} \left\{g(\xi) - \mathrm{d}f(\xi)/\mathrm{d}\xi\right\} \exp(\alpha\, sign(\dot{\varepsilon})(\xi - \varepsilon))\, \mathrm{d}\xi.$$

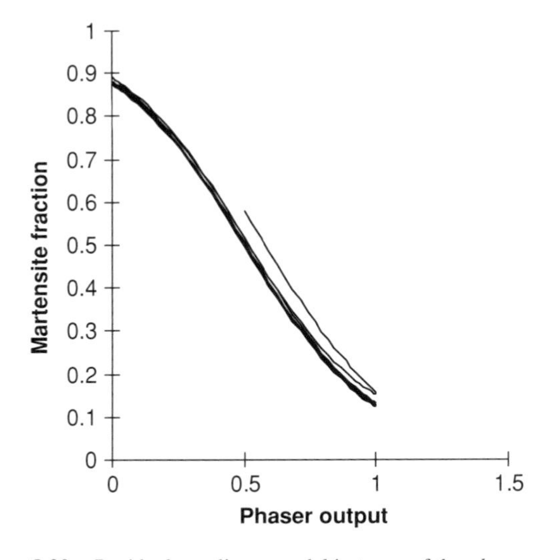

Figure 8.20 Residual non-linear model in terms of the *phaser* output

The function $I(\varepsilon, \varepsilon_0)$ satisfies the relationship

$$\mathrm{d}I(\varepsilon, \varepsilon_0)\big/\mathrm{d}\varepsilon = -\alpha\, sign(\dot{\varepsilon})\, I(\varepsilon, \varepsilon_0). \tag{8.40}$$

Thus, it follows that

$$\frac{\mathrm{d}}{\mathrm{d}\varepsilon}\sigma(\varepsilon, \dot{\varepsilon}) = \frac{\mathrm{d}}{\mathrm{d}\varepsilon} f(\varepsilon) - \alpha\, sign(\dot{\varepsilon}) \{\sigma(\varepsilon, \dot{\varepsilon}) - f(\varepsilon)\}. \tag{8.41}$$

Hence for non-zero values of the parameter α, the modulus can exhibit a variety of non-linear and hysteretic behaviours. To cast the constitutive relationship in the Duhem form, the above relationship may be expressed as

$$\frac{\mathrm{d}}{\mathrm{d}t}\sigma(\varepsilon, \dot{\varepsilon}) = \frac{(1 + sign(\dot{\varepsilon}))}{2}\left(\frac{\mathrm{d}}{\mathrm{d}\varepsilon} f(\varepsilon)\dot{\varepsilon} - \alpha\dot{\varepsilon}\{\sigma(\varepsilon, \dot{\varepsilon}) - f(\varepsilon)\}\right)$$
$$+\frac{(1 - sign(\dot{\varepsilon}))}{2}\left(\frac{\mathrm{d}}{\mathrm{d}\varepsilon} f(\varepsilon)\dot{\varepsilon} + \alpha\dot{\varepsilon}\{\sigma(\varepsilon, \dot{\varepsilon}) - f(\varepsilon)\}\right). \tag{8.42}$$

The model is completely defined when the function $f(\varepsilon)$ representing the static constitutive, albeit non-linear, constitutive relationship and the parameter α representing the extension to the dynamic situation are specified. Thus, the above generalization of the constitutive relationships represents a quasi-static generalization of the non-linear constitutive relationships when $\dot{\varepsilon} = 0$ ($\alpha = 0$), as the model only represents sufficiently slow time domain variations of ε. However, this is adequate for most smart actuators based on piezoelectric or ferromagnetic materials or on SMAs.

8.7 Shape Memory Alloy Actuators: Architecture and Model Structure

SMA actuators based on thin films, ribbons and wires have been proposed by several researchers (Miyazaki *et al.*, 1997; Seward *et al.*, 1999; Oh *et al.*, 2001; Rediniotis *et al.*, 2002b). The principle of operation of typical SMA tube actuator is illustrated in Figure 8.21a. It is expected that distributed arrays of these actuators would be employed to bend a catheter in a controlled manner. A typical example is illustrated in Figure 8.21b. Employing such a principle, a prototype of an actively controlled actuator was built to demonstrate the feasibility of the concept. Thus, the control of these actuators is of primary importance.

8.7.1 Simulation and Inverse Modelling of Shape Memory Alloy Actuators

To fully understand the input–output features of the open-loop system, it was simulated using MATLAB®/Simulink®. The basic modules employed in the simulation of the open-loop system are illustrated in Figure 8.22a. Figure 8.22b illustrates the non-linear nature of the input–output behaviour of the constituent modules.

The inverse model may now be constructed from the complete heating, cooling and phase transformation model, which may be expressed as

$$\tau\frac{\partial T}{\partial t} + T = T_c, \quad \xi = \xi_{PT}(\sigma, T). \tag{8.43}$$

Thus, the inverse model is given by

$$T = \xi_{PT}^{-1}(\sigma, \xi), \quad T_c = \tau\frac{\partial T}{\partial t} + T \tag{8.44}$$

and has the structure of a proportional-differential (PD) controller.

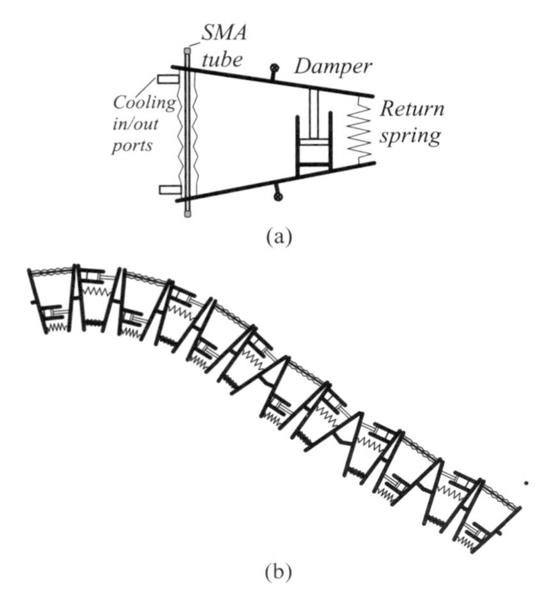

Figure 8.21 SMA tube actuator: (a) a single cell in an SMA tube actuator and (b) a string of SMA actuators employed to actuate a control surface

To invert the transformation from austenite to martensite we assume that $\xi_0(\sigma_{init}, T_{init}) = 0$, so it follows that

$$T = M_{f\sigma} + \frac{(M_{s\sigma} - M_{f\sigma})}{\pi}\cos^{-1}\left(2\xi^{A\to M}(\sigma, T) - 1\right), \tag{8.45}$$

where the temperature is assumed to be in the range that limits the domain of the cosine function to $(0, \pi)$. Similarly, to invert the transformation from martensite to austenite, we assume that $\xi_0(\sigma_{init}, T_{init}) = 1$, so

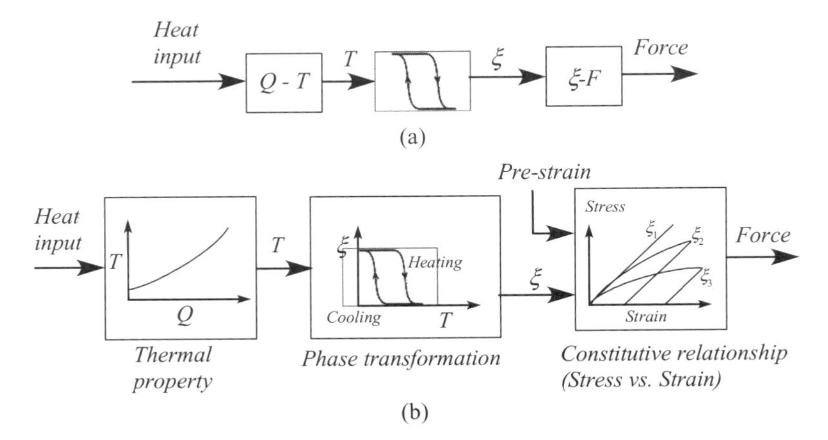

Figure 8.22 (a) Simplified block diagram of an SMA actuator and (b) modelling the non-linear force actuator

it follows that

$$T = A_{s\sigma} + \tfrac{(A_{f\sigma} - A_{s\sigma})}{\pi} \cos^{-1}\left(2\xi^{M\to A}(\sigma, T) - 1\right). \tag{8.46}$$

To invert the constitutive relationship, we assume that ξ_S is independent of ξ and rearrange the equation to solve for ξ to obtain

$$\xi = \frac{\sigma - \sigma_0 - E_A(\varepsilon - \varepsilon_L \xi_S) + (E_A + \xi_0(E_M - E_A))(\varepsilon_0 - \varepsilon_L \xi_{S0})}{(E_M - E_A)(\varepsilon - \varepsilon_L \xi_S) + (\Theta_M - \Theta_A)(T - T_0)}$$

$$- \frac{\Theta_A(T - T_0)}{(E_M - E_A)(\varepsilon - \varepsilon_L \xi_S) + (\Theta_M - \Theta_A)(T - T_0)}. \tag{8.47}$$

The module that plays a primary role in the design of controller is the hysteretic subsystem relating the martensitic fraction to the temperature. This system can be modelled by a variety of hysteresis models such as the Ishlinskii hysteresis model, the Preisach model, the Duhem model and several others based on physically meaningful modelling of the $B - H$ hysteresis in ferromagnetic materials, such as the Jiles–Atherton model. The latter is, however, equivalent to the Duhem model. The Duhem model is our preferred choice for representing the hysteresis in an SMA actuator. It is interesting to observe that the Duhem model is completely invertible.

8.7.2 Control of Shape Memory Alloy Actuators

The design of a controller for an SMA actuator is usually based on a model to predict the response of the actuator to a given temperature increment, a thermal model to compute the temperature change in the device and a continuum-mechanical model to predict the martensite fraction on the SMA (see, for example, Mihálcz, 2001). The model-based design approach is particularly important for certain applications such as catheters (Dumont and Kuehl, 2004; Langelaar and van Keulen, 2004). Certain key considerations are important in the controller design for such actuators. In an SMA actuator, stressing, heating or unloading can recover only about 8.5% of the strain. Strain above the limiting value will remain as a permanent plastic deformation. Overheating should be strictly avoided, as it results in the material losing its shape memory property. Thus, the operating temperature for shape memory devices must not move significantly away from the transformation range, or else the shape memory characteristics may be altered and consequently must be controlled quite precisely. Moreover, the system must acquire the commanded temperatures very rapidly for the shape memory effect to remain prominent and stationary. While a *NiTi* SMA may be deformed at a temperature below A_s, the maximum deformation is limited by the intrinsic strain tolerance of the material. These performance requirements essentially dictate the choice of the control system design parameters such as the cost functions and maximum allowable excursions in the model states about the nominal values.

The method and control approach adopted in various simulations is different to the methods published so far in the literature (Kudva *et al.*, 1996; Oh *et al.*, 2001; Rediniotis *et al.*, 2002a; Ma and Song, 2003; Dutta *et al.*, 2005; Song *et al.*, 2005). The presence of the hysteresis in the input–output model can be dealt with by the Duhem model, and the controller is designed around the inverse model of the system dynamics. The basic closed-loop control technique for the regulation of hysteretic servo-actuators is the proportional-integral-derivative control. Actual implementation could be based on pulse width modulation, which was also simulated. In this case, the controller output is just two valued independent pulse modulation controls employed for heating and the cooling, ensuring that the switching in of one excludes the operation of the other. The application is to actuating a typical elastic member. The required force command is computed as for a robot manipulator by the computed torque approach, and inverse models of the constitutive relationships, phase transformation hysteresis and the temperature–heat input relationship are employed to compute the desired and actual heat input. The error in the heat input/extracted is then employed to drive a Proportional-Integral-Derivative (PID) controller and a pulse-width pulse-frequency type modulator consisting of a lag filter in series with a Schmidt trigger with a unity feedback loop around them. The output is a pulse command sequence to the flow control valves, which adjusts the pulse width and/or the pulse frequency. It was found by simulations that the PID control algorithm was most appropriate to simulate and evaluate a typical non-electrically heated and cooled SMA force actuator.

Exercises

1. Show that the energy loss per cycle due to hysteresis in a piezoelectric actuator is proportional to the area of the hysteresis loop.

2.

(i) Calculate the energy per unit volume in J/m^3 expended during one full hysteresis cycle of a magnetic material having a rectangular hysteresis loop. Assume for a magnetic field intensity of $H_c = 200$ A/m that the saturated magnetic induction B_s=1.4 T (Tesla). Note that 1 T = 1 Tesla = 1 kg/(s C) and 1 A = 1 Ampere = 1 C/s.
(ii) Hysteresis losses are important in designing electrical transformers. The material from part (i) is used for constructing a transformer operating at 60 Hz. Assume that the material is completely saturated during each cycle of the magnetic field and that the core has a mass of 10 kg and a density of 7 g/cm^3. Estimate the power in watts dissipated during the operation of this transformer.
(iii) An *Fe–Si* alloy with $H_c = 8$ A/m and $B_s = 2$ T is used as an alternative material to construct the another transformer. Find the hysteresis loss in this case.

3. Consider a single degree of freedom vibrating system with an element characterized by Bouc–Wen hysteresis acting in parallel with linear spring and damper, as shown in Figure 8.23. Additionally, the mass is assumed to be excited by a sinusoidal force. The static displacement may be assumed to be unity, the natural frequency in the absence of the hysteresis is given as 1000 rad/s, the damping ratio in the absence of hysteresis is given as 0.5 and $b_w/m = 0.1$. The constants in the Bouc–Wen hysteresis model are assumed to be $\alpha = \beta = \gamma = 0.5$, $n = 2$.
The equations of motion are

$$\frac{\mathrm{d}}{\mathrm{d}t}\begin{bmatrix} x(t) \\ \dot{x}(t) \end{bmatrix} = \begin{bmatrix} 0 & 1 \\ -k/m & -c/m \end{bmatrix}\begin{bmatrix} x(t) \\ \dot{x}(t) \end{bmatrix} + \begin{bmatrix} 0 \\ 1/m \end{bmatrix}(f(t) - b_w z(t))$$

$$\dot{z}(t) = \alpha\dot{x}(t) - \alpha\beta|\dot{x}(t)||z(t)|^{n-1}z(t) - \alpha\gamma\dot{x}(t)|z(t)|^n.$$

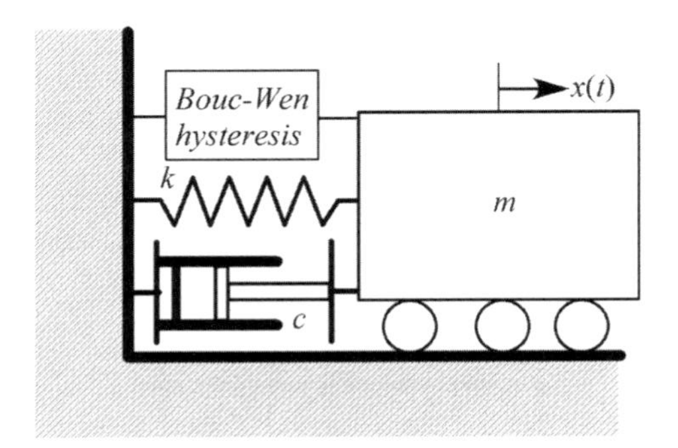

Figure 8.23 Single degree of freedom hysteretic system

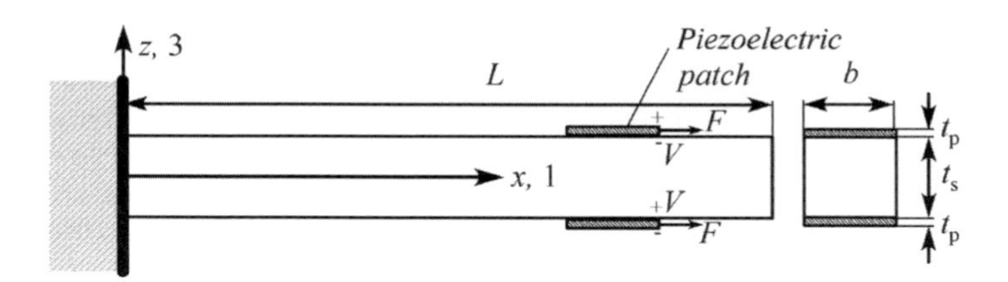

Figure 8.24 Cantilever beam excited symmetrically by a bimorph configuration

Simulate the above system in MATLAB/Simulink and obtain plots of the response $x(t)$ as a function of the input $f(t)$.

4. Open-loop piezo-actuators exhibit hysteresis in their dielectric and electromagnetic large-signal behaviour. Hysteresis is based on crystalline polarization effects and molecular effects within the piezoelectric material. Two piezoelectric patches are attached to the flat bar, as shown in Figure 8.24. Assume that actuation of the patches results only in the axial motion of the bar.

(i) Given the Young's modulus and thickness of the beam, and the actuator patch materials as E_s andE_p, and t_s andt_p, respectively, show from first principles that the strain ε of the bar with only external loads present is

$$\varepsilon = (F/b)/(2E_pt_p + E_st_s).$$

Hence show that the strain response is analogous to the response of two springs in parallel subjected to a single external force, as shown in Figure 8.25.

(ii) Show from first principles that the strain of the bar with only piezoelectric patch actuation present is

$$\varepsilon = (2E_pt_p)/(2E_pt_p + E_st_s)d_{31}(V/t_p).$$

Hence show that the strain response can be modelled as the displace response of the two springs in parallel, with one of them undergoing displacement excitation, as shown in Figure 8.26.

(iii) Obtain an expression for the combined strain when both inputs are present and explain how the presence of hysteresis can be modelled using the Bouc–Wen model.

5. Consider the Tan–Balas–Krishnaprasad model and show how the hysteresis in a piezoelectric material can be represented. Assume the model output represents the polarization that may be represented as a

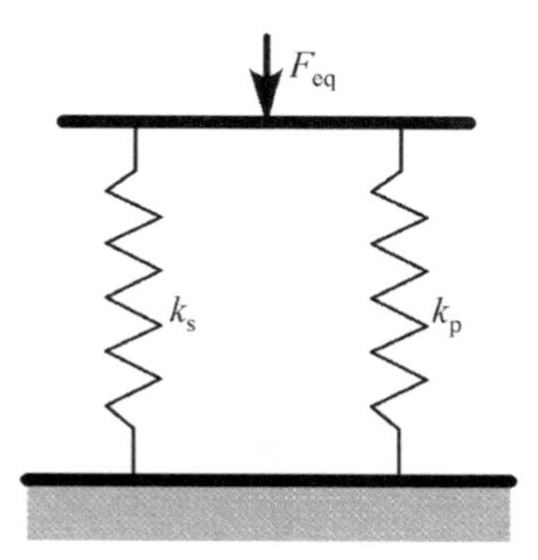

Figure 8.25 Analogy of two springs in parallel

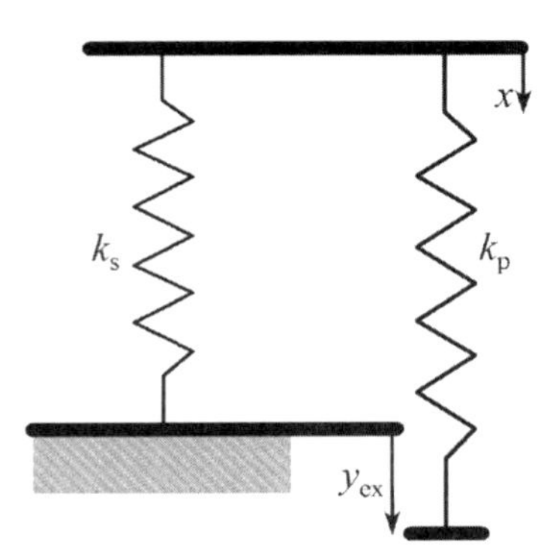

Figure 8.26 Two spring analogy with piezoelectric patch excitation

hysteron and a function of the internal state. The model input may be assumed to be the applied voltage while the internal state is the electric field. Identify the essential parameters representing the model. [Hint: Establish the internal state dynamics in terms of the electric displacement D, which is the eliminated using the relationship $D(t) = \varepsilon_0 E(t) + P(t)$, where ε_0 is the permittivity of free space.]

6. Consider the Tan–Balas–Krishnaprasad model and show how the hysteresis in an SMA can be represented. Assume that the model output represents the temperature-dependent martensite to austenite fraction. The model input may be assumed to be the commanded temperature while the internal state is the actual temperature. Identify the essential parameters representing the model.

7. The closed-loop representation of a Nitinol wire actuator based on the Tan–Balas–Krishnaprasad model is shown in Figure 8.27. The stroke of the actuator varies from 0 mm in a fully martensite state to 4 mm in the austenite state. The phase transition temperatures are given in Table 8.1.

(i) The key to simulating the closed-loop system requires the simulation of the Preisach hysteron. Making suitable assumptions of linear phase transformation relationships within specific temperature ranges, simulate a Preisach hysteron.

(ii) Close the loop around the hysteresis model and simulate the closed-loop system. Hence or otherwise obtain the optimum parameters for the controller. How does the time constant of the lead filter compare with the time constant of the lag filter model of the heating circuit?

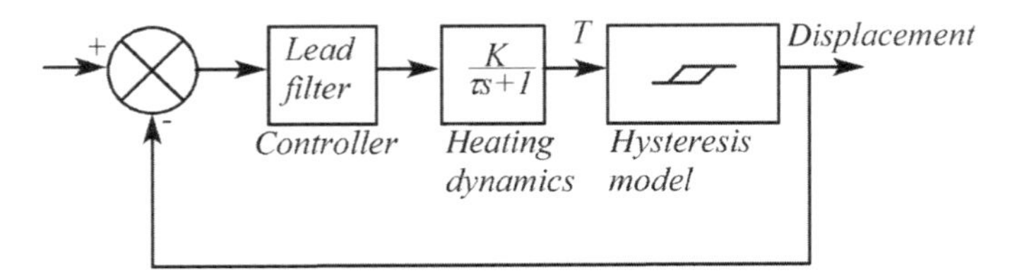

Figure 8.27 Closed-loop control of an SMA actuator

Table 8.1 Phase transition temperatures (°C) of Nitinol

Alloy	M_s	M_f	A_s	A_f
Nitinol	46.1	32.2	67.2	85

Table 8.2 Phase transition temperatures (°C) of *NiTi*-based alloys

Alloy	M_s	M_f	A_s	A_f
Nitinol strip	18	9	35	49
Nitinol wire	46	32	67	85
TiNi mat	60	52	71	77
$TiNi_{0.95}$	70	60	108	113

8. A number of *NiTi*-based SMAs are compared as potential materials for designing displacement actuators. The approximate phase transition temperatures of four of these materials are given in Table 8.2.
Identify the material with worst hysteresis characteristics and the one that is capable of delivering the highest resolution. Explain the reasons for your choices.

9. It is proposed to simulate the Nitinol wire-based displacement actuator considered in exercise 7 using a Bouc–Wen-like hysteresis model to represent the temperature–martensite fraction hysteresis, as illustrated in Figure 8.28. The hysteresis model (unscaled) is based on the model of a typical *B–H* curve. The model is described by the following equations:

$$\dot{x} = \alpha \times i - \beta \times x|i| - \gamma \times i|x|, \quad z = 1 - \tanh(i - 0.5x), \quad \alpha = 2.8, \quad \beta = 1.5, \quad \gamma = 3\beta.$$

(i) After appropriate scaling simulate the closed dynamics of the actuator and obtain the optimum parameters for the controller.
(ii) Show that the above model is equivalent to a Duhem model.

10.

(i) A Nitinol wire actuator is constructed from a bundle of 10 wires each of 0.4 mm diameter and length 200 mm. In a test, each wire is initially strained in the martensite phase by 2% at a temperature below M_f and rigidly fixed at the two ends. The wire is heated beyond the austenite finish temperature A_f. Estimate the force generated.
(ii) Each wire in the actuator is attached to a rigid body that is restrained by a stiff spring with a stiffness of 10 N/mm. Estimate the displacement of the rigid body as well as the strain and force developed in each wire.
(iii) Ignoring all hysteresis and non-linear effects design a controller for the closed-loop control of the actuator. Assume the heating dynamics to be represented by a first-order lag.
(iv) Introduce the hysteresis in the martensite fraction–temperature relationship and optimize the controller for closed-loop performance.

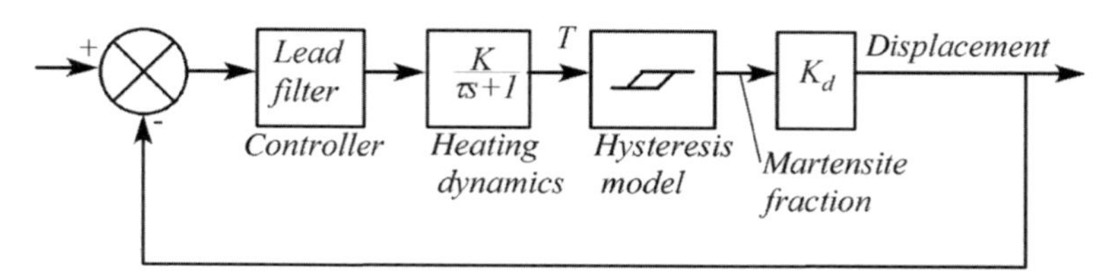

Figure 8.28 Closed-loop control of an SMA actuator with Bouc–Wen hysteresis model in the forward path

References

Bekker, A. and Brinson, L. C. (1997) Thermo-induced transformation in a prestressed 1-D SMA polycrystalline body: phase diagram kinetics approach, *Journal of the Mechanics and Physics of Solids*, 45, 949–988.

Brinson, L.C. (1993) One dimensional constitutive behaviour of shape memory alloys: thermomechanical derivation with non-constant material functions, *Journal of Intelligent Material Systems and Structures*, 4, 229–242.

Brokate, M. and Sprekels, J. (1996) *Hysteresis and Phase Transitions*, Springer-Verlag, Berlin, Heidelberg, New York.

Coleman, B. D. and Hodgdon, M. L. (1986) On a class of constitutive relations for ferromagnetic hysteresis, *International Journal of Engineering Science*, 24(6), 897–919.

Croft, D., Shed, G. and Devasia, S. (2001) Creep, hysteresis, and vibration compensation for piezoactuators: atomic force microscopy application, *Journal of Dynamic Systems, Measurement, and Control*, 123(35), 35–43.

Cruz-Hernáandez, J. M. and Hayward, V. (2005) Position stability for phase control of the Preisach hysteresis model, *Transactions of the CSME*, 29(2), 129–142.

DeCastro, J. A., Melcher, K. J., and Noebe, R. D. (2005) System-level design of a shape memory alloy actuator for active clearance control in the high pressure turbine, AIAA Paper 2005–3988, NASA/TM—2005-213834.

DeCastro, J.A., Melcher, K.J., Noebe, R.D. and Gaydosh, D. (2006) Development of a numerical model for high temperature shape memory alloys, NASA/TM—2006-214356.

Dumont, G. and Kuehl, C. (2004) A dynamical training and design simulator for active catheters, *International Journal of Advanced Robotic Systems*, 1(4), 245–250, ISSN 1729–8806.

Dutta, S. M., Ghorbel, F. H. and Dabney, J. B. (2005) Modeling and Control of a Shape Memory Alloy Actuator, Paper no. WeA01-4, Proceedings of the 2005 IEEE International Symposium on Intelligent Control, Limassol, Cyprus.

Hodgdon, M. L. (1988a) Application of a theory of ferromagnetic hysteresis, *IEEE Transactions on Magnetics*, 24(1), 218–221.

Hodgdon, M. L. (1988b) Mathematical theory and calculations of magnetic hysteresis curves, *IEEE Transactions on Magnetics*, 24(6), 3120–3122.

Jiles, D. and Atherton, D. (1984) Theory of ferromagnetic hysteresis, *Journal of Applied Physics*, 55(6), 2115–2120.

Kozek, M. and Gross, B. (2005) Identification and inversion of magnetic hysteresis for sinusoidal magnetization, *IJOE International Journal on Online Engineering* (www.i-joe.org), 1–10.

Kudva, J. N., Lockyer, A. J. and Appa, K. (1996) Adaptive Aircraft Wing, AGARD-LS-205, Paper 10.

Langelaar, M. and van Keulen, F. (2004) Modeling of a Shape Memory Alloy Active Catheter, 45th AIAA/ASME/ASCE/AHS/ASC Structures, Structural Dynamics & Materials Conference, 19–22 April 2004, Palm Springs, California.

Liang, C. and Rogers, C.A. (1990) One-dimensional thermomechanical constitutive relations for shape memory material, *Journal of Intelligent Material Systems and Structures*, 1, 207–234.

Ma, N. and Song, G. (2003) *Control of Shape Memory Alloy Actuator Using Pulse Width Modulation, Smart Material Structures*, Vol. 12, pp. 712–719, Bristol: Institute of Physics Publishing.

Meurer, T., Qu, J. and Jacobs, L. J. (2002) Wave propagation in nonlinear and hysteretic media – a numerical study, *International Journal of Solids and Structures*, 39, 5585–5614.

Mihálcz, I. (2001) Fundamental characteristics and design method for nickel-titanium shape memory alloy, *Periodica Polytechnica Series Mechanical Engineering*, 45 (1), 75–86.

Miyazaki, S., Nomura, K., Ishida, A. and Kajiwara, S. (1997) Recent developments in sputter-deposited Ti-Ni-base shape memory alloy thin films, *Journal de Physique IV France*, 7(Col. C5), 275–280.

Oh, J. T., Park, H. C. and Hwang, W. (2001) Active shape control of a double-plate structure using piezoceramics and SMA wires, *Smart Materials and Structures*, 10, 1100–1106.

Pare, T. E. (2000) *Passivity Based Analysis and the Control of Nonlinear Systems*, Ph. D. Dissertation, Department of Mechanical Engineering, Stanford University, November, 2000.

Prahlad, H. and Chopra, I. (2001) Comparative evaluation of shape memory alloy constitutive models with experimental data, *Journal of Intelligent Material Systems and Structures*, 12, 383–395.

Rediniotis, O. K., Lagoudas, D. C., Jun, H. Y. and Allen, R. D. (2002a) *Fuel-Powered Compact SMA Actuator*, Aerospace Engineering Department, Texas A&M University, College Station, TX.

Rediniotis, O. K., Wilson, L. N., Lagoudas D. C. and Khan, M. M. (2002b) Development of a shape-memory-alloy actuated biomimetic hydrofoil, *Journal of Intelligent Material Systems and Structures*, 13(1), 35–49.

Seward, K. P., Krulevitch, P., Ackler, H. D. and Ramsey, P. B. (1999) A New Mechanical Characterization Method for Microactuators Applied to Shape Memory Films, Transducers '99, *The 10th International Conference on Solid-State Sensors and Actuators*, Senda, Japan.

Song, G., Zhao, J., Zhou, X. and De Abreu-García, J. A. (2005) Tracking control of a piezoceramic actuator with hysteresis compensation using inverse Preisach model, *IEEE/ASME Transactions on Mechatronics*, 10(2), 198-209.

Tan, X. and Baras, J. S. (2004) Modeling and control of hysteresis in magnetostrictive actuators, *Automatica*, 40, 1469–1480.

Vedantam, S. (2000) Constitutive Modelling of Cu-Al-Ni Shape Memory Alloys, Ph. D. Dissertation, Massachusetts Institute of Technology, Boston, Massachusetts, USA.

Visintin, A. (1994) *Differential Models of Hysteresis*, Springer-Verlag, Berlin, Heidelberg, New York.

9

Controller Design for Flexible Structures

9.1 Introduction to Controller Design

The design of high-performance controllers for flexible structures leads to a number of challenging non-linear control synthesis problems. The requirements of the load to inertia ratio, high efficiency, robustness, disturbance rejection and matched fast enough dynamic response have to be met by any practical and viable control system. Flexible structures subjected to time-dependent loads are characterized by deformations over a wide band of frequencies. Such systems are essentially systems with an infinite number of degrees of freedom. Although models with a finite number of modes representing the dynamics may be developed and applied to design control laws for such linear systems, it is suspected that they lead to very rigorous restrictions and that there is possibly an unnecessary wastage of control effort. In order to develop truly optimal high-performance actuators and drives, it is essential that the broadband problem is dealt with rather than employing a 'finite-state' model. In linear control, it is customary to regard most of the synthesis techniques based on finite state models as validation tools.

9.2 Controller Synthesis for Structural Control

There are essentially two classes of methods of control law synthesis for finite-state plants. The first class of methods is based on performance metric optimization while the second is based on Lyapunov stability. Each class of methods has its own set of advantages and disadvantages.

The first approach to controller synthesis for a flexible structure is based on performance metric optimization. The performance metric optimization-based controller synthesis technique has its roots in the optimal feedback control of a linear system where a control law is sought to minimize a quadratic performance index. The solution of the problem naturally led to Hamiltonian system of equations, which are generalizations of the Hamiltonian equations that arise in connection with the dynamics of particles, and the solution to the equations may be conveniently expressed in terms of the solution of a non-linear matrix differential equation, the Riccati equation. However, the realization that such an optimization is not adequate led to the development of the H_∞ controller synthesis technique.

In controller synthesis, it is essential to establish a trade-off between two opposing and competing requirements involving stability robustness and disturbance rejection. Thus, the performance index involves multi-objective optimization, and the philosophy of H_∞ optimization is based on differential game theory. The H_∞ optimal controller minimizes the worst case gain of the system in order to ensure that both requirements are adequately satisfied. Here the underlying principle is that minimizing the H_∞ gain with respect to the control improves stability and that the disturbance always tends to maximize it. Thus, the problem is one ensuring optimal trade-offs. The solution is characterized by an equilibrium point, which is known as a 'saddle point', with the dynamics being essentially non-linear. The linear H_∞ controller synthesis technique for a linear plant involves optimizing quadratic performance trade-offs. Here again there are two classes of problems involving two different situations. In the first case, it is assumed that all plant and disturbance states are available for feedback, while in the second case only the measurements are assumed to be available for feedback. In the latter case, an H_∞ estimator is used to estimate the states of the system and the optimal control input.

9.2.1 Problems Encountered in Structural Control: Spillover, Model Uncertainty, Non-causal Compensators and Sensor Noise

Lightly damped structures are characterized by a large number of normal modes and are therefore difficult to control. It is generally easy to model the first few modes quite accurately. However for the design of high-integrity controllers, it is often necessary to be able to model a large number of the normal modes of the structure. There are a number of problems that are typical to structural control and that are encountered when one is designing and implementing a feedback controller for flexible structures. Practical controller design for flexible structures involves addressing these problems effectively and avoiding any undesirable responses that may be caused by them.

First, there is the issue of spillover. When one is designing a structural controller with the intention of controlling a few of the normal modes, the control actuators must not only be located so they will influence these modes but also explicitly satisfy the requirement that they do not influence a significant number of the remaining normal modes. Otherwise one is faced with the problem of controller spillover that results from the control input exciting one of the other normal modes that the designer did not intend to control in the first place. Similarly, when one is sensing the normal modes, it is important to explicitly ensure that only the amplitudes of the normal modes earmarked to be controlled are measured and that a significant number of the remaining modal amplitudes are not. If not, one would be faced with problem of observation spillover that results from the measurements containing information about those modes that were not intended to be controlled. When the measured signals are fed back to the controller, the closed loop will significantly influence modes other than those that were intended to be controlled, triggering a number of unexpected responses. It is important to be able to control the first few modes of a structure without in any way destabilizing or, in some cases, disturbing higher-frequency modes due to spillover. Spillover-free design of the controller is essential to avoid spillover problems that are commonly encountered in the design of controllers for flexible structures.

One structural feedback control concept that avoids spillover is direct velocity feedback with an array of distributed but co-located sensors and actuators. With co-located sensors, it is generally easier to ensure that the closed-loop system has essentially similar properties as the open loop in the sense that the structure is passive in terms of energy flow, i.e. it does not generate energy within but only conserves or dissipates it. With the use of distributed sensors and velocity feedback it is relatively easy to augment the damping in all the structural modes. However, such methods treat all modes on an equal footing, as they do not generally damp those that are relatively less stable. Such an approach is referred to as *low authority control*. Controller design methods can focus on stabilizing the less stable modes, and such methods are based on the use of estimators or are specially designed to be robust. They generally do not co-locate the sensors and actuators but ensure stabilization within a frequency band. Such methods are generally classed as *high authority controllers*.

The second important feature is the uncertainties involved in the representation of the dynamics of the structure. Often a few important normal modes are ignored because they are perceived as not being important. For example, when one is controlling a wing structure, it is often the practice to ignore modes associated with the engine pods and pylons. Yet these may contribute secondary normal modes, which fall within the bandwidth of the controller and therefore should not be ignored.

Often when a control system compensator is designed in the frequency domain, it may be non-causal. A non-causal compensator is one that requires inputs of future behaviour, and these are not available in practice. For example, when one is compensating a system with a time delay, a popular method of compensation is the Smith predictor, which is essentially non-causal when implemented in the continuous time domain, as it requires inputs from the future to construct a filter that can 'undo' the delay introduced by the plant. One must then construct the optimal causal approximation to the non-causal compensator. H_∞ control provides a feasible solution to the problem of causal approximation.

The last feature that is encountered in the control of flexible structures is the relatively high levels of sensor noise that one must contend with while implementing the controller. Often sensor noise must be filtered out, but in many situations linear filters are inadequate for the task and some form of non-linear filtering must be adopted.

9.2.2 Concepts of Stability

The first and most important concept of stability is loosely termed Lyapunov stability. Formally it is referred to as *global uniform asymptotic stability* (see, for example, M. Vidyasagar, Chapter 5, pp. 135–269). It relies on several concepts, notable amongst them being:

1. stability, which requires that initial conditions due to perturbations near an equilibrium point result in responses that approach that point;
2. *global* stability implying stability in the large as against *local* stability near the equilibrium point;
3. *uniformness* to eliminate the dependence on the starting time;
4. *attractivity*, which considers a set of initial conditions in a region around the equilibrium point and requires the resulting responses to approach it;
5. *asymptotic stability*, which implies both stability and attractivity.

Global uniform asymptotic stability is defined as:

the equilibrium $\mathbf{x} = \mathbf{0}$ is globally uniformly asymptotically stable if (1) it is uniformly stable and (2) for each pair of positive numbers M and ε with arbitrarily large and ε arbitrarily small, there exists a finite number $T = T(M, \varepsilon)$ such that

$$\|\mathbf{x}_0\| < M, \quad t_0 \geq 0 \tag{9.1}$$

implies

$$\|\mathbf{x}(t_0 + t, t_0, \mathbf{x}_0)\| < \varepsilon, \quad \forall t \geq T(M, \varepsilon). \tag{9.2}$$

In the above definition, $\|\cdot\|$ refers to the norm or more specifically finite 2-norm. Let $p \geq 1$ be a real number, then the finite p-norm is defined as

$$\|\mathbf{x}\|_p \equiv \left(\sum_{i=1}^{n} |x_i|^p \right)^{\frac{1}{p}}. \tag{9.3}$$

Note that for $p = 2$ we get the Euclidean norm, which is also denoted as $\|\cdot\|$.

There are in fact other concepts of stability that do not rely on the notion of states. Two forms of bounded input/bounded output (BIBO) stability are applied extensively for controller synthesis. These are

1. BIBO stability: A system is BIBO stable if for every bounded input, $|u_i(t)| < M_1$ for all t and for all i, the output is bounded, $|y_j(t)| < M_2$ for all t and for all j, provided that the initial conditions are zero.
2. L_2 stability/L_p stability: A system is L_2 stable (L_p stable, $p = 1, 2, 3, \ldots, \infty$) if all the inputs and signals with finite 2-norms (p-norms, $p = 1, 2, 3, \ldots, \infty$), if the system *gain* is bounded, i.e.

$$\frac{\|y(t)\|_2}{\|u(t)\|_2} < M_{2-norm}, \tag{9.4}$$

$$\left(\frac{\|y(t)\|_p}{\|u(t)\|_p} < M_{p-norm},\ p = 1, 2, 3, \ldots, \infty\right). \tag{9.5}$$

Internal stability: Clearly the above definition of stability based on input/output behaviour may be inadequate when a BIBO-stable system still has internal signals that are unbounded. Thus, there is a need for internal stability as well.

Input to state stability is one definition of stability that will ensure that internal stability is guaranteed. A system is input to state stable when there exist functions β and γ such that for every initial condition $x0$ and every input u, such that its finite ∞-norm is bounded, every response solution $x(t)$ starting from $t = 0$ satisfies

$$|x(t)| \leq \beta\left(|x_0|, t\right) + \gamma\left(\|u\|_\infty\right) \tag{9.6}$$

for all $t \geq 0$, where $\|.\|_\infty$is the finite ∞-norm of the response, defined as

$$\|\mathbf{x}\|_\infty = \max_i |x_i| \quad \text{(the maximum absolute row)}. \tag{9.7}$$

A well-known theorem concerning the ∞-norms in the context of input–output systems shows that it represents the maximum gain over all possible 2-norm inputs, i.e.

$$\|\mathbf{G}\|_\infty = \sup_{u \neq 0} \frac{\|y\|_2}{\|u\|_2}. \tag{9.8}$$

The ratio of the input/output finite 2-norms is known as the H_∞-norm of the transfer function.

When $|x_0| = 0$ the requirement for input to state stability reduces to

$$\|x(t)\|_\infty \leq \gamma_0\left(\|u\|_\infty\right), \tag{9.9}$$

which is equivalent to L_∞ stability. Ensuring input to state stability means computing β and γ, and Lyapunov functions are the main tools for achieving this.

It is important to emphasize that the above concepts may be applied to the *nominal* system, i.e. stability of the systems free from modelling errors or disturbances or the *actual* system, including the various physical uncertainties.

The stability analysis of the asymptotic systems was pioneered by A. M. Lyapunov in his thesis that was published more than 100 years ago (Parks, 1992). Lyapunov's method not only is a basic tool in the analysis of system stability but has also emerged as a very valuable tool for the synthesis of controllers for non-linear systems. Lyapunov divided the problem of stability analysis of non-linear systems into two groups. The first group of problems involved non-linear systems that could be either solved exactly or reduced by some means to a linear system. These included not only exact methods but also, in some cases, approximate techniques where the stability of the linearized system yields useful properties about the stability of equilibrium of the non-linear system.

In the case of a linear system, it is possible to analyse the stability of the solution without having to derive elaborate general solutions to the problem. Methods of evaluating stability, such as the Routh–Hurwitz method and Routh's tabulation, allow the control system designer to establish bounds on important system gains to guarantee stability. For the second group of stability problems, Lyapunov was motivated by similar considerations and he sought to develop similar methods applicable to the analysis of the stability of motion. The Lagrangian method, for the analysis of the stability of static equilibrium, where the potential energy function is required to be a minimum, is an example of such a technique.

Lyapunov defined the so-called Lyapunov function as a generalization of the potential energy that could yield the stability boundaries without the necessity of having to solve the governing equations of motion. For this reason, the method is sometimes referred to as the direct method.

The choice of the Lyapunov function is an important step in Lyapunov analysis. Not only should the function be positive definite in terms of its variables and satisfy a condition of radial unboundedness, but it must satisfy the condition that its time rate of change along every admissible trajectory of motion is always negative. Such a function is not necessarily unique and criteria for stability based on a particular Lyapunov function are in general sufficient but not necessary. It is important to observe that should a particular function fail to show that the system under consideration is stable or unstable, there is no assurance that another Lyapunov function cannot be found that results in adequate stability criteria for the system. Moreover, when such a function is found and certain stability limits are established, there is no indication that they are not overly conservative.

One of the most common ways to the study the stability of a non-linear dynamic system described by a set of non-linear differential equations is to find a Lyapunov function, denoted by $V(\mathbf{x})$, which is radially unbounded and positive definite but with a negative definite time derivative (i.e. $V(\mathbf{x}) > 0 \forall \mathbf{x} \in \mathbf{\Omega} \subset R^N$ and $V(\mathbf{x}) = 0$ when $\mathbf{x} \equiv \mathbf{0}$). Thus, $V(\mathbf{x})$ is a function on the state space, which is bounded from below and monotonically decreasing along every non-constant continuous trajectory. The above definition may be extended to the case when $V = V(\mathbf{x}, t)$ is a function of time as well.

Consider a dynamical system described by a set of coupled differential equations

$$\dot{\mathbf{x}} = f(\mathbf{x}). \tag{9.10}$$

Lyapunov's theorem may be stated as: if for a given system of differential equations there exists a Lyapunov function, i.e. it is radially unbounded and positive definite and satisfies the relationship

$$\frac{\mathrm{d}V(\mathbf{x})}{\mathrm{d}t} = \sum_{j=1}^{n} \frac{\mathrm{d}V(\mathbf{x})}{\mathrm{d}x_j}\frac{\mathrm{d}x_j}{\mathrm{d}t} = \frac{\mathrm{d}V(\mathbf{x})}{\mathrm{d}\mathbf{x}} f(\mathbf{x}) < 0, \tag{9.11}$$

then the system is globally uniformly asymptotically stable, as the trajectories necessarily converge to the equilibrium point $\mathbf{x} = \mathbf{0}$. The concept is easily extended to the non-autonomous case.

Thus, the existence of a Lyapunov function is itself a necessary and sufficient condition for the stability of a linear system. Classical Lyapunov stability theory is in reality L_1 stability but is also equivalent to L_2 stability.

Consider the construction of a suitable Lyapunov function for a linear system. The linear system is assumed to be described by the state–space model

$$\dot{\mathbf{x}}(t) = \mathbf{A}\mathbf{x}(t). \tag{9.12}$$

A suitable Lyapunov function is assumed to be

$$V(\mathbf{x}) = \mathbf{x}^T(t)\mathbf{P}(t)\mathbf{x}(t), \tag{9.13}$$

where $\mathbf{P}(t)$ is a positive semi-definite matrix.

Then, the time derivative of the Lyapunov function along the trajectories of the system must satisfy

$$\frac{dV(\mathbf{x})}{dt} = \sum_{j=1}^{n} \frac{dV(\mathbf{x})}{dx_j}\frac{dx_j}{dt} = \frac{dV(\mathbf{x})}{d\mathbf{x}}\mathbf{A}\mathbf{x}(t) < 0. \tag{9.14}$$

To satisfy this requirement, we assume that the time derivative of the Lyapunov function along the trajectories of the system must be

$$\frac{dV(\mathbf{x})}{dt} \equiv \dot{V}(\mathbf{x}) \equiv -\mathbf{x}^T(t)\mathbf{Q}(t)\mathbf{x}(t), \tag{9.15}$$

where $\mathbf{Q}(t)$ is a positive definite matrix. This fact is stated as a linear matrix inequality (LMI) as

$$\mathbf{Q}(t) > 0. \tag{9.16}$$

However, the time derivative of the Lyapunov function along the trajectories of the system is

$$\dot{V}(\mathbf{x}) = \dot{\mathbf{x}}^T(t)\mathbf{P}(t)\mathbf{x}(t) + \mathbf{x}^T(t)\dot{\mathbf{P}}(t)\mathbf{x} + \mathbf{x}^T(t)\mathbf{P}(t)\dot{\mathbf{x}}(t).$$

Using equation (9.12),

$$\dot{V}(\mathbf{x}) = \mathbf{x}^T(t)\mathbf{A}^T\mathbf{P}(t)\mathbf{x}(t) + \mathbf{x}^T(t)\dot{\mathbf{P}}(t)\mathbf{x} + \mathbf{x}^T(t)\mathbf{P}(t)\mathbf{A}\mathbf{x}(t).$$

Thus, the time derivative of the Lyapunov function along the trajectories of the system must satisfy

$$\dot{V}(\mathbf{x}) = \mathbf{x}^T(t)\mathbf{A}^T\mathbf{P}(t)\mathbf{x}(t) + \mathbf{x}^T(t)\dot{\mathbf{P}}(t)\mathbf{x} + \mathbf{x}^T(t)\mathbf{P}(t)\mathbf{A}\mathbf{x}(t) = -\mathbf{x}^T(t)\mathbf{Q}(t)\mathbf{x}(t). \tag{9.17}$$

Hence, the matrix $\mathbf{P}(t)$ must satisfy the constraint

$$\dot{\mathbf{P}}(t) + \mathbf{A}^T\mathbf{P}(t) + \mathbf{P}(t)\mathbf{A} = -\mathbf{Q}(t). \tag{9.18}$$

Since $\mathbf{Q}(t)$ is a positive definite matrix, the constraint on $\mathbf{P}(t)$ may also be stated as an LMI in the form

$$\dot{\mathbf{P}}(t) + \mathbf{A}^T\mathbf{P}(t) + \mathbf{P}(t)\mathbf{A} < 0. \tag{9.19}$$

If one further assumes that $\mathbf{P}(t) = \mathbf{P}$ is assumed to be time invariant, the constraint equation may be expressed as

$$\mathbf{A}^T\mathbf{P} + \mathbf{P}\mathbf{A} = -\mathbf{Q}. \tag{9.20}$$

Equation (9.20) is the celebrated Lyapunov equation, which may be solved if $\mathbf{A}$ can be diagonalized. Thus when $\mathbf{A} = \mathbf{V}\Lambda\mathbf{V}^{-1}$,

$$\mathbf{P} = \left(\mathbf{V}^{-1}\right)^T \bar{\mathbf{P}}\mathbf{V}^{-1}, \tag{9.21}$$

where i, jth element of $\bar{\mathbf{P}}$ is given in terms of i, jth element of $\bar{\mathbf{Q}}$ by $\bar{P}_{ij} = -\bar{Q}_{ij}/(\lambda_{ii} + \lambda_{jj})$, where $\bar{\mathbf{Q}} = \mathbf{V}^T\mathbf{Q}\mathbf{V}$. It is also customary to choose $\mathbf{Q} = \mathbf{I}$.

There is an important consequence that follows the existence of a Lyapunov function, which is known as the *bounded real lemma*. Given a system in state-space form with

$$\dot{\mathbf{x}}(t) = \mathbf{A}\mathbf{x}(t) + \mathbf{B}\mathbf{u}(t), \quad \mathbf{x}(0) = 0, \quad y(t) = \mathbf{C}\mathbf{x}(t) \tag{9.22}$$

and a suitable Lyapunov function

$$V(\mathbf{x}) = \mathbf{x}^T(t)\mathbf{P}(t)\mathbf{x}(t), \tag{9.23}$$

where $\mathbf{P}(t)$ is a positive semi-definite matrix and

$$\dot{V}(\mathbf{x}) = \mathbf{w}^T(t)\mathbf{w}(t) - \gamma^2\mathbf{y}^T(t)\mathbf{y}(t) < 0 \quad \text{for all } \mathbf{x} \text{ and } \mathbf{w}, \tag{9.24}$$

one can show by integrating equation (9.24) that the root mean square gain does not exceed γ, i.e.

$$\int_0^\infty \mathbf{y}^T(t)\mathbf{y}(t)\mathrm{d}t \le \gamma^2 \int_0^\infty \mathbf{u}^T(t)\mathbf{u}(t)\mathrm{d}t. \tag{9.25}$$

Another important consequence that is extremely useful in controller synthesis arises in closed-loop systems where $\mathbf{P}(t)$ and the inverse of $\mathbf{P}(t)$ can be decomposed into

$$\mathbf{P}(t) = \begin{bmatrix} \mathbf{Y} & \mathbf{N} \\ \mathbf{N}^T & \mathbf{V} \end{bmatrix} \text{ and } \mathbf{P}(t)^{-1} = \begin{bmatrix} \mathbf{X} & \mathbf{M} \\ \mathbf{M}^T & \mathbf{U} \end{bmatrix}. \tag{9.26}$$

since

$$\begin{bmatrix} \mathbf{X} & \mathbf{M} \\ \mathbf{M}^T & \mathbf{U} \end{bmatrix}\begin{bmatrix} \mathbf{Y} & \mathbf{N} \\ \mathbf{N}^T & \mathbf{V} \end{bmatrix} = \begin{bmatrix} \mathbf{I} & \mathbf{0} \\ \mathbf{0} & \mathbf{I} \end{bmatrix}, \tag{9.27a}$$

$$\begin{bmatrix} \mathbf{Y} & \mathbf{N} \\ \mathbf{N}^T & \mathbf{V} \end{bmatrix}\begin{bmatrix} \mathbf{X} & \mathbf{I} \\ \mathbf{M}^T & \mathbf{0} \end{bmatrix} = \begin{bmatrix} \mathbf{I} & \mathbf{Y} \\ \mathbf{0} & \mathbf{N}^T \end{bmatrix} \tag{9.27b}$$

and it follows that

$$\mathbf{XY} + \mathbf{MN}^T = \mathbf{I} \tag{9.28}$$

and

$$\begin{bmatrix} \mathbf{X} & \mathbf{I} \\ \mathbf{M}^T & \mathbf{0} \end{bmatrix}^T \begin{bmatrix} \mathbf{Y} & \mathbf{N} \\ \mathbf{N}^T & \mathbf{V} \end{bmatrix}\begin{bmatrix} \mathbf{X} & \mathbf{I} \\ \mathbf{M}^T & \mathbf{0} \end{bmatrix} = \begin{bmatrix} \mathbf{X} & \mathbf{M} \\ \mathbf{I} & \mathbf{0} \end{bmatrix}\begin{bmatrix} \mathbf{I} & \mathbf{Y} \\ \mathbf{0} & \mathbf{N}^T \end{bmatrix} = \begin{bmatrix} \mathbf{X} & \mathbf{I} \\ \mathbf{I} & \mathbf{Y} \end{bmatrix} > \mathbf{0}, \text{ i.e. } \mathbf{XY} - \mathbf{I} > 0. \tag{9.29}$$

9.2.3 *Passive Controller Synthesis*

The concept of Lyapunov stability could be exploited to design controllers for flexible structures. To set the scene for the analysis and design of controllers, we first consider a flexible structure with the dynamics described by a linear system

$$\dot{\mathbf{x}}(t) = \mathbf{Ax}(t) + \mathbf{Bu}(t), \tag{9.30}$$

where $\mathbf{u}(t)$ is a control input vector to the system. If one assumes that

$$\mathbf{u}(t) = -\mathbf{K}_c\mathbf{x}(t), \tag{9.31}$$

the dynamics of the closed-loop system is described by

$$\dot{\mathbf{x}}(t) = (\mathbf{A} - \mathbf{BK}_c)\mathbf{x}(t). \tag{9.32}$$

If a suitable Lyapunov function for the closed-loop system is assumed to be

$$V(\mathbf{x}) = \mathbf{x}^T(t)\mathbf{P}(t)\mathbf{x}(t), \tag{9.33}$$

then the matrix $\mathbf{P}(t)$ must satisfy the constraint

$$\dot{\mathbf{P}}(t) + (\mathbf{A} - \mathbf{BK}_c)^T\mathbf{P}(t) + \mathbf{P}(t)(\mathbf{A} - \mathbf{BK}_c) = -\mathbf{Q}(t). \tag{9.34}$$

Since the controller gain matrix $\mathbf{K}_c$ is yet to be determined, one must choose both $\mathbf{K}_c$ and $\mathbf{P}(t)$ to satisfy the above constraint. To see how this may be applied to synthesize a passive structural controller, consider the dynamics of a flexible structure in the form

$$\mathbf{M}\ddot{\mathbf{x}}(t) + \mathbf{C}\dot{\mathbf{x}}(t) + \mathbf{Kx}(t) = \mathbf{u}(t), \tag{9.35}$$

where $\mathbf{x}(t)$ is an $n \times 1$ vector of the displacement degrees of freedom, $\mathbf{u}(t)$ is an $n \times 1$ vector of control forces generated by the actuators, $\mathbf{M}$ is an $n \times n$ mass matrix associated with the flexible structure, $\mathbf{C}$ is an $n \times n$ damping matrix associated with the structure and $\mathbf{K}$ is the associated $n \times n$ stiffness matrix. The dynamics may be expressed in state-space form as

$$\begin{bmatrix} \mathbf{M} & \mathbf{0} \\ \mathbf{0} & \mathbf{M} \end{bmatrix} \frac{\mathrm{d}}{\mathrm{d}t} \begin{bmatrix} \mathbf{x} \\ \dot{\mathbf{x}} \end{bmatrix} = \begin{bmatrix} \mathbf{0} & \mathbf{M} \\ -\mathbf{K} & -\mathbf{C} \end{bmatrix} \begin{bmatrix} \mathbf{x} \\ \dot{\mathbf{x}} \end{bmatrix} + \begin{bmatrix} \mathbf{0} \\ \mathbf{I} \end{bmatrix} \mathbf{u} \tag{9.36a}$$

and reduced to

$$\frac{\mathrm{d}}{\mathrm{d}t} \begin{bmatrix} \mathbf{x} \\ \dot{\mathbf{x}} \end{bmatrix} = \begin{bmatrix} \mathbf{0} & \mathbf{I} \\ -\mathbf{M}^{-1}\mathbf{K} & -\mathbf{M}^{-1}\mathbf{C} \end{bmatrix} \begin{bmatrix} \mathbf{x} \\ \dot{\mathbf{x}} \end{bmatrix} + \begin{bmatrix} \mathbf{0} \\ \mathbf{M}^{-1} \end{bmatrix} \mathbf{u}. \tag{9.36b}$$

One assumes that the passive controller takes the form

$$\mathbf{u}(t) = -\Delta\mathbf{K}\mathbf{x}(t) - \Delta\mathbf{C}\dot{\mathbf{x}}, \tag{9.37}$$

where $\Delta\mathbf{K}$ and $\Delta\mathbf{C}$ are symmetric matrices representing modifications to the structure's stiffness and damping matrices. We have tacitly assumed that the structure's mass matrix is unchanged, but this is generally not the case in practice. This aspect will be considered in a latter section. The closed-loop dynamics is now given by

$$\frac{\mathrm{d}}{\mathrm{d}t} \begin{bmatrix} \mathbf{x} \\ \dot{\mathbf{x}} \end{bmatrix} = -\begin{bmatrix} \mathbf{0} & -\mathbf{I} \\ \mathbf{M}^{-1}(\mathbf{K}+\Delta\mathbf{K}) & \mathbf{M}^{-1}(\mathbf{C}+\Delta\mathbf{C}) \end{bmatrix} \begin{bmatrix} \mathbf{x} \\ \dot{\mathbf{x}} \end{bmatrix}. \tag{9.38}$$

If one assumes that the matrix $\mathbf{P}(t)$ is now given by the partitioned matrix

$$\mathbf{P}(t) = \begin{bmatrix} \mathbf{P}_{11} & \mathbf{P}_{12} \\ \mathbf{P}_{21} & \mathbf{P}_{22} \end{bmatrix}, \tag{9.39}$$

then the matrix $\mathbf{P}(t)$ must satisfy the constraint

$$\begin{aligned} &\begin{bmatrix} \dot{\mathbf{P}}_{11} & \dot{\mathbf{P}}_{12} \\ \dot{\mathbf{P}}_{21} & \dot{\mathbf{P}}_{22} \end{bmatrix} - \begin{bmatrix} \mathbf{0} & (\mathbf{K}+\Delta\mathbf{K})\mathbf{M}^{-1} \\ -\mathbf{I} & (\mathbf{C}+\Delta\mathbf{C})\mathbf{M}^{-1} \end{bmatrix} \begin{bmatrix} \mathbf{P}_{11} & \mathbf{P}_{12} \\ \mathbf{P}_{21} & \mathbf{P}_{22} \end{bmatrix} \\ &- \begin{bmatrix} \mathbf{P}_{11} & \mathbf{P}_{12} \\ \mathbf{P}_{21} & \mathbf{P}_{22} \end{bmatrix} \begin{bmatrix} \mathbf{0} & -\mathbf{I} \\ \mathbf{M}^{-1}(\mathbf{K}+\Delta\mathbf{K}) & \mathbf{M}^{-1}(\mathbf{C}+\Delta\mathbf{C}) \end{bmatrix} = -\begin{bmatrix} \mathbf{Q}_{11} & \mathbf{Q}_{12} \\ \mathbf{Q}_{21} & \mathbf{Q}_{22} \end{bmatrix}. \end{aligned} \tag{9.40}$$

The components of the matrix $\mathbf{P}(t)$ must satisfy the constraints

$$\dot{\mathbf{P}}_{11}(t) - (\mathbf{K}+\Delta\mathbf{K})\mathbf{M}^{-1}\mathbf{P}_{21}(t) - \mathbf{P}_{21}^{T}\mathbf{M}^{-1}(\mathbf{K}+\Delta\mathbf{K}) = -\mathbf{Q}_{11}, \tag{9.41a}$$

$$\dot{\mathbf{P}}_{22}(t) + \mathbf{P}_{21}^{T}(t) + \mathbf{P}_{21}(t) - (\mathbf{C}+\Delta\mathbf{C})\mathbf{M}^{-1}\mathbf{P}_{22}(t) - \mathbf{P}_{22}\mathbf{M}^{-1}(\mathbf{C}+\Delta\mathbf{C}) = -\mathbf{Q}_{22} \tag{9.41b}$$

$$\dot{\mathbf{P}}_{21}(t) + \mathbf{P}_{11}(t) - (\mathbf{C}+\Delta\mathbf{C})\mathbf{M}^{-1}\mathbf{P}_{21}(t) - \mathbf{P}_{22}\mathbf{M}^{-1}(\mathbf{K}+\Delta\mathbf{K}) = -\mathbf{Q}_{21}. \tag{9.41c}$$

In particular, if $\mathbf{P}(t) = \mathbf{P}$ is assumed to be time invariant, the constraint equations (9.41) may be expressed as

$$(\mathbf{K}+\Delta\mathbf{K})\mathbf{M}^{-1}\mathbf{P}_{21} + \mathbf{P}_{21}^{T}\mathbf{M}^{-1}(\mathbf{K}+\Delta\mathbf{K}) = \mathbf{Q}_{11}, \tag{9.42a}$$

$$(\mathbf{C}+\Delta\mathbf{C})\mathbf{M}^{-1}\mathbf{P}_{22} + \mathbf{P}_{22}\mathbf{M}^{-1}(\mathbf{C}+\Delta\mathbf{C}) = \mathbf{Q}_{22} + \mathbf{P}_{21}^{T} + \mathbf{P}_{21}, \tag{9.42b}$$

$$\mathbf{P}_{11} = (\mathbf{C}+\Delta\mathbf{C})\mathbf{M}^{-1}\mathbf{P}_{21} + \mathbf{P}_{22}\mathbf{M}^{-1}(\mathbf{K}+\Delta\mathbf{K}) - \mathbf{Q}_{21}. \tag{9.42c}$$

Given the modifications to the structure's stiffness and damping matrices, $\Delta\mathbf{K}$ and $\Delta\mathbf{C}$, the above three equations (9.42) may be solved for $\mathbf{P}_{21}$, $\mathbf{P}_{22}$ and $\mathbf{P}_{11}$. Stability is then guaranteed if

$$\begin{bmatrix} \mathbf{P}_{11} & \mathbf{P}_{12} \\ \mathbf{P}_{21} & \mathbf{P}_{22} \end{bmatrix} > \mathbf{0}. \tag{9.43}$$

Although the above is not a synthesis technique, it is actually quite handy to check in practice if the structural modifications that have been made to the flexible structural system are adequate to enhance stability. For purposes of synthesis, the multi-objective synthesis method based on LMIs is most suitable and will be discussed in a latter section.

An interesting feature of most uncontrolled flexible structures is that they are usually Hamiltonian in nature or could be ideålized as Hamiltonian systems by ignoring the dissipative forces that may be present. We discuss the simplest problem of obtaining the optimal control of a Hamiltonian and a near-Hamiltonian system. Given the system model by the set of n governing system of equations in Hamiltonian form

$$J\dot{\mathbf{x}} = \frac{\mathrm{d}H}{\mathrm{d}\mathbf{x}}^T - \begin{bmatrix} \mathbf{F} \\ \mathbf{0} \end{bmatrix} - \begin{bmatrix} \mathbf{B} \\ \mathbf{0} \end{bmatrix}\mathbf{u}, \quad \mathbf{x} = \begin{bmatrix} \mathbf{q} \\ \mathbf{p} \end{bmatrix}, \quad J = \begin{bmatrix} 0 & -\mathbf{I} \\ \mathbf{I} & 0 \end{bmatrix}, \quad J^T = J^{-1} = -J, \tag{9.44}$$

where J is given by the symplectic form, H is the Hamiltonian function in terms of the generalized displacements $\mathbf{q}$ and the generalized momenta $\mathbf{p}$, defined in terms of the Lagrangian by the equations

$$H = \mathbf{p}^T\dot{\mathbf{q}} - L, \quad \mathbf{p} = \frac{\partial L}{\partial \dot{\mathbf{q}}}^T, \quad L = T - V, \tag{9.45}$$

where T and V are the kinetic and potential energies. $\mathbf{F}$ is the vector of normalized external forces that cannot be derived from the potential function V, and $\mathbf{Bu}$ are the normalized control forces.

It is often desired to choose the control $\mathbf{u} = \mathbf{u}^*$ to minimize the performance or cost functional,

$$J = \frac{1}{2}\int_0^\infty \left(\mathbf{x}^T\mathbf{Q}\mathbf{x} + \mathbf{u}^T\mathbf{R}\mathbf{u}\right)\mathrm{d}t = \int_0^\infty \mathrm{L}(\mathbf{x}, \mathbf{u})\mathrm{d}t. \tag{9.46}$$

The matrices $\mathbf{Q}$ and $\mathbf{R}$ are assumed to be symmetric, and $\mathbf{R}$ is assumed to be invertible. To find $\mathbf{u}$, we augment the cost functional and minimize

$$J_{augmented} = \int_0^\infty \left[\mathrm{L}(\mathbf{x}, \mathbf{u}) + \lambda^T\left(\frac{\mathrm{d}H}{\mathrm{d}\mathbf{x}}^T - \begin{bmatrix} \mathbf{F} \\ \mathbf{0} \end{bmatrix} - \begin{bmatrix} \mathbf{B} \\ \mathbf{0} \end{bmatrix}\mathbf{u} - J\dot{\mathbf{x}}\right)\right]\mathrm{d}t. \tag{9.47}$$

Minimizing the augmented cost functional with respect to the n-dimensional multiplier λ^T, which is also known as the co-state vector, results in the given system model equations. An approach to solving this augmented minimization problem defines another Hamiltonian $H_{\mathrm{OC}}$where H_{OC} is defined by the relationship

$$H_{\mathrm{OC}} = \int_0^\infty \mathrm{H}(\mathbf{x}, \lambda, \mathbf{u}^*)\mathrm{d}t, \tag{9.48}$$

where

$$\mathrm{H}(\mathbf{x}, \lambda, \mathbf{u}^*) = \min_u \left[\mathrm{L}(\mathbf{x}, \mathbf{u}) + \lambda^T\left(\frac{\mathrm{d}H}{\mathrm{d}\mathbf{x}}^T - \begin{bmatrix} \mathbf{F} \\ \mathbf{0} \end{bmatrix} - \begin{bmatrix} \mathbf{B} \\ \mathbf{0} \end{bmatrix}\mathbf{u}\right)\right]. \tag{9.49}$$

In defining $\mathrm{H}(\mathbf{x}, \lambda, \mathbf{u}^*)$, an n-dimensional multiplier or co-state vector λ^T has been introduced. The multiplier raises the dimensionality of the problem to $2n$ but allows the conditions for the optimal $\mathbf{u}^*$ to be obtained. Integrating the last term in the integral defining the augmented cost functional by parts and minimizing the augmented cost functional with respect to $\mathbf{u}$ and $\mathbf{x}$, the conditions for the optimal $\mathbf{u}^*$ are obtained as

$$\partial\mathrm{H}(\mathbf{x}, \lambda, \mathbf{u})/\partial\mathbf{u} = 0, \Rightarrow \mathbf{u}^* = -\mathbf{R}^{-1}\left[\mathbf{B}^T\mathbf{0}\right]\lambda \tag{9.50a}$$

and

$$\dot{\lambda}^T J = -\frac{\partial\mathrm{H}^T(\mathbf{x}, \lambda, \mathbf{u})}{\partial\mathbf{x}} = -\lambda^T\left[\frac{\mathrm{d}^2 H}{\mathrm{d}\mathbf{x}^2} - \begin{bmatrix} \dfrac{\mathrm{d}\mathbf{F}}{\mathrm{d}\mathbf{x}} \\ 0 \end{bmatrix}\right] - \mathbf{x}^T\mathbf{Q}. \tag{9.50b}$$

It is customary to introduce a further transformation $\lambda = \mathbf{P}x$, so

$$\mathbf{u}^* = -\mathbf{R}^{-1}\left[\mathbf{B}^T\mathbf{0}\right]\mathbf{P}\mathbf{x}. \tag{9.51}$$

If we let $\mathbf{H} = \mathrm{d}^2 H/\mathrm{d}\mathbf{x}^2$, then from equation (9.50),

$$J\dot{\lambda} = \left[\mathbf{H} - \left[\frac{\mathrm{d}\mathbf{F}^T}{\mathrm{d}\mathbf{x}} 0\right]\right]\lambda + \mathbf{Q}\mathbf{x}. \tag{9.52}$$

When $\mathbf{F}=\mathbf{0}$ and $\mathbf{Q}=\mathbf{0}$, $J\dot{\lambda} = \mathbf{H}\boldsymbol{\lambda}$. If we assume that $\boldsymbol{\lambda} = \dot{\mathbf{x}}$,

$$\dot{\boldsymbol{\lambda}} = \ddot{\mathbf{x}} = \frac{\mathrm{d}}{\mathrm{d}t} J^{-1}\frac{\mathrm{d}H}{\mathrm{d}\mathbf{x}}^T = J^{-1}\frac{\mathrm{d}^2 H}{\mathrm{d}\mathbf{x}^2}\dot{\mathbf{x}},$$

and this results in

$$J\dot{\lambda} = \frac{\mathrm{d}^2 H}{\mathrm{d}\mathbf{x}^2}\dot{\mathbf{x}} = \frac{\mathrm{d}^2 H}{\mathrm{d}\mathbf{x}^2}\boldsymbol{\lambda} = \mathbf{H}\boldsymbol{\lambda} \tag{9.53a}$$

and

$$\mathbf{u}^* = -\mathbf{R}^{-1}\left[\mathbf{B}^T\mathbf{0}\right]\dot{\mathbf{x}} = -\mathbf{R}^{-1}\mathbf{B}^T\dot{\mathbf{q}}. \tag{9.53b}$$

This result is quite physically reasonable, as it implies that the application of optimal control theory to a strictly Hamiltonian system results in a controller that requires a feedback directly proportional to the velocity vector. The feedback is a special case of the control law (equation (9.51)) that increases the damping in the system that is already not unstable.

Furthermore, when $\mathbf{R}=\mathbf{0}$ and $\mathbf{u}=\mathbf{0}$,

$$\mathrm{H} = \frac{dH}{d\mathbf{x}}\dot{\mathbf{x}} = \frac{dH}{dt}.$$

Hence it follows that

$$H_{\mathrm{OC}} \equiv H. \tag{9.54}$$

It follows therefore that the two Hamiltonians in this case are identical. This is an important conclusion since the Hamiltonian H_{OC} may be viewed as a generalization of the classical Hamiltonian H. It must be noted, however, that although H_{OC} is a constant, it does not represent the total energy and hence the system under consideration is not a conservative system. Furthermore, the general equation for the multiplier vector $\boldsymbol{\lambda}$,

$$J\dot{\lambda} = \left[\mathbf{H} - \left[\frac{\mathrm{d}\mathbf{F}^T}{\mathrm{d}\mathbf{x}} 0\right]\right]\lambda + \mathbf{Q}\mathbf{x} = \partial H^T(\mathbf{x}, \lambda, \mathbf{u})/\partial\mathbf{x},\ \ \mathbf{u}^* = -\mathbf{R}^{-1}\left[\mathbf{B}^T\mathbf{0}\right]\lambda, \tag{9.55}$$

is a linear equation in the co-state vector $\boldsymbol{\lambda}$. Together with the governing system of equations, the pair of equation sets

$$J\dot{\mathbf{x}} = \frac{\mathrm{d}H}{\mathrm{d}\mathbf{x}}^T - \begin{bmatrix}\mathbf{F}\\ \mathbf{0}\end{bmatrix} - \begin{bmatrix}\mathbf{B}\\ \mathbf{0}\end{bmatrix}\mathbf{u},\ \ J\dot{\lambda} = \left[\mathbf{H} - \left[\frac{\mathrm{d}\mathbf{F}^T}{\mathrm{d}\mathbf{x}} 0\right]\right]\lambda + \mathbf{Q}\mathbf{x} \tag{9.56}$$

constitute a Hamiltonian system of equations. When the Hamiltonian H is a homogeneous quadratic function, $\mathbf{u}$ and $\mathbf{P}$ (equation (9.51)) are obtained by a method involving eigenvalue–eigenvector decomposition. This aspect is discussed in the next section.

9.2.4 *Active Controller Synthesis and Compensation*

The classical time-domain method of synthesizing a controller for a linear time-invariant system is based on shaping the dynamic response of the system by suitably altering the eigenvalues (and eigenvectors in the case of multiple control inputs). Consider a flexible structure excited by a single control actuator and assume that the dynamics may be represented by the set of state-space equations

$$\dot{\mathbf{x}}(t) = \mathbf{A}\mathbf{x}(t) + \mathbf{B}u(t), \tag{9.30}$$

where $u(t)$ is a scalar control input to the system. The open-loop characteristic polynomial is given by

$$\det|\mathbf{A} - s\mathbf{I}| \equiv s^n + a_1 s^{n-1} \ldots a_{n-1}s + a_n = 0. \tag{9.57}$$

If one assumes that

$$u(t) = -\mathbf{K}\mathbf{x}(t), \tag{9.58}$$

the dynamics of the closed-loop system is described by

$$\dot{\mathbf{x}}(t) = (\mathbf{A} - \mathbf{B}\mathbf{K})\mathbf{x}(t). \tag{9.59}$$

The control gains are chosen so that the desired closed-loop characteristic polynomial is assumed to be given by

$$\det|\mathbf{A} - \mathbf{B}\mathbf{K} - s\mathbf{I}| \equiv s^n + c_1 s^{n-1} \ldots c_{n-1}s + c_n = 0. \tag{9.60}$$

If it is assumed that the desired closed-loop characteristic polynomial may be expressed in terms of the desired closed-loop modal natural frequencies ω_{di}, $i = 1, 2, 3, \ldots n$, and the desired closed-loop damping ratios ζ_{di}, $i = 1, 2, 3, \ldots n$, then the coefficients c_i, $i = 1, 2, 3, \ldots n$ may be obtained from the relationship

$$\prod_{i=1}^{n} \left(s^2 + 2\zeta_{di}\omega_{di}s + \omega_{di}^2\right) \equiv s^n + c_1 s^{n-1} \ldots c_{n-1}s + c_n = 0. \tag{9.61}$$

It is desired to design a regulator so the flexible structure has the desirable damping and frequency response characteristics. One approach to designing such a regulator is to first design a full state feedback control law in order to alter all the characteristic roots or pole locations and place them at preferred locations. This is done by transforming the system by the state transformation

$$\mathbf{x}(t) = \mathbf{T}\mathbf{z}(t). \tag{9.62}$$

The corresponding system matrices are required to take the form

$$\mathbf{A}' = \mathbf{T}^{-1}\mathbf{A}\mathbf{T} = \begin{bmatrix} 0 & 1 & \ldots & 0 & 0 \\ 0 & 0 & \ldots & 0 & 0 \\ \ldots & \ldots & \ldots & \ldots & \ldots \\ 0 & 0 & \ldots & 0 & 1 \\ -a_n & -a_{n-1} & \ldots & a_2 & a_1 \end{bmatrix}, \quad \mathbf{B}' = \mathbf{T}^{-1}\mathbf{B} = \begin{bmatrix} 0 \\ 0 \\ \ldots \\ 0 \\ 1 \end{bmatrix}. \tag{9.63}$$

The transformation is given by

$$\mathbf{T} = \mathbf{C}_O\mathbf{C}_T^{-1}, \tag{9.64}$$

where $\mathbf{C}_O$ is the controllability matrix of the original system and $\mathbf{C}_T$is the controllability matrix of the transformed system and is defined as

$$\mathbf{C}_O = [\,\mathbf{B} \quad \mathbf{A}\mathbf{B} \quad \ldots \quad \mathbf{A}^{n-2}\mathbf{B} \quad \mathbf{A}^{n-1}\mathbf{B}\,] \tag{9.65}$$

and

$$\mathbf{C}_T = [\,\mathbf{B}' \quad \mathbf{A}'\mathbf{B}' \quad \ldots \quad \mathbf{A}'^{n-2}\mathbf{B}' \quad \mathbf{A}'^{n-1}\mathbf{B}'\,]. \tag{9.66}$$

The control law may be expressed as

$$u(t) = -\mathbf{Kx}(t) = -\mathbf{KTz}(t). \tag{9.67}$$

The desired closed-loop system matrix in terms of the transformed states is then given as

$$\mathbf{A}'_c = \mathbf{A}' - \mathbf{B}'\mathbf{KT} = \begin{bmatrix} 0 & 1 & \dots & 0 & 0 \\ 0 & 0 & \dots & 0 & 0 \\ \dots & \dots & \dots & \dots & \dots \\ 0 & 0 & \dots & 0 & 1 \\ -c_n & -c_{n-1} & \dots & c_2 & c_1 \end{bmatrix}. \tag{9.68}$$

It follows that the transformed control gains are given by

$$\mathbf{KT} = \begin{bmatrix} c_n - a_n & c_{n-1} - a_{n-1} & \dots & c_2 - a_2 & c_1 - a_1 \end{bmatrix}. \tag{9.69}$$

The required control gains are then given by

$$\mathbf{K} = \mathbf{C}_T \mathbf{C}_O^{-1} \begin{bmatrix} c_n - a_n & c_{n-1} - a_{n-1} & \dots & c_2 - a_2 & c_1 - a_1 \end{bmatrix}. \tag{9.70}$$

The difficulty with this type of control law is that it assumes that all the states of the system are available for feedback. In practice, this is not the case and some form of filtering must be provided for to compensate for the fact that all the states of the system are not actually measured or are available for feedback.

This aspect can be elegantly dealt with if one adopts the so-called *linear-quadratic-Gaussian* (LQG) approach. There are two aspects to the LQG controller design. The first is the *linear quadratic regulator* (LQR), which allows a full state controller to be synthesized, and second is the *linear quadratic estimator* (LQE), which allows some or all of the states of the system that are not actually measured or are available for feedback to be compensated for.

To formulate the LQR problem, the plant representing the dynamics of the flexible structure is assumed to be described by the state–space model

$$\dot{\mathbf{x}}(t) = \mathbf{Ax}(t) + \mathbf{Bu}(t), \quad \mathbf{z}(t) = \mathbf{Cx}(t), \tag{9.71}$$

where $\mathbf{z}(t)$ is assumed to be the controlled output. The control $\mathbf{u} = \mathbf{u}^*$ is chosen so as to minimize the performance or cost functional

$$J = \frac{1}{2} \int_0^\infty \left(\mathbf{x}^T \mathbf{Q} \mathbf{x} + \mathbf{u}^T \mathbf{R} \mathbf{u} \right) \mathrm{d}t. \tag{9.72}$$

The matrices $\mathbf{Q}$ and $\mathbf{R}$ are assumed to be symmetric and $\mathbf{R}$ is assumed to be invertible. Furthermore, $\mathbf{Q} = \mathbf{C}^T \mathbf{C}$. This approach was discussed in the preceding section and following the methods outlined there the solution may be expressed as

$$\begin{bmatrix} \dot{\mathbf{x}} \\ \dot{\boldsymbol{\lambda}} \end{bmatrix} = \mathbf{H} \begin{bmatrix} \mathbf{x} \\ \boldsymbol{\lambda} \end{bmatrix}, \quad \mathbf{u}^* = -\mathbf{R}^{-1}\mathbf{B}^T \boldsymbol{\lambda}, \tag{9.73}$$

where the matrix $\mathbf{H}$ is known as the Hamiltonian matrix and is

$$\mathbf{H} = \begin{bmatrix} \mathbf{A} & -\mathbf{BR}^{-1}\mathbf{B}^T \\ -\mathbf{Q} & -\mathbf{A}^T \end{bmatrix}. \tag{9.74}$$

It is customary to introduce a further transformation $\boldsymbol{\lambda} = \mathbf{P}x$ so the constant symmetric positive semi-definite matrix $\mathbf{P}$ satisfies the equation

$$\begin{bmatrix} \mathbf{P} & -\mathbf{I} \end{bmatrix} \mathbf{H} \begin{bmatrix} \mathbf{I} \\ \mathbf{P} \end{bmatrix} = \begin{bmatrix} \mathbf{P} & -\mathbf{I} \end{bmatrix} \begin{bmatrix} \mathbf{A} & -\mathbf{BR}^{-1}\mathbf{B}^T \\ -\mathbf{Q} & -\mathbf{A}^T \end{bmatrix} \begin{bmatrix} \mathbf{I} \\ \mathbf{P} \end{bmatrix} = 0. \tag{9.75}$$

Hence, we obtain a single equation for $\mathbf{P}$ known as the *algebraic Riccati equation* and given by

$$\mathbf{PA} + \mathbf{A}^T\mathbf{P} + \mathbf{Q} - \mathbf{PBR}^{-1}\mathbf{B}^T\mathbf{P} = 0. \tag{9.76}$$

The optimal control may be expressed as

$$\mathbf{u}^* = -\mathbf{R}^{-1}\mathbf{B}^T\mathbf{Px} \equiv -\mathbf{Kx}, \text{ where } \mathbf{K} = \mathbf{R}^{-1}\mathbf{B}^T\mathbf{P}. \tag{9.77}$$

Thus, the optimal control gain matrix $\mathbf{K}$ may be obtained provided the *algebraic Riccati equation* can be solved for the matrix $\mathbf{P}$. To solve for the unique positive semi-definite matrix $\mathbf{P}$, we consider the stable eigenvalues and the corresponding eigenvectors of the matrix $\mathbf{H}$ such that

$$\mathbf{H}\begin{bmatrix}\mathbf{P}_1\\ \mathbf{P}_2\end{bmatrix} = \begin{bmatrix}\mathbf{P}_1\\ \mathbf{P}_2\end{bmatrix}\Lambda_{\mathbf{H}}. \tag{9.78}$$

Then, provided $\det|\mathbf{P}_1| \neq 0$,

$$\mathbf{P} = \mathbf{P}_2\mathbf{P}_1^{-1}. \tag{9.79}$$

The state feedback control law (9.4) results in a closed-loop system of the form

$$\dot{\mathbf{x}}(t) = (\mathbf{A} - \mathbf{BK})\mathbf{x}(t),\ \ \mathbf{K} = \mathbf{R}^{-1}\mathbf{B}^T\mathbf{P}. \tag{9.80}$$

A crucial property of LQR controller design is that this closed loop is asymptotically stable (i.e. all the eigenvalues of $\mathbf{A} - \mathbf{BK}$ have negative real parts) as long as the following two conditions hold:

1. The pair $\mathbf{A}$, $\mathbf{B}$ is controllable; i.e. the controllability matrix

$$\mathbf{C}_C = \begin{bmatrix}\mathbf{B}\ \mathbf{AB}\ \ldots\ \mathbf{A}^{n-2}\mathbf{B}\ \mathbf{A}^{n-1}\mathbf{B}\end{bmatrix}$$

is of full rank.

2. The pair $\mathbf{A}$, $\mathbf{C}$ is observable when we regard $\mathbf{z}$ as the sole output.

The LQE employs the concept of an *observer* to reconstruct all the states of the system assuming that a measurement of a linear combination of states is available. The dynamics of a flexible structure could be expressed in state space, which is assumed to be in the form

$$\dot{\mathbf{x}} = \mathbf{Ax} + \mathbf{B}u,\ \mathbf{y} = \mathbf{Hx}. \tag{9.81}$$

One may construct an observer-based compensator, which satisfies the equation

$$\dot{\mathbf{z}} = \mathbf{Fz} + \mathbf{Gy} + \mathbf{M}u, \tag{9.82}$$

where the observer states asymptotically tend to a linear combination of the states of the system model, which may be expressed by the relationship

$$\underset{t\to+\infty}{\text{limit}}\ \mathbf{z} = \bar{\mathbf{z}} \equiv \mathbf{Sx}, \tag{9.83}$$

so

$$\underset{t\to+\infty}{\text{limit}}\,(\mathbf{z} - \mathbf{Sx}) = 0. \tag{9.84}$$

The feedback control law relating the scalar input u to a linear combination of an auxiliary input u', the states of the observer $\mathbf{z}$ and the measurements $\mathbf{y}$ is defined as

$$u = \left(u' - \mathbf{Ez} - \mathbf{Ly}\right). \tag{9.85}$$

The control input asymptotically tends to a linear combination of the auxiliary input u' and the states of the system model, which may be expressed by the relationship

$$\lim_{t\to+\infty} u = \lim_{t\to+\infty} \left(u' - \mathbf{Ez} - \mathbf{Ly}\right) = \lim_{t\to+\infty} \left(u' - \mathbf{Kx}\right). \tag{9.86}$$

The equations governing the parameter matrices of the observer, **F**, **G** and **M**, the transformation matrix **S**, and the control law parameter vectors **E** and **L**, in terms of **A**, **B**, **H** and **K**, may be obtained.

Let the observer error be defined as

$$\mathbf{e} = \mathbf{z} - \mathbf{Sx}. \tag{9.87}$$

Hence,

$$\dot{\mathbf{e}} = \dot{\mathbf{z}} - \mathbf{S}\dot{\mathbf{x}} = \mathbf{F}(\mathbf{z} - \mathbf{Sx}) + \mathbf{FSx} - \mathbf{SAx} + \mathbf{Gy} + (\mathbf{M} - \mathbf{SB})\,\mathbf{u} \tag{9.88}$$

or

$$\dot{\mathbf{e}} = \dot{\mathbf{z}} - \mathbf{S}\dot{\mathbf{x}} = \mathbf{Fe} + (\mathbf{FS} - \mathbf{SA} + \mathbf{GC})\,\mathbf{x} + (\mathbf{M} - \mathbf{SB})\,\mathbf{u}. \tag{9.89}$$

Assuming that the observer error satisfies

$$\dot{\mathbf{e}} = \mathbf{Fe}, \tag{9.90}$$

where its characteristic equation $\det|\mathbf{sI} - \mathbf{F}| = 0$ has asymptotically stable roots

$$(\mathbf{FS} - \mathbf{SA} + \mathbf{GH}) = \mathbf{0},\ \ (\mathbf{M} - \mathbf{SB}) = \mathbf{0}. \tag{9.91}$$

Since

$$u = u' - \mathbf{Kx} = u' - \mathbf{Ez} - \mathbf{Ly},$$
$$-\mathbf{Kx} = -\mathbf{Ez} - \mathbf{Ly} = -\mathbf{ESx} - \mathbf{LHx},$$

and it follows that

$$\mathbf{ES} + \mathbf{LH} = \mathbf{K}. \tag{9.92}$$

The complete set of observer equations is

$$(\mathbf{FS} - \mathbf{SA} + \mathbf{GH}) = \mathbf{0},\ \ (\mathbf{M} - \mathbf{SB}) = \mathbf{0} \text{ and } \mathbf{ES} + \mathbf{LH} = \mathbf{K}. \tag{9.93}$$

In the special case when the transformation matrix **S** is a unit matrix, one has a full order observer. In this case,

$$\mathbf{F} = \mathbf{A} - \mathbf{GH},\ \ \mathbf{M} = \mathbf{B},\ \ \mathbf{E} = \mathbf{K} \text{ and } \mathbf{L} = \mathbf{0}. \tag{9.94}$$

The observer dynamics may now be expressed in terms of the equation

$$\dot{\mathbf{z}} = \mathbf{Az} + \mathbf{G}\,(\mathbf{y} - \mathbf{Hz}) + \mathbf{B}u. \tag{9.95}$$

The observer error dynamics is then determined by

$$\dot{\mathbf{e}} = (\mathbf{A} - \mathbf{GH})\,\mathbf{e}. \tag{9.96}$$

In general, the output **y** is affected by measurement noise and the process dynamics are also affected by disturbance. In light of this, a more reasonable model for the process is

$$\dot{\mathbf{x}}(t) = \mathbf{Ax}(t) + \mathbf{Bu}(t) + \mathbf{B}_{\mathbf{w}}\mathbf{w}(t),\ \ \mathbf{y} = \mathbf{Hx} + \mathbf{v}(t), \tag{9.97}$$

where $\mathbf{w}(t)$ is the process disturbance noise and $\mathbf{v}(t)$ is the measurement disturbance noise. To reconstruct all of the states of the system, one may assume that full order observer is employed. However, as the measurements and the process are corrupted by the presence of noise, the observer can only at best estimate the states and the estimator is assumed to be of the form

$$\dot{\hat{\mathbf{x}}} = \mathbf{A}\hat{\mathbf{x}} + \mathbf{G}(\mathbf{y} - \mathbf{H}\hat{\mathbf{x}}) + \mathbf{B}u, \tag{9.98}$$

where $\hat{\mathbf{x}}$ is an estimate of the state vector $\mathbf{x}$. The observer error dynamics is now given by

$$\dot{\mathbf{e}} = (\mathbf{A} - \mathbf{GH})\mathbf{e} + \mathbf{B}_w\mathbf{w} - \mathbf{Gv}, \ \ \mathbf{e} = \mathbf{x} - \hat{\mathbf{x}}. \tag{9.99}$$

One could define a new process noise variable $\mathbf{n}$ as the weighted sum of $\mathbf{w}(t)$ and $\mathbf{v}(t)$ as

$$\mathbf{n} = \mathbf{B}_w\mathbf{w} - \mathbf{Gv}. \tag{9.100}$$

The observer error dynamics may be expressed as

$$\dot{\mathbf{e}} = (\mathbf{A} - \mathbf{GH})\mathbf{e} + \mathbf{n}. \tag{9.101}$$

One makes the assumption that both $\mathbf{w}(t)$ and $\mathbf{v}(t)$ are broadband zero-mean Gaussianly distributed white noise random processes, so their autocorrelation functions may be expressed in terms of the Dirac delta function $\delta(t - \tau)$ while their power spectra may be assumed to be constants. Since $\mathbf{n}$ is the weighted sum of $\mathbf{w}(t)$ and $\mathbf{v}(t)$, which are independent of each other, the noise correlation matrix is given by

$$E\left(\mathbf{n}(t)\mathbf{n}^T(\tau)\right) = \left(\mathbf{B}_w\mathbf{W}\mathbf{B}_w^T + \mathbf{GVG}^T\right)\delta(t - \tau), \tag{9.102}$$

where $E(\cdot)$ is the expectation operator and $\mathbf{W}$ and $\mathbf{V}$ are the magnitudes of the power spectra of $\mathbf{w}(t)$ and $\mathbf{v}(t)$, respectively, over the frequency band of the error dynamics. The time-domain evolution of the error dynamics may itself be expressed in terms of state transition matrix of the error dynamics in the absence of the noise $\Phi(t, t_0)$ as

$$\mathbf{e}(t) = \Phi(t, t_0)\mathbf{e}(t_0) + \int_{t_0}^{t} \Phi(t, \tau)\mathbf{n}(\tau)\,\mathrm{d}\tau. \tag{9.103}$$

Thus, the correlation matrix of the error dynamics may be expressed in terms of the covariance matrix $\mathbf{S}(t)$ of the error at time t. Thus,

$$E\left(\mathbf{e}(t)\mathbf{e}^T(\tau)\right) \equiv \mathbf{R}_\mathbf{e}(t, \tau) = \mathbf{S}(t)\Phi^T(\tau, t), \ \tau \geq t, \tag{9.104}$$

$$\mathbf{S}(t) = \Phi(t, t_0)\mathbf{S}(t_0)\Phi^T(t, t_0) + \int_{t_0}^{t} \Phi(t, \tau)\left(\mathbf{B}_w\mathbf{W}\mathbf{B}_w^T + \mathbf{GVG}^T\right)\Phi^T(t, \tau)\,\mathrm{d}\tau. \tag{9.105}$$

Differentiating the expression for the covariance matrix $\mathbf{S}(t)$ and noting that $\Phi(t, t_0)$ satisfies

$$\frac{\mathrm{d}}{\mathrm{d}t}\Phi(t, t_0) = (\mathbf{A} - \mathbf{GH})\Phi(t, t_0), \tag{9.106}$$

one may show that

$$\dot{\mathbf{S}}(t) = (\mathbf{A} - \mathbf{GH})\mathbf{S}(t) + \mathbf{S}(t)(\mathbf{A} - \mathbf{GH})^T + \left(\mathbf{B}_w\mathbf{W}\mathbf{B}_w^T + \mathbf{GVG}^T\right). \tag{9.107}$$

At this stage, it is important to note that the matrix gain $\mathbf{G}$ is yet to be found. On approach to find the matrix gain $\mathbf{G}$ is to minimize the performance metric

$$J_e = \lim_{t\to\infty} E\left(\left\|\mathbf{e}^T(t)\mathbf{e}(t)\right\|\right). \tag{9.108}$$

The solution to this minimization problem may be shown to be

$$\mathbf{G} = \mathbf{SH}^T\mathbf{V}^{-1}. \tag{9.109}$$

Substituting for $\mathbf{G}$ in the equation for the covariance matrix $\dot{\mathbf{S}}(t)$ and assuming that $t \to \infty$ in steady state,

$$\mathbf{AS} + \mathbf{SA}^T + \mathbf{B}_w\mathbf{W}\mathbf{B}_w^T - \mathbf{SH}^T\mathbf{V}^{-1}\mathbf{HS} = 0. \tag{9.110}$$

The above is also an algebraic Riccati equation and is a dual of the one we have already encountered in solving the LQR problem. It is associated with the solution to the LQE problem and is often referred to as the estimator algebraic Riccati equation to distinguish it from the former, which is known as the control algebraic Riccati equation. A positive semi-definite solution may be obtained by eigenvalue–eigenvector analysis of the corresponding Hamiltonian matrix in exactly the same way as was discussed in the case of the control algebraic Riccati equation.

Thus, the problem of having to provide compensation for the absence of all the states in the measurements is now solved. The closed-loop system may now be constructed for system model, which is now governed by the equation set

$$\begin{aligned} \dot{\mathbf{x}}(t) &= \mathbf{A}\mathbf{x}(t) + \mathbf{B}\mathbf{u}(t) + \mathbf{B}_{\mathrm{w}}\mathbf{w}(t) \\ \mathbf{y}(t) &= \mathbf{H}\mathbf{x}(t) + \mathbf{v}(t), \ \ \mathbf{z}(t) = \mathbf{C}\mathbf{x}(t). \end{aligned} \tag{9.111}$$

The control law is defined by feeding back a linear combination of the *estimated* states rather than the actual states. Thus, the control law is defined as

$$\begin{aligned} &\mathbf{u} = -\mathbf{R}^{-1}\mathbf{B}^{T}\mathbf{P}\hat{\mathbf{x}} \equiv -\mathbf{K}\hat{\mathbf{x}}, \ \ \dot{\hat{\mathbf{x}}} = \mathbf{A}\hat{\mathbf{x}} + \mathbf{G}\left(\mathbf{y} - \mathbf{H}\hat{\mathbf{x}}\right) + \mathbf{B}\mathbf{u}, \\ &\mathbf{G} = \mathbf{S}\mathbf{H}^{T}\mathbf{V}^{-1}. \end{aligned} \tag{9.112}$$

P and **S**, respectively, satisfy the control and estimator algebraic Riccati equations. The closed-loop dynamics is therefore defined in terms of the error vector as

$$\begin{aligned} \dot{\mathbf{x}}(t) &= (\mathbf{A} - \mathbf{B}\mathbf{K})\mathbf{x}(t) + \mathbf{B}\mathbf{K}\mathbf{e} + \mathbf{B}_{\mathrm{w}}\mathbf{w}(t), \\ \dot{\mathbf{e}} &= (\mathbf{A} - \mathbf{G}\mathbf{H})\,\mathbf{e} + \mathbf{B}_{w}\mathbf{w} - \mathbf{G}\mathbf{v}. \end{aligned} \tag{9.113}$$

One can easily observe that the error dynamics is decoupled from the dynamics of the flexible structure's state vector $\mathbf{x}(t)$. This is the celebrated *separation* principle and provides a basis for using the estimated states for feedback in place of the actual states of plant, which in this case is the flexible structure. However, any uncertainties in one's knowledge of the plant model, which manifests as errors in the plant parameters, would mean that the error dynamics would no longer be decoupled from the plant's state vector. Yet we have been able to demonstrate the basic design of a dynamic and active feedback controller without the need for making measurements of all the states in the state vector.

9.2.5 Reduced-order Modelling: Balancing

One of the many major problems that arise during the design of controller for a flexible structure is one of size. It is an important practical issue that must be addressed, especially when one is designing a controller for a flexible structure capable of being dynamically active in several modes simultaneously. Irrespective of the number of states that are included in the model, the ability to control a particular state depends on the degree of coupling between the state in question and the control input and the measurement. If the state is only weakly coupled to the control input and to the measurement, it is almost impossible to control. Thus, it is quite pointless to include it in the mathematical model of flexible structure, particularly if it is already stable. Under these circumstances, it is desirable to truncate the model and delete all those states that are weakly coupled to the control input and to the measurement. Balancing based on the controllability and observability grammians, which provide a means for quantifying the coupling between the states and the control input and between the states and the measurement, is a technique that facilitates model truncation and the deletion of states that are weakly coupled to the control input and to the measurement.

Considering linear system in its standard form (equation (9.71)), the controllability grammian is defined as

$$\mathbf{P}_C = \int_0^{\infty} \exp(\mathbf{A}t)\,\mathbf{B}\mathbf{B}^T \exp\left(\mathbf{A}^T t\right) \mathrm{d}t. \tag{9.114a}$$

It is the solution of the Lyapunov equation,

$$\mathbf{A}\mathbf{P}_C + \mathbf{P}_C\mathbf{A}^T + \mathbf{B}\mathbf{B}^T = 0. \tag{9.114b}$$

Thus, it can be interpreted as the steady-state state covariance matrix when the system is subject to a white noise input with the spectral density equal to an identity matrix. The relative magnitude of the variance is indicative of relative degree of coupling with the input.

The observability grammian is defined as

$$\mathbf{P}_O = \int_0^{\infty} \exp\left(\mathbf{A}^T t\right) \mathbf{C}\mathbf{C}^T \exp(\mathbf{A}t)\, \mathrm{d}t. \tag{9.115a}$$

It is the solution of the Lyapunov equation,

$$\mathbf{A}^T\mathbf{P}_O + \mathbf{P}_O\mathbf{A} + \mathbf{C}^T\mathbf{C} = 0. \tag{9.115b}$$

Thus, it can be interpreted as the steady-state state covariance matrix when the adjoint system, defined as

$$\dot{\mathbf{x}}(t) = \mathbf{A}^T\mathbf{x}(t) + \mathbf{C}^T\mathbf{u}(t), \tag{9.116}$$

is subject to a white noise input with the spectral density equal to an identity matrix. The relative magnitude of the variance is indicative of relative degree of coupling with the input.

When a similarity transformation $\mathbf{x}(t) = \mathbf{T}\mathbf{z}(t)$ is applied to the state of the system, the matrix $\mathbf{A}$ is transformed to $\mathbf{T}^{-1}\mathbf{A}\mathbf{T}$, $\mathbf{B}$ is transformed to $\mathbf{T}^{-1}\mathbf{B}$, $\mathbf{C}$ is transformed to $\mathbf{C}\mathbf{T}$, the controllability grammian $\mathbf{P}_C$ is transformed to $\mathbf{T}^{-1}\mathbf{P}_C\mathbf{T}^{T^{-1}}$ and the observability grammian $\mathbf{P}_O$ is transformed to $\mathbf{T}^T\mathbf{P}_O\mathbf{T}$. To define a balanced realization of the plant, the task is to find a similarity transformation, which will transform the controllability grammian and the observability grammian to diagonal form and also make them both equal to each other. If the elements on the diagonals of these matrices are arranged in decreasing magnitude, it means that not only are the states equally coupled to the control inputs and the measurements, but they are also ordered with the states that are maximally coupled appearing first and the states that are minimally coupled appearing last. Thus, the transformation to a balanced realization would facilitate model truncation and the deletion of states that are weakly coupled to the control input and to the measurement.

To find such a similarity transformation, the controllability grammian $\mathbf{P}_C$ is first decomposed to the form

$$\mathbf{P}_C = \mathbf{R}^T\mathbf{R}.$$

The triple matrix product $\mathbf{R}\mathbf{P}_O\mathbf{R}^T$ is then constructed, and it is decomposed by the process singular value decomposition to $\mathbf{R}\mathbf{P}_O\mathbf{R}^T = \mathbf{U}\Sigma^2\mathbf{U}^T$. The similarity transformation that would simultaneously diagonalize $\mathbf{P}_C$ and $\mathbf{P}_O$ and also make them equal to each other is then given by $\mathbf{T} = \mathbf{R}^T\mathbf{U}\Sigma^{-1/2}$. Thus, the construction of a balanced realization facilitated the construction of reduced-order models, which could the used for the purpose of controller synthesis.

A popular approach in structural vibration control is modal control. It is based on the fact that vibration of a structure is almost always dominated by a small sub-set of its vibration modes. This makes it possible to construct a reduced model by retaining the relevant sub-set of dynamics of the structure by using limited sub-set of the modes to represent the structure's displacements and velocities. The approach is based on the use of modal truncation to reduce the order of dynamic system and design a controller in the modal space. By controlling the modal responses of dominant modes, vibration of the real structure can be controlled. However, since the control system is designed to work in modal space, model parameters (modal displacement, modal velocity, etc.) are needed for the controller to compute control force. These modal parameters cannot be measured directly, and transformation from physical measurement to modal space is needed. Once the states to be deleted in the transformed balanced domain are identified, one could associate a set of modal displacement amplitudes and the corresponding modal velocities with those states that are weakly coupled to the control input and the measurement. This finite set of modal displacement amplitudes and the corresponding modal velocities may then be deleted from the original model.

9.2.6 *Zero-spillover Controller Synthesis*

The use of a reduced-order model for the purpose of controller synthesis is responsible for a problem that is peculiar to flexible structures. To illustrate this problem effectively, we consider a general vibrating flexible structure with n degrees of freedom, m force actuators, k displacement sensors and l velocity sensors. The dynamics is assumed to be described by the equations of motion

$$\mathbf{M}\ddot{\mathbf{x}}(t) + \mathbf{C}\dot{\mathbf{x}}(t) + \mathbf{K}\mathbf{x}(t) = \mathbf{B}_2\mathbf{u}(t), \ \mathbf{y}_1 = \mathbf{H}_1\mathbf{x}, \ \mathbf{y}_2 = \mathbf{H}_2\dot{\mathbf{x}}, \tag{9.117}$$

where $\mathbf{x}(t)$ is an $n \times 1$ vector of the displacement degrees of freedom, $\mathbf{u}(t)$ is an $m \times 1$ vector of control forces generated by the actuators, $\mathbf{y}_1$ is a $k \times 1$ vector of measured displacements assumed to be disturbance free, $\mathbf{y}_2$ is an $l \times 1$ vector of measured disturbance free velocities, $\mathbf{M}$ is an $n \times n$ mass matrix associated with the flexible structure, $\mathbf{C}$ is an $n \times n$ damping matrix associated with the structure, $\mathbf{K}$ is the associated $n \times n$ stiffness matrix, $\mathbf{B}_2$ is the $n \times m$ actuator force distribution matrix, and $\mathbf{H}_1$ and $\mathbf{H}_2$ relate the measured displacements and velocities to the displacement degrees of freedom. For equation (9.117), the corresponding eigen-problem is defined by

$$\mathbf{K}\mathbf{\Phi} = \mathbf{M}\mathbf{\Phi}\Omega^2, \tag{9.118}$$

where $\mathbf{\Phi}$ is an $n \times n$ normal mode matrix, assumed to be normalized with respect the mass matrix, i.e. $\mathbf{\Phi}^T\mathbf{M}\mathbf{\Phi} = \mathbf{I}$, and $\mathbf{\Omega}^2 = \mathbf{\Phi}^T\mathbf{K}\mathbf{\Phi}$ is an $n \times n$ diagonal matrix with entries ω_{ni}^2, $i = 1, 2, 3, \ldots n$, ω_{ni} being the ith natural frequency of the structure. It is assumed that matrix $\mathbf{C}$ is assumed to be a linear combination of the mass matrix and the stiffness matrix, so it satisfies $\mathbf{\Phi}^T\mathbf{C}\mathbf{\Phi} = 2\mathbf{Z}\mathbf{\Omega}$, where $\mathbf{Z}$ is an $n \times n$ diagonal matrix with entries ζ_i, $i = 1, 2, 3, \ldots n$, ζ_i being the damping ratio in the ith normal mode. Assuming that system (9.117) is transformed to normal mode coordinates by the transformation $\mathbf{x} = \mathbf{\Phi}\mathbf{q}$, $\mathbf{q}$ being an $n \times 1$ vector of modal amplitudes, one obtains the transformed equations of motion given by

$$\ddot{\mathbf{q}}(t) + 2\mathbf{Z}\mathbf{\Omega}\dot{\mathbf{q}}(t) + \mathbf{\Omega}^2\mathbf{q}(t) = \mathbf{\Phi}^T\mathbf{B}_2\mathbf{u}(t), \ \mathbf{y}_1 = \mathbf{H}_1\mathbf{\Phi}\mathbf{q}, \ \mathbf{y}_2 = \mathbf{H}_2\mathbf{\Phi}\dot{\mathbf{q}}. \tag{9.119}$$

It is now assumed that the intention is to construct a reduced-order model with just first j modes, with $j < n$ and possibly even $j < m$. The system is therefore partitioned and written as

$$\begin{aligned} &\ddot{\mathbf{q}}_i(t) + 2\mathbf{Z}_i\mathbf{\Omega}_i\dot{\mathbf{q}}_i(t) + \mathbf{\Omega}_i^2\mathbf{q}_i(t) = \mathbf{\Phi}_i^T\mathbf{B}_2\mathbf{u}(t), \ i = 1, 2, \\ &\mathbf{y}_1 = \mathbf{H}_1\left(\mathbf{\Phi}_1\mathbf{q}_1 + \mathbf{\Phi}_2\mathbf{q}_2\right), \end{aligned} \tag{9.120}$$

where

$$\mathbf{q} = \begin{bmatrix} \mathbf{q}_1 \\ \mathbf{q}_2 \end{bmatrix}, \ \mathbf{\Phi} = \begin{bmatrix} \mathbf{\Phi}_1 & \mathbf{\Phi}_2 \end{bmatrix}, \ \mathbf{\Omega} = \begin{bmatrix} \mathbf{\Omega}_1 & \mathbf{0} \\ \mathbf{0} & \mathbf{\Omega}_2 \end{bmatrix}, \ \mathbf{Z} = \begin{bmatrix} \mathbf{Z}_1 & \mathbf{0} \\ \mathbf{0} & \mathbf{Z}_2 \end{bmatrix},$$

$\mathbf{q}_1$ being a $j \times 1$ vector of the first j modal amplitudes and $\mathbf{q}_2$ being an $(n - j) \times 1$ vector of the remaining modal amplitudes. If it is desired that we delete the modal amplitude vector $\mathbf{q}_2$, it is necessary to ensure that $\mathbf{\Phi}_2^T\mathbf{B}_2 = \mathbf{0}$ and that $\mathbf{H}_i\mathbf{\Phi}_2 = \mathbf{0}$, $i = 1, 2$. When the first condition is violated, i.e. $\mathbf{\Phi}_2^T\mathbf{B}_2 \neq \mathbf{0}$, any control input applied to the force actuators would also excite the modal amplitude vector $\mathbf{q}_2$, and this is not desirable. This situation is referred to as *control spillover*. When the second condition is violated, i.e. $\mathbf{H}_i \Phi_2 \neq \mathbf{0}$, $i = 1$ or 2, the measurement would be contaminated by the presence of components of the modal amplitude vector $\mathbf{q}_2$, which when fed back to the force actuators would necessarily couple $\mathbf{q}_1$ and $\mathbf{q}_2$. This again is not desirable and is referred to as *observation spillover*. While balanced realizations ensure to minimize spillover, they do not eliminate the spillover altogether unless both the conditions for spillover-free control are explicitly met, i.e. $\mathbf{\Phi}_2^T \mathbf{B}_2 = \mathbf{0}$ and $\mathbf{H}_i \Phi_2 = \mathbf{0}$, $i = 1, 2$. The control spillover condition may be met by locating the actuators at the nodal points of the modes, which are not to be excited by the control forces. The observation spillover condition may be met by filtering the measurements. If the frequency response of the modal amplitude vector $\mathbf{q}_2$ is in a contiguous high-frequency band, measurements could be filtered by a low-pass filter with appropriate passband characteristics. Alternatively, a bank of notch filters, with a notch-like frequency response characteristic at resonance frequencies of the unwanted modes, may be used to filter out the unwanted responses in the measurements.

Once the conditions for the absence of spillover are met, the model equations for the first j modes may be cast in state-space form and the LQG methodology enunciated in the previous section can be applied. Such a controller is referred to as an H_2 controller, which is one that internally stabilizes the plant and minimizes the H_2-norm of transfer matrix T_{zw} from disturbance w to the controlled output z. The equations defining the complete H_2 controller may be succinctly stated as follows: Given a state-space system

$$\dot{\mathbf{x}}(t) = \mathbf{A}\mathbf{x}(t) + \mathbf{B}_1\mathbf{w}(t) + \mathbf{B}_2\mathbf{u}(t), \;\; \mathbf{z}(t) = \mathbf{C}_1\mathbf{x}(t) + \mathbf{D}_{12}\mathbf{u}(t),$$
$$\mathbf{y}(t) = \mathbf{C}_2\mathbf{x}(t) + \mathbf{D}_{21}\mathbf{w}(t), \tag{9.121}$$

the plant may be represented in matrix form as

$$\begin{bmatrix} \dot{\mathbf{x}} \\ \mathbf{z} \\ \mathbf{y} \end{bmatrix} = \begin{bmatrix} \mathbf{A} & \mathbf{B}_1 & \mathbf{B}_2 \\ \mathbf{C}_1 & \mathbf{0} & \mathbf{D}_{12} \\ \mathbf{C}_2 & \mathbf{D}_{21} & \mathbf{0} \end{bmatrix} \begin{bmatrix} \mathbf{x} \\ \mathbf{w} \\ \mathbf{u} \end{bmatrix} \tag{9.122}$$

and the plant's transfer matrix representation has the form

$$\mathbf{G} = \begin{bmatrix} \mathbf{A} & \mathbf{B}_1 & \mathbf{B}_2 \\ \mathbf{C}_1 & \mathbf{0} & \mathbf{D}_{12} \\ \mathbf{C}_2 & \mathbf{D}_{21} & \mathbf{0} \end{bmatrix}. \tag{9.123}$$

When the system matrices meet the following conditions:

1. $(\mathbf{A}, \mathbf{B}_2)$ is 'stabilizable' and $(\mathbf{C}_2, \mathbf{A})$ is 'detectable'; a system is said to be stabilizable if there exists a feedback of a weighted linear combination of the states such that the closed loop is asymptotically stable with all eigenvalues in the open left-half plane and a linear system with a set of measurements of some of the states is said to be detectable if the remaining states can be reconstructed from the measurements;
2. $\mathbf{D}_{12}$ has full row rank, $\mathbf{D}_{21}$ has full column rank;
3. For all ω, the matrix

$$\begin{bmatrix} \mathbf{A} - j\omega\mathbf{I} & \mathbf{B}_2 \\ \mathbf{C}_1 & \mathbf{D}_{12} \end{bmatrix}$$

has full column rank;

4. For all ω, the matrix

$$\begin{bmatrix} \mathbf{A} - j\omega\mathbf{I} & \mathbf{B}_1 \\ \mathbf{C}_2 & \mathbf{D}_{21} \end{bmatrix}$$

has full row rank;

there exists a unique optimal dynamic controller

$$\begin{bmatrix} \dot{\hat{\mathbf{x}}} \\ \mathbf{u} \end{bmatrix} = \mathbf{K}(s) \begin{bmatrix} \hat{\mathbf{x}} \\ \mathbf{y} \end{bmatrix}, \tag{9.124}$$

with the transfer matrix given by

$$\mathbf{K}(s) = \begin{bmatrix} \mathbf{A}_c & -\mathbf{L}_2 \\ \mathbf{F}_2 & \mathbf{0} \end{bmatrix}, \tag{9.125}$$

where $\mathbf{A}_c$ is the closed-loop dynamics matrix given by, $\mathbf{A}_c = \mathbf{A} + \mathbf{B}_2\mathbf{F}_2 + \mathbf{L}_2\mathbf{C}_2$, $\mathbf{F}_2$ and $\mathbf{L}_2$ are the controller and observer gains given by $\mathbf{F}_2 = -\left(\mathbf{B}_2^T\mathbf{X} + \mathbf{D}_{12}^T\mathbf{C}_1\right)$, $\mathbf{L}_2 = -\left(\mathbf{Y}\mathbf{C}_2^T + \mathbf{B}_1\mathbf{D}_{21}^T\right)$, $\mathbf{X}$ and $\mathbf{Y}$ being the positive semi-definite matrix solutions of the control and observer algebraic Riccati equations with respective Hamiltonian matrices

$$\mathbf{H}_C = \begin{bmatrix} \mathbf{A} - \mathbf{B}_2\mathbf{D}_{12}^T\mathbf{C}_1 & -\mathbf{B}_2\mathbf{B}_2^T \\ -\mathbf{C}_1^T\left(\mathbf{I} - \mathbf{D}_{12}\mathbf{D}_{12}^T\right)\mathbf{C}_1 & -\mathbf{A}^T + \mathbf{C}_1^T\mathbf{D}_{12}\mathbf{B}_2^T \end{bmatrix}, \tag{9.126a}$$

$$\mathbf{H}_O = \begin{bmatrix} \mathbf{A}^T - \mathbf{C}_2^T\mathbf{D}_{21}^T\mathbf{B}_1 & -\mathbf{C}_2^T\mathbf{C}_2 \\ -\mathbf{B}_1\left(\mathbf{I} - \mathbf{D}_{21}^T\mathbf{D}_{21}\right)\mathbf{B}_1^T & -\mathbf{A} + \mathbf{B}_1\mathbf{D}_{21}^T\mathbf{C}_2 \end{bmatrix}. \tag{9.126b}$$

The first condition guarantees the stabilizability of the closed loop by output feedback using an observer, and the last two conditions ensure that $\mathbf{X}$ and $\mathbf{Y}$ exist and are the positive semi-definite matrix solutions of the control and observer algebraic Riccati equations. The H_2-norm minimizing controller involves finding a control law $\mathbf{u}(s) = \mathbf{K}(s)\mathbf{y}(s)$ over the set of all stabilizing controllers that minimizes the H_2-norm of the $\mathbf{w} \to \mathbf{z}$ transfer function $\mathbf{T}_{\mathbf{zw}}$,

$$\mathbf{T}_{\mathbf{zw}} = \mathbf{G}_{\mathbf{zu}}(s)\mathbf{K}(s)\left(\mathbf{I} - \mathbf{G}_{\mathbf{yu}}(s)\mathbf{K}(s)\right)^{-1}\mathbf{G}_{\mathbf{yw}}(s) + \mathbf{G}_{\mathbf{zw}}(s), \tag{9.127}$$

where the transfer function $\mathbf{G}(s)$ defines the relationship between the outputs $\mathbf{z}$ and $\mathbf{y}$ and inputs $\mathbf{w}$ and $\mathbf{u}$ as

$$\begin{bmatrix} \mathbf{z} \\ \mathbf{y} \end{bmatrix} = \mathbf{G}(s)\begin{bmatrix} \mathbf{w} \\ \mathbf{u} \end{bmatrix} \equiv \begin{bmatrix} \mathbf{G}_{\mathbf{zw}}(s) & \mathbf{G}_{\mathbf{zu}}(s) \\ \mathbf{G}_{\mathbf{yw}}(s) & \mathbf{G}_{\mathbf{yu}}(s) \end{bmatrix}\begin{bmatrix} \mathbf{w} \\ \mathbf{u} \end{bmatrix}. \tag{9.128}$$

The H_2-norm of a transfer function $\mathbf{T}$ is defined as

$$\|\mathbf{T}\|_2 = \left(\frac{1}{2}\int_{-\infty}^{\infty} trace\left[\mathbf{T}^*(j\omega)\,\mathbf{T}(j\omega)\right]\mathrm{d}\omega\right)^{\frac{1}{2}}. \tag{9.129}$$

The minimum H_2-norm of the $\mathbf{w} \to \mathbf{z}$ transfer function $\mathbf{T}_{\mathbf{zw}}$ is given by

$$\min \|\mathbf{T}_{zw}\|_2^2 = \|\mathbf{G}_c\mathbf{L}_2\|_2^2 + \left\|\mathbf{C}_1\mathbf{G}_f\right\|_2^2, \tag{9.130}$$

where the transfer matrices are given by

$$\mathbf{G}_c(s) = \begin{bmatrix} \mathbf{A} + \mathbf{B}_2\mathbf{F}_2 & \mathbf{I} \\ \mathbf{C}_1 + \mathbf{D}_{12}\mathbf{F}_2 & \mathbf{0} \end{bmatrix}, \quad \mathbf{G}_f(s) = \begin{bmatrix} \mathbf{A} + \mathbf{L}_2\mathbf{C}_2 & \mathbf{B}_1 + \mathbf{L}_2\mathbf{D}_{21} \\ \mathbf{I} & \mathbf{0} \end{bmatrix}. \tag{9.131}$$

9.3 Optimal Control Synthesis: H_∞ and Linear Matrix Inequalities

9.3.1 The Basis for Performance Metric Optimization-based Controller Synthesis

To first explain the basis for H_∞ controller synthesis as well as Lyapunov synthesis methods, it is important to consider some of the main concepts of optimization. The H_∞ controller synthesis technique not only requires the control law to render the system Lyapunov stable but also requires it to be optimal in a certain sense. Before considering the general non-linear case, we shall briefly consider linear H_∞ control, which is emerging as fundamental paradigm in linear control theory. The H_∞-norm (L_2 induced norm) can be related to a number of practical metrics in control engineering, such as the Nyquist distance, which are extremely relevant in model matching, model reduction, robust control design and tracking. Hence, the H_∞ approach is recognized as being of fundamental importance in numerous applications. The basic problem is to find a controller that stabilizes the closed-loop system internally while attenuating the worst case disturbances in the outputs to a prespecified level. The early approaches to H_∞ control were based on frequency-domain methods, spectral factorization, the Nehari approximation problem and the Navalina–Pick algorithm. However, the problem could be equivalently formulated in the time domain.

The plant is assumed to be described by the state-space model

$$\dot{\mathbf{x}}(t) = \mathbf{A}\mathbf{x}(t) + \mathbf{B}\mathbf{u}(t) + \mathbf{B}_{\mathbf{w}}\mathbf{w}(t), \tag{9.132}$$

where $\mathbf{u}(t)$ is the control input and $\mathbf{w}(t)$ is the disturbance input. The objective is to minimize a performance measure with respect to the control input of the form

$$J_{worst\ case} = \max_{w(t)} J\left\{\mathbf{x}(t), \mathbf{u}(t), \mathbf{w}(t)\right\}, \tag{9.133}$$

where $J_{worst\ case}$ represent the worst case performance measure with $\mathbf{w}(t) = \mathbf{w}_c(t)$, i.e.

$$J_{worst\ case} = J\left\{\mathbf{x}(t), \mathbf{u}(t), \mathbf{w}_c(t)\right\} \tag{9.134}$$

and $\mathbf{x}(t)$ satisfies

$$\dot{\mathbf{x}}(t) = \mathbf{A}\mathbf{x}(t) + \mathbf{B}\mathbf{u}(t) + \mathbf{B}_{\mathbf{w}}\mathbf{w}(t). \tag{9.132}$$

Thus, the problem reduces to a constrained min–max problem that may be stated as

$$J^* = J_{optimum} = \min_{u(t)}\left\{\max_{w(t)} J\left\{\mathbf{x}(t), \mathbf{u}(t), \mathbf{w}(t)\right\}\right\},$$

where $\mathbf{x}(t)$ satisfies the constraint equations given by equations (9.132).

The solution is a 'saddle point' of the objective function, as it satisfies the inequalities

$$J\left\{\mathbf{x}(t), \mathbf{u}^*(t), \mathbf{w}(t)\right\} \leq J\left\{\mathbf{x}^*(t), \mathbf{u}^*(t), \mathbf{w}_c(t)\right\} \tag{9.135}$$

and

$$J\left\{\mathbf{x}^*(t), \mathbf{u}^*(t), \mathbf{w}_c(t)\right\} \leq J\left\{\mathbf{x}(t), \mathbf{u}(t), \mathbf{w}_c(t)\right\}. \tag{9.136}$$

The problem can be thought of as a game with two non-cooperating participants: the control systems designer who seeks a control input to minimize the performance measure and the environment that is seeking a disturbance that will maximize the same measure. Such games are referred to as *differential games*, as the dynamics of the game is governed by differential equations. Another interesting interpretation of the two-person non-cooperative game is as a zero-sum differential game with quadratic performance criteria. The term '*zero-sum*' refers to the fact while one player of the game wants maximize the payoff or performance measure the other wants to minimize it. Lagrangian multipliers may be used to convert the constrained min–max problem into an unconstrained one. Thus,

$$J^* = \min_{u(t)}\left\{\max_{w(t)} J_{uc}\left\{\mathbf{x}(t), \mathbf{u}(t), \mathbf{w}(t)\right\}\right\}, \tag{9.137}$$

where

$$J_{uc} = J\left\{\mathbf{x}(t), \mathbf{u}(t), \mathbf{w}(t)\right\} + \int_0^{t_f} \mathbf{p}^T(t)\left\{\mathbf{A}\mathbf{x}(t) + \mathbf{B}\mathbf{u}(t) + \mathbf{B}_{\mathbf{w}}\mathbf{w}(t) - \dot{\mathbf{x}}(t)\right\} \mathrm{d}t.$$

The first step in the optimization process is to select an appropriate performance measure. Furthermore it is assumed that all states and disturbance inputs are available for feedback.

In order to construct a suitable performance measure, an input/output model of the plant is assumed to be of the form

$$\dot{\mathbf{x}}(t) = \mathbf{A}\mathbf{x}(t) + \mathbf{B}\mathbf{u}(t) + \mathbf{B}_{\mathrm{w}}\mathbf{w}(t),\ \ \mathbf{y}(t) = \mathbf{C}\mathbf{x}(t) + \mathbf{D}\mathbf{u}(t), \tag{9.138}$$

where the output is normalized such that

$$\mathbf{D}^T\mathbf{D} = \mathbf{R} \text{ and } \mathbf{D}^T\mathbf{C} = \mathbf{0};$$

$\mathbf{R}$ is a positive definite and diagonal matrix.

The performance measure is

$$J\left\{\mathbf{x}(t), \mathbf{u}(t), \mathbf{w}(t)\right\} = \|\mathbf{y}(t)\|_2^2 - \gamma^2\|\mathbf{w}(t)\|_2^2, \tag{9.139}$$

where $\|\cdot\|_2$ is the finite 2-norm. An equivalent performance measure is given by

$$J\left\{\mathbf{x}(t), \mathbf{u}(t), \mathbf{w}(t)\right\} = \frac{\|\mathbf{y}(t)\|_2}{\|\mathbf{w}(t)\|_2} - \gamma. \tag{9.140}$$

In fact, we may define an equivalent optimization problem in the frequency domain as

$$\min_{u(t)}\left\{\sup_{\|\mathbf{w}(t)\|_2 \neq 0}\frac{\|\mathbf{y}(t)\|_2}{\|\mathbf{w}(t)\|_2}\right\} < \gamma. \tag{9.141}$$

However, the quantity

$$\sup_{\|\mathbf{w}(t)\|_2 \neq 0}\frac{\|\mathbf{y}(t)\|_2}{\|\mathbf{w}(t)\|_2} = \left\|\mathbf{G}_{\mathrm{y/w}}\right\|_{\infty} \tag{9.142}$$

is the disturbance input/output ∞-norm. The optimization problem may be expressed as

$$\min_{u(t)}\left\{\left\|\mathbf{G}_{\mathrm{y/w}}\right\|_{\infty}\right\} < \gamma. \tag{9.143}$$

Thus, equivalent solutions of the H_∞ control problem may be obtained both in the time domain and in the frequency domain.

Based on standard techniques in linear optimal control, the time-domain solution may be written as

$$\begin{bmatrix}\dot{\mathbf{x}}\\ \dot{\mathbf{p}}\end{bmatrix} = \begin{bmatrix}\mathbf{A} & -\left(\mathbf{B}\mathbf{R}^{-1}\mathbf{B}^T - \dfrac{1}{\gamma^2}\mathbf{B}_{\mathrm{w}}\mathbf{B}_{\mathrm{w}}^T\right)\\ -\mathbf{C}^T\mathbf{C} & -\mathbf{A}^{\mathrm{T}}\end{bmatrix}\begin{bmatrix}\mathbf{x}\\ \mathbf{p}\end{bmatrix}, \tag{9.144}$$

with

$$\mathbf{x}(t_0) = \mathbf{x}_0,\ \ \mathbf{p}\left(t_f\right) = \mathbf{0},\ \ \mathbf{u}(t) = -\mathbf{R}^{-1}\mathbf{B}^T\mathbf{p}(t) \text{ and } \mathbf{w}(t) = \frac{1}{\gamma^2}\mathbf{B}_{\mathrm{w}}^T\mathbf{p}(t).$$

Making the substitution $\mathbf{p}(t) = \mathbf{P}(t)\mathbf{x}(t)$, the optimal solution may be shown to satisfy

$$\mathbf{u}(t) = -\mathbf{R}^{-1}\mathbf{B}^T\mathbf{P}(t)\mathbf{x}(t), \tag{9.145}$$

where $\mathbf{P}(t)$ satisfies the Riccati equation

$$\dot{\mathbf{P}} = -\mathbf{P}\mathbf{A} - \mathbf{A}^T\mathbf{P} - \mathbf{C}^T\mathbf{C} + \mathbf{P}\left(\mathbf{B}\mathbf{R}^{-1}\mathbf{B}^T - \frac{1}{\gamma^2}\mathbf{B}_{\mathrm{w}}\mathbf{B}_{\mathrm{w}}^T\right)\mathbf{P}, \tag{9.146}$$

where $\mathbf{P}\left(t_f\right) = \mathbf{0}$.

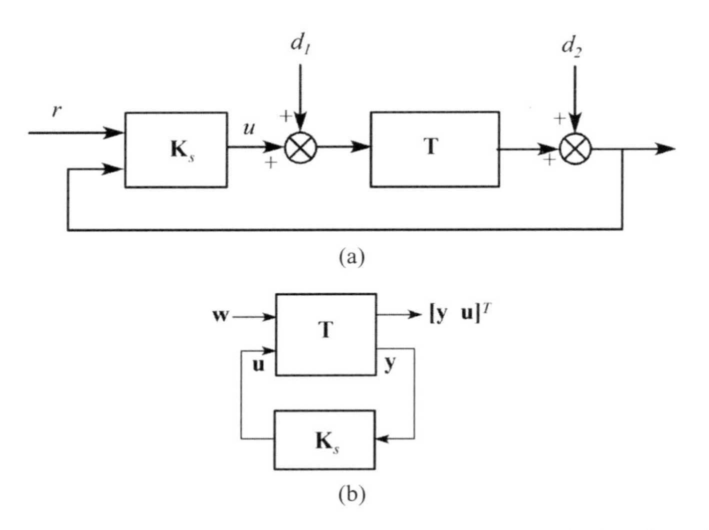

Figure 9.1 (a) Typical feedback configuration and (b) standard feedback configuration

It is important to note that

$$V(\mathbf{x}) = \mathbf{x}^T(t)\mathbf{P}(t)\mathbf{x}(t) \tag{9.147}$$

is a Lyapunov function.

In the frequency-domain referring to the typical feedback configuration illustrated in Figure 9.1a, the disturbance input/output ∞-norm is given by

$$\left\| \mathbf{G}_{\mathbf{y}/\mathbf{w}} \right\|_\infty = \left\| \begin{matrix} \mathbf{SG} & \mathbf{S} \\ \mathbf{K}_s\mathbf{SG} & \mathbf{K}_s\mathbf{S} \end{matrix} \right\|_\infty, \tag{9.148}$$

where $\mathbf{S} = (\mathbf{I} - \mathbf{GK}_s)^{-1}$. Equivalently in terms of the transfer function $\mathbf{T}$ in Figure 9.1b,

$$\left\| \mathbf{G}_{\mathbf{y}/\mathbf{w}} \right\|_\infty = \left\| \mathbf{T}_{11} + \mathbf{T}_{12}\mathbf{K}_s\left(\mathbf{I} - \mathbf{T}_{22}\mathbf{K}_s\right)^{-1}\mathbf{T}_{21} \right\|_\infty. \tag{9.149}$$

The transfer function $\mathbf{T}$ may be conveniently expressed in terms of the lower fractional transformation. Thus if

$$\begin{bmatrix} \mathbf{z} \\ \mathbf{y} \end{bmatrix} = \mathbf{P}\begin{bmatrix} \mathbf{w} \\ \mathbf{u} \end{bmatrix} = \begin{bmatrix} \mathbf{P}_{11} & \mathbf{P}_{12} \\ \mathbf{P}_{21} & \mathbf{P}_{22} \end{bmatrix}\begin{bmatrix} \mathbf{w} \\ \mathbf{u} \end{bmatrix}, \quad \mathbf{u} = \mathbf{Ky}, \tag{9.150}$$

$\mathbf{y}$ may be expressed as

$$\mathbf{y} = [\mathbf{I} - \mathbf{P}_{22}\mathbf{K}]^{-1}\mathbf{P}_{21}\mathbf{w}. \tag{9.151}$$

Hence, the output z is

$$\mathbf{z} = \left\lfloor \mathbf{P}_{11} + \mathbf{P}_{12}\mathbf{K}\left[\mathbf{I} - \mathbf{P}_{22}\mathbf{K}\right]^{-1}\mathbf{P}_{21} \right\rfloor \mathbf{w} = \mathsf{F}_l\left(\mathbf{P}, \mathbf{K}\right)\mathbf{w} \tag{9.152a}$$

or

$$\mathbf{z} = \left\lfloor \mathbf{P}_{11} + \mathbf{P}_{12}\left[\mathbf{K}^{-1} - \mathbf{P}_{22}\right]^{-1}\mathbf{P}_{21} \right\rfloor \mathbf{w} = \mathsf{F}_l\left(\mathbf{P}, \mathbf{K}\right)\mathbf{w}. \tag{9.152b}$$

Similarly, the upper fractional transformation corresponding to the transfer function

$$\begin{bmatrix} \mathbf{w} \\ \mathbf{u} \end{bmatrix} = \begin{bmatrix} \mathbf{S}_{11} & \mathbf{S}_{12} \\ \mathbf{S}_{21} & \mathbf{S}_{22} \end{bmatrix}\begin{bmatrix} \mathbf{z} \\ \mathbf{y} \end{bmatrix}, \quad \text{with } \mathbf{z} = \mathbf{Gw}, \tag{9.153}$$

i.e.

$$\mathbf{z} = \lfloor \mathbf{G}^{-1} - \mathbf{S}_{11} \rfloor \mathbf{S}_{12}\mathbf{y} \tag{9.154}$$

is

$$\mathbf{u} = \lfloor \mathbf{S}_{22} + \lfloor \mathbf{G}^{-1} - \mathbf{S}_{11} \rfloor \mathbf{S}_{12} \rfloor \mathbf{y} = \mathsf{F}_{\mathsf{u}}(\mathbf{S}, \mathbf{G})\,\mathbf{y}. \tag{9.155}$$

Two important theorems in control theory facilitate the use of the linear fractional transformation representations. In the first instance, consider the transfer function relating **w** and **u** to **z** and **y**. The controller $\mathbf{u} = \mathbf{K}\mathbf{y}$ stabilizes the transfer function **P** if and only if it stabilizes the transfer function $\mathbf{P}_{22}$ as the two transfer functions share the same characteristic polynomial. In the second instance, consider the transfer function relating **z** and **y** to **w** and **u**. The closed-loop transfer function is well posed and stable (internally) for all $\|\mathbf{G}\|_\infty \leq 1$ if and only if.$\|\mathbf{S_{22}}\|_\infty \leq 1$ This is known as the *small gain theorem*.

In the case of a linear system, it is always possible to factorize the transfer function into a matrix ratio of two matrix transfer functions, both of which are stable. This factorization is known as coprime factorization. Coprime factorization is an important step, as it leads to the Youla–Kucera parameterization and the Bezout identity, which permits the controller to be parameterized in such a way that the solution of the optimal control problem can be reduced to a parameter optimization problem. This reduction of the controller design problem to a parameter optimization problem is an important step and leads naturally to the H_∞ design problem. It can be shown that the solution to the optimization problem also reduces to the solution of the Riccati equation presented earlier.

The need to achieve a prespecified level of disturbance attenuation has led to a number of new reformulations of the H_∞ problem in terms of dynamic games, which has in turn led to new results concerning the structure of H_∞ controllers. One important feature of linear H_∞ controllers is that it is possible to design a robust controller when prescribed levels of uncertainties are present in the model dynamics.

There is also a dual to the equations (9.145) and (9.146) representing the optimal H_∞ filtering problem. In the optimal H_∞ filtering problem, the performance measure

$$J\{\mathbf{y}_m(t), \hat{\mathbf{y}}(t), \mathbf{w}(t)\} = \|\mathbf{y}_m(t) - \hat{\mathbf{y}}(t)\|_2^2 - \gamma^2 \|\mathbf{w}(t)\|_2^2 \tag{9.156}$$

is minimized where $\mathbf{y}_m(t)$ is the measurement and $\hat{\mathbf{y}}(t)$ is the estimate of the measurement. An important feature of the H_∞ approach is that all results pertaining to the optimal filter can be easily obtained by recognizing the duality between the optimal controller and the optimal filter.

9.3.2 Optimal H_∞ Control: Problem Definition and Solution

The H_∞-norm minimizing controller minimizes the worst case of the ratio between the H_2-norm of the H_∞ performance assessment output $\mathbf{z}_\infty = \mathbf{T}_{H_\infty}\mathbf{x}$ and the input **x**, which is

$$\left\|\mathbf{T}_{H_\infty}(j\omega)\right\|_\infty = \sup_{\omega\in R}\left[\max_{x\in C^n}\left[\frac{\left\|\mathbf{T}_{H_\infty}(j\omega)\,\mathbf{x}\right\|_2}{\|\mathbf{x}\|_2}\right]\right], \tag{9.157}$$

where the H_2 and H_∞-norms are, respectively, defined as

$$\|\mathbf{T}\|_2 = \left(\frac{1}{2}\int_{-\infty}^{\infty} trace\left[\mathbf{T}^*(j\omega)\,\mathbf{T}(j\omega)\right]\mathrm{d}\omega\right)^{\frac{1}{2}}, \tag{9.158a}$$

$$\|\mathbf{T}\|_\infty = \sup_{\omega}\left[\sigma_{\max}\left(\mathbf{T}(j\omega)\right)\right]. \tag{9.158b}$$

The H_∞-norm bounding but sub-optimal controller may also be stated in the alternative form. The plant is assumed to be in a state-space form

$$\dot{\mathbf{x}} = \mathbf{A}\mathbf{x} + \mathbf{B}_1\mathbf{w} + \mathbf{B}_2\mathbf{u},\ \ \mathbf{z} = \mathbf{C}_1\mathbf{x} + \mathbf{D}_{11}\mathbf{w} + \mathbf{D}_{12}\mathbf{u},$$
$$\mathbf{y} = \mathbf{C}_2\mathbf{x} + \mathbf{D}_{21}\mathbf{w} + \mathbf{D}_{22}\mathbf{u}. \tag{9.159}$$

The plant may be represented in matrix form as

$$\begin{bmatrix} \dot{\mathbf{x}} \\ \mathbf{z} \\ \mathbf{y} \end{bmatrix} = \begin{bmatrix} \mathbf{A} & \mathbf{B}_1 & \mathbf{B}_2 \\ \mathbf{C}_1 & \mathbf{D}_{11} & \mathbf{D}_{12} \\ \mathbf{C}_2 & \mathbf{D}_{21} & \mathbf{D}_{22} \end{bmatrix} \begin{bmatrix} \mathbf{x} \\ \mathbf{w} \\ \mathbf{u} \end{bmatrix} \tag{9.160}$$

and the plant's transfer matrix representation has the form

$$\mathbf{G} = \begin{bmatrix} \mathbf{A} & \mathbf{B}_1 & \mathbf{B}_2 \\ \mathbf{C}_1 & \mathbf{D}_{11} & \mathbf{D}_{12} \\ \mathbf{C}_2 & \mathbf{D}_{21} & \mathbf{D}_{22} \end{bmatrix}. \tag{9.161}$$

When the system matrices meet the following conditions:

1. $(\mathbf{A}, \mathbf{B}_1)$ is controllable and $(\mathbf{C}_1, \mathbf{A})$ is observable;
2. $(\mathbf{A}, \mathbf{B}_2)$ is controllable and $(\mathbf{C}_2, \mathbf{A})$ is observable;
3. $\mathbf{D}_{12}^T \begin{bmatrix} \mathbf{C}_1 & \mathbf{D}_{12} \end{bmatrix} = \begin{bmatrix} \mathbf{0} & \mathbf{I} \end{bmatrix}, \begin{bmatrix} \mathbf{B}_1 \\ \mathbf{D}_{21} \end{bmatrix} \mathbf{D}_{21}^T = \begin{bmatrix} \mathbf{0} \\ \mathbf{I} \end{bmatrix}.$

Let the $\mathbf{w} \to \mathbf{z}$ transfer function be defined as $\mathbf{T}_{\mathbf{zw}}(s)$ with its H_∞-norm equal to $||\mathbf{T}_{\mathbf{zw}}(s)||_\infty$. The problem is to find a control law, $\mathbf{u}(s) = \mathbf{K}(s)\mathbf{y}(s)$, such that the H_∞-norm is bounded as $||\mathbf{T}_{\mathbf{zw}}||_\infty < \gamma$, where $\gamma \in R$. By noting the duality between the optimal controller and optimal estimator, the solution may be conveniently stated in terms of the positive semi-definite solutions of two algebraic Riccati equations

$$\mathbf{X}_\infty\mathbf{A} + \mathbf{A}^T\mathbf{X}_\infty + \mathbf{X}_\infty(\mathbf{B}_1\mathbf{B}_1^T/\gamma^2 - \mathbf{B}_2\mathbf{B}_2^T)\mathbf{X}_\infty + \mathbf{C}_1^T\mathbf{C} = \mathbf{0} \tag{9.162a}$$

and

$$\mathbf{A}\mathbf{Y}_\infty + \mathbf{Y}_\infty\mathbf{A}^T + \mathbf{Y}_\infty(\mathbf{C}_1^T\mathbf{C}_1/\gamma^2 - \mathbf{C}_2^T\mathbf{C}_2)\mathbf{Y}_\infty + \mathbf{B}_1\mathbf{B}_1^T = \mathbf{0}. \tag{9.162b}$$

Provided, $\mathbf{X}_\infty$ and $\mathbf{Y}_\infty$ satisfy the condition

$$\begin{bmatrix} \gamma\mathbf{Y}_\infty^{-1} & \mathbf{I}_n \\ \mathbf{I}_n & \gamma\mathbf{X}_\infty^{-1} \end{bmatrix} > 0, \tag{9.163}$$

there exists a unique optimal dynamic controller

$$\begin{bmatrix} \dot{\hat{\mathbf{x}}} \\ \mathbf{u} \end{bmatrix} = \mathbf{K}(s) \begin{bmatrix} \hat{\mathbf{x}} \\ \mathbf{y} \end{bmatrix}, \tag{9.164}$$

with the transfer matrix given by

$$\mathbf{K}(s) = \begin{bmatrix} \mathbf{A}_\infty & -\mathbf{Z}_\infty\mathbf{L}_\infty \\ \mathbf{F}_\infty & \mathbf{0} \end{bmatrix}, \tag{9.165}$$

where

$$\mathbf{A}_\infty = \mathbf{A} + \gamma^{-2}\mathbf{B}_1\mathbf{B}_1^T\mathbf{X}_\infty + \mathbf{B}_2\mathbf{F}_\infty + \mathbf{Z}_\infty\mathbf{L}_\infty\mathbf{C}_2,\ \ \mathbf{F}_\infty = -\mathbf{B}_2^T\mathbf{X}_\infty,$$
$$\mathbf{L}_\infty = -\mathbf{Y}_\infty\mathbf{C}_2^T \text{ and } \mathbf{Z}_\infty = (\mathbf{I} - \gamma^{-2}\mathbf{Y}_\infty\mathbf{X}_\infty)^{-1}.$$

The closed-loop dynamics of the plant state and estimator may be expressed as

$$\frac{\mathrm{d}}{\mathrm{d}t}\begin{bmatrix}\mathbf{x}\\ \hat{\mathbf{x}}\end{bmatrix}=\begin{bmatrix}\mathbf{A} & \mathbf{B}_2\mathbf{F}_\infty\\ -\mathbf{Z}_\infty\mathbf{L}_\infty\mathbf{C}_2 & \mathbf{A}_\infty\end{bmatrix}\begin{bmatrix}\mathbf{x}\\ \hat{\mathbf{x}}\end{bmatrix}+\begin{bmatrix}\mathbf{B}_1\\ -\mathbf{Z}_\infty\mathbf{L}_\infty\mathbf{D}_{21}\end{bmatrix}\mathbf{w},$$

$$\mathbf{z}=\begin{bmatrix}\mathbf{C}_1\\ \mathbf{D}_{12}\mathbf{F}_\infty\end{bmatrix}\begin{bmatrix}\mathbf{x}\\ \hat{\mathbf{x}}\end{bmatrix}+\mathbf{0}\cdot\mathbf{w}. \tag{9.166}$$

The closed-loop transfer matrix is represented as

$$\mathbf{T}_{zw}(s)=\begin{bmatrix}\mathbf{A}_c & \mathbf{B}_c\\ \mathbf{C}_c & \mathbf{0}\end{bmatrix},\ \gamma>0,\ ||\mathbf{T}_{zw}||_\infty<\gamma,\ \mathbf{P}_c\geq 0, \tag{9.167a}$$

where $\mathbf{P}_c$ satisfies

$$\mathbf{P}_c\mathbf{A}_c+\mathbf{A}_c^T\mathbf{P}_c+\mathbf{P}_c\mathbf{B}_c\mathbf{B}_c^T\mathbf{P}_c/\gamma^2+\mathbf{C}_c^T\mathbf{C}_c=0, \tag{9.167b}$$

with

$$\mathbf{P}_c=\gamma^2\begin{bmatrix}\mathbf{Y}_\infty^{-1} & -\mathbf{Y}_\infty^{-1}\mathbf{Z}_\infty^{-1}\\ -\left(\mathbf{Z}_\infty^T\right)^{-1}\mathbf{Y}_\infty^{-1} & \mathbf{Y}_\infty^{-1}\mathbf{Z}_\infty^{-1}\end{bmatrix}. \tag{9.167c}$$

Further details may be found in Zhou and Doyle (1998).

9.3.3 *Optimal Control Synthesis: Linear Matrix Inequalities*

Given the following linear dynamic system in state-space form,

$$\dot{\mathbf{x}}=\mathbf{A}\mathbf{x}+\mathbf{B}\mathbf{w},\ \mathbf{z}=\mathbf{C}\mathbf{x}+\mathbf{D}\mathbf{w}, \tag{9.168}$$

where $\mathbf{x}$ is system state vector, $\mathbf{w}$ is an exogenous disturbance signal and $\mathbf{z}$ is a controlled output signal. The system matrices $(\mathbf{A}\ \mathbf{B}\ \mathbf{C}\ \mathbf{D})$ are constant matrices of appropriate dimensions. For a prescribed scalar $\gamma>0$, we define the performance index by

$$J(\mathbf{w})=\int_0^\infty\left(\mathbf{z}^T\mathbf{z}-\gamma^2\mathbf{w}^T\mathbf{w}\right)\mathrm{d}t \tag{9.169}$$

Then, it follows that $J(\mathbf{w})<0$, for all non-zero $\mathbf{w}$, if and only if there exists a symmetric positive definite matrix $\mathbf{P}>\mathbf{0}$ to satisfy the LMI

$$\begin{bmatrix}\mathbf{AP}+\mathbf{PA}^T & \mathbf{PC}^T & \mathbf{B}\\ \mathbf{CP} & -\mathbf{I} & \mathbf{D}\\ \mathbf{B}^T & \mathbf{D}^T & -\gamma^2\mathbf{I}\end{bmatrix}<0, \tag{9.170}$$

where the symmetric positive matrix $\mathbf{P}$ is usually called as Lyapunov function matrix. The Schur complement equivalence allows equation (9.170) to be reduced to an equivalent of constraints. Thus, if $\mathbf{P}$ and $\mathbf{R}$ are symmetric matrices, the condition

$$\begin{bmatrix}\mathbf{P} & \mathbf{S}\\ \mathbf{S}^T & \mathbf{R}\end{bmatrix}>\mathbf{0} \tag{9.171}$$

is equivalent to $\mathbf{R}>\mathbf{0}$ and $\mathbf{P}-\mathbf{SR}^{-1}\mathbf{S}^T>\mathbf{0}$.

Thus, equation (9.170) is equivalent to

$$\begin{bmatrix}\mathbf{AP}+\mathbf{PA}^T+\mathbf{BB}^\mathbf{T}/\gamma^2 & \mathbf{PC}^T+\mathbf{BD}^T/\gamma^2\\ \mathbf{CP}+\mathbf{DB}^\mathbf{T}/\gamma^2 & -\mathbf{I}+\mathbf{DD}^T/\gamma^2\end{bmatrix}<0, \tag{9.172}$$

which is then equivalent to solving the algebraic Riccati-like equation for $\mathbf{P}$,

$$\mathbf{AP}+\mathbf{PA}^T+\mathbf{BB}^\mathbf{T}/\gamma^2+\left(\mathbf{PC}^T+\mathbf{BD}^T/\gamma^2\right)\left(\mathbf{I}-\mathbf{DD}^T/\gamma^2\right)^{-1}\left(\mathbf{CP}+\mathbf{DB}^\mathbf{T}/\gamma^2\right)<0. \tag{9.173}$$

Pre- and post-multiplying by $\mathbf{P}^{-1}$ results in

$$\begin{aligned}&\mathbf{P}^{-1}\mathbf{A}+\mathbf{A}^T\mathbf{P}^{-1}+\mathbf{P}^{-1}\mathbf{BB}^\mathbf{T}\mathbf{P}^{-1}/\gamma^2+\\&\left(\mathbf{C}^T+\mathbf{P}^{-1}\mathbf{BD}^T/\gamma^2\right)\left(\mathbf{I}-\mathbf{DD}^T/\gamma^2\right)^{-1}\left(\mathbf{C}+\mathbf{DB}^\mathbf{T}\mathbf{P}^{-1}/\gamma^2\right)<0.\end{aligned} \tag{9.174}$$

Denoting the last term of equation (9.174) by $\mathbf{Q}$, the algebraic Riccati equation-like structure of the equation is apparent. Moreover, both the H_∞-norm and H_2-norm minimizing controllers could be expressed in terms of LMIs.

To solve the H_∞-norm minimizing controller, as before the plant is assumed to be in a state-space form, with the plant's transfer matrix representation having the form

$$\mathbf{G}=\begin{bmatrix}\mathbf{A} & \mathbf{B}_1 & \mathbf{B}_2\\ \mathbf{C}_1 & \mathbf{D}_{11} & \mathbf{D}_{12}\\ \mathbf{C}_2 & \mathbf{D}_{21} & \mathbf{D}_{22}\end{bmatrix}. \tag{9.175}$$

Given a proper rational controller in the form

$$\begin{bmatrix}\dot{\hat{\mathbf{x}}}\\ \mathbf{u}\end{bmatrix}=\begin{bmatrix}\hat{\mathbf{A}} & \hat{\mathbf{B}}\\ \hat{\mathbf{C}} & \hat{\mathbf{D}}\end{bmatrix}\begin{bmatrix}\hat{\mathbf{x}}\\ \mathbf{y}\end{bmatrix}\equiv\hat{\mathbf{G}}\begin{bmatrix}\hat{\mathbf{x}}\\ \mathbf{y}\end{bmatrix}, \tag{9.176}$$

the closed-loop system matrix is

$$\mathbf{G}_C\equiv\begin{bmatrix}\mathbf{A}_C & \mathbf{B}_C\\ \mathbf{C}_C & \mathbf{D}_C\end{bmatrix}=\begin{bmatrix}\mathbf{A}_0 & \mathbf{B}_0\\ \mathbf{C}_0 & \mathbf{D}_0\end{bmatrix}+\begin{bmatrix}\bar{\mathbf{B}} & \mathbf{0}\\ \mathbf{0} & \bar{\mathbf{D}}_{12}\end{bmatrix}\begin{bmatrix}\hat{\mathbf{A}} & \hat{\mathbf{B}}\\ \hat{\mathbf{C}} & \hat{\mathbf{D}}\end{bmatrix}\begin{bmatrix}\bar{\mathbf{C}} & \mathbf{0}\\ \mathbf{0} & \bar{\mathbf{D}}_{21}\end{bmatrix}, \tag{9.177}$$

where

$$\begin{bmatrix}\mathbf{A}_0 & \mathbf{B}_0\\ \mathbf{C}_0 & \mathbf{D}_0\end{bmatrix}=\begin{bmatrix}\begin{bmatrix}\mathbf{A} & \mathbf{0}\\ \mathbf{0} & 0*\hat{\mathbf{A}}\end{bmatrix} & \begin{bmatrix}\mathbf{B}_1\\ \mathbf{0}\end{bmatrix}\\ \begin{bmatrix}\mathbf{C}_1 & \mathbf{0}\end{bmatrix} & \mathbf{D}_{11}\end{bmatrix},$$

$$\begin{bmatrix}\bar{\mathbf{B}} & \mathbf{0}\\ \mathbf{0} & \bar{\mathbf{D}}_{12}\end{bmatrix}=\begin{bmatrix}\begin{bmatrix}\mathbf{0} & \mathbf{B}_2\\ \mathbf{I} & \mathbf{0}\end{bmatrix} & \begin{bmatrix}\mathbf{0} & \mathbf{0}\\ \mathbf{0} & \mathbf{0}\end{bmatrix}\\ \begin{bmatrix}\mathbf{0} & \mathbf{0}\end{bmatrix} & \begin{bmatrix}\mathbf{0} & \mathbf{D}_{12}\end{bmatrix}\end{bmatrix},$$

$$\begin{bmatrix}\bar{\mathbf{C}} & \mathbf{0}\\ \mathbf{0} & \bar{\mathbf{D}}_{21}\end{bmatrix}=\begin{bmatrix}\begin{bmatrix}\mathbf{0} & \mathbf{I}\\ \mathbf{C}_2 & \mathbf{0}\end{bmatrix} & \begin{bmatrix}\mathbf{0}\\ \mathbf{0}\end{bmatrix}\\ \begin{bmatrix}\mathbf{0} & \mathbf{0}\\ \mathbf{0} & \mathbf{0}\end{bmatrix} & \begin{bmatrix}\mathbf{0}\\ \mathbf{D}_{21}\end{bmatrix}\end{bmatrix}.$$

From the bounded real lemma, there exists a positive definite Lyapunov function

$$V(\mathbf{x})=\mathbf{x}^T(t)\mathbf{P}^{-1}(t)\mathbf{x}(t),\ \ \mathbf{P}^{-1}>\mathbf{0} \tag{9.178}$$

such that

$$\dot{V}(\mathbf{x})=\mathbf{w}^T(t)\mathbf{w}(t)-\gamma^2\mathbf{y}^T(t)\mathbf{y}(t)<0 \tag{9.179}$$

for all $\mathbf{x}$ and $\mathbf{w}$. Substituting the closed-loop system equations into inequality (9.179) and rearranging it, the problem of finding a control law $\mathbf{u}(s)=\mathbf{K}(s)\mathbf{y}(s)$, such that the H_∞-norm is bounded as $||\mathbf{T_{zw}}||_\infty<\gamma$, where $\gamma\in R$, i.e. the H_∞ sub-optimal control problem, is equivalent to the existence of a solution to the inequality for $\mathbf{P} > \mathbf{0}$,

$$\begin{bmatrix} \mathbf{PA}_C + \mathbf{A}_C^T\mathbf{P} & \mathbf{PB}_C & \mathbf{C}_C^T \\ \mathbf{B}_C^T\mathbf{P} & -\gamma\mathbf{I} & \mathbf{D}_C^T \\ \mathbf{C}_C & \mathbf{D}_C & -\gamma\mathbf{I} \end{bmatrix} < 0 \quad \text{or}$$

$$\begin{bmatrix} \mathbf{PA}_C + \mathbf{A}_C^T\mathbf{P} + \mathbf{C}_C^T\mathbf{C}_C/\gamma & \mathbf{PB}_C + \mathbf{C}_C^T\mathbf{D}_C/\gamma \\ \mathbf{B}_C^T\mathbf{P} + \mathbf{D}_C^T\mathbf{C}_C/\gamma & -\gamma\mathbf{I} + \mathbf{D}_C^T\mathbf{D}_C/\gamma \end{bmatrix} < 0. \tag{9.180a}$$

which is equivalent to

$$\mathbf{PA}_C + \mathbf{A}_C^T\mathbf{P} + \mathbf{C}_C^T\mathbf{C}_C/\gamma + \left(\mathbf{PB}_C + \mathbf{C}_C^T\mathbf{D}_C/\gamma\right)\left(\gamma\mathbf{I} - \mathbf{D}_C^T\mathbf{D}_C/\gamma\right)^{-1}\left(\mathbf{B}_C^T\mathbf{P} + \mathbf{D}_C^T\mathbf{C}_C/\gamma\right) < 0. \tag{9.180b}$$

To solve the problem, one solves two LMI inequalities:

$$\begin{bmatrix} \mathbf{AR} + \mathbf{RA}^T & \mathbf{RC}_1^T & \mathbf{B}_1 \\ \mathbf{C}_1\mathbf{R} & -\gamma\mathbf{I} & \mathbf{D}_{11} \\ \mathbf{B}_1^T & \mathbf{D}_{11}^T & -\gamma\mathbf{I} \end{bmatrix} < 0, \tag{9.181a}$$

$$\begin{bmatrix} \mathbf{SA} + \mathbf{A}^T\mathbf{S} & \mathbf{SB}_1 & \mathbf{C}_1^T \\ \mathbf{B}_1^T\mathbf{S} & -\gamma\mathbf{I} & \mathbf{D}_{11}^T \\ \mathbf{C}_1 & \mathbf{D}_{11} & -\gamma\mathbf{I} \end{bmatrix} < 0, \tag{9.181b}$$

for **R** and **S**, provided

$$\begin{bmatrix} \mathbf{R} & \mathbf{I} \\ \mathbf{I} & \mathbf{S} \end{bmatrix} \geq \mathbf{0}, \ \text{let } \mathbf{X} = \gamma\mathbf{R}^{-1}, \ \mathbf{Y} = \gamma\mathbf{S}^{-1}. \tag{9.182}$$

Furthermore, we make the simplifying assumptions that $\mathbf{D}_{11} = \mathbf{0}$,

$$\mathbf{D}_{12}^T\left[\mathbf{D}_{12} \quad \mathbf{C}_1\right] = \left[\mathbf{I} \quad \mathbf{0}\right] \quad \text{and} \quad \mathbf{D}_{21}\left[\mathbf{D}_{21}^T \quad \mathbf{B}_1^T\right] = \left[\mathbf{I} \quad \mathbf{0}\right]. \tag{9.183}$$

The matrices **X** and **Y** satisfy

$$\mathbf{XA} + \mathbf{A}^T\mathbf{X} + \mathbf{X}(\mathbf{B}_1\mathbf{B}_1^T/\gamma^2 - \mathbf{B}_2\mathbf{B}_2^T)\mathbf{X} + \mathbf{C}_1^T\mathbf{C} < \mathbf{0} \tag{9.184a}$$

and

$$\mathbf{AY} + \mathbf{YA}^T + \mathbf{Y}(\mathbf{C}_1^T\mathbf{C}_1/\gamma^2 - \mathbf{C}_2^T\mathbf{C}_2)\mathbf{Y} + \mathbf{B}_1\mathbf{B}_1^T < \mathbf{0}. \tag{9.184b}$$

The solution for **P** may be

$$\mathbf{P} = \gamma^2\begin{bmatrix} \mathbf{Y}^{-1} & -\mathbf{Y}^{-1}\mathbf{Z}^{-1} \\ -\left(\mathbf{Z}^T\right)^{-1}\mathbf{Y}^{-1} & \mathbf{Y}^{-1}\mathbf{Z}^{-1} \end{bmatrix}, \tag{9.185}$$

with

$$\mathbf{Z} = (\mathbf{I} - \gamma^{-2}\mathbf{YX})^{-1}. \tag{9.186}$$

To find the solution for **P** in the general case ($\mathbf{D}_{11} \neq \mathbf{0}$), note that it may be expressed as

$$\mathbf{P} = \begin{bmatrix} \mathbf{S} & \mathbf{N} \\ \mathbf{N}^T & \mathbf{V} \end{bmatrix} \quad \text{and} \quad \mathbf{P}^{-1} = \begin{bmatrix} \mathbf{R} & \mathbf{M} \\ \mathbf{M}^T & \mathbf{U} \end{bmatrix}. \tag{9.187}$$

Hence,

$$\mathbf{MN}^T = \mathbf{I} - \mathbf{RS} \text{ and } \mathbf{M}^T\mathbf{N} = \mathbf{I} - \mathbf{UV}. \tag{9.188}$$

Two full column rank matrices $\mathbf{N}$ and $\mathbf{M}$ can then be computed from the first of the above two identities. The solution for $\mathbf{P}$ is then obtained from

$$\mathbf{P} = \begin{bmatrix} \mathbf{S} & \mathbf{I} \\ \mathbf{N}^T & \mathbf{0} \end{bmatrix} \begin{bmatrix} \mathbf{I} & \mathbf{R} \\ \mathbf{0} & \mathbf{M}^T \end{bmatrix}^{-1}. \tag{9.189}$$

The LMI representation is convenient for us to analyse and synthesize nominal control performance for a linear time-invariant system. The LMI representation also facilitates the design of a control system that can potentially meet several LMI-type constraints. Given several LMI constraints of the form

$$\mathbf{F}_i(\mathbf{x}) = \mathbf{x}^T\mathbf{Q}_i\mathbf{x} + \mathbf{S}_i^T\mathbf{x} + \mathbf{x}\mathbf{S}_i + \mathbf{R}_i \geq \mathbf{0},\ i = 0, 1, 2, 3, \ldots n, \tag{9.190}$$

the condition $\mathbf{F}_0(\mathbf{x}) \geq \mathbf{0}$ such that $\mathbf{F}_i(\mathbf{x}) \geq \mathbf{0},\ i = 1, 2, 3, \ldots n$, holds provided there exist non-negative coefficients $\lambda_i i = 1, 2, 3, \ldots n$, which is satisfied for all $\mathbf{x}$

$$\mathbf{F}_0(\mathbf{x}) - \sum_{i=1}^{n} \lambda_i \mathbf{F}_i(\mathbf{x}) \geq \mathbf{0}. \tag{9.191}$$

Numerical procedures for finding a solution that meets several LMI-type constraints have been developed. Thus, one must be able to solve the control design problem by meeting an appropriately chosen LMI constraint and by appropriate choice of the performance parameters the multiple LMI constraints may be met.

The H_∞ controller minimizes the worse case of the ratio between the L_2 norms of input and output signals. It can effectively ensure that the plant tracks the steady operating design set point. The H_∞ performance is convenient to enforce robustness against model uncertainty and to express frequency-domain specifications such as bandwidth and low-frequency gain. The H_2 controller can effectively limit the output fluctuations and thus regulate the output. It minimizes the effects of stochastic inputs such as measurement noise and random disturbance. Furthermore, restricting the location of the closed-loop poles within a sector and to the right of the minimum damping vertical in the complex plane ensures a desirable time response and closed-loop damping. A multi-objective H_∞/H_2 controller may be designed to recover the advantages of both the H_∞ and the H_2 controllers as well as meet the pole placement constraints. The above considerations imply that the following performance and pole placement constraints are essential for the operation of the closed-loop system:

1. The H_∞ control objective: following the bounded real lemma, the H_∞ performance $||\mathbf{T}_{\mathbf{zw}}(s)||_\infty < \gamma$ is equivalent to the following matrix inequality:

$$\begin{bmatrix} \mathbf{PA}_C + \mathbf{A}_C^T\mathbf{P} & \mathbf{PB}_C & \mathbf{C}_C^T \\ \mathbf{B}_C^T\mathbf{P} & -\gamma\mathbf{I} & \mathbf{D}_C^T \\ \mathbf{C}_C & \mathbf{D}_C & -\gamma\mathbf{I} \end{bmatrix} < 0. \tag{9.192a}$$

2. The H_2 control objective: the H_2 objective for the closed-loop system, the performance $||\mathbf{T}_{\mathbf{zw}}(s)||_2 < \nu$ if and only if $\mathbf{D}_C = 0$ and the following matrix inequalities hold with the auxiliary variable $\mathbf{Q}$ satisfying *Trace* $(\mathbf{Q}) < \nu^2$:

$$\begin{bmatrix} \mathbf{PA}_C + \mathbf{A}_C^T\mathbf{P} & \mathbf{PB}_C \\ \mathbf{B}_C^T\mathbf{P} & -\mathbf{I} \end{bmatrix} < 0, \begin{bmatrix} \mathbf{P} & \mathbf{C}_C^T \\ \mathbf{C}_C & \mathbf{Q} \end{bmatrix} > 0. \tag{9.192b}$$

3. The constrained pole placement objective: the region of interest for our control purposes is the set $S(\alpha, r, \theta)$ of complex numbers $x + jy$ such that $x < -\alpha < 0$, $|x + jy| < r$, and bounded by the

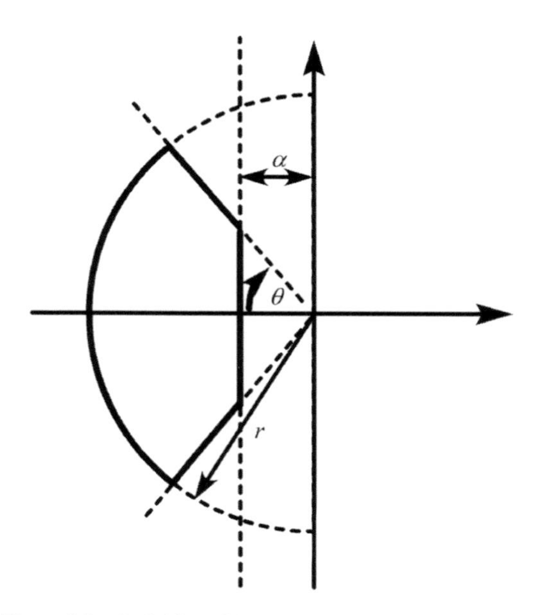

Figure 9.2 Definition of the pole placement region $S(\alpha, r, \theta)$

lines $y = \pm x \tan\theta$, as shown in Figure 9.2. The conditions of the poles of the closed-loop system matrix, lying in the region $S(\alpha, r, \theta)$, are characterized by the following matrix inequalities (Chilali *et al.*, 1999):

$$\mathbf{PA}_C + \mathbf{A}_C^T\mathbf{P} + 2\alpha\mathbf{P} < 0, \tag{9.192c}$$

$$\begin{bmatrix} -r\mathbf{P} & \mathbf{A}_C^T\mathbf{P} \\ \mathbf{PA}_C & -r\mathbf{P} \end{bmatrix} < 0, \tag{9.192d}$$

$$\begin{bmatrix} \left(\mathbf{PA}_C + \mathbf{A}_C^T\mathbf{P}\right)\sin\theta & -\left(\mathbf{PA}_C - \mathbf{A}_C^T\mathbf{P}\right)\cos\theta \\ \left(\mathbf{PA}_C - \mathbf{A}_C^T\mathbf{P}\right)\cos\theta & \left(\mathbf{PA}_C + \mathbf{A}_C^T\mathbf{P}\right)\sin\theta \end{bmatrix} < 0. \tag{9.192e}$$

The actual design of the controller may be achieved numerically by using the MATLAB®/Simulink® robust control toolbox. Jeng *et al.* (2005) have succinctly summarized the LMI-based design approach. For further details on the use and application of LMIs, the reader is referred to the book by Boyd *et al.* (1994).

9.4 Optimal Design of Structronic Systems

Smart structures (or active structures as they are also referred to in the structural dynamics and control literature) are now a distinct possibility as compared to passive structures. The availability of low-cost micromechanical actuators and sensors, using a range of new active materials such as active plastics, piezoelectric crystals and ceramics, shape memory alloys and new breed of semiconductor materials, which can be integrated into layered elastic structures, has led to the development of smart structures. The availability of these fibre-integrated or wafer-integrated smart structures has in turn spawned the possibility of precise control of structural behaviour or actively adapting the structure's dynamics and its behaviour to a continuously changing environment.

There is also the distinct possibility of adapting the structure's response to different disturbances or set points. Thus, there is a breed of emerging structronic systems that may be defined as control-integrated or control-configured structures. Associated with structronic systems are a number of problems that are peculiar to the design and development of these systems. First of these is the issue of the robustness of these actively controlled structures, which are difficult to model precisely. The second is the issue of the optimal distribution of the sensors and actuators. Finally, there is the issue of the overall optimal design of practical structronic systems. The issue of robustness of the actively controlled structure is considered in the next section. To illustrate the problems associated with the location of sensors and actuators for such actively controlled structures, the design of an actively controlled clamp is considered in detail in Section 9.4.2. The design of a piezo-ceramic patch-integrated smart plate structure is considered in Section 9.4.3. In Section 9.4.4 we consider the robust stabilization of the classical problem of a fluttering panel. The examples will serve to highlight some of the key issues to consider in the design and implementation of such systems.

9.4.1 Optimal Robust Design of Controlled Structures

So far in the design of the controller for achieving the desired performance, no mention has been made of the problems of dealing with the uncertainties in a typical structure. Although uncertainties are always present in any system, they are particularly important in the design of controllers for smart structures, as in these systems one can expect significant levels of uncertainties while establishing their physical properties due to the nature of the processes involved in their manufacture. At this stage it is therefore important to consider the tools necessary to establish the robustness of a particular controller in the presence of uncertainties.

Control theorists have introduced a representation of closed-loop systems to facilitate such analysis, which is a generalization of the representation presented earlier in Figure 9.1(b). This representation is shown in Figure 9.3(a). The inputs and outputs are arranged into three classes of inputs and three classes

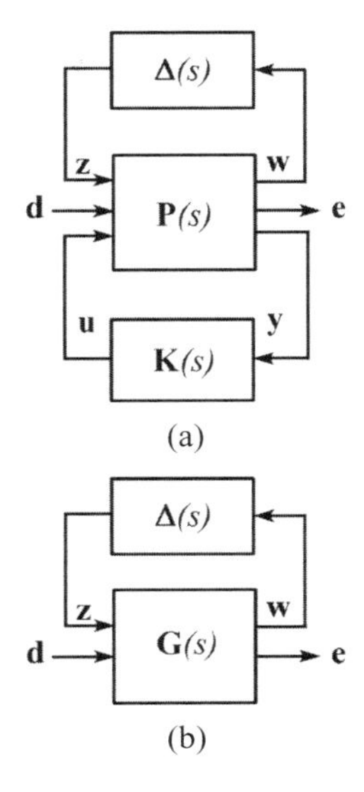

Figure 9.3 (a) Input–output representation of a plant, controller and uncertainties and (b) input–output representation of a controlled plant, with uncertainties

of outputs. The input in the lower end is the controlling input, while the corresponding outputs at the lower end are the measurements available for feedback. The inputs in the middle are the typical inputs that the plant is used for, including disturbances, while corresponding outputs on the right are the errors in the outputs when compared with the standard desired responses. The inputs at the top correspond to inputs that are essential to represent the uncertainties in defining the plant. These uncertainties are defined by means of a feedback loop linking the corresponding output at the top by means of a gain matrix, which represents the uncertainties. Since we have assumed that the design of the controller is completed, it could be absorbed into the plant and this process results in a simplified diagram, which is illustrated in Figure 9.3(b).

The inputs **d**, the perturbation Δ and the output error **e** are all assumed to be normalized to unity with all weightings and scaling factors absorbed into the generalized plant structure **P**. This arrangement results in standardized conditions for robust stability and performance expressed in terms of certain measures of the structured uncertainties denoted by μ. Although the connection of the uncertainty is standardised and is bounded, real uncertainties are characterised by further constraints. Uncertainty models are said to "structured" when they must satisfy additional constraints on the parameters. A structured uncertainty is a more general form of uncertainty than an unstructured uncertainty. For further details on the definitions of uncertainties are discussed by Burl (1993). For an interesting discussion of the μ synthesis methodology, the reader is referred to the paper by Moser (1993).

The first and most important step prior to the analysis involves the creation of the model in Figures 9.3 and 9.4 with the uncertainties represented as an outer feedback loop. This involves a process known amongst control theorists as 'pulling out the deltas' and will now be discussed briefly.

Consider the dynamics of a flexible structure in the form

$$\mathbf{M}\ddot{\mathbf{x}}(t) + \mathbf{C}\dot{\mathbf{x}}(t) + \mathbf{K}\mathbf{x}(t) = \mathbf{F}_C\mathbf{u}(t) + \mathbf{F}_D\mathbf{d}(t), \tag{9.193}$$

where $\mathbf{x}(t)$ is an $n \times 1$ vector of the displacement degrees of freedom, $\mathbf{u}(t)$ is an $n \times 1$ vector of control forces generated by the actuators, $\mathbf{d}(t)$ is an $n \times 1$ vector of disturbance forces, $\mathbf{M}$ is an $n \times n$ mass matrix associated with the flexible structure, $\mathbf{C}$ is an $n \times n$ damping matrix associated with the structure, $\mathbf{K}$ is the associated $n \times n$ stiffness matrix, $\mathbf{F}_C$ is the control force distribution vector and $\mathbf{F}_D$ is the disturbance force distribution vector. The disturbance forces are often said to be *exogenous* as they are generated external to the plant. The dynamics may be expressed in state-space form as

$$\begin{bmatrix} \mathbf{M} & \mathbf{0} \\ \mathbf{0} & \mathbf{M} \end{bmatrix} \frac{\mathrm{d}}{\mathrm{d}t} \begin{bmatrix} \mathbf{x} \\ \dot{\mathbf{x}} \end{bmatrix} = \begin{bmatrix} \mathbf{0} & \mathbf{M} \\ -\mathbf{K} & -\mathbf{C} \end{bmatrix} \begin{bmatrix} \mathbf{x} \\ \dot{\mathbf{x}} \end{bmatrix} + \begin{bmatrix} \mathbf{F}_D & \mathbf{F}_C \end{bmatrix} \begin{bmatrix} \mathbf{d} \\ \mathbf{u} \end{bmatrix} \tag{9.194}$$

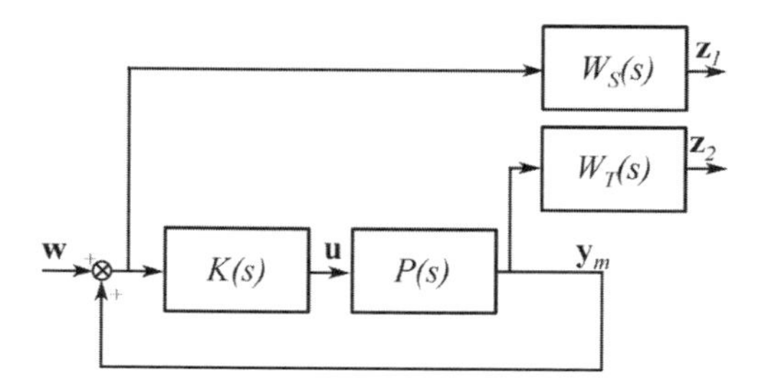

Figure 9.4 Traditional representation of the plant, the controller, the measurement ym and performance assessing outputs z_1 and z_2

and reduced to

$$\frac{\mathrm{d}}{\mathrm{d}t}\begin{bmatrix}\mathbf{x}\\ \dot{\mathbf{x}}\end{bmatrix}=\begin{bmatrix}\mathbf{0} & \mathbf{I}\\ -\mathbf{M}^{-1}\mathbf{K} & -\mathbf{M}^{-1}\mathbf{C}\end{bmatrix}\begin{bmatrix}\mathbf{x}\\ \dot{\mathbf{x}}\end{bmatrix}+\begin{bmatrix}\mathbf{0}\\ \mathbf{M}^{-1}\end{bmatrix}\begin{bmatrix}\mathbf{F}_D & \mathbf{F}_C\end{bmatrix}\begin{bmatrix}\mathbf{d}\\ \mathbf{u}\end{bmatrix}$$

$$\equiv \mathbf{A}\begin{bmatrix}\mathbf{x}\\ \dot{\mathbf{x}}\end{bmatrix}+\mathbf{B}_1\begin{bmatrix}\mathbf{d}\\ \mathbf{u}\end{bmatrix}+\mathbf{B}_2\begin{bmatrix}\mathbf{d}\\ \mathbf{u}\end{bmatrix}. \quad (9.195)$$

For a system with uncertain parameters,

$$\frac{\mathrm{d}}{\mathrm{d}t}\begin{bmatrix}\mathbf{x}\\ \dot{\mathbf{x}}\end{bmatrix}=\begin{bmatrix}\mathbf{0} & \mathbf{I}\\ -(\mathbf{M}+\Delta\mathbf{M})^{-1}(\mathbf{K}+\Delta\mathbf{K}) & -(\mathbf{M}+\Delta\mathbf{M})^{-1}(\mathbf{C}+\Delta\mathbf{C})\end{bmatrix}\begin{bmatrix}\mathbf{x}\\ \dot{\mathbf{x}}\end{bmatrix}$$

$$+\begin{bmatrix}\mathbf{0}\\ (\mathbf{M}+\Delta\mathbf{M})^{-1}\end{bmatrix}\begin{bmatrix}\mathbf{F}_D+\Delta\mathbf{F}_D & \mathbf{F}_C+\Delta\mathbf{F}_D\end{bmatrix}\begin{bmatrix}\mathbf{d}\\ \mathbf{u}\end{bmatrix}$$

$$=(\mathbf{A}+\Delta\mathbf{A})\begin{bmatrix}\mathbf{x}\\ \dot{\mathbf{x}}\end{bmatrix}+\begin{bmatrix}\mathbf{B}_1+\Delta\mathbf{B}_1 & \mathbf{B}_2+\Delta\mathbf{B}_2\end{bmatrix}\begin{bmatrix}\mathbf{d}\\ \mathbf{u}\end{bmatrix}. \quad (9.196a)$$

Hence,

$$\Delta\mathbf{A}=\begin{bmatrix}\mathbf{0} & \mathbf{I}\end{bmatrix}^T\times\lfloor -(\mathbf{M}+\Delta\mathbf{M})^{-1}(\mathbf{K}+\Delta\mathbf{K})-\mathbf{M}^{-1}\mathbf{K} \quad -(\mathbf{M}+\Delta\mathbf{M})^{-1}(\mathbf{C}+\Delta\mathbf{C})-\mathbf{M}^{-1}\mathbf{C}\rfloor \quad (9.196b)$$

and

$$\Delta\mathbf{B}_1=\begin{bmatrix}\mathbf{0}\\ \mathbf{I}\end{bmatrix}\left((\mathbf{M}+\Delta\mathbf{M})^{-1}(\mathbf{F}_D+\Delta\mathbf{F}_D)-\mathbf{M}^{-1}\mathbf{F}_D\right), \quad (9.196c)$$

$$\Delta\mathbf{B}_2=\begin{bmatrix}\mathbf{0}\\ \mathbf{I}\end{bmatrix}\left((\mathbf{M}+\Delta\mathbf{M})^{-1}(\mathbf{F}_C+\Delta\mathbf{F}_C)-\mathbf{M}^{-1}\mathbf{F}_C\right). \quad (9.196d)$$

Given the above structure in the uncertainty matrices $\mathbf{A}$, $\mathbf{B}_1$ and $\mathbf{B}_2$, the entire system transfer matrix is now considered and expressed as

$$\bar{\mathbf{G}}+\Delta\bar{\mathbf{G}}=\begin{bmatrix}\mathbf{A}+\Delta\mathbf{A} & \mathbf{B}_1+\Delta\mathbf{B}_1 & \mathbf{B}_2+\Delta\mathbf{B}_2\\ \mathbf{C}_1+\Delta\mathbf{C}_1 & \mathbf{D}_{11}+\Delta\mathbf{D}_{11} & \mathbf{D}_{12}+\Delta\mathbf{D}_{12}\\ \mathbf{C}_2+\Delta\mathbf{C}_2 & \mathbf{D}_{21}+\Delta\mathbf{D}_{21} & \mathbf{D}_{22}+\Delta\mathbf{D}_{22}\end{bmatrix}. \quad (9.197a)$$

Furthermore, it is assumed that $\Delta\bar{\mathbf{G}}$ may be expressed as

$$\Delta\bar{\mathbf{G}}=\begin{bmatrix}\Delta\mathbf{A} & \Delta\mathbf{B}_1 & \Delta\mathbf{B}_2\\ \Delta\mathbf{C}_1 & \Delta\mathbf{D}_{11} & \Delta\mathbf{D}_{12}\\ \Delta\mathbf{C}_2 & \Delta\mathbf{D}_{21} & \Delta\mathbf{D}_{22}\end{bmatrix}=\begin{bmatrix}\sum_{i=1}^{k}\delta_i\hat{\mathbf{A}}_i & \sum_{i=1}^{k}\delta_i\hat{\mathbf{B}}_{1i} & \sum_{i=1}^{k}\delta_i\hat{\mathbf{B}}_{2i}\\ \sum_{i=1}^{k}\delta_i\hat{\mathbf{C}}_{1i} & \sum_{i=1}^{k}\delta_i\hat{\mathbf{D}}_{11i} & \sum_{i=1}^{k}\delta_i\hat{\mathbf{D}}_{12i}\\ \sum_{i=1}^{k}\delta_i\hat{\mathbf{C}}_{2} & \sum_{i=1}^{k}\delta_i\hat{\mathbf{D}}_{21i} & \sum_{i=1}^{k}\delta_i\hat{\mathbf{D}}_{22i}\end{bmatrix}. \quad (9.197b)$$

Define

$$\Delta_p=diag\left\{\delta_1\mathbf{I}_{q_1} \quad \delta_2\mathbf{I}_{q2} \quad \cdots \quad \delta_k\mathbf{I}_{qk}\right\}, \quad (9.198)$$

where q_i denotes, for each i, the rank of the perturbation transfer matrix,

$$\mathbf{P}_i=\begin{bmatrix}\hat{\mathbf{A}}_i & \hat{\mathbf{B}}_{1i} & \hat{\mathbf{B}}_{2i}\\ \hat{\mathbf{C}}_{1i} & \hat{\mathbf{D}}_{11i} & \hat{\mathbf{D}}_{12i}\\ \hat{\mathbf{C}}_{2i} & \hat{\mathbf{D}}_{21i} & \hat{\mathbf{D}}_{22i}\end{bmatrix}. \quad (9.199)$$

Then $\mathbf{P}_i$ and $\delta_i\mathbf{P}_i$ could be expressed as

$$\mathbf{P}_i = \begin{bmatrix} \mathbf{L}_i \\ \mathbf{W}_{1i} \\ \mathbf{W}_{2i} \end{bmatrix} \begin{bmatrix} \mathbf{R}_i \\ \mathbf{Z}_{1i} \\ \mathbf{Z}_{2i} \end{bmatrix}^*, \quad \delta_i\mathbf{P}_i = \begin{bmatrix} \mathbf{L}_i \\ \mathbf{W}_{1i} \\ \mathbf{W}_{2i} \end{bmatrix} \begin{bmatrix} \delta_i \mathbf{I}_{qi} \end{bmatrix} \begin{bmatrix} \mathbf{R}_i \\ \mathbf{Z}_{1i} \\ \mathbf{Z}_{2i} \end{bmatrix}^*.$$

Thus, $\Delta\mathbf{G}$ may be expressed as

$$\Delta\bar{\mathbf{G}} = \begin{bmatrix} \mathbf{L}_1 & \cdots & \mathbf{L}_k \\ \mathbf{W}_{11} & \cdots & \mathbf{W}_{1k} \\ \mathbf{W}_{21} & \cdots & \mathbf{W}_{2k} \end{bmatrix} \begin{bmatrix} \delta_1 \mathbf{I}_{q_1} & \mathbf{0} & \mathbf{0} \\ \mathbf{0} & \ddots & \mathbf{0} \\ \mathbf{0} & \mathbf{0} & \delta_1 \mathbf{I}_{qk} \end{bmatrix} \begin{bmatrix} \mathbf{R}_1^* & \mathbf{Z}_{11}^* & \mathbf{Z}_{21}^* \\ \cdots & \cdots & \cdots \\ \mathbf{R}_k^* & \mathbf{Z}_{1k}^* & \mathbf{Z}_{2k}^* \end{bmatrix}. \tag{9.200}$$

The transfer matrix may now be augmented as

$$\bar{\mathbf{G}}_{\mathbf{G}+\Delta} = \begin{bmatrix} \mathbf{D}_{33} & \mathbf{C}_3 & \mathbf{D}_{31} & \mathbf{D}_{32} \\ \mathbf{B}_3 & \mathbf{A} & \mathbf{B}_1 & \mathbf{B}_2 \\ \mathbf{D}_{13} & \mathbf{C}_1 & \mathbf{D}_{11} & \mathbf{D}_{12} \\ \mathbf{D}_{23} & \mathbf{C}_2 & \mathbf{D}_{21} & \mathbf{D}_{22} \end{bmatrix}, \tag{9.201a}$$

where

$$\begin{bmatrix} \mathbf{D}_{33} \\ \mathbf{B} \\ \mathbf{D}_{13} \\ \mathbf{D}_{23} \end{bmatrix} = \begin{bmatrix} & \mathbf{0} & \\ \mathbf{L}_1 & \cdots & \mathbf{L}_k \\ \mathbf{W}_{11} & \cdots & \mathbf{W}_{1k} \\ \mathbf{W}_{21} & \cdots & \mathbf{W}_{2k} \end{bmatrix}, \tag{9.201b}$$

$$\begin{bmatrix} \mathbf{C}_3 & \mathbf{D}_{31} & \mathbf{D}_{32} \end{bmatrix} = \begin{bmatrix} \mathbf{R}_1^* & \mathbf{Z}_{11}^* & \mathbf{Z}_{21}^* \\ \cdots & \cdots & \cdots \\ \mathbf{R}_k^* & \mathbf{Z}_{1k}^* & \mathbf{Z}_{2k}^* \end{bmatrix}. \tag{9.201c}$$

Thus, it now follows that the system transfer matrix may be expressed as the upper fractional transform of $\bar{\mathbf{G}}_{\mathbf{G}+\Delta}$ and Δ_p

$$\bar{\mathbf{G}} + \Delta\bar{\mathbf{G}} = \mathsf{F}_\mathsf{U}\left(\bar{\mathbf{G}}_{\mathbf{G}+\Delta},\ \Delta_p\right). \tag{9.202}$$

It then follows that the system transfer function may be expressed as

$$\begin{bmatrix} \mathbf{e} \\ \mathbf{y} \end{bmatrix} = \mathbf{G}_{io} \begin{bmatrix} \mathbf{d} \\ \mathbf{u} \end{bmatrix}, \tag{9.203}$$

where $\mathbf{G}_{io}$ may be expressed in terms of the upper fractional transform and a set of integrators as

$$\mathbf{G}_{io} = \mathsf{F}_\mathsf{U}\left(\mathsf{F}_\mathsf{U}\left(\bar{\mathbf{G}}_{\mathbf{G}+\Delta},\ \Delta_p\right), \mathbf{I}/s\right). \tag{9.204}$$

Thus, closing only the second loop with the integrators, and the lower loop with a negative feedback controller, $\mathbf{u} = -\mathbf{K}(s)\mathbf{y}$, we obtain the relationship between the uncertainty and disturbance inputs and the corresponding outputs as

$$\begin{bmatrix} \mathbf{w} \\ \mathbf{e} \end{bmatrix} = \mathbf{G}(s) \begin{bmatrix} \mathbf{z} \\ \mathbf{d} \end{bmatrix}, \tag{9.205}$$

$$\mathbf{G}(s) = \mathsf{F}_\mathsf{l}\left(\mathbf{P}(s), -\mathbf{K}(s)\right), \tag{9.206}$$

where

$$\mathbf{P}(s) = \mathsf{F}_\mathsf{U}\left(\mathbf{G}_{\mathbf{G}+\Delta}, \mathbf{I}/s\right), \ \mathbf{G}_{\mathbf{G}+\Delta} = \begin{bmatrix} \mathbf{A} & \mathbf{B}_3 & \mathbf{B}_1 & \mathbf{B}_2 \\ \mathbf{C}_3 & \mathbf{D}_{33} & \mathbf{D}_{31} & \mathbf{D}_{32} \\ \mathbf{C}_1 & \mathbf{D}_{13} & \mathbf{D}_{11} & \mathbf{D}_{12} \\ \mathbf{C}_2 & \mathbf{D}_{23} & \mathbf{D}_{21} & \mathbf{D}_{22} \end{bmatrix}. \tag{9.207}$$

The input–output relationship between **e** and **d** may be expressed in terms of the upper fractional transform as

$$\mathbf{e} = \mathsf{F}_\mathsf{U}\left(\bar{\mathbf{G}}, \Delta_p\right)\mathbf{d}.$$

However, the conditions for stability, performance and robustness are usually represented in terms of the properties of the transfer function matrix $\mathbf{G}(s)$. The feedback is then internally stable provided the poles of the closed-loop system, i.e. the roots of the equation

$$\det\left(\mathbf{I} + \mathbf{P}_{22}(s)\mathbf{K}(s)\right) = 0, \tag{9.208}$$

all have negative real parts. In the multi-input–multi-output case, the determinant equation (9.206) provides a convenient test for stability. The condition for the poles of the closed-loop system can be simplified for the case of single-input–single-output systems when there are no pole-zero cancellations between the plant and controller. In this case,

$$\det\left(\mathbf{I} + \mathbf{P}_{22}(s)\mathbf{K}(s)\right) = \left(1 + P_{22}(s)K(s)\right) = 0 \tag{9.209}$$

reduces to a scalar equation, the return difference equation. Additionally, the poles of the open-loop system are cancelled and do not appear in the roots of the return difference equation. Thus, the closed-loop system in this case is internally stable if and only if all the roots of the return difference equation have negative real parts.

In the absence of uncertainty, $\boldsymbol{\Delta}_p$, the nominal performance objectives are expressed in terms of

$$\left\|\mathbf{G}_{22}\left(j\omega\right)\right\|_\infty \equiv \sup_\omega\left[\sigma_{\max}\left(\mathbf{G}_{22}\left(j\omega\right)\right)\right] < 1, \tag{9.210}$$

which relates the H_2-norm of **e** to the H_2-norm of **d**. In practice, the use of scalings and weightings is necessary to represent and normalize the varying frequency and spatial content of input and output sets.

Consider plant perturbations $\boldsymbol{\Delta}_p$ that could destabilize a nominally stable plant. Robust stability is guaranteed for the unstructured uncertainty if only if the following condition holds:

$$\left\|\mathbf{G}_{11}\left(j\omega\right)\right\|_\infty < 1 \text{ for all } \boldsymbol{\Delta}_p,\ \sigma_{\max}(\boldsymbol{\Delta}_p) \equiv \bar{\sigma}(\boldsymbol{\Delta}_p) < 1. \tag{9.211}$$

In the case of problems associated with the control of flexible structures, the uncertainty consists of parameter variations and multiple norm-bounded perturbations that represent the unmodelled dynamics of the system. Parameter variations often arise due to changes in inertial and stiffness properties and represent changes in the coefficients of the equations describing the flexible structural system. Unfortunately, the norm-bounded test is insufficient and inadequate in dealing with robust performance and realistic models of plant uncertainty, which can be said to possess a structure.

To represent bounded structured uncertainty, the structured singular value (SSV) is defined and the function μ are used to develop necessary and sufficient conditions. The function μ is defined as

$$\mu(\mathbf{G}) = \frac{1}{\min\limits_{\boldsymbol{\Delta}_p\in\boldsymbol{\Delta}}\left\{\bar{\sigma}(\boldsymbol{\Delta}_p)\middle|\det\left(\mathbf{I} + \mathbf{G}\Delta_p\right) = 0 \text{ for all } \mathbf{G}\right\}} \tag{9.212}$$

unless no $\boldsymbol{\Delta}_p \in \boldsymbol{\Delta}$ makes the determinant of $\mathbf{I} + \mathbf{G}\boldsymbol{\Delta}_p$ zero, in which case $\mu(\mathbf{G})$ is also set to be zero. With this definition of the structured singular value μ, the conditions for robust stability condition have been derived. Robust stability is guaranteed if and only if

$$\left\|\mathbf{G}_{11}\left(j\omega\right)\right\|_\mu \equiv \sup_\omega\left[\mu\left(\mathbf{G}_{11}\left(j\omega\right)\right)\right] \le 1 \text{ for all } \boldsymbol{\Delta}_p \in \boldsymbol{\Delta}_\mathbf{B} \tag{9.213}$$

where the subscript **B** is used to denote the fact that $\boldsymbol{\Delta}_p$ belongs to the set of all uncertainties within the unit ball. It may be noted that, in contrast with $\bar{\sigma}(\boldsymbol{\Delta}_p)$, the value of $\mu(\mathbf{G})$ is dependent on $\mathbf{G}_{11}\left(j\omega\right)$ as well as on the structure of perturbations. Since $\mathbf{G}_{11}\left(j\omega\right)$ is a function of frequency, $\mu(\mathbf{G})$ is taken as the largest value over all frequencies. The value of $\mu(\mathbf{G})$ is an SSV that specifies the norm of the smallest matrix Δ_p (in a family of plausible Δ_ps), which destabilizes the closed-loop synthesis problem. When $\mu(\mathbf{G}) > 1$ means that the controller will either not achieve the specified performance for the modelled uncertainties or not be stable for the particular uncertainties, or a combination of these violations of the requirements of performance and stability. A value of $\mu(\mathbf{G}) \leq 1$ ensures that the controller will meet the specified performance for the given uncertainties.

When designing a control system, the issue of primary importance is performance of the system in the presence of uncertainty. Uncertainties in structural systems may be modelled as additive, which are not directly dependent on the magnitude of the plant transfer function, as multiplicative representing errors in modal frequencies, modal damping ratios and modal participation factors or as parametric variations in the mass, stiffness and damping coefficients of nominal model. The μ-analysis methodology may be applied in all three cases, provided the performance of the controller is analysed with both the perturbation and the disturbances occurring simultaneously. To consider the issue of robust performance, i.e. performance with perturbations and disturbances occurring simultaneously, the following conditions must be satisfied:

$$\begin{aligned} &\mathsf{F_U}\left(\bar{\mathbf{G}}, \mathbf{\Delta}_p\right) \text{ is stable and } \left\|\mathsf{F_U}\left(\bar{\mathbf{G}}, \mathbf{\Delta}_p\right)\right\|_\infty \leq 1 \text{ for all } \mathbf{\Delta}_p \in \mathbf{\Delta_B}, \\ &\left\|\mathbf{G}_{11}\left(j\omega\right)\right\|_\mu \equiv \sup_\omega \left[\mu\left(\mathbf{G}_{11}\left(j\omega\right)\right)\right] \leq 1. \end{aligned} \tag{9.214}$$

The computation of $\mu(\mathbf{G})$ can be prohibitively expensive even for examples with moderately large numbers of degrees of freedom. Rather than computing $\mu(\mathbf{G})$, a practical alternative is to compute an upper and a lower bound to it. An upper bound for $\mu(\mathbf{G})$ allows one to establish a sufficient condition for the non-singularity of $\mathbf{I} + \mathbf{G}\mathbf{\Delta}_p$ for all possible uncertainties under consideration. From the upper bound, one can estimate a lower bound of the margin for the robustness. A lower bound for $\mu(\mathbf{G})$ allows one to establish a sufficient condition for the singularity of $\mathbf{I} + \mathbf{G}\mathbf{\Delta}_p$ for all possible uncertainties under consideration. The bounds on $\mu(\mathbf{G})$ are computed based on the extreme cases of the structure of $\mathbf{\Delta}_p$. In general, the structure of $\mathbf{\Delta}_p$ consists of repeated scalar blocks and full matrix blocks. $\mu(\mathbf{G})$ can be computed exactly from the upper bound if the structure of Δ_p corresponds to

$$2b_\mathrm{s} + b_\mathrm{f} \leq 3, \tag{9.215}$$

where b_s is the number of repeated scalar blocks and b_f is the number of full blocks. In this case, it is possible to express $\mu(\mathbf{G})$ as

$$\mu(\mathbf{G}) = \inf_{\mathbf{D}} \bar{\sigma}\left(\mathbf{D}\mathbf{G}\mathbf{D}^{-1}\right). \tag{9.216}$$

The transformation $\mathbf{DGD}^{-1}$ represents a scaling of the inputs and outputs of $\mathbf{G}$, which does not change the value of $\mu(\mathbf{G})$. In addition, since $\mu(\mathbf{G})$ can be computed exactly just as a maximum singular value ($\bar{\sigma}$) after appropriate scaling, the methods developed for H_∞ optimal control can be used to optimize $\mu(\mathbf{G})$. MATLAB's robust control toolbox provides number of m-files that are extremely useful in designing a robust controller for a flexible structure. These are summarized in Table 9.1.

Using MATLAB's toolboxes μ-analysis can be combined with H_∞ optimal control to produce μ-synthesis, which provides H_∞ performance in the presence of structured uncertainty. The scaling matrices $\mathbf{D}$ and $\mathbf{D}^{-1}$ are used to parameterize the structure of $\mathbf{\Delta}_p$ over the frequency range. The problem could be reformulated as an H_∞-norm minimization of

$$\left\|\mathbf{D}\mathsf{F_l}\left(\mathsf{F_U}\left(\mathbf{G}_{\mathbf{G}+\mathbf{\Delta}}, \mathbf{I}/s\right), -\mathbf{K}(s)\right)\mathbf{D}^{-1}\right\|_\infty \leq 1. \tag{9.217}$$

The above norm minimization problem is known as **D–K** iteration. As the **D–K** name implies, the μ-synthesis approach is to iterate between the **D** and **K** until the solution converges. However, there is no guarantee that it will converge. Although the method is not proven to produce a global minimum or to converge, the results widely published in the literature have demonstrated its usefulness in assisting a control system designer in picking an optimal and robust controller design.

Table 9.1 Summary of H_2 and H_∞ control and μ-synthesis (via **D–K** iteration) MATLAB robust control toolbox design commands

Function	Description
augw	Augments plant weights for mixed-sensitivity control design
h2syn	H_2 controller synthesis
h2hinfsyn	Mixed H_2/H_∞ controller synthesis
hinfsyn	H_∞ controller synthesis
loopsyn	H_∞ loop-shaping controller synthesis
ltrsyn	Loop-transfer recovery controller synthesis
mixsyn	H_∞ mixed-sensitivity controller synthesis
ncfsyn	H_∞ normalized co-prime factor controller synthesis
dksyn	Synthesis of a robust controller via μ-synthesis
dkitopt	Create a dksyn options object
drawmag	Interactive mouse-based sketching and fitting tool
fitfrd	Fit scaling frequency response data with LTI model
fitmagfrd	Fit scaling magnitude data with stable, minimum-phase model

9.4.2 *Optimum Placement and Co-location of the Sensor and Actuators: The Active Clamp*

One of the simplest applications of using feedback-controlled smart structures is in the design of an active clamp. In many situations, one is faced with need to actively bolster a strut or clamp a simply supported structure for a certain time period. An active strut is generally bolstered by an additionally controllable longitudinal tensile force that is provided by a piezoelectric stack actuator. A piezoelectric stack actuator consists of assemblage of identical piezoelectric discs that provide a cumulative axial actuation force along the common axis of the discs.

Active feedback-controlled smart structures could be effectively employed to emulate the effect of clamping the structure. Generally when a structure is clamped, moments are applied to it at the boundary. The same effect can be simulated by applying a force to the structure where the displacement is a maximum. In this section, the design of a feedback controller to emulate an active clamp is discussed, as are some of the practical aspects of implementing such a controller.

Consider a moderately thick plate that is simply supported along all its edges. Assuming that both the in-plane and the rotary inertia effects may be neglected, the transverse displacement $w(x, y)$ of a rectangular, isotropic, thin plate undergoing moderately large deflections may be expressed as

$$\nabla^4\Phi + Eh\left(\frac{\partial^2 w}{\partial x^2}\frac{\partial^2 w}{\partial y^2} - \left(\frac{\partial^2 w}{\partial x\partial y}\right)^2\right) = 0, \tag{9.218}$$

$$D\nabla^4 w - \frac{\partial^2\Phi}{\partial y^2}\frac{\partial^2 w}{\partial x^2} - \frac{\partial^2\Phi}{\partial x^2}\frac{\partial^2 w}{\partial y^2} + 2\frac{\partial^2\Phi}{\partial x\partial y}\frac{\partial^2 w}{\partial x\partial y} + \rho h\frac{\partial^2 w}{\partial t^2} = q, \tag{9.219}$$

with

$$D = \frac{Eh^3}{12\left(1-\nu^2\right)},$$

where D is the plate's flexural rigidity, E the Young's modulus, ν the Poisson ratio, q the distributed loading that is assumed to be the sum of a control pressure loading and a demanded or disturbance pressure loading, h the plate thickness, ρ the mass density and $\Phi(x, y)$ the Airy stress function. The in-plane stress resultants are

$$N_{xx} = \frac{\partial^2\Phi}{\partial y^2},\ N_{yy} = \frac{\partial^2\Phi}{\partial x^2},\ N_{xy} = -\frac{\partial^2\Phi}{\partial x\partial y}. \tag{9.220}$$

For a simply supported plate, the transverse displacement may be expressed in terms of the dominant first mode as

$$w_{ss}(x, y, t) = q_{ss}(t)h \cos\left(\frac{\pi x}{a}\right) \cos\left(\frac{\pi y}{b}\right), \tag{9.221}$$

where q_{ss} is the non-dimensional amplitude of motion at the origin of the reference frame that is located at the geometric centre of the plate. The lengths of the sides of the plate parallel to the x and y axes are a and b, respectively. Since the plate is moment free along its edges, the boundary conditions are

$$M_{xx} = D\left(w_{,xx} + \nu w_{,yy}\right) = 0 \text{ along } x = \pm a/2, \tag{9.222a}$$

$$M_{yy} = D\left(w_{,yy} + \nu w_{,xx}\right) = 0 \text{ along } y = \pm b/2. \tag{9.222b}$$

Substituting the assumed solution for $w(x, y)$ into the equation for the Airy stress function $\Phi(x, y)$, one could write the general solution for stress function as the sum of a complimentary solution $\Phi_c(x, y)$ and a particular integral $\Phi_p(x, y)$.

$$\Phi(x, y) = \Phi_c(x, y) + \Phi_p(x, y). \tag{9.223}$$

The particular integral $\Phi_p(x, y)$ is given by

$$\Phi_p(x, y) = -\frac{3}{8} D\left(1 - \nu^2\right) r^2 q^2 \left(\cos\left(\frac{2\pi x}{a}\right) + \frac{1}{r^4}\cos\left(\frac{2\pi y}{b}\right)\right), \tag{9.224}$$

where the aspect ratio of the plate is given by $r = a/b$.

The complimentary solution, $\Phi_c(x, y)$ may be expressed as

$$\Phi_c(x, y) = \bar{N}_{xx}\frac{y^2}{2} + \bar{N}_{yy}\frac{x^2}{2} - \bar{N}_{xy}xy, \tag{9.225}$$

where $\bar{N}_{xx}$, $\bar{N}_{yy}$ and $\bar{N}_{xy}$ are arbitrary constants that are determined by the application of the in-plane boundary conditions. If the edges are assumed to be movable, the averaged in-plane stresses could be assumed to be zero; thus, $\bar{N}_{xx} = 0$, $\bar{N}_{yy} = 0$ and $\bar{N}_{xy} = 0$. If the edges are assumed to be immovable, $\bar{N}_{xy} = 0$ along the edges and the averaged longitudinal strains normal to the edges must be equal to zero. Thus,

$$\int_S \varepsilon_{xx} \mathrm{d}y\mathrm{d}z = \int_S \left\{\frac{1}{E}\left(f_{,yy}(x, y) - \nu\Phi_{,xx}(x, y)\right) - \frac{1}{2}w_{,x}^2\right\} \mathrm{d}y\mathrm{d}z = 0 \tag{9.226a}$$

along $x = \pm a/2$,

$$\int_S \varepsilon_{yy} \mathrm{d}y\mathrm{d}z = \int_S \left\{\frac{1}{E}\left(\Phi_{,xx}(x, y) - \nu\Phi_{,yy}(x, y)\right) - \frac{1}{2}w_{,y}^2\right\} \mathrm{d}y\mathrm{d}z = 0 \tag{9.226b}$$

along $y = \pm b/2$ and $f_{,xy}(x, y) = 0$. Thus, the arbitrary constants $\bar{N}_{xx}$ and $\bar{N}_{yy}$ are given by

$$\bar{N}_{xx} = -\frac{3\pi^2}{2a^2} D\left(1 + \nu r^2\right) q^2, \ \bar{N}_{yy} = -\frac{3\pi^2}{2a^2} D\left(r^2 + \nu\right) q^2. \tag{9.227}$$

A similar approach may be adopted in the case of a rectangular plate clamped on all edges. In this case, the transverse displacement may be expressed in terms of the dominant assumed first mode as

$$w_{cc}(x, y, t) = q_{cc}(t)h \frac{\left(1 + \cos\left(\frac{2\pi x}{a}\right)\right)\left(1 + \cos\left(\frac{2\pi y}{b}\right)\right)}{4}, \tag{9.228}$$

where q_{cc} is the non-dimensional amplitude of motion at the origin of the reference frame that is located at the geometric centre of the plate.

The particular integral $\Phi_p(x, y)$ for the case of a clamped plate is given by

$$\Phi_p(x, y) = -\frac{3}{8} D \left(1 - \nu^2\right) r^2 q^2 \bar{\Phi}_p(x, y), \tag{9.229}$$

where

$$\begin{aligned}\bar{\boldsymbol{\Phi}}_p(x, y) = {} & \cos\left(\frac{2\pi x}{a}\right) + \frac{1}{r^4}\cos\left(\frac{2\pi y}{b}\right) + \frac{1}{16}\left(\cos\left(\frac{4\pi x}{a}\right) + \frac{1}{r^4}\cos\left(\frac{4\pi y}{b}\right)\right) \\ & + \frac{2}{\left(1+r^2\right)^2}\cos\left(\frac{2\pi x}{a}\right)\cos\left(\frac{2\pi y}{b}\right) + \frac{2}{\left(4+r^2\right)^2}\cos\left(\frac{4\pi x}{a}\right)\cos\left(\frac{2\pi y}{b}\right) \\ & + \frac{2}{\left(1+4r^2\right)^2}\cos\left(\frac{2\pi x}{a}\right)\cos\left(\frac{4\pi y}{b}\right). \end{aligned} \tag{9.230}$$

If the edges are assumed to be movable, the averaged in-plane stresses could be assumed to be zero; thus, $\bar{N}_{xx} = 0$, $\bar{N}_{yy} = 0$ and $\bar{N}_{xy} = 0$. If the edges are assumed to be immovable, $\bar{N}_{xy} = 0$ along the edges and the transverse displacements and slopes normal to the boundary at the edges must be equal to zero. Thus,

$$w_{cc}(x, y, t) = w_{cc,x}(x, y, t) = 0 \text{ along } x = \pm a/2, \tag{9.231a}$$

$$w_{cc}(x, y, t) = w_{cc,y}(x, y, t) = 0 \text{ along } y = \pm b/2 \tag{9.231b}$$

and $f_{,xy}(x, y) = 0$. Thus, the arbitrary constants $\bar{N}_{xx}$ and $\bar{N}_{yy}$ are given by

$$\bar{N}_{xx} = -\frac{9\pi^2}{8a^2} D \left(1 + \nu r^2\right) q^2, \tag{9.232}$$

$$\bar{N}_{yy} = -\frac{9\pi^2}{8a^2} D \left(r^2 + \nu\right) q^2. \tag{9.233}$$

With the solution for $f(x, y)$ determined, one could obtain the equation for the transverse displacement $w(x, y)$ by substituting the assumed solution into the governing equation and forcing the error to be orthogonal to the assumed first mode. Thus, assuming that the plate is subjected to a control pressure distribution $p(x, y, t)$ and a demanded pressure distribution $p^d(x, y, t)$, one obtains for the simply supported plate

$$\ddot{q}_{ss} + \omega_{0ss}^2 q_{ss} + \beta_{ss} q_{ss}^3 = p_{ss}(t) + p_{ss}^d(t), \tag{9.234}$$

with

$$\begin{aligned} \omega_{0ss}^2 &= \frac{D\pi^4}{\rho h b^4}\left(1 + \frac{1}{r^2}\right)^2, \quad \beta_{ss} = \frac{D\pi^4}{\rho h b^4 r^4}\left(\frac{3}{2}\beta^c + \left(1 - \nu^2\right)\beta_{ss}^p\right), \\ \beta_{ss}^p &= \frac{3}{4}\left(1 + r^4\right), \quad \beta^c = 1 + \nu r^2 + r^2\left(r^2 + \nu\right) \\ p_{ss}(t) &= \frac{4}{\rho h a b}\int_S p(x, y, t)\cos\left(\frac{\pi x}{a}\right)\cos\left(\frac{\pi y}{b}\right)\mathrm{d}x\mathrm{d}y, \\ p_{ss}^d(t) &= \frac{4}{\rho h a b}\int_S p^d(x, y, t)\cos\left(\frac{\pi x}{a}\right)\cos\left(\frac{\pi y}{b}\right)\mathrm{d}x\mathrm{d}y \end{aligned}$$

and

$$m_{cc}\ddot{q}_{cc} + \omega_{0cc}^2 q_{cc} + \beta_{cc} q_{cc}^3 = p_{cc}(t) + p_{cc}^d(t), \tag{9.235}$$

with

$$m_{cc} = \frac{9}{16},\ \omega_{0cc}^2 = \frac{D\pi^4}{\rho h b^4}\left(3 + \frac{2}{r^2} + \frac{3}{r^4}\right),$$

$$\beta_{cc} = \frac{D\pi^4}{\rho h b^4 r^4}\left(\frac{3}{2}\beta^c + \left(1 - \nu^2\right)\beta_{cc}^p\right),$$

$$\beta_{cc}^p = \frac{4}{3}\left(\frac{17}{16}\left(1 + r^4\right) + \frac{2r^4}{\left(1 + r^2\right)^2} + \frac{r^4}{2\left(4 + r^2\right)^2} + \frac{r^4}{2\left(1 + 4r^2\right)^2}\right),$$

$$p_{cc}(t) = \frac{4}{\rho hab}\int_S p(x, y, t)\frac{\left(1 + \cos\left(\frac{2\pi x}{a}\right)\right)\left(1 + \cos\left(\frac{2\pi y}{b}\right)\right)}{4}\mathrm{d}x\mathrm{d}y,$$

$$p_{cc}^d(t) = \frac{4}{\rho hab}\int_S p^d(x, y, t)\frac{\left(1 + \cos\left(\frac{2\pi x}{a}\right)\right)\left(1 + \cos\left(\frac{2\pi y}{b}\right)\right)}{4}\mathrm{d}x\mathrm{d}y.$$

In general, if in the case of the simply supported plate the control pressure mode is assumed to be

$$p_{ss}(t) = u + Kp_{cc}(t) \tag{9.236}$$

and

$$u = -\left(Km_{cc} - 1\right)\ddot{q}_{ss} - \left(K\omega_{0cc}^2 - \omega_{0ss}^2\right)q_{ss} - \left(K\beta_{cc} - \beta_{ss}\right)q_{ss}^3, \tag{9.237}$$

then q_{ss} satisfies

$$m_{cc}\ddot{q}_{ss} + \omega_{0cc}^2 q_{ss} + \beta_{cc}q_{ss}^3 = p_{cc}(t) + \frac{1}{K}p_{ss}^d(t). \tag{9.238}$$

Furthermore, when

$$\frac{1}{K}p_{ss}^d(t) = p_{cc}^d(t), \tag{9.239}$$

$$m_{cc}\ddot{q}_{ss} + \omega_{0cc}^2 q_{ss} + \beta_{cc}q_{ss}^3 = p_{cc}(t) + p_{cc}^d(t). \tag{9.240}$$

Thus, choosing

$$K = p_{ss}^d(t)/p_{cc}^d(t), \tag{9.241}$$

it is possible to force the transverse displacement amplitude of a simply supported plate in the first mode to satisfy the same equation as the transverse displacement amplitude of a plate clamped along all edges in the first mode. Thus, the closed-loop system emulates the stiffness characteristics of a clamped plate, at least in the first mode.

On the other hand, if $p^d(x, y, t)$ is a disturbance pressure distribution that must be attenuated, the controller gain may be chosen to be relatively large within a feasible range of values, so the closed system not only emulates the stiffness characteristics of a clamped plate but also attenuates the disturbance pressure in the process.

To implement the controller implied by equation (9.237), it is essential that the modal acceleration amplitude is measured. The rate and position may then be obtained by integrating once and twice, respectively. The controller may be considered to be a PI^2 controller. The simulation and assessment of the closed system in the presence of disturbances and noise are left as an exercise for the reader.

There a few pertinent issues that must be stressed if one wishes to implement such a controller. The plant model in this is non-linear, although it is still a single degree of freedom system. To implement the controller, one must obtain accurate estimates of the amplitude of the first mode q_{ss}. While the sensor location is extremely important to obtain accurate state estimates, the Kalman and H_∞ filters discussed earlier may be wholly inadequate for the purpose and it may be necessary to modify the filter to account for the non-linearity. The methodologies of the extended Kalman filter and the unscented Kalman filter are suitable candidates. However, a complete discussion of these filters is beyond the scope of this chapter. Furthermore, it may be essential to augment the plate model to include higher modes and estimate the higher-mode amplitudes as well to obtain an accurate estimate of the first dominant mode.

The second important aspect is the need to carefully ensure that only the first mode is excited. One must explicitly ensure that the next three or four modes are not excited. Thus, a large number of PZT patches must be used and the control signal distributed to these patches proportionately so only the first mode is excited and the next three or four modes are not excited. This is done by methods discussed in Chapter 3 to reduce controller spillover and will be discussed further in the next example. Moreover, the actual location would be different for transverse force actuators, which must be concentrated near the plate's centre, and for moment generators, which must be located closer to the edges. A judicious combination of both these types of actuators will facilitate the choice of a minimum number of actuators.

A third feature of the closed-loop controller is the need for independent acceleration and displacement feedback. In practice, one may also have to introduce rate feedback to ensure that the closed-loop structure is adequately damped. In the above analysis, it was implicitly assumed that there was adequate structural damping already present and this was capable of providing the desired stability margins. In practice, it may also not be possible to achieve the full stiffness characteristics of a clamped plate. However, the controller is definitely capable of substantially emulating the stiffness characteristics of a clamped plate.

9.4.3 Optimal Controller Design Applied to Smart Composites

To illustrate the optimal design of laminated or patched smart composites, an archetypal example will be considered. It is a uniform isotropic plate with proportional damping excited by a distribution of piezoelectric PZT patches with classical displacement sensing. However, the methodology could also be applied to a uniform orthotropic plate structure with proportional damping excited by a distribution of piezoelectric PZT patches with co-located and integrated piezoelectric sensing. For more details on the methods of modelling smart composites, the reader is referred to the papers by Gabbert *et al.* (2000a,b) and Gabbert *et al.* (2002) and the thesis by Liang (1997).

Consider a uniform isotropic thin plate of thickness h excited on its top and bottom surfaces by a distribution of M piezoelectric in-plane isotropic $\left(\bar{e}_{31}^{p} = \bar{e}_{32}^{p}\right)$ PZT bimorph patch actuators. Assuming that the voltage input to each patch is

$$V_m = E_{33} t_{p_m}, \tag{9.242}$$

the equation of motion for the transverse displacement may be expressed, by modifying the equations derived in Chapter 7, as

$$\left(\left(D^s + \sum_{m=1}^{M} \Psi_m(x,y) D_m^p\right)\nabla^4 - \bar{\rho}\left(h + 2\sum_{m=1}^{M} t_{p_m}\Psi_m(x,y)\right)\omega^2\right) w(x,y)$$
$$= q(x,y) + \bar{e}_{31}^{p} \sum_{m=1}^{M} V_m\left(h + t_{p_m}\right)\nabla^2 \boldsymbol{\Psi}_m(x,y). \tag{7.20}$$

The thickness of the mth piezoelectric patch is t_{p_m}. Assuming the laminate to be symmetric about the middle surface and integrating the density over the total thickness of the laminate h_t, the averaged density is expressed as

$$\bar{\rho} = \int_{-h_t/2}^{h_t/2} \rho \mathrm{d}z / h_t. \tag{9.243}$$

The flexural rigidity of the mth piezoelectric patch defined relative to the middle surface of the plate is D_m^p. All other variables are defined in Chapter 7, Section 7.1.6. The plate is assumed to be simply supported along all its edges. The transverse displacement of the plate is assumed to be given in terms of summation of assumed modes as

$$w(x, y, t) = h \sum_{n=1}^{N} P_n(t)\phi_n(x, y) = h \sum_{i=1}^{I} \sum_{j=1}^{J} P_n(t) \sin\left(\frac{i\pi x}{a}\right) \sin\left(\frac{j\pi y}{b}\right), \tag{9.244}$$

where $N = I \times J$ and $n = (i - 1)\, J + j$. Hence, one could obtain the equation for the coefficients $P_n(t)$ in the assumed solution for transverse displacement $w(x, y)$ by substituting the assumed solution into the governing equation and forcing the error to be orthogonal to the assumed modes. Thus, assuming that the only external loading present is due to the control inputs, and integrating the terms associated with piezoelectric control inputs by parts, one obtains the differential equations

$$\sum_{k=1}^{N} m_{nk} \frac{\mathrm{d}^2 P_k(t)}{\mathrm{d}^2 t} + s_{nk} P_k(t) = \bar{e}_{31}^p \sum_{m=1}^{M} g_{nm} V_m(t), \tag{9.245}$$

where

$$m_{nk} = \int\limits_S \bar{\rho} \left(h + 2\sum_{m=1}^{M} t_{p_m} \mathbf{\Psi}_m(x, y)\right) \phi_k(x, y)\phi_n(x, y)\mathrm{d}x\mathrm{d}y,$$

$$s_{nk} = \int\limits_S \left(D^s + \sum_{m=1}^{M} \mathbf{\Psi}_m(x, y) D_m^p\right) \phi_n(x, y)\nabla^4\phi_k(x, y)\mathrm{d}x\mathrm{d}y,$$

$$g_{nm} = \int\limits_S \left(h + t_{p_m}\right) \mathbf{\Psi}_m(x, y)\nabla^2\phi_n(x, y)\mathrm{d}x\mathrm{d}y.$$

In matrix form, one obtains

$$\mathbf{M}\frac{\mathrm{d}^2}{\mathrm{d}t^2}\mathbf{P}(t) + \mathbf{KP}(t) = \bar{e}_{31}^p \mathbf{GV}. \tag{9.246}$$

From the solution of the eigenvalue problem $\mathbf{K}x = \lambda \mathbf{M}x$, one obtains the matrix of eigenvectors $\mathbf{T}$, the equations (9.246) may be diagonalized as

$$\mathbf{T}^{-1}\mathbf{MT}\frac{\mathrm{d}^2}{\mathrm{d}t^2}\tilde{\mathbf{P}}(t) + \mathbf{T}^{-1}\mathbf{KT}\tilde{\mathbf{P}}(t) = \bar{e}_{31}^p \mathbf{FV},\ \ \mathbf{F} = \mathbf{T}^{-1}\mathbf{G}. \tag{9.247}$$

If one requires that the control inputs only influence the first few modes only, it may be assumed that

$$Q_n(t) \equiv \bar{e}_{31}^p \sum_{m=1}^{M} f_{nm} V_m(t) = u_n(t),\ \ \text{for } n = 1 \ldots N_c \tag{9.248a}$$

$$\text{and}\quad Q_n(t) \equiv \bar{e}_{31}^p \sum_{m=1}^{M} f_{nm} V_m(t) = 0,\ \ n = N_c \ldots N. \tag{9.248b}$$

Only the first N modes are considered in the analysis. The vector of required exciting forces may be expressed in matrix form as

$$\mathbf{Q} = \bar{e}_{31}^p \mathbf{FV} = \left[\mathbf{I}\ \mathbf{0}\right]^T \mathbf{u}. \tag{9.249}$$

assuming that the vector-applied voltages $\mathbf{V}$ may be expressed, in terms of a weighting matrix $\mathbf{W}$ and an auxiliary set of control inputs $\mathbf{u}$, as $\mathbf{V} = \mathbf{Wu}$. Thus, the voltage inputs to the *PZT* patches are constructed as a weighted linear combinations of the control inputs. A possible optimum solution for the weighting matrix $\mathbf{W}$ may be obtained as

$$\mathbf{W} = \left(\mathbf{F}^T\mathbf{F}\right)^{-1} \frac{1}{\bar{e}_{31}^p}\mathbf{F}^T \begin{bmatrix} \mathbf{I} \\ \mathbf{0} \end{bmatrix}. \tag{9.250}$$

Consequently,

$$\mathbf{Q} = \mathbf{F}\left(\mathbf{F}^T\mathbf{F}\right)^{-1}\mathbf{F}^T\left[\mathbf{I}\ \mathbf{0}\right]^T\mathbf{u}. \tag{9.251}$$

The equations of motion may be expressed in matrix notation as

$$\mathbf{T}^{-1}\mathbf{M}\mathbf{T}\frac{\mathrm{d}^2}{\mathrm{d}t^2}\tilde{\mathbf{P}}(t) + \mathbf{T}^{-1}\mathbf{K}\mathbf{T}\tilde{\mathbf{P}}(t) = \mathbf{F}\left(\mathbf{F}^T\mathbf{F}\right)^{-1}\mathbf{F}^T\left[\mathbf{I}\ \mathbf{0}\right]^T\mathbf{u}. \tag{9.252}$$

In the above equation, $\mathbf{T}^{-1}\mathbf{M}\mathbf{T}$ is the $N \times N$ diagonal generalized mass matrix, $\mathbf{T}^{-1}\mathbf{K}\mathbf{T}$ is the $N \times N$ diagonal stiffness matrix and $\mathbf{u}$ is the vector of N_c control inputs, which predominantly influences the first N_c modes of vibration. It may be noted that the control inputs influence the other modes as well, and this leads to the problem of *control spillover*, particularly when one wants to construct a reduced-order model. Equations (9.252) are in exactly same form as equation (3.133) in Chapter 3. However, when the number of patches is sufficiently large, the equations of motion are expressed as

$$\mathbf{T}^{-1}\mathbf{M}\mathbf{T}\frac{\mathrm{d}^2}{\mathrm{d}t^2}\tilde{\mathbf{P}}(t) + \mathbf{T}^{-1}\mathbf{K}\mathbf{T}\tilde{\mathbf{P}}(t) = \begin{bmatrix} \mathbf{I} \\ \mathbf{0} \end{bmatrix}\mathbf{u}, \tag{9.253}$$

and the problem of control spillover is avoided. To include the influence of structural damping, equations (9.253) are modified and written as

$$\mathbf{T}^{-1}\mathbf{M}\mathbf{T}\frac{\mathrm{d}^2}{\mathrm{d}t^2}\tilde{\mathbf{P}}(t) + \mathbf{T}^{-1}\mathbf{K}\mathbf{T}\left(g\frac{\mathrm{d}}{\mathrm{d}t} + 1\right)\tilde{\mathbf{P}}(t) = \begin{bmatrix} \mathbf{I} \\ \mathbf{0} \end{bmatrix}\mathbf{u}. \tag{9.254}$$

The parameter g is the structural damping coefficient. It is further assumed that the transverse displacement of the plate is measured at a number of discrete locations. Thus, the measurements are expressed as

$$y_l(t) = w\left(x_l, y_l, t\right) = h\sum_{n=1}^{N} P_n(t)\phi_n\left(x_l, y_l\right),\ l = 1\ldots L. \tag{9.255}$$

When the measurements are made using identical piezoelectric patches that are sensitive to the curvature, driven by a constant voltage V_p, they may be expressed as

$$y_l(t) = \mu\sum_{n=1}^{N} P_n(t)\int_{S_l} \Omega_l(x, y)\nabla^2\phi_n(x, y)\mathrm{d}x\mathrm{d}y,\ l = 1\ldots L, \tag{9.256}$$

where

$$\mu = e_{31}\left(h + t_p\right)ht_pE_{33} \equiv e_{31}\left(h + t_p\right)hV_p.$$

In matrix form, in either case, these may be expressed in terms of the transformed modal displacement amplitudes $\tilde{\mathbf{P}}(t)$ as

$$\mathbf{Y} = \mathbf{H}\tilde{\mathbf{P}}(t). \tag{9.257}$$

When it is desired to sense a finite number of modal amplitudes and a large number of sensor outputs are available, these may be combined, so $\mathbf{H}$ may be expressed as

$$\mathbf{H} = \left[\mathbf{I}\quad \mathbf{0}\right]. \tag{9.258}$$

This aspect is quite similar to the manner in which the control inputs were combined and was discussed in Chapter 3.

It is important to add that in designing a controller it is essential to include a number of disturbances and unmodelled effects. These include measurement noise, sensor dynamics, process disturbances, process noise and noise dynamics. However, before discussing these effects and designing an archetypal controller, the open-loop response of the structure will be briefly discussed. Considering the undamped case and eliminating the transformed modal displacement amplitudes $\tilde{\mathbf{P}}(t)$ from the equations of motion and measurements by taking Laplace transforms, the input-to-output transfer function may be expressed as

$$\tilde{\mathbf{Y}}(s) = \mathbf{G}(s)\tilde{\mathbf{U}}(s),\ \ \mathbf{G}(s) = \mathbf{H}\left(\mathbf{T}^{-1}\mathbf{M}\mathbf{T}s^2 + \mathbf{T}^{-1}\mathbf{K}\mathbf{T}\right)^{-1}\begin{bmatrix}\mathbf{I}\\ \mathbf{0}\end{bmatrix}. \tag{9.259}$$

When the Laplace transform variable s is assumed to be purely imaginary, i.e. $s = \omega\sqrt{-1}$, the transfer function $\mathbf{G}\,(i\omega)$ represents the frequency response function and can be expanded into an infinite summation in terms of the residues evaluated at the poles of the transfer function $\mathbf{G}(s)$. The expansion may expressed as

$$\mathbf{G}(s) = \mathbf{H}\left\{\sum_{k=1}^{\infty}\frac{\mathbf{R}_k}{\left(s^2+\omega_n^2\right)}\right\}\begin{bmatrix}\mathbf{I}\\ \mathbf{0}\end{bmatrix}, \tag{9.260}$$

where $\mathbf{R}_k$ are the positive definite residue matrices. It may be observed that, if in place of the piezoelectric patch actuators, a set of force actuators were co-located at the same points as the sensors, the transfer function may be expressed as

$$\mathbf{G}_c(s) = \mathbf{H}\left\{\sum_{k=1}^{\infty}\frac{\mathbf{R}_k}{\left(s^2+\omega_n^2\right)}\right\}\mathbf{H}^T. \tag{9.261}$$

It is customary to truncate the above expansion after a finite but large number of terms, and in this case one has

$$\mathbf{G}_c(s) \approx \mathbf{H}\left\{\sum_{k=1}^{K}\frac{\mathbf{R}_k}{\left(s^2+\omega_n^2\right)}\right\}\mathbf{H}^T. \tag{9.262}$$

The structure now has all the properties of a passive electrical network. The property of passivity has a number of important consequences, the most import one being that passive systems do not generate any energy. Moreover, passive systems can never be unstable unless they are interacting with another system that is not passive. Finally, the transfer function of a single-input–single-output system with pure imaginary poles and a co-located actuator and sensor has the form

$$G_c(s) = K_G\frac{\prod_{k=1}^{K}\left(s^2+z_n^2\right)}{\prod_{k=1}^{K}\left(s^2+\omega_n^2\right)},\ \ \omega_0 < z_0 < \omega_1 < z_1 < \omega_2 < z_2\cdots. \tag{9.263}$$

The issue of optimal location and distribution of the actuator has been discussed at length by Seeger and Gabbert (2003). A related problem is the optimal placement and distribution of the sensors used to complement the actuators that is determined by the requirement that the measurements must be adequate to be able to determine the mode shapes used in modelling the structure. The implementation of filters for the determination of the mode shapes is discussed by Meirovitch and Baruh (1985).

In this section, MATLAB functions such as *h2syn.*, *hinfsyn.m* and *h2hinfsyn.m* in the robust control toolbox are employed to design a controller for a typical jet engine axial compressor. The robust control toolbox has been revised by Balas *et al.* (2004) by integrating the more recent LMIs (Gahinet *et al.*, 1995) and the μ-analysis and μ-synthesis toolboxes with it. Many robust control analysis and synthesis routines provided by MATLAB based on powerful general-purpose functions for solving a class of convex non-linear programming problems are known as LMIs. The LMI-based controller synthesis methods are implemented by robust control toolbox functions that evaluate worst case performance, as well and functions like *h2syn.*, *hinfsyn.m* and *h2hinfsyn.m*. In the last case, the mixed sensitivity optimization can also be performed and the ∞-norm $\|\mathbf{T}_1\|_\infty$ of a weighted linear combination of such as the sensitivity $S(s)$ and the complementary sensitivity $T(s)$ is minimized. The resulting regulator seems to also possess the structure of a notch filter, with notch frequency located slightly above the structural resonance frequency. H_∞ control design is mainly concerned with frequency-domain performance and does not guarantee good transient responses of the closed-loop system. H_2 control gives more suitable performance with regard to closed-loop transient responses. Combining H_2 and H_∞ control objectives in a controller, where one must optimize the H_∞-norm $\|\mathbf{T}_1\|_\infty$ subject to a constraint that the H_2-norm satisfies, $\|\mathbf{T}_2\|_2 < g$, is desirable but finding a solution to the mixed H_∞/H_2 sub-optimal controller synthesis problem is often only possible if the controllers obtained by H_∞ and H_2 controller synthesis are relatively close to each other. Since our interest here involves applications where flexible structures interact with other non-passive systems, it will generally not be assumed that the structure being controlled is a passive system. However, when the interacting structure is definitely known to be passive, one could exploit some of the special features of the system to design robust controllers. The example considered in this section is one such case, and the design of H_∞ and H_2 controllers is illustrated.

Reconsider the equations of motion of general uniform thin plate excited by piezoelectric patch actuators with a number of displacement sensors distributed on the surface of the plate. To design an H_∞ and an H_2 controller, a set of representative disturbance models for the flexible plate structure are chosen to represent the unmodelled dynamics in the actuators in terms of the first natural frequency of the structure. Thus,

$$W_{d1} = \frac{0.1\omega_1}{s + 2\omega_1}, \quad W_{d2} = \frac{0.1\omega_1}{s + 0.5\omega_1}, \quad W_{d3} = 0.01 + \frac{0.09\omega_1}{s + \omega_1} \tag{9.264}$$

The measurement noise represents the higher frequency noise in the displacement measurement.

$$W_{mn} = \frac{s + 0.1\omega_1}{0.05s + \omega_1} = 20 - \frac{398\omega_1}{s + 20\omega_1}. \tag{9.265}$$

The H_∞ performance weight W_{11} is chosen so as to disable the controller from tracking the noise and high-frequency disturbances. The H_∞ performance weight W_{12} is chosen to attenuate high frequencies in the control signal.

$$\begin{aligned} W_{11} &= W_S = \frac{0.1s + \omega_1}{10s + \omega_1} = 0.01 + \frac{0.099\omega_1}{s + 0.1\omega_1}, \\ W_{12} &= W_T = \frac{0.01s + 0.001\omega_1}{0.1s + \omega_1} = 0.1 - \frac{0.99\omega_1}{s + 10\omega_1}. \end{aligned} \tag{9.266}$$

For a closed-loop system, the H_∞-norm equal to $||\mathbf{T}_{\mathbf{zw}}(s)||_\infty$ may be expressed as

$$||\mathbf{T}_{\mathbf{zw}}(s)||_\infty = \left\| \begin{matrix} W_S(s)\mathbf{S}(s) \\ W_T(s)\mathbf{T}(s) \end{matrix} \right\|_\infty < 1, \tag{9.267}$$

where $\mathbf{S}(s)$ is the sensitivity transfer function and $\mathbf{T}(s) = \mathbf{I} - \mathbf{S}(s)$ is the complementary sensitivity transfer function with

$$\mathbf{S}(s) = (\mathbf{I} + \mathbf{G}(s))^{-1}.$$

If one assumes that the basic plant is modelled by the state-space representation as

$$\dot{\mathbf{x}}_p = \mathbf{A}_p\mathbf{x}_p + \mathbf{B}_p\mathbf{u}, \ \mathbf{y}_\mathbf{o} = \mathbf{H}_p\mathbf{x}_p, \tag{9.268}$$

the disturbance dynamics is modelled to represent a dynamic system driven by white noise inputs as

$$\dot{\mathbf{x}}_d = \mathbf{A}_d \mathbf{x}_d + \mathbf{w}, \tag{9.269}$$

where

$$\mathbf{A}_d = \begin{bmatrix} -2 & 0 & 0 & 0 \\ 0 & -0.5 & 0 & 0 \\ 0 & 0 & -1 & 0 \\ 0 & 0 & 0 & -20 \end{bmatrix} \omega_1.$$

The plant dynamics, including the disturbance states, is

$$\dot{\mathbf{x}}_p = \mathbf{A}_p \mathbf{x}_p + \mathbf{A}_{pd} \mathbf{x}_d + \mathbf{B}_p \mathbf{u} + \mathbf{B}_d \mathbf{w}, \tag{9.270a}$$

$$\mathbf{y}_m = \mathbf{H}_p \mathbf{x}_p + \mathbf{I}\begin{bmatrix} 0 & 0 & 0 & -398 \end{bmatrix} \mathbf{x}_d + \mathbf{I}\begin{bmatrix} 0 & 0 & 0 & 20 \end{bmatrix} \mathbf{w}, \tag{9.270b}$$

where

$$\mathbf{A}_{pd} = \mathbf{B}_p \begin{bmatrix} 1 \\ 1 \\ \vdots \\ 1 \end{bmatrix} \begin{bmatrix} 0.1 & 0.1 & 0.09 & 0 \end{bmatrix}, \ \mathbf{B}_d = \mathbf{B}_p \begin{bmatrix} 1 \\ 1 \\ \vdots \\ 1 \end{bmatrix} \begin{bmatrix} 0 & 0 & 0.01 & 0 \end{bmatrix}.$$

The complete plant model is then expressed as

$$\dot{\mathbf{x}}_c = \mathbf{A}_c \mathbf{x}_c + \mathbf{B}_c \mathbf{u} + \mathbf{B}_{cd} \mathbf{w}, \ \mathbf{y}_\mathbf{o} = \mathbf{C}_{oc} \mathbf{x}_c, \ \mathbf{y}_m = \mathbf{C}_c \mathbf{x}_c + \mathbf{D}_c \mathbf{w}, \tag{9.271}$$

where

$$\mathbf{x}_c = \begin{bmatrix} \mathbf{x}_p \\ \mathbf{x}_d \end{bmatrix}, \ \mathbf{A}_c = \begin{bmatrix} \mathbf{A}_p & \mathbf{A}_{pd} \\ \mathbf{0} & \mathbf{A}_d \end{bmatrix}, \ \mathbf{B}_c = \begin{bmatrix} \mathbf{B}_p \\ \mathbf{0} \end{bmatrix}, \ \mathbf{B}_{cd} = \begin{bmatrix} \mathbf{B}_d \\ \mathbf{I}_{4\times4} \end{bmatrix}, \ \mathbf{C}_{oc} = \begin{bmatrix} \mathbf{H}_p & \mathbf{0} \end{bmatrix},$$

$$\mathbf{C}_c = \begin{bmatrix} \mathbf{H}_p & \begin{bmatrix} 1 \\ \vdots \\ 1 \end{bmatrix} \begin{bmatrix} 0 \cdots -398 \end{bmatrix} \end{bmatrix}, \ \mathbf{D}_c = \begin{bmatrix} 1 \\ \vdots \\ 1 \end{bmatrix} \begin{bmatrix} 0 & 0 & 0 & 20 \end{bmatrix}.$$

The traditional block diagram representation of the plant, the controller, the measurement $\mathbf{y}_m$ and performance assessing outputs $\mathbf{z}_{11}$ and $\mathbf{z}_{12}$ is shown in Figure 9.4.

An alternative representation leading to the lower LFT representation is shown in Figures 9.5(a) and 9.5(b). In particular, the transfer function W_T could be incorporated in the plant as shown in Figure 9.5(c) and the augmented plant may be expressed as

$$\dot{\mathbf{x}}_a = \mathbf{A}_a \mathbf{x}_a + \mathbf{B}_a \mathbf{u} + \mathbf{B}_{ad} \mathbf{w}, \ \begin{bmatrix} \mathbf{z}_{12} & \mathbf{y}_m \end{bmatrix}^T = \mathbf{C}_a \mathbf{x}_a + \mathbf{D}_a \mathbf{w}. \tag{9.272}$$

By a similar process, one could combine the plant with the transfer function W_S. However, the use of the MATLAB robust control toolbox function *sysic.m* also facilitates this and is the preferred option. The problem is to find a control law $\mathbf{u}(s) = \mathbf{K}(s)\mathbf{y}(s)$ such that the H_∞-norm is bounded as $||\mathbf{T}_{\mathbf{zw}}||_\infty < \gamma$, where $\gamma \in R$. The solution may be conveniently stated in terms of the solutions of two algebraic Riccati equations (Glover and Doyle, 1988; Doyle *et al.*, 1989).

Gahinet and Apkarian (1994) have shown that the same sub-optimal problem may also be stated in terms of LMIs, provided the controller is also defined as a dynamic controller. Khargonekar and Rotea (1991) have defined the mixed H_2/H_∞-norm optimization of a linear dynamic controller, and this again can be cast in LMI format. Although the exact solution to the problem is known to exist (Sznaier *et al.*, 2000), our experience indicated that finding a solution to this problem is not easy.

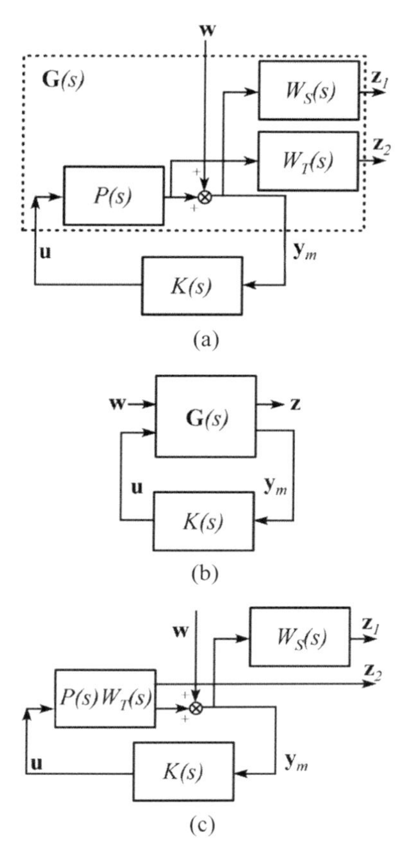

Figure 9.5 (a) Alternative representation of plant, measurement and controller, (b) linear fractional transformation representation of feedback system, (c) incorporating the weighting function in the plant

In the example considered in this section, the output for assessing H_2 performance is chosen so the optimization cost function represents the traditional LQG performance indices. Thus, the plant in a state-space form is expressed in the form

$$\dot{\mathbf{x}} = \mathbf{A}\mathbf{x} + \mathbf{B}_1\mathbf{w} + \mathbf{B}_2\mathbf{u}, \tag{9.273a}$$

$$\mathbf{z}_1 = \mathbf{C}_1\mathbf{x} + \mathbf{D}_{11}\mathbf{w} + \mathbf{D}_{12}\mathbf{u}, \tag{9.273b}$$

$$\mathbf{z}_2 = \mathbf{C}_2\mathbf{x} + \mathbf{D}_{21}\mathbf{w} + \mathbf{D}_{22}\mathbf{u}, \tag{9.273c}$$

$$\mathbf{y} = \mathbf{C}_3\mathbf{x} + \mathbf{D}_{31}\mathbf{w} + \mathbf{D}_{32}\mathbf{u}. \tag{9.273d}$$

The representation facilitates the use of the MATLAB robust control toolbox. The synthesis is demonstrated with a typical numerical example.

Consider an isotropic rectangular plate of aspect ratio 2 and assume that it is patched by an array of 25 piezo-ceramic PZT actuators. The ratio of patch thickness to the plate thickness is assumed to be 0.5. The initial model is constructed using 6×6 modes. The inputs to the patches are configured so that only the first two modes are excited and the sensor outputs are configured so that the modal amplitudes of the

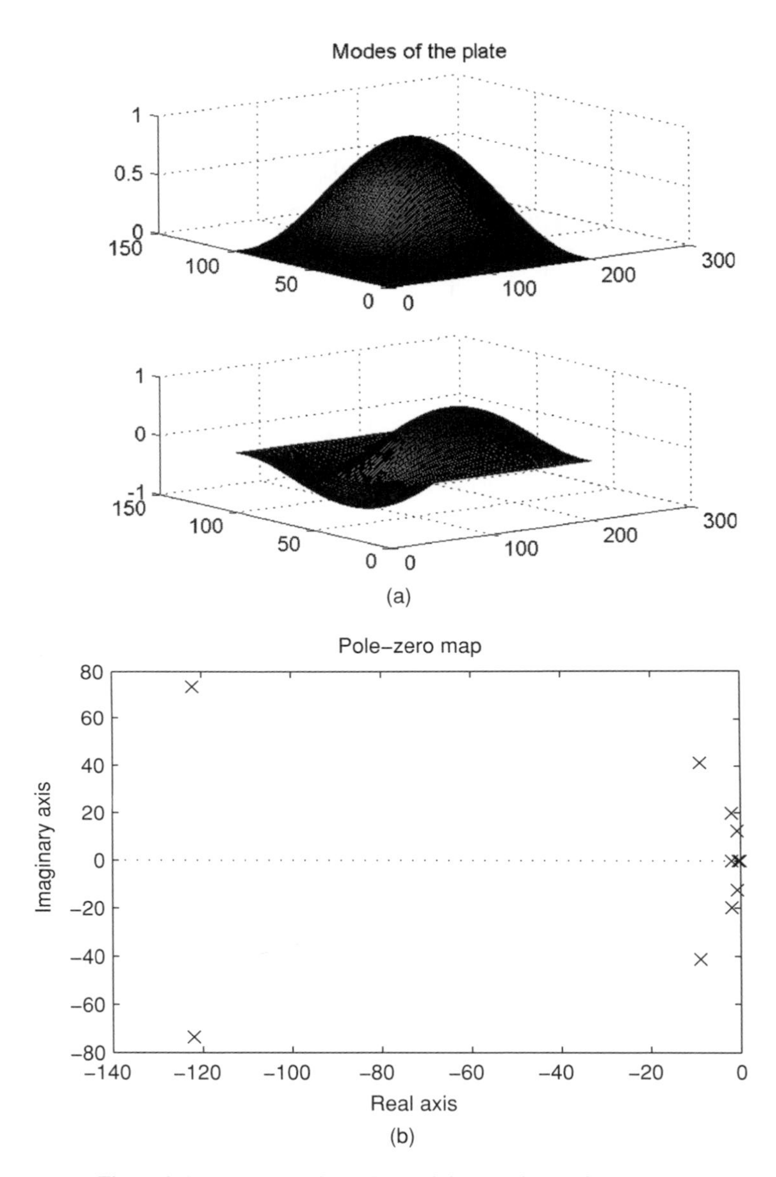

Figure 9.6 (a) Modes of the plate and (b) open-loop pole-zero map

first two modes are measured. The patch locations are chosen such that as many as 23 modes neither are excited nor contribute to the measurement, so spillover effects will be minimized.

The first two modes of the plate are illustrated in Figure 9.6(a). A reduced eighth-order state-space model is then constructed by grammian balancing of the complete model, which has 72 states. Small and equal magnitudes of viscous and structural damping are assumed in the form of Rayleigh damping. Appropriate models are assumed for the process and measurement of noise as well as for defining the H_∞ performance. The open-loop pole-zero map is illustrated in Figure 9.6(b).

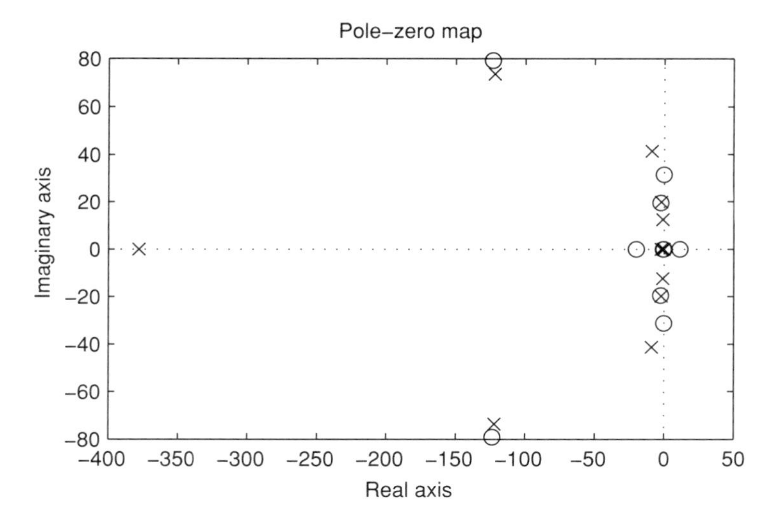

Figure 9.7 Closed-loop pole-zero map with the H_2 controller in place

The controller is designed both by H_2 and H_∞ performance optimization. The closed-loop pole-zero map with the H_2 and H_∞ controllers in place are illustrated in Figures 9.7 and 9.8, respectively.

At first sight, it seems as if both the controllers have not significantly changed the locations of the closed-loop poles. A closer examination would reveal that both the controllers introduce a significant number of zeros and additional poles, some located on the real axis and some close to the real axis. The zeros are absent in the open-loop pole-zero map. Moreover, the zeros are almost always close to the open-loop poles, indicating that the controllers function as multi-dimensional notch filters. Notch filters generally attenuate the responses at the frequencies of the notch. When the frequencies are designed to match the structural frequencies, the notch filters are particularly capable in filtering the structural

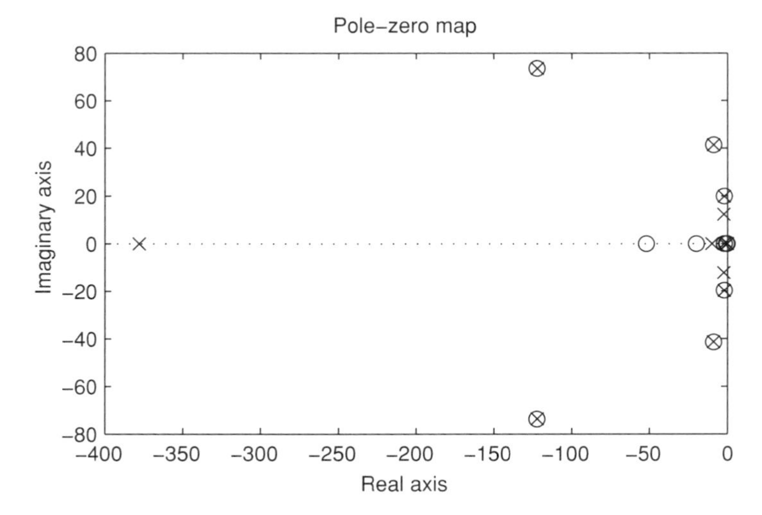

Figure 9.8 Closed-loop pole-zero map with the H_∞ controller in place

oscillatory modes. The zeros introduced by the controller are generally not too close to the poles at low frequencies and for modes with low damping. Over these frequencies, the controller acts as a band-stop filter, attenuating the response over wider bands than at higher frequencies where the attenuation has a characteristic notch structure.

Comparing the two controllers, one observes that in the case of the H_2 controller one or more of the zeros lie in the right half of the complex plane. Control filters with right half plane zeros are characterized by a non-minimum phase response and can be notoriously transparent to noise or disturbances. An interesting feature of the H_∞ controller is the absence of zeros in the right half of the complex plane. Furthermore, the characteristics of the closed-loop response with the H_∞ controller could be shaped depending on the nature of the desired performance by shaping the performance shaping filters W_S and W_T.

Although the plate was assumed to be isotropic, the controller design *m*-file was design was written, so controllers could be synthesized for any laminated composite. For most realistic composite plates with the appropriate distribution of piezoelectric patches, similar results could be obtained. Also, the initial high-order model could be set up by employing a finite element model, although in this case the model order reduction must be performed in several steps in order to ensure that the reduced-order model captures all the significant features of the structure being controlled.

Kar *et al.* (2000a,b) have discussed the bending and torsional and multi-mode vibration control of a flexible plate structure using an H_∞-based robust control law. Nonami and Sivrioglu (1996) have considered the active vibration control of a structural system using LMI-based mixed H_2/H_∞ state and output feedback control with non-linearities considered in the model.

9.4.4 Optimal Robust Stabilization of Smart Structures

To highlight the issue of robustness of the controller design, we consider the classic problem of flutter of a structural panel exposed to supersonic or hypersonic flow on one side. The supersonic or hypersonic panel flutter problem is an archetypal example of a non-linear continuous elastic system subjected to a combination of dissipative and circulatory forces (non-dissipative forces that cannot be derived from a potential energy function). There are a number of interesting features of the panel flutter problem. First, the presence of non-uniform damping amongst the normal modes due to the addition of aerodynamic requires that the limiting situations must be considered carefully. Otherwise, the so-called apparent destabilization of damping may be observed.

Second, the structural panel with aerodynamic loading possess a number of equilibrium states, thus allowing for different bifurcation modes of the equilibrium states. The reason for this feature is the fact that multiple parameters are present in the non-linear equilibrium relationships corresponding to the in-plane force resultants and the dynamic pressure of the aerodynamic flow. Thus, one or more of the buckled states could remain stable and, depending on the design requirements, this could have a significant impact on the closed-loop controller design. Recent studies of this problem have indicated that the dynamics of panel flutter could be chaotic in the domain of instability. In this section, we shall briefly discuss the design of an optimal controller for such a panel by using a finite-dimensional model and highlight the need for robustness of the closed controller.

Consider a flat rectangular panel with sides a and b and thickness h, and simply supported along all its edges. The Cartesian reference frame is such that the sides of the panel are $x = 0$ and $x = a$, and $y = 0$ and $y = a$. The panel is assumed to be performing moderately large oscillations. The governing equations for the transverse displacement $w(x, y)$ are

$$D\nabla^4 w - \frac{\partial^2 \mathbf{\Phi}}{\partial y^2}\frac{\partial^2 w}{\partial x^2} - \frac{\partial^2 \mathbf{\Phi}}{\partial x^2}\frac{\partial^2 w}{\partial y^2} + 2\frac{\partial^2 \mathbf{\Phi}}{\partial x \partial y}\frac{\partial^2 w}{\partial x \partial y} + \rho h \frac{\partial^2 w}{\partial t^2} = p_a(x, y, t)$$

$$\nabla^4 \mathbf{\Phi} + Eh\left(\frac{\partial^2 w}{\partial x^2}\frac{\partial^2 w}{\partial y^2} - \left(\frac{\partial^2 w}{\partial x \partial y}\right)^2\right) = 0, \; D = \frac{Eh^3}{12\left(1-\nu^2\right)}, \tag{9.273}$$

where D is the plate's flexural rigidity, E the Young's modulus, ν the Poisson ratio, $p_a(x, y, t)$ the distributed aerodynamic loading, h the plate thickness, ρ the plate's mass density and $\boldsymbol{\Phi}(x, y)$ the Airy stress function. The in-plane stress resultants are

$$N_{xx} = \frac{\partial^2 \boldsymbol{\Phi}}{\partial y^2}, \ N_{yy} = \frac{\partial^2 \boldsymbol{\Phi}}{\partial x^2}, \ N_{xy} = -\frac{\partial^2 \boldsymbol{\Phi}}{\partial x \partial y}. \tag{9.274}$$

The supersonic aerodynamic linearized pressure loading is assumed to be given by

$$p_a(x, y, t) = p_s + p_0(x, y, t) + \frac{2q_\infty}{\beta}\left(\frac{\partial w}{\partial x} + \frac{\beta^2 - 1}{\beta^2 U_\infty}\frac{\partial w}{\partial t}\right),$$

$$q_\infty = \rho_\infty U_\infty^2/2, \ \beta = \sqrt{M^2 - 1}, \tag{9.275}$$

where q_∞, ρ_∞ and U_∞ are the undisturbed free stream dynamic pressure, flow density and flow velocity far away from the panel, p_s is the uniform pressure acting on the panel that is only a function of time, p_0 is that component of the spatial pressure distribution that is independent of the transverse displacement and M is the free stream Mach number. In the case when $M \gg 2$, it may be assumed that $\beta \to M$ and $\beta^2 - 1 \to M^2$.

The transverse displacement is assumed to be a linear combination of pairs of assumed modes of the form

$$w_j(x, y, t) = h\left(P_{1j}(t)\sin\left(\frac{\pi x}{a}\right) + P_{2j}(t)\sin\left(\frac{2\pi x}{a}\right)\right)\sin\left(\frac{j\pi y}{b}\right), \quad j = 1, \ 2. \tag{9.276}$$

The particular solution for the corresponding Airy stress function $\boldsymbol{\Phi}_{jp}(x, y)$ is given by

$$\boldsymbol{\Phi}_{jp}(x, y) = -3D\left(1 - \nu^2\right)\mu^2\left(\bar{\boldsymbol{\Phi}}_{0p}(x, y) + \bar{\boldsymbol{\Phi}}_{jp}(x, y)\cos\left(\frac{2j\pi y}{b}\right)\right), \tag{9.277}$$

with

$$\bar{\boldsymbol{\Phi}}_{0p}(x, y) = -P_{1j}P_{2j}\cos(\alpha x) + \frac{P_{1j}^2}{8}\cos(2\alpha x) + \frac{P_{1j}P_{2j}}{14}\cos(3\alpha x) + \frac{P_{2j}^2}{32}\cos(4\alpha x)$$

and

$$\bar{\boldsymbol{\Phi}}_{jp}(x, y) = \frac{P_{1j}^2 + 4P_{2j}^2}{8\mu^4} + \frac{9P_{1j}P_{2j}}{\left(1 + 4\mu^2\right)^2}\cos(\alpha x) - \frac{P_{1j}P_{2j}}{\left(9 + 4\mu^2\right)^2}\cos(3\alpha x),$$

where

$$\alpha = \pi/a \text{ and } \mu = ja/b.$$

Proceeding in exactly in the same manner as in Section 9.4.1, one could derive the governing equations for the dynamics of the model amplitudes P_{1j} and P_{2j}. These are given by

$$\frac{d^2 P_{1j}}{d\tau^2} + \sqrt{\frac{\gamma\lambda}{M}}\frac{dP_{1j}}{d\tau} + \left(M_1 + N_1 P_{1j}^2 + QP_{2j}^2\right)P_{1j} - \frac{8\lambda}{3}P_{2j} = \frac{8}{\pi^2 j}\left((-1)^j - 1\right)\bar{p}_s + p_{10}, \tag{9.278a}$$

$$\frac{d^2 P_{2j}}{d\tau^2} + \sqrt{\frac{\gamma\lambda}{M}}\frac{dP_{2j}}{d\tau} + \left(M_2 + N_2 P_{2j}^2 + QP_{1j}^2\right)P_{2j} + \frac{8\lambda}{3}P_{1j} = p_{20}, \tag{9.278b}$$

where

$$\tau = \alpha^2 t\sqrt{\frac{D}{\rho h}},\ \gamma = \frac{\rho_\infty a}{\rho h},\ \lambda = \frac{2qa^3}{\pi^4 \beta D},\ \bar{p}_s = \frac{p_s}{Dh\alpha^4},$$

$$p_{i0}(t) = \frac{4}{D\alpha^4 h^2 ab}\int\limits_S p_0(x, y, t)\frac{\partial w}{\partial C_{ij}}\mathrm{d}x\mathrm{d}y \quad i = 1,\ 2.$$

$$M_k = \left(k^2 + \mu^2\right)^2 - \left(k^2 N_{xx} + \mu^2 N_{yy}\right),\ N_k = 12\left(1 - \nu^2\right)\left(k^4 + \mu^4\right)/16,$$
$$k = 1,\ 2,\ N_{xx} = \bar{N}_{xx}/D\alpha^2,\ N_{yy} = \bar{N}_{yy}/D\alpha^2$$

and

$$Q = 12\left(1 - \nu^2\right)\left\{\frac{\left(1 + \mu^4\right)}{4} + \frac{8\mu^4}{16\left(1 + 4\mu^2\right)^2} + \frac{\mu^4}{16\left(9 + 4\mu^2\right)^2}\right\}.$$

The very special structure of the above equations may be noted. This is due to the fact that the assumed loading is linear and, although it is not conservative in nature, the non-linear terms are entirely from the dynamics of the plate structure, which by itself is a conservative system. First, it is possible in principle to patch the panel with a distribution of PZT actuators so that both modes are independently controlled. Moreover, when the non-linear terms are ignored, one could construct a optimal controller based on the LQR method. This would correspond to a particular solution of a Riccati equation, which could also be considered to be a Lyapunov function. If one could modify this Lyapunov function so that the complete non-linear system is adequately stable, then a robust controller that is capable of delivering the desired closed-loop performance could be successfully designed. This is the underpinning principle of the design of robust control system, when applied to an uncertain system.

In this situation, it is possible to exploit the special features of the non-linear equations and design a robust controller by standard methods. Consider the governing equations for the amplitudes P_{1j} and P_{2j}, and express them as

$$\frac{\mathrm{d}^2 P_{1j}}{\mathrm{d}\tau^2} + \sqrt{\frac{\gamma\lambda}{M}}\frac{dP_{1j}}{d\tau} + M_1 P_{1j} + K_j\left(N_1 + Q\kappa_j\right)P_{1j} - \frac{8\lambda}{3}P_{2j} = \frac{8}{\pi^2 j}\left((-1)^j - 1\right)\bar{p}_s + p_{10}, \quad (9.279a)$$

$$\frac{\mathrm{d}^2 P_{2j}}{\mathrm{d}\tau^2} + \sqrt{\frac{\gamma\lambda}{M}}\frac{dP_{2j}}{d\tau} + M_2 P_{2j} + K_j\left(N_2\kappa_j + Q\right)P_{2j} + \frac{8\lambda}{3}P_{1j} = p_{20}, \quad (9.279b)$$

where

$$K_j = P_{1j}^2 \text{ and } \kappa_j = \left|P_{2j}/P_{1j}\right|^2$$

are both positive. If one assumes that the structure is in a state of near-normal mode operation or operating with an almost constant modal amplitude ratio, the ratio κ_j may be determined from the ratio of the modes of the linear small-amplitude vibrations in the vicinity of the operating point. The problem is now transformed into the case of a linear parameter varying system. Moreover, one could now design a feedback controller using a distribution of PZT patches, assuming that it is possible to excite the first two modal amplitudes P_{1j} and P_{2j}.

Thus, the equations for designing the controller may be expressed as

$$\frac{\mathrm{d}^2 P_{1j}}{\mathrm{d}\tau^2} + \sqrt{\frac{\gamma\lambda}{M}}\frac{\mathrm{d}P_{1j}}{\mathrm{d}\tau} + M_1 P_{1j} + K_j\left(N_1 + Q\kappa_j\right)P_{1j} - \frac{8\lambda}{3}P_{2j} = u_1, \quad (9.280a)$$

$$\frac{\mathrm{d}^2 P_{2j}}{\mathrm{d}\tau^2} + \sqrt{\frac{\gamma\lambda}{M}}\frac{\mathrm{d}P_{2j}}{\mathrm{d}\tau} + M_2 P_{2j} + K_j\left(N_2\kappa_j + Q\right)P_{2j} + \frac{8\lambda}{3}P_{1j} = u_2, \quad (9.280b)$$

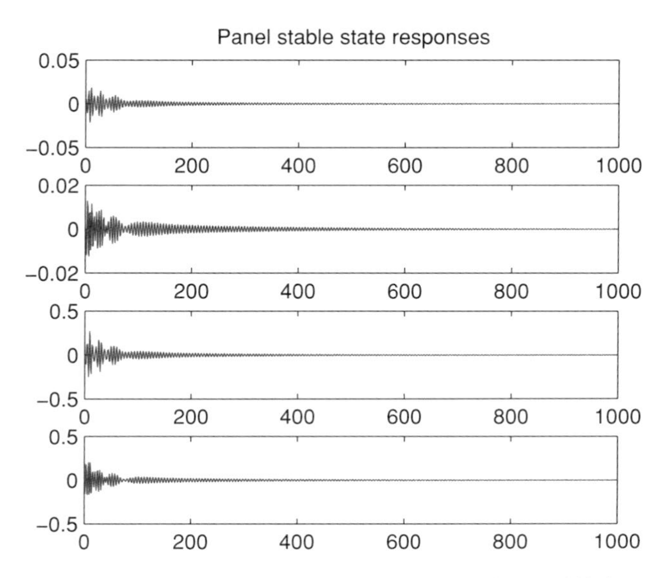

Figure 9.9 Typical asymptotically stable panel state responses. The time frame (1000 time steps) corresponds to 0.92 s

where u_1 and u_2 are the control inputs to the two degrees of freedom. Typical open-loop state responses for the case of asymptotically stable rectangular panel in supersonic flow ($M = 5$), with uniform, light, in-plane loading around the panel, are shown in Figure 9.9. When the loading is increased to a critical value such that the panel is in a state of neutral stability, the corresponding state responses in non-dimensional time P_{1j}, P_{2j}, $\mathrm{d}P_{1j}/\mathrm{d}\tau$ and $\mathrm{d}P_{2j}/\mathrm{d}\tau$ are shown in Figure 9.10.

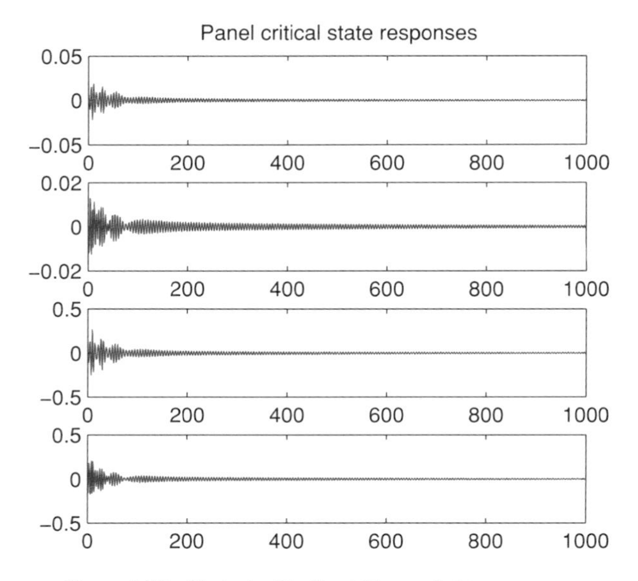

Figure 9.10 Typical critically stable panel state responses

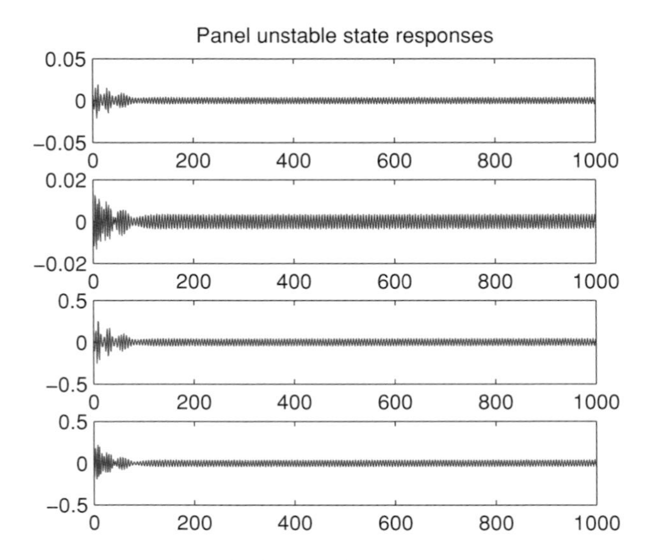

Figure 9.11 Typical unstable panel state responses

When the loading is increased beyond the critical value, the corresponding state responses, which have a limit-cycle-like structure, are shown in Figure 9.11. An LQR is designed for $K_j = 0.0005$ and $\kappa_j = 0.2$, for the typical unstable case shown in Figure 9.11. The negative feedback control gains are given by

$$K_{cl} = \begin{bmatrix} 0.0399 & 0.4821 & 0.5936 & 0.1342 \\ -0.1626 & 0.0377 & 0.1342 & .5917 \end{bmatrix}. \tag{9.281}$$

The feedback control gains are all relatively reasonable in magnitude. The corresponding closed-loop responses are shown in Figure 9.12. Because $K_j = 0.0005$ is reasonably small, the robustness of the LQR could be established and verified numerically. The control gains indicate that the implementation of the closed-loop controller is a feasible exercise and therefore it is possible to suppress the occurrence of panel flutter at the design point and ensure adequate closed-loop performance. The actual implementation of the controller by using a distribution of PZT actuators is not discussed here but can be relatively easily realized in the same manner as discussed in earlier sections. The distribution of actuators and sensors used must be configured to avoid control and observation spillover. This is essential to ensure that closed-loop system has the desired features of robustness.

Before concluding this section, it is worth mentioning a number of computations that were not performed here and should be done prior to implementation of such a controller. It is often necessary to assess and validate the closed-loop controllers by including realistic, additional dynamic modes of the structure. The performance of the controller can be demonstrated by considering the full dynamic response of the system. The results may be plotted in the frequency domain by employing such tools as Bode plots and Nyquist diagrams. This aspect was deliberately avoided as the background material for the analysis and interpretation of stability margins such as the gain and phase margin, as obtained from such plots, was not covered in Chapter 3. These aspects, which are related to the validation of control system designs, are generally covered to a greater depth in textbooks dedicated to control system design and validation.

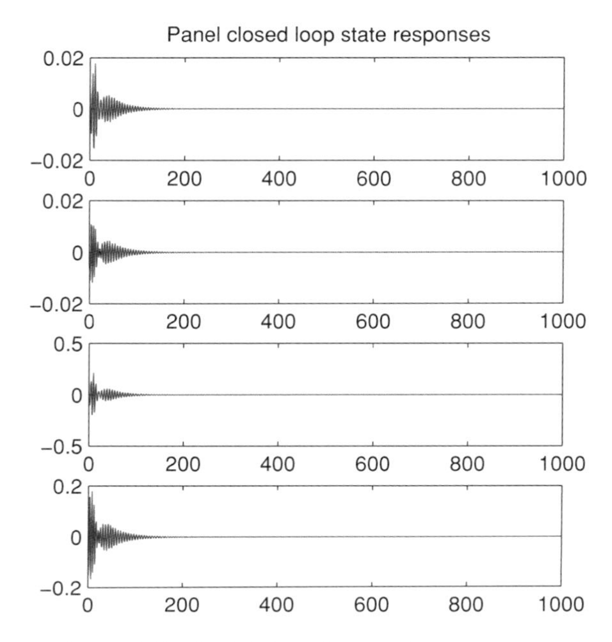

Figure 9.12 Typical closed-loop panel state responses

9.5 Design of an Active Catheter

While there are a number of applications of smart structures and smart control of structural vibrations and instabilities such as the active control of machine tool chatter, shape control of an aerofoil, shape control of a heart valve, flow control over aerofoils to minimize drag and control and stabilization of bio-dynamic fluid flows, in this section the application of the theory developed to the design of an active catheter is briefly discussed. Smart structures are already employed in the design of a variety of cardio-vascular applications, and the design of active catheters may be considered as an emerging technological application, as several prototype catheters are currently under development. Yet there is still considerable room for improvements in the proposed designs of active catheters and in innovation, as there are still a number of design issues that must be addressed.

A catheter is a thin flexible plastic tube inserted into the body to permit introduction or withdrawal of fluids or implantable devices. The catheter is particularly useful to a cardiologist in the diagnosis and treatment of various cardiac conditions. During angioplasty, a cardiologist inserts a catheter directly into an artery, gently working its way up to the heart. The cardiologist is able to visualize the catheter by concurrent magnetic resonance imaging (MRI) scans and, through this procedure, the heart, blood vessels and pressure, and any blockages or anomalies. Subsequently, he/she can clear the passgeways in arteries or expand them using a ballooning stent, which is inserted via the catheter. It is useful to have a catheter that is automatically tracked and also controlled so that it can be targeted to specific locations that can be visualized on the MRI scan.

In designing an active catheter, a number of key design issues must first be addressed. The first is the issue of providing for automatic tracking. Continuous tracking of both the position and the orientation and visualization of the catheter in real time is essential. The second is the provision of actuation, the method actuation, the location and number of actuators, and the authority. The operator must retain control of the whole automated process, but there is the issue of the extent of the operator's authority and the extent to which the automatic control system should be able to exercise its authority during the deployment of the catheter. Many active catheter designs are currently under development, and these employ a range of actuation methods such as pneumatic (compressed air), micro-electro mechanical actuators, SMAs and ionic-polymer metal composites as they possess attractive characteristics including large strain, low operation voltage and the potential for miniaturization. Most of these actuators are plagued by non-linear hysteresis, and the control system must be designed to compensate for it. In some of these designs, the whole of the inserted catheter may be actively controlled, while in others it is just the tip of the catheter. In such designs, complete automatic control of motion in two or even in three dimensions is a desirable feature.

Tracking of an inserted catheter may be performed by incorporating three or more inductively coupled radio-frequency (RF) micro-coils in the body of the catheter and using a special pulse sequence to localize the coils in 3D space. The operation is similar to a GPS navigation system used in aircraft or cars. Thus, when multiple coils are incorporated, one can determine the orientation of the device by localizing the three coils, followed by an automatic scan plane adjustment. However, including micro-coils in the body of the catheter alters its flexibility, which is not often desirable. An alternative to using RF coils is to use image-based navigation techniques.

To design the controller, it is best to simulate the open- and closed-loop dynamics of the catheter. The finite element-based modelling of the catheter may be done by employing a curved beam finite element (see exercise 3 of Chapter 7). Cubic spline-based curve fitting techniques may be used to recover the geometry of each element from the images and compared with the desired geometry based on the operator's commands. The closed-loop control may be exercised by a continuous 'snake' actuator. Such automatically controlled catheters with appropriate image-based tracking are currently under development, and several such devices should be available quite commonly in the very near future.

9.6 Modelling and Control of Machine Tool Chatter

Continuous structural systems with an infinite number of degrees of freedom could be modelled by a set of discrete masses, interconnected or supported by a set of springs and with a very large number of degrees of freedom. One classic feature of a class of structural systems is the fact that they are characterized by component sub-systems with identical natural frequencies. Thus, when coupled together they vibrate with a large number of natural frequencies. Rather than being dispersed from each other, the natural frequencies are essentially close to each other, but the normal modes of vibration corresponding to each of them are substantially different. Thus, they have all the characteristic features of band-pass electrical networks.

The classical method of modelling structures with repeated identical structural components and periodic excitation is that they lead to systems of differential with periodic coefficients. Such systems cannot be dealt with as systems with constant coefficients but only as systems with periodic coefficients. The analysis is generally based on the theory of Floquet (see also Section 6.6.2) and naturally leads to wave-like solutions, with wavelengths related to the characteristic distance between two repeated components. When additional conditions are imposed at an interface boundary, the wave-like motions will reflect to form new wave-like envelopes of the vibration modes. The interactions of the envelope motion with the natural modes or imposed vibration can either stabilize or destabilize the structure and must be analysed carefully.

Such distributed systems also often exhibit certain features of periodicity in the mass and stiffness properties that could be efficiently exploited to compute the natural frequencies and the normal mode shapes. Typical examples are the vibration of turbo-machinery such as a mid-shrouded forward fan and low-pressure compressor blade–rotor assemblies, end-shrouded blade–rotor assemblies and other similar systems and assemblies. An approach based on *wave-based control* is one recently evolving method of

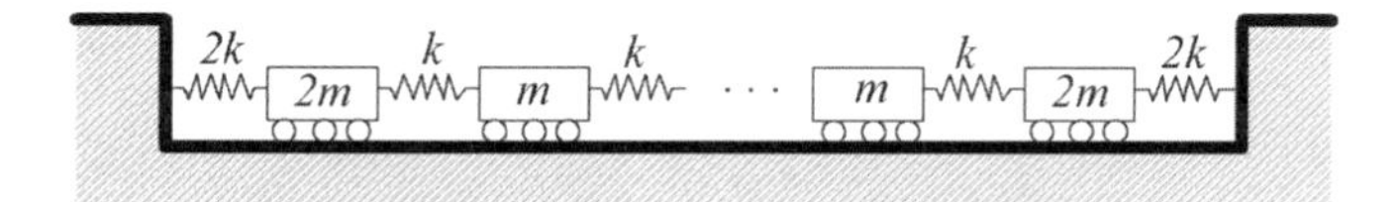

Figure 9.13 A continuous structure modelled by a large number of discrete masses and springs

control synthesis for such systems with repeated components, and the reader is referred to the paper by O'Connor and Fumagalli (2009). The wave-based approach interprets the motion of the structure as two-way mechanical waves, which enter and leave the structure. Those entering and leaving the structure at the actuator system interface are particularly relevant. Measuring the two-way wave motion at the actuator–structure interface and then controlling it by imposing additional forces and motion generated by the actuator are the bases for the design of a wave-based controller. The strategy is quite similar in principle to that adopted in transmission lines by attaching wave-absorbing impedances at certain points along the transmission line. Thus, the wave-based control is based on using an actuator capable of launching another wave, which along with its reflection attenuates the uncontrolled structural vibration. Another approach of controller synthesis for systems with repeated components relies on clustering modes with similar phase and is referred to as *cluster control*, as discussed by Tanaka (2009). The cluster control approach is particularly suitable for attenuating acoustic response where a response to an acoustic disturbance is cancelled by superposing an out-of-phase signal generated by a controller. Such control systems generally require the use of feed-forward controllers, which have not been discussed in this text. One common method of analysis of systems with repeated identical components is illustrated with a typical example.

Consider the system illustrated below with a large number of particles with equal masses and connected together by springs to two spring–mass systems attached to the two end walls, as shown in Figure 9.13. The end masses have twice the mass of each of the other masses, which are equal. The springs at both ends are twice as stiff as each of the other springs, which are equal to each other.

The equations of motion are

$$m\begin{bmatrix} 2 & 0 & 0 & 0 & \cdots & 0 & 0 \\ 0 & 1 & 0 & 0 & \cdots & 0 & 0 \\ \cdots & \cdots & \cdots & \cdots & & & \\ 0 & 0 & 0 & 0 & \cdots & 1 & 0 \\ 0 & 0 & 0 & 0 & \cdots & 0 & 2 \end{bmatrix}\begin{bmatrix} \ddot{x}_1 \\ \ddot{x}_2 \\ \cdots \\ \ddot{x}_{n-1} \\ \ddot{x}_n \end{bmatrix}$$

$$+k\begin{bmatrix} 3 & -1 & 0 & 0 & 0 & \cdots & 0 & 0 \\ -1 & 2 & -1 & 0 & 0 & \cdots & 0 & 0 \\ \cdots & \cdots & \cdots & \cdots & \cdots & \cdots & \cdots & \\ 0 & 0 & 0 & 0 & \cdots & -1 & 2 & -1 \\ 0 & 0 & 0 & 0 & \cdots & 0 & -1 & 3 \end{bmatrix}\begin{bmatrix} x_1 \\ x_2 \\ \cdots \\ x_{n-1} \\ x_n \end{bmatrix} = \begin{bmatrix} 0 \\ 0 \\ \cdots \\ 0 \\ 0 \end{bmatrix}, \tag{9.282}$$

which may be expressed in scalar form as

$$2m\ddot{x}_1(t) + 3kx_1(t) - kx_2(t) = 0, \tag{9.283a}$$

$$m\ddot{x}_r(t) + 2kx_r(t) - kx_{r-1}(t) - kx_{r+1}(t) = 0, r = 2, 3, \cdots, n-1 \tag{9.283b}$$

and

$$2m\ddot{x}_n(t) + 3kx_n(t) - kx_{n-1}(t) = 0. \tag{9.283c}$$

Assuming normal mode oscillations, we have

$$\left[-2m\omega^2 + 3k\right]x_1 - kx_2 = 0 \tag{9.284}$$

Hence, it follows that

$$x_1 = \frac{k}{3k - 2m\omega^2} x_2 = \frac{1}{3 - 2\dfrac{\omega^2}{\omega_0^2}} x_2, \tag{9.285}$$

where $\omega_0^2 = k/m$. If we denote $\omega^2/\omega_0^2 = \lambda$,

$$x_1 = \frac{1}{3 - 2\lambda} x_2. \tag{9.286}$$

Similarly, the last equation may be written as

$$x_n = \frac{1}{3 - 2\lambda} x_{n-1}. \tag{9.287}$$

The remaining equations are

$$\left[-m\omega^2 + 2k\right] x_r - kx_{r+1} - kx_{r-1} = 0. \tag{9.288}$$

Hence,

$$x_{r+1} - \left(2 - \frac{\omega^2}{\omega_0^2}\right) x_r + x_{r-1} = 0 \tag{9.289}$$

or

$$x_{r+1} - (2 - \lambda) x_r + x_{r-1} = 0, r = 2, 3, \cdots, n. \tag{9.290}$$

If we introduce the shift operator

$$\mathbf{E}x_r = x_{r+1}, \tag{9.291}$$

we may rewrite the above equation as

$$\left(\mathbf{E}^2 - (2 - \lambda)\mathbf{E} + 1\right) x_{r-1} = 0. \tag{9.292}$$

This equation is satisfied if

$$\left(\mathbf{E}^2 - (2 - \lambda)\mathbf{E} + 1\right) = 0 \quad \text{or} \quad \text{if } x_{r-1} = 0. \tag{9.293}$$

The latter solution is a trivial solution and is not considered. The former is the desired characteristic equation

$$\mathbf{E}^2 - (2 - \lambda)\mathbf{E} + 1 = 0. \tag{9.294}$$

The roots of this equation are

$$\mathbf{E}_{1,2} = 1 - \frac{\lambda}{2} \pm i\sqrt{1 - \left(1 - \frac{\lambda}{2}\right)^2} = \cos(\theta) \pm i\,\sin(\theta), \tag{9.295}$$

where

$$\cos(\theta) = 1 - \frac{\lambda}{2} \quad \text{or} \quad \lambda = 1 - 2\cos(\theta).$$

Thus, the roots $\mathbf{E}_{1,2}$ may be expressed as $\mathbf{E_1} = \exp(\theta)$ and $\mathbf{E}_2 = \exp(-\theta)$.
The general solution for the difference equation may be written as

$$x_r = \mathrm{B}\alpha^r, \tag{9.296}$$

where α satisfies the equation $\alpha^2 - (2 - \lambda)\alpha + 1 = 0$. Since this equation is the same as the equation for **E,** the solution for xr may be written as

$$x_r = \mathrm{B}_1 \cos(r\theta) + \mathrm{B}_2 \sin(r\theta). \tag{9.297}$$

Since this solution must also satisfy the first and the last equation, one may substitute it in the first and last equation and obtain

$$\mathrm{B}_1 \left(\cos(\theta) - \bar{\lambda}\cos(2\theta)\right) + \mathrm{B}_2 \left(\sin(\theta) - \bar{\lambda}\sin(2\theta)\right) = 0 \tag{9.298a}$$

and

$$\mathrm{B}_1 \left(\cos(n\theta) - \bar{\lambda}\cos((n+1)\theta)\right) + \mathrm{B}_2 \left(\sin(n\theta) - \bar{\lambda}\sin((n+1)\theta)\right) = 0, \tag{9.298b}$$

where $\bar{\lambda} = 1/(3 - 2\lambda)$. Since both these equations must be satisfied, the determinant

$$\begin{vmatrix} \left(\cos(\theta) - \bar{\lambda}\cos(2\theta)\right) & \left(\sin(\theta) - \bar{\lambda}\sin(2\theta)\right) \\ \left(\cos(n\theta) - \bar{\lambda}\cos((n+1)\theta)\right) & \left(\sin(n\theta) - \bar{\lambda}\sin((n+1)\theta)\right) \end{vmatrix} = 0. \tag{9.299}$$

The above determinant is a transcendental equation in θ, which must be solved numerically for a given n. However, if n is large then this equation reduces to

$$\sin(n\theta) = 0. \tag{9.300}$$

Hence, $n\theta = j\pi, j = 1, 2, 3, \ldots, n$ and

$$\lambda = 2\left(1 - \cos\left(\frac{j\pi}{n}\right)\right), \; j = 1, 2, 3 \cdots n. \tag{9.301}$$

Thus, the squares of the natural frequencies are

$$\omega_n^2 = 2\frac{k}{m}\left(1 - \cos\left(\frac{j\pi}{n}\right)\right), \; j = 1, 2, 3 \cdots n. \tag{9.302}$$

Since

$$\mathrm{B}_2 = -\mathrm{B}_1 \frac{\left(\cos\left(\dfrac{j\pi}{n}\right) - \dfrac{1}{3-2\lambda}\cos\left(2\dfrac{j\pi}{n}\right)\right)}{\left(\sin\left(\dfrac{j\pi}{n}\right) - \dfrac{1}{3-2\lambda}\sin\left(2\dfrac{j\pi}{n}\right)\right)}, \tag{9.302}$$

the mode shapes are given by

$$x_r = \mathrm{B}_1 \left[\cos(r\theta) - \frac{\left(\cos\left(\dfrac{j\pi}{n}\right) - \dfrac{1}{3-2\lambda}\cos\left(2\dfrac{j\pi}{n}\right)\right)}{\left(\sin\left(\dfrac{j\pi}{n}\right) - \dfrac{1}{3-2\lambda}\sin\left(2\dfrac{j\pi}{n}\right)\right)}\sin(r\theta)\right]. \tag{9.303}$$

In the case of $n = 5$, the above formula for the natural frequencies gives

$$\omega_n^2 = 2\omega_0^2\left(1 - \cos(j36^\circ)\right), \; j = 1, 2, 3, 4, 5. \tag{9.304}$$

Thus, $\omega_1^2 = 0.382\ \omega_0^2, \omega_2^2 = 1.382\ \omega_0^2, \omega_3^2 = 2.618\ \omega_0^2, \omega_4^2 = 3.618\ \omega_0^2$ and $\omega_5^2 = 4\ \omega_0^2$. The exact frequencies are $\omega_1^2 = 0.3515\ \omega_0^2, \omega_2^2 = \omega_0^2, \omega_3^2 = 1.6\ \omega_0^2, \omega_4^2 = 2.5\ \omega_0^2$ and $\omega_5^2 = 3.5419\ \omega_0^2$. Thus, the approximate frequencies are upper bounds for the exact frequencies.

9.6.1 Stability Analysis of Machine Tool Chatter

Machine tool chatter is the regenerative vibration associated with machine tools involving a cutting tool and the rotating workpiece. It is an unstable relative vibration between the workpiece and the cutting tool and is a frequent problem in machining operations. It occurs because of the presence of large cutting forces at the interface between the cutting tool and workpiece as well as lagged feedback from the workpiece to the cutting tool. These cutting tool forces have all the features of the output of a mechanical band-pass filter of the type discussed in the preceding section. Since machine tool chatter is extremely detrimental to the machine tool, as it not only promotes damage and tool wear but also adversely affects the surface finish of the workpiece, many methods have been suggested to mathematically model and analyse its occurrence. Based on these models and analysis, methods of eliminating chatter by controlling the regenerative feedback loops have been tried and tested. In this section we propose a simple model of the phenomenon based on the representation of the cutting tool forces as a signal involving time lags and apply the techniques introduced in the last section to approximately determine the natural frequencies of vibration. Several non-linear effects such as backlash and hysteresis, which are known to be present, are ignored in this simple analysis.

The equation of motion is based on a simple mass, spring and damper model excited by lagged feedback. Thus, it is represented as

$$m\ddot{x}(t) + c\dot{x}(t) + kx(t) = kb\left(x\left(t - T\right) - x(t)\right), \tag{9.305}$$

where $T = 2\pi/\Omega_r$, m, c and k refer to the mass damping and stiffness of the deformable component in the cutting process, x is the displacement of that deformation, b is a dimensionless ratio relating the stiffness associated with the depth of metal removal to the stiffness of the cutting tool, T is the time delay associated with the delayed feedback associated with workpiece and Ω_r is the tool passing frequency. To reduce the equation of motion to a standard form, we introduce a dimensionless time variable given by

$$\tau = \Omega_r t, \tag{9.306}$$

and resulting equation of motion is

$$y''(\tau) + \zeta y'(\tau) + \omega^2 y(\tau) = \omega^2 b\left(y\left(\tau - 2\pi\right) - y\left(\tau\right)\right), \tag{9.307}$$

where

$$\omega^2 = \frac{k}{m\Omega_r^2} = \frac{1}{\Omega^2},\ y\left(\tau\right) = x(t),\ \zeta = \frac{c}{2\sqrt{mk}} \quad \text{and} \quad ()' = \frac{\mathrm{d}}{\mathrm{d}\tau}.$$

The new independent coordinate τ is the revolution angle and the dimensionless rotation speed is $\Omega = 1/\omega$. We also define a lag operator Δ_T by the equation $\Delta_T f\left(\tau\right) = f\left(\tau - T\right)$ and a differential operator D by the equation

$$D = \frac{\mathrm{d}}{\mathrm{d}\tau} = ()'. \tag{9.308}$$

Thus, in terms of these operators the equation of motion is

$$\left[D^2 + \zeta D + \omega^2 - \omega^2 b\left(\Delta_{2\pi} - 1\right)\right] y\left(\tau\right) = 0. \tag{9.309}$$

If we let $y\left(\tau\right) = \mathrm{Re}\left(\exp\left[i\theta\tau\right]\right)$, the equation reduces to

$$\left[\theta^2 + i\zeta\theta + \omega^2 - \omega^2 b\left(\exp\left[i2\pi\theta\right] - 1\right)\right] \times \mathrm{Re}\left(\exp\left[i\theta\tau\right]\right) = 0. \tag{9.310}$$

To solve equation (9.309), we expand $y\left(\tau\right)$ in a Fourier series given by

$$y\left(\tau\right) = \mathrm{Re}\left(\exp\left[i\theta\tau\right] \sum_{n=-N}^{n=N} C_n \exp\left[i\frac{n\tau}{2}\right]\right), \tag{9.311}$$

where N is chosen to ensure that the solution is of adequate accuracy. Substituting the solution in equation (9.311) for $y(\tau)$ and setting the sum of all the coefficients of like harmonics to be zero results in a set of $2N+1$ linear homogeneous algebraic equations given by

$$C_n F\left(\theta + \frac{n}{2}\right) = 0, n = -N, \ldots, 0, \ldots, N, \tag{9.312}$$

where

$$F(\theta) = \left[\theta^2 + i\zeta\theta + \omega^2 - \omega^2 b\left(\exp\left[i2\pi\theta\right] - 1\right)\right]. \tag{9.313}$$

Thus, the system of equations is uncoupled and the characteristic frequencies are obtained by setting the coefficient of C_n to zero. Thus, the characteristic roots are obtained by solving the equation

$$F\left(\theta + \frac{n}{2}\right) = 0, n = -N, \ldots, 0, \ldots, N, \tag{9.314}$$

which is equivalent to solving the equation

$$F(\theta) = 0$$

or

$$F(\theta) = \left[\theta^2 + i\zeta\theta + \omega^2 - \omega^2 b\left(\exp\left[i2\pi\theta\right] - 1\right)\right] = 0. \tag{9.315}$$

Because this equation is not really a polynomial equation, it has an infinity of roots. It follows that the system has infinite degrees of freedom.

Assuming that the parameter b is small, we may find the principal natural frequencies as $b \to 0$. Thus, the characteristic equation is written as

$$\theta^2 + i\left(\zeta\theta - \omega^2 b\sin(2\pi\theta)\right) + \omega^2 - \omega^2 b\left(\cos(2\pi\theta) - 1\right) = 0 \tag{9.316}$$

and further approximations of $\sin(2\pi\theta)$ and $\cos(2\pi\theta) - 1$ must be made to obtain the roots of the equation.

An alternative approach is to approximate the time delay term by the Padé approximant given by

$$\exp\left[i\theta T\right] \approx \frac{4 + i2T\theta + (iT\theta)^2}{4 - i2T\theta + (iT\theta)^2}, \tag{9.317}$$

where T is the time delay.

For our purposes, we shall assume that $\sin(2\pi\theta) \approx 2\pi\theta$ and $\cos(2\pi\theta) - 1 \approx 0$. It follows that the frequency equation is

$$\theta^2 + i\left(\zeta - 2\pi\omega^2 b\right)\theta + \omega^2 = 0. \tag{9.318}$$

Observe the similarity between equations (9.318) and (9.294). The frequency shows that the system oscillates harmonically when

$$\omega^2 = \frac{k}{m\Omega_r^2} = \frac{\zeta}{2\pi b}, \tag{9.319}$$

i.e. when the rotation speed is

$$\Omega_r = \sqrt{\frac{2\pi bk}{m\zeta}}. \tag{9.320}$$

Although approximate, the equation gives a practical estimate for the rotation speed at which there is a possibility of the onset of chatter vibration and is based on a formula presented in a recent paper by Minis and Yanushevsky (1993) on the subject. A discussion of the conditions under which the formula is applicable is well beyond the scope of this section and may be found in a paper on the subject of machine tool chatter by Segalman and Butcher (2000). A more extensive analysis generally involves the generation of a stability lobe diagram for the dynamic system depicting a set of rotation speeds at which chatter is significant and the inclusion of the non-linear effects of backlash and hysteresis as well as additional broadband periodic excitation. The absence of chatter is then assessed from the stability lobes diagram.

9.6.2 Feedback Control of Machine Tool Chatter

For purposes of controller synthesis, the dynamic model may be re-stated in terms of the dynamic forces acting on the machine tool in the form

$$\mathbf{F}(t) = (1/2)aK_t\mathbf{AG}(D)\left(\mathbf{F}(t) - \mathbf{F}(t - T)\right) + \mathbf{F}_c(t) \tag{9.321}$$

or as

$$\mathbf{F}(t) = (1/2)aK_t\mathbf{AG}(D)\left(\mathbf{I} - \exp(-DT)\right)\mathbf{F}(t) + \mathbf{F}_c(t), \tag{9.322}$$

where $\mathbf{F}(t)$ is the dynamic force vector acting on the cutting tool, a is the axial depth of cut, K_t is the tangential cutting force coefficient, $\mathbf{A}(t)$ is the immersion-dependent directional cutting force coefficient matrix, which could in general be a periodic function of time satisfying the condition $\mathbf{A}(t + T) = \mathbf{A}(t)$, $\mathbf{G}(D)$ is the direction-dependent and frequency-dependent transfer function relating the static and dynamic cutting force vector, T is the inter-tooth time of passage and $\mathbf{F}_c(t)$ is the control force vector acting on the cutting tool. Equation (9.322) is analogous to equation (9.305), but in terms of the force vector. Further information on the formulation of the force equations for a generic case, including the application of Floquet theory to machine tool chatter, may be found in the paper by Minis and Yanushevsky (1993).

In many applications to machine tool chatter, the immersion-dependent directional cutting force coefficient matrix $\mathbf{A}$ may be approximated by a constant coefficient matrix. The problem can be reformulated as a closed-loop control problem and the control law defining the control force vector acting on the cutting tool $\mathbf{F}_c(t)$ may be synthesized using standard techniques of control law synthesis. Thus, if the control law takes the form

$$\mathbf{F}_c(t) = -\mathbf{K}_{cl}(D)\mathbf{F}(t),$$

it follows that

$$\mathbf{F}(t) = (1/2)aK_t\left(\mathbf{I} + \mathbf{K}_{cl}(D)\right)^{-1}\mathbf{AG}(D)\left(1 - \exp(-DT)\right)\mathbf{F}(t).$$

The open- and closed-loop characteristic equations may be, respectively, expressed as

$$\det\left|\mathbf{I} - (1/2)aK_t\left(1 - \exp(-DT)\right)\mathbf{AG}(D)\right| = 0$$

and

$$\det\left|\mathbf{I} - (1/2)aK_t\left(1 - \exp(-DT)\right)\left(\mathbf{I} + \mathbf{K}_{cl}(D)\right)^{-1}\mathbf{AG}(D)\right| = 0.$$

The control law synthesis problem now reduces to finding a suitable control law $\mathbf{K}_{cl}(D)$ to eliminate chatter within the operational speed range of the machine tool. In practice, the control force vector acting on the cutting tool $\mathbf{F}_c(t)$ is generated by a distribution of force actuators, such as piezoelectric ceramic (PZT) film actuators, attached to the cutting tool. The objective is for the control force vector to generate relative motion between the cutting tool and workpiece so as to compensate and reduce the open-loop relative vibration. Typically, a closed-loop stability lobes diagram is numerically established. The effectiveness of the controller is then assessed on the basis of the stability lobes diagram. A complete discussion of the synthesis of the feedback controller for suppressing regenerative chatter is beyond the scope of this chapter, and more details may be found in the literature on the active control of chatter in machine tools.

Exercises

1. The concept of an elastica was first put forward by Bernoulli in the 17th century and extended by Euler in 1744. The concept of an elastica refers to the modelling of a thin wire-like structure by an elastic curve, the elastica, capable of non-linear large-scale deflections and rotations. The solution for the curve of the elastica is particularly useful in studying buckling and other related stability problems.

Its formulation in a plane involves two geometrical constraint equations, two equations for the force equilibrium in two orthogonal directions, an equation for the moment equilibrium and a final constitutive equation relating the applied bending moment to the slope (Plaut and Virgin, 2009). In this example, we shall apply the elastica to a control problem with a number of practical applications.

Consider a uniform, thin, flexible, inextensible, elastic strip, with length S, bending rigidity EI, cross-sectional area A, mass per unit length m, weight per unit length mg and mass moment of inertia J. The elastric strip is assumed to be incapable of being deformed in shear. The effect of weight, transverse and axial inertia forces and rotary inertia are considered in the formulation. Only the in-plane motion is considered, while all damping is neglected.

To define the geometry of the elastica, the height of the strip is assumed to be H and the arc length of a typical mass point from the left end is assumed to be s. The horizontal and vertical coordinates of a typical mass point are assumed to be $x(s, t)$ and $y(s, t)$, respectively, while the angle of rotation is assumed to be $\theta(s, t)$, where t is the time. The angle with the horizontal at $s = 0$ is denoted by α.

The bending moment $M(s, t)$ is assumed to be positive if it is counter-clockwise on a positive face. The horizontal force $P(s, t)$ is assumed to be positive in compression, while the vertical force $Q(s, t)$ is positive if it is downward on a positive face.

2. Consider a rectangular plate $30 \times 20 \times 0.01$ cm with the Young's modulus and Poisson's ratio, respectively, equal to 30×10^{10} Pa and 0.2. The plate is simply supported along two of its shorter edges and free along the other two edges. The plate is modelled as 15×10 plate elements based or Kirchhoff's plate theory. The plate is actuated by a distribution of 24 PZT patches where each patch covers an area of 2×2 cm.

(i) Choose appropriate locations for bonding the PZT patches on both sides of the plate and assume that it can be electrically actuated anti-symmetrically. The patches are networked, so only three independent controls are available to excite the first three normal modes while at least the next seven normal modes are not excited.

Obtain the transformation relating the control inputs to the inputs to the patch actuators.

(ii) Assume that 32 displacement sensors have been bonded to the plate at various locations. The sensors are interconnected, so only the modal amplitudes of the first three modes are measured while at least the modal amplitudes of the next seven modes do not corrupt the measurements. Obtain the transformation relating the sensor outputs to the three desired modal amplitudes.

(iii) Design a proportional-derivative H_2 optimal controller, assuming that both the $\mathbf{Q}$ and $\mathbf{R}$ weighting matrices in the performance index are diagonal, equal to unity and equal to each other. State the proportional and derivative gain matrices after normalizing the largest element in each of them to unity.

(iv) Plot the open- and closed-loop poles and comment on your design.

(v) The plate is subjected to sinusoidally varying transverse point load of constant amplitude at its geometric centre. Plot and compare the open- and closed-loop frequency responses over a suitable range of frequencies. Hence or otherwise comment on the performance of the controller in the presence of the disturbance.

3.

(i) Reconsider the previous exercise. Assume that you would now like to minimize the number of patches and sensors used. What are the minimum numbers of patches and sensors that are needed to design the controller, meeting the above specification?

(ii) By carefully creating only the most active transducers identify the locations of the sensors and actuator patches.

(iii) Redesign the controller and plot the modified closed-loop poles. Comment on any significant differences in the location of the closed-loop poles.

4. Reconsider exercise 1: In order to cut costs it is decided to use the same piezoelectric patches both as actuators and as sensors to measure the plate curvature. The sensors and actuators are now co-located.

Redesign the controller and compare with the results obtained in exercises 1 and 2. Comment on any significant differences in the performance of the controller.

5.

(i) Reconsider exercise 1 and assume that five PZT patches and 32 displacement sensors are available to implement the controller. Redesign the controller to minimize spillover effects.
(ii) If the number of sensors is also reduced to five, is it feasible to design a controller avoiding spillover effects? Design a controller to minimize spillover effects and investigate if any spillover is present by investigating the response of the first 10 modes.
(iii) Explain how the spillover effect could be suppressed. Hence or otherwise design a low-pass filter to process the sensor outputs to eliminate observation spillover.

6. Consider a simply supported deep beam, where both shear and rotary inertia may be considered important. For certain applications, it is necessary to actively and effectively clamp both ends of the beam by applying suitable feedback control forces along the length of the beam.

(i) Design a generic controller capable of clamping a simply supported beam so its first three modes have the same natural frequencies as those of a beam clamped at both ends.
(ii) How many actuators and sensors would you need if it is essential to avoid spillover effects in the next three modes?
(iii) If only a maximum of four actuators are available, investigate the effects of spillover on the closed-loop frequency response. Assume that

$$\frac{J}{mL^2} = \frac{I}{AL^2} = 0.06, \; \frac{E}{KG} = 1.88.$$

(You may find it convenient to use the finite element model developed in exercise 2 of Chapter 7 or use a finite element analysis package such as ABACUS.)

7.

(i) A non-uniform horizontal bar is supported on two springs, one of which is to the left of the centre of gravity and at a distance l_1 from it, while the second is to the right of the centre of gravity and at a distance l_2 from it. The spring constants are k_1 and k_2, respectively. The mass of the bar is m, while the moment of inertia about the centre of gravity is I_{cg}.

Consider a reference frame with coordinates x pointing down and θ denoting clockwise rotation of the bar about the centre of gravity and show that the governing equations of motion are

$$\begin{bmatrix} m & 0 \\ 0 & I_{cg} \end{bmatrix}\begin{bmatrix} \ddot{x} \\ \ddot{\theta} \end{bmatrix} + \begin{bmatrix} k_1 + k_2 & (k_2 l_2 - k_1 l_1) \\ (k_2 l_2 - k_1 l_1) & (k_2 l_2^2 + k_1 l_1^2) \end{bmatrix}\begin{bmatrix} x \\ \theta \end{bmatrix} = m\begin{bmatrix} g \\ 0 \end{bmatrix}.$$

(ii) Assume that the origin of the coordinate frame is moved to the left of the centre of gravity by a distance x_{cg} to a point O, so the two springs are at distances l_3 and l_4 from the origin. The moment of inertia about the origin is I_o.
(a) Show that the governing equations of motion are

$$\begin{bmatrix} m & mx_{cg} \\ mx_{cg} & I_o \end{bmatrix}\begin{bmatrix} \ddot{x} \\ \ddot{\theta} \end{bmatrix} + \begin{bmatrix} k_1 + k_2 & (k_2 l_4 - k_1 l_3) \\ (k_2 l_4 - k_1 l_3) & (k_2 l_4^2 + k_1 l_3^2) \end{bmatrix}\begin{bmatrix} x \\ \theta \end{bmatrix} = m\begin{bmatrix} g \\ 0 \end{bmatrix}.$$

(b) There exists a point E, at a distance e to the right of the origin, where an applied external force causes only a pure downward translation and an applied external moment causes a pure rotation, i.e. the translation and rotational static displacements are uncoupled and it follows that

$$k_2(l_4 - e) - k_1(l_3 + e) = 0.$$

Furthermore the bar is subjected to an upward external force L_e and an anti-clockwise external moment M_e at the origin O.

Show that the equations of motion may be expressed as

$$\begin{bmatrix} 1 & x_{cg} \\ x_{cg} & r_o^2 \end{bmatrix} \begin{bmatrix} \ddot{x} \\ \ddot{\theta} \end{bmatrix} + \omega_T^2 \begin{bmatrix} 1 & e \\ e & l_e^2 \end{bmatrix} \begin{bmatrix} x \\ \theta \end{bmatrix} = \begin{bmatrix} g \\ 0 \end{bmatrix} - \begin{bmatrix} L_e \\ M_e \end{bmatrix},$$

where

$$\omega_T^2 = (k_2 + k_1)/m, \; mr_o^2 = I_o \text{ and } l_e^2 = \left(k_2 l_4^2 + k_1 l_3^2\right)/(k_2 + k_1).$$

(c) Show that the characteristic equation may be expressed as

$$\left(\frac{\omega^2}{\omega_T^2}\right)^2 - \left(\frac{\omega^2}{\omega_T^2}\right)\left(1 + \frac{l_e^2 - e^2}{r_0^2 - x_{cg}^2} + \frac{\left(x_{cg} - e\right)^2}{r_0^2 - x_{cg}^2}\right) + \frac{l_e^2 - e^2}{r_0^2 - x_{cg}^2} = 0$$

or as

$$\left(\frac{\omega^2}{\omega_T^2} - 1\right)\left(\frac{\omega^2}{\omega_T^2} - \frac{l_e^2 - e^2}{r_0^2 - x_{cg}^2}\right) - \left(\frac{\omega^2}{\omega_T^2}\right)\frac{\left(x_{cg} - e\right)^2}{r_0^2 - x_{cg}^2} = 0.$$

Hence discuss the consequence of this result, i.e. the extent of the coupling between the two modes of oscillation depends on the distance of the point E, representing the point where system is statically uncoupled, and the centre of gravity C where the system is dynamically coupled.

8. An aircraft wing is modelled as a non-uniform flat plate of mass m as shown in Figure 9.14. The stiffness of the wing is idealized and represented by two springs k_A and k_B. The chord of the wing is assumed to be $2b$ and all the distances along the wing are non-dimensionalized by the semi-chord b. The origin of the coordinate system is assumed to be located at the elastic centre of the cross-section, which is assumed to be located at EA.

The line joining the elastic centres of the various cross-sections of the wing is assumed to be a straight line and referred to as the *elastic axis*. The elastic axis is assumed to be located at a distance ba from the mid-chord and the centre of mass is assumed to be located at a distance $bx\alpha$ from the elastic axis. The mass moment of inertia of the wing model about the elastic axis is $I\alpha$.

(i) Show that the non-dimensional distance of the elastic axis from mid chord satisfies the relationship

$$a = \frac{k_B a_2 - k_A a_1}{k_B + k_A}.$$

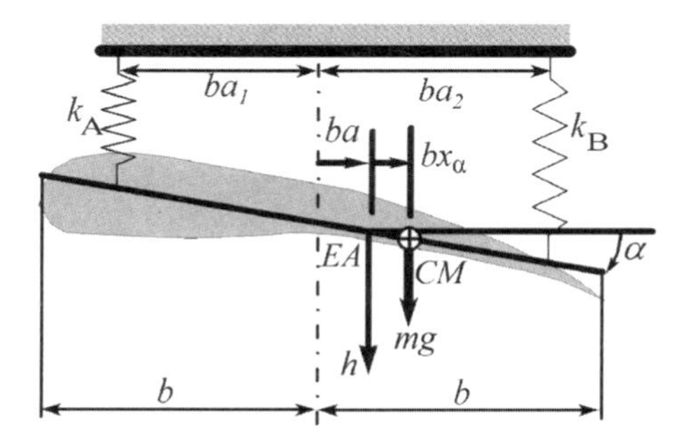

Figure 9.14 Typical section idealization of an aircraft wing

Hence, show that the equivalent translation stiffness is

$$k_h = k_B + k_A$$

and the equivalent torsional stiffness is

$$k_\alpha = k_2 b^2 \left(a_2^2 - a^2\right) + k_1 b^2 \left(a_1^2 - a^2\right).$$

(ii) Show that the potential energy and the kinetic energy are given by

$$V = \frac{1}{2}k_h h^2 + \frac{1}{2}k_\alpha \alpha^2, \; T = \frac{1}{2}m\left(\dot{h} + bx_\alpha\dot{\alpha}\right)^2 + \frac{1}{2}\left(I_\alpha - m\left(bx_\alpha\right)^2\right)\dot{\alpha}^2.$$

(iii) It is proposed to control the bending-torsion vibrations of an aircraft wing prototype by using a distribution of piezoelectric PZT patch actuators to generate a pure bending moment and a pure torsional moment. Considering the typical section of the wing, identify the precise locations of the PZT patch actuators relative to the elastic centre.

Derive the equations of motion of the typical section idealization.

9. Two masses, $M = 1$ kg and $m = 0.2$ kg, are attached to a slender uniform light rod of length $L = 1$ m to form a dumbbell. The dumbbell is supported by two springs of stiffness $k1 = 1000$ N/m and $k2 = 2000$ N/m, which are attached to it at a distance $L/4$ from each end of the rod. Two dashpots with damping coefficients $c1 = 0.25$ N s/m and $c2 = 0.25$ N s/m are attached to each of the masses. Two transverse white noise disturbance forces with magnitudes $d1 = 0.2$ and $d_2 = 0.1$ and two transverse control forces $u1$ and u_2 are assumed to act on the two masses, respectively, in the directions of the transverse displacements of the masses. Assuming that both the transverse disturbance forces are measured and that the measurements are corrupted by white noise disturbances of magnitudes $v1 = 0.1$ and $v_2 = 0.2$, design a feedback controller to optimize the H_∞-norm of the transverse displacement output labelled z_{inf} in Figure 9.15.

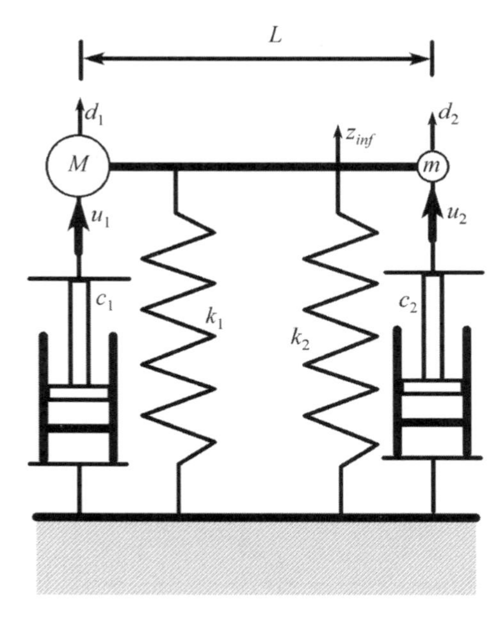

Figure 9.15 Transverse vibration model of an aircraft wing section

10. Consider the bending torsion vibrations of a large aspect ratio cantilevered rectangular box beam of flexural stiffness EI, length L and mass m. The origin of the coordinate frame is to the left of the centre of gravity at a distance x_{cg} at a point O. The moment of inertia of the beam cross-section about the origin is I_o, the polar second moment of the area of cross-section is J and the modulus of rigidity is G. The transverse displacement is w and the clockwise torsional displacement is θ. The differential equation of bending-torsion vibration is given by

$$\begin{bmatrix} m & mx_{cg} \\ mx_{cg} & I_o \end{bmatrix}\begin{bmatrix} \ddot{w} \\ \ddot{\theta} \end{bmatrix} + \begin{bmatrix} EI\partial^4 w/\partial x^4 \\ -GJ\partial^2\theta/\partial x^2 \end{bmatrix} = m\begin{bmatrix} g \\ 0 \end{bmatrix}.$$

(i) Assume appropriate mode shapes for the bending and torsional modes and establish a multi-mode vibration model of the box beam. Determine the first two bending and first two torsional modes and natural frequencies in terms of appropriate non-dimensional parameters.
(ii) Choose appropriate locations for bonding the PZT patches on both sides of the plate and assume that it can be electrically actuated anti-symmetrically. The patches are networked such that only two independent controls are available to excite the first two bending modes and two independent controls are available to excite the first two torsion modes, while at least the next three bending and the next three torsion modes are not excited.

Obtain the transformation relating the control inputs to the inputs to the patch actuators.
(iii) Assume that 16 displacement sensors have been bonded to the plate at various locations. The sensors are interconnected such that only the modal amplitudes of the first two bending and the first two torsion modes are measured, while at least the modal amplitudes of the next three bending and the next three torsion modes do not corrupt the measurements. Obtain the transformation relating the sensor outputs to the three desired modal amplitudes.
(iv) Assume appropriate measurement and process disturbances and design an H_∞ optimal controller to minimize the H_∞-norm of the transverse displacement at the 3/4; chord point from the leading edge.
(v) If the number of sensors and actuators is also reduced to five, is it feasible to design an H_∞ controller avoiding spillover effects? Design an H_∞ controller to minimize spillover effects and investigate if any spillover is present by investigating the response of the first 10 modes.

References

Balas, G. J., Packard, A. K., Safonov M. G. and Chiang, R. Y. (2004) Next generation of tools for robust control, Proc. American Control Conf., *Boston, MA*, June 30–July 2, 2004.

Boyd, S. L., Ghaoui, L. E., Feron, E. and Balakrishnan, V. (1994) *Linear Matrix Inequalities in System and Control Theory,* Vol. 15 of *SIAM Studies in Applied Mathematics*, SIAM, Philadelphia, PA, USA.

Burl, Jeffrey B. (1999) *Linear Optimal Control: H_2 and H_∞ methods*, Menlo Park; Addison Wesley Longman, Inc.

Chilali, M., Gahinet, P. and Apkarian, P. (1999) Robust pole placement in LMI region, *IEEE Transactions on Automatic Control*, 44(12), 2257–2269.

Doyle, J. C., Glover, K. Khargonekar P. P. and Francis B. (1989) State-space solutions to the standard H_2 and H_∞ control problems, *IEEE Transactions on Automatic Control*, 34, 831–847.

Gabbert, U., Berger, H., Köppe, H. and Cao X. (2000a) On modelling and analysis of piezoelectric smart structures by the finite element method, *Applied Mechanics and Engineering*, 5(1), 127–142.

Gabbert, U., Köppe, H., Fuchs, K. and Seeger, F. (2000b) Modelling of smart composites controlled by thin piezoelectric fibers, in Varadan, V.V. (ed.), *Mathematics and Control in Smart Structures*, Bellingham WA USA: International Society for Optical Engineering, SPIE Proceedings Series, Vol. 3984, pp. 2–11.

Gabbert, U., Trajkov, T. N. and Köppe, H. (2002) *Modelling, Control and Simulation of Piezoelectric Smart Structures using Finite Element Method and Optimal Control, The Scientific Journal Facta*

Universitatis, (University of Niš, Serbia.) Series: Mechanics, Automatic Control and Robotics, Vol. 3, No. 12, pp. 417–430.
Gahinet, P. and Apkarian, P. (1994) A linear matrix inequality approach to H_∞ control, *International Journal of Robust and Nonlinear Control*, 4, 421–448.
Gahinet, P., Nemirovski, A., Laub, A. J. and Chilali, M., (1995) *LMI Toolbox for use with MATLAB, User Guide*, Vol. 1, The Mathworks Inc., MA, USA.
Glover, K. and Doyle, J. C. (1988) State space formulae for all stabilizing controllers that satisfy an H_∞-norm bound and relations to risk sensitivity, *Systems & Control Letters*, 11, 167–172.
Jeng, Y-F., Chen, P. C. and Chang, Y.-H. (2005) Design of a class of extended dynamic linear controller via LMI approach, *Journal of the Chinese Institute of Engineers*, 28(3), 423–432.
Kar, I. N., Miyakura, T. and Seto, K. (2000a) Bending and torsional vibration control of a flexible plate structure using H_∞-based robust control law, *IEEE Transactions on Control Systems Technology*, 8, 545–553.
Kar, I. N., Seto, K. and Doi, F. (2000b) Multimode vibration control of a flexible structure using H_∞-based robust control, *IEEE/ASME Transactions on Mechatronics*, 5, 23–31.
Khargonekar, P. P. and Rotea, M. A. (1991) Mixed H_2/H_∞ control: a convex optimization approach, *IEEE Transactions on Automatic Control*, 36(7), 824–837.
Liang, X. (1997) Dynamic Response of Linear/Nonlinear Laminated Structures Containing Piezoelectric Laminas, Ph. D. Dissertation, Virginia Polytechnic Institute and State University, VA, USA.
Meirovitch, L. and Baruh, H. (1985) The implementation of modal filters for control of structures, *Journal of Guidance, Control, and Dynamics*, 8, 707–716.
Moser, A. N. (1993) Designing controllers for flexible structures with H_∞/μ-synthesis, *IEEE Control Systems Magazine*, 13(2), 79–89.
Minis, I. and Yanushevsky, R. (1993) A new theoretical approach for the prediction of machine tool chatter in milling, *Journal of Engineering for Industry*, 115, 1–8.
Nonami, K. and Sivrioglu, S. (1996) Active vibration control using LMI-based mixed H_2/H_∞ state and output feedback control with nonlinearity, Proceedings of the 35th IEEE Conference on Decision and Control, Kobe, Japan.
O'Connor, W. J. and Fumagalli, A. (2009) Refined wave based control applied to nonlinear, bending and slewing flexible systems, *Journal of Applied Mechanics*, 76(4), 041005-1:9.
Parks, P. C. (1992) A M Lyapunov's stability theory - 100 years on, *IMA Journal of Mathematical Control and Information*, 9(4), 275–303.
Plaut R. H. and Virgin, L. N. (2009) Vibration and snap-through of bent elastica strips subjected to end rotations, *Journal of Applied Mechanics*, 76 (041011) pp. 041011-1:7
Seeger, F. and Gabbert, U. (2003) Optimal placement of distributed actuators for a controlled smart elastic plate, *PAMM Proceedings of the Applied Mathematics and Mechanics*, 2, 262–263.
Segalman, D. J. and Butcher, E. A. (2000) Suppression of regenerative chatter via impedance modulation, *Journal of Vibration and Control*, 6, 243–256.
Sznaier, M., Rotstein, H., Bu, J. and Sideris, A. (2000) An exact solution to continuous-time mixed H_2/H_∞ control problems, *IEEE Transactions On Automatic Control*, 45(11), 2095–2101.
Tanaka, N. (2009) Cluster control of distributed-parameter structures, *International Journal of Acoustics and Vibration*, 14(1), 24–34.
Vidyasagar M.. (1993) *Nonlinear systems analysis* (2nd ed.), Prentice-Hall, Inc. Upper Saddle River, NJ, USA.
Zhou, K. and Doyle, J. C. (1998) *Essentials of Robust Control*, Upper Saddle River, NJ: Prentice-Hall.

Index

Dynamics of Smart Structures Dr. Ranjan Vepa